CYANOBACTERIAL LIFESTYLE AND ITS APPLICATIONS IN BIOTECHNOLOGY

CYANOBACTERIAL LIFESTYLE AND ITS APPLICATIONS IN BIOTECHNOLOGY

Edited by

PRASHANT KUMAR SINGH

Department of Biotechnology, Mizoram University (A Central University), Pachhunga University College Campus, Aizawl, Mizoram, India

MARÍA F. FILLAT

Departamento de Bioquímica y Biología Molecular y Celular, Universidad de Zaragoza, Zaragoza, Spain

AJAY KUMAR

Agriculture Research Organization (ARO), The Volcani Center, Rishon-Lezion, Israel

ELSEVIER

ACADEMIC PRESS

An imprint of Elsevier

elsevier.com/books-and-journals

Academic Press is an imprint of Elsevier
125 London Wall, London EC2Y 5AS, United Kingdom
525 B Street, Suite 1650, San Diego, CA 92101, United States
50 Hampshire Street, 5th Floor, Cambridge, MA 02139, United States
The Boulevard, Langford Lane, Kidlington, Oxford OX5 1GB, United Kingdom

British Library Cataloguing-in-Publication Data
A catalogue record for this book is available from the British Library

Library of Congress Cataloging-in-Publication Data
A catalog record for this book is available from the Library of Congress

ISBN: 978-0-323-90634-0

For Information on all Academic Press publications visit our website at https://www.elsevier.com/books-and-journals

Publisher: Andre G. Wolff
Acquisitions Editor: Glyn Jones
Editorial Project Manager: Barbara Makinster
Production Project Manager: Swapna Srinivasan
Cover Designer: Matthew Limbert

Contents

11. Cyanobacteria as biostimulants in the paddy fields

Prashant Kumar Singh, Ajay Kumar, Mukesh Kumar Yadav, Kushlendra, Savita Singh

12. Molecular characterization of local cyanobacterial isolates using *16S rRNA*, *rpo*B, and *nif*H biomarkers

Nidhi Kumari, Lal Chand Rai

13. Cyanobacteria-mediated heavy metal and xenobiotics bioremediation

Shweta Pandey, Sumit Kumar Dubey, Arun Kumar Kashyap, Buddhi Prakash Jain

14. Cyanobacteria: as a promising candidate for nanoparticles synthesis

Dnyaneshwar K. Kulal, Dinesh N. Navale, Santosh W. Zote, Prasanna B. Ranade, Pramod K. Kalambate

Contributors

Rafael Ambrosio Instituto de Investigacines en Biodiversidad y Biotecnología (INBIOTEC-CONICET), Mar del Plata, Argentina; Fundación para Investigaciones Biológicas Aplicadas, Argentina

Rocío C. Arce Instituto de Biología Molecular y Celular de Rosario (IBR-UNR/CONICET), Facultad de Ciencias Bioquímicas y Farmacéuticas, Universidad Nacional de Rosario (UNR), Rosario, Argentina

Punuri Jayasekhar Babu Department of Biotechnology, Mizoram University (A Central University), Pachhunga University College Campus, Aizawl, Mizoram, India

María V. Barja Centre for Research in Agricultural Genomics, Barcelona, Spain

William Carrie Department of Biotechnology, Mizoram University, Mizoram, India

Liansangmawii Chhakchhuak Department of Biotechnology, Mizoram University (A Central University), Pachhunga University College Campus, Aizawl, Mizoram, India

Kaushal Kishore Choudhary Ram Jaipal College, Chapra, Bihar, India

Leonardo Curatti Instituto de Investigaciones en Biodiversidad y Biotecnología (INBIOTEC-CONICET), Mar del Plata, Argentina; Fundación para Investigaciones Biológicas Aplicadas, Argentina

Purbajyoti Deka Department of Biotechnology, Mizoram University, Mizoram, India

Mariana Demarchi Instituto de Biología Molecular y Celular de Rosario (IBR-UNR/CONICET), Facultad de Ciencias Bioquímicas y Farmacéuticas, Universidad Nacional de Rosario (UNR), Rosario, Argentina

Sumit Kumar Dubey Department of Microbiology, D. L. S. Postgraduate College, Bilaspur, Chhattisgarh, India

María F. Fillat Departamento de Bioquímica y Biología Molecular y Celular, Universidad de Zaragoza, Zaragoza, Spain

Sandip Ghuge Agriculture Research Organization, The Volcani Center, Rishon-Lezion, Israel

Rajan Kumar Gupta Department of Botany, Institute of Science, Banaras Hindu University, Varanasi, Uttar Pradesh, India

Vaishali Gupta Government College, Bichhua, Chhindwara, Madhya Pradesh, India

Jose A. Hernandez Department of Biochemistry and Molecular Genetics, College of Graduate Studies, Midwestern University, Glendale, AZ, United States; Arizona College of Osteopathic Medicine, Midwestern University, Glendale, AZ, United States

Buddhi Prakash Jain Department of Zoology, Mahatma Gandhi Central University, Motihari, Bihar, India

Naveen Chandra Joshi Amity Institute of Microbial Technology (AIMT), Amity University, Noida, Uttar Pradesh, India

Pramod K. Kalambate State Key Laboratory of Materials Processing and Die & Mould Technology, School of Materials Science and Engineering, Huazhong University of Science and Technology, Wuhan, Hubei, PR China

Arun Kumar Kashyap Department of Biotechnology, Govt. E Raghavendra Rao Postgraduate Science College, Bilaspur, Chhattisgarh, India

Kaushalendra Department of Zoology, Mizoram University (A Central University), Pachhunga University College Campus, Aizawl, Mizoram, India

Galyna Kufryk Grand Canyon University, Phoenix, AZ, United States

Reena Kujur Department of Botany, Institute of Science, Banaras Hindu University, Varanasi, Uttar Pradesh, India

Dnyaneshwar K. Kulal Ramnarain Ruia College, Mumbai, Maharashtra, India

Ajay Kumar Agriculture Research Organization (ARO), The Volcani Center, Rishon-Lezion, Israel

Anil Kumar Agriculture Research Organization, The Volcani Center, Rishon-Lezion, Israel

Nidhi Kumari Department of Biological Sciences, Indian Institute of Science Education and Research Mohali, Mohali, Punjab, India; Centre for Advanced Study in Botany, Institute of Science, Banaras Hindu University, Varanasi, Uttar Pradesh, India

Rosie Lalmuanpuii Department of Botany, Mizoram University, (A Central University), Aizawl, Mizoram, India

Esther Lalnunmawii Department of Botany, Mizoram University, (A Central University), Aizawl, Mizoram, India

Lalrokimi Department of Biotechnology, Mizoram University, Mizoram, India

Anabella F. Lodeyro Instituto de Biología Molecular y Celular de Rosario (IBR-UNR/CONICET), Facultad de Ciencias Bioquímicas y Farmacéuticas, Universidad Nacional de Rosario (UNR), Rosario, Argentina

Bendangchuchang Longchar Department of Life Science, Mizoram University (A Central University), Pachhunga University College Campus, Aizawl, Mizoram, India

Gajanan T. Mehetre Department of Biotechnology, Mizoram University, Mizoram, India

Arpan Modi Agriculture Research Organization, The Volcani Center, Rishon-Lezion, Israel

José Miguel Mulet Instituto de Biología Molecular y Celular de Plantas, Universidad Politécnica de Valencia-CSIC, Valencia, Spain

Mauro Do Nascimiento Instituto de Investigaciones en Biodiversidad y Biotecnología (INBIOTEC-CONICET), Mar del Plata, Argentina; Fundación para Investigaciones Biológicas Aplicadas, Argentina

Dinesh N. Navale Vivekanand Education Society's College of Arts, Science & Commerce, Mumbai, Maharashtra, India

Hugo Guillermo Jimenez Pacheco Universidad Católica de Santa María, Arequipa, Perú

Kapil Deo Pandey Department of Botany, Institute of Science, Banaras Hindu University, Varanasi, Uttar Pradesh, India

Shweta Pandey Govt VYT PG Autonomous College, Durg, Chhattisgarh, India

M. Luisa Peleato Departamento de Bioquímica y Biología Molecular y Celular, Universidad de Zaragoza, Zaragoza, Spain

Avinash Chandra Rai Agriculture Research Organization, The Volcani Center, Rishon-Lezion, Israel

Lal Chand Rai Centre for Advanced Study in Botany, Institute of Science, Banaras Hindu University, Varanasi, Uttar Pradesh, India

Lilia Mary Miranda Ramos Universidad Nacional de San Agustín de Arequipa, Arequipa, Perú

Prasanna B. Ranade Vivekanand Education Society's College of Arts, Science & Commerce, Mumbai, Maharashtra, India

Lara Sanchez Rizza Instituto de Investigaciones en Biodiversidad y Biotecnología (INBIOTEC-CONICET), Mar del Plata, Argentina; Fundación para Investigaciones Biológicas Aplicadas, Argentina

Violeta C. Sein-Echaluce Departamento de Bioquímica y Biología Molecular y Celular, Universidad de Zaragoza, Zaragoza, Spain

Rahul Prasad Singh Department of Botany, Institute of Science, Banaras Hindu University, Varanasi, Uttar Pradesh, India

Sandeep Kumar Singh Department of Botany, Institute of Science, Banaras Hindu University, Varanasi, Uttar Pradesh, India

Garima Singh Department of Botany, Mizoram University (A Central University), Pachhunga University College Campus, Aizawl, Mizoram, India

Avinash Singh Department of Biochemistry and Molecular Pharmacology, New York University, New York, USA

Prashant Kumar Singh Department of Biotechnology, Mizoram University (A Central University), Pachhunga University College Campus, Aizawl, Mizoram, India

Bhim Pratap Singh Department of Agriculture and Environmental Sciences, National Institute of Food Technology Entrepreneurship & Management (NIFTEM), Sonepat, Haryana, India

Sandeep Kumar Singh Department of Botany, Banaras Hindu University, Varanasi, Uttar Pradesh, India

Savita Singh Department of Botany, Babu Shivnath Agrawal College, Mathura, Uttar Pradesh, India

Naveen Kumar Verma Government College, Bichhua, Chhindwara, Madhya Pradesh, India

Paula Vicino Instituto de Biología Molecular y Celular de Rosario (IBR-UNR/CONICET), Facultad de Ciencias Bioquímicas y Farmacéuticas, Universidad Nacional de Rosario (UNR), Rosario, Argentina

Wenjing Wang Department of Biology and Food Science, Shangqiu Normal University, Shangqiu, The People Republic of China

Mukesh Kumar Yadav Department of Biotechnology, Mizoram University (A Central University), Pachhunga University College Campus, Aizawl, Mizoram, India

Priya Yadav Department of Botany, Institute of Science, Banaras Hindu University, Varanasi, Uttar Pradesh, India

Santosh W. Zote Department of Chemistry, Sathaye College, Mumbai, Maharashtra, India

Zothanpuia Department of Biotechnology / Life sciences, Mizoram University, Pachhunga University College, Aizawl, Mizoram, India

About the editors

Dr. Prashant Kumar Singh

Dr. Prashant Kumar Singh is currently working in the Department of Biotechnology, Mizoram University (A Central University), Pachhunga University College Campus, Aizawl, Mizoram, India. He completed his PhD from the Department of Botany (Centre of Advanced Studies in Botany), Institute of Sciences, Banaras Hindu University, Varanasi, India. Dr. Singh worked as a Foreign Expert cum Visiting Researcher at Henan University, China, and Visiting Scientist at Agriculture Research Organization (ARO)—the Volcani Centre, Israel. His research focuses on plant and cyanobacterial stress biology rounding around the omics approaches. His current research area is to decipher the cyanobacterial and crop plants adaptation at high altitudes using omics approaches. He authored more than 70 articles, book chapters, and one book entitled "Advances in Cyanobacterial Biology," Elsevier, USA. Dr. Singh is also awarded the Young Scientist Award, early carrier research grant award, lifetime members of various scientific societies, and a reviewer of various peer-reviewed journals.

Prof. María F. Fillat

Prof. María F. Fillat is a Full Professor in the Department of Biochemistry and Molecular Cell Biology (University of Zaragoza) and Leader of the group "Genetic regulation and physiology of cyanobacteria." During the last decades, her work has focused on the functional characterization of the ferric uptake regulator proteins in cyanobacteria, including their regulation, participation in regulatory networks, their relationship with redox homeostasis, and their role in the synthesis of cytotoxic microcystin. She completed her PhD in Sciences from the University of Zaragoza and worked as postdoctoral fellow at the University of Utrecht (The Netherlands). She also visited other laboratories at the University of Konstanz (Germany), Emory University (Atlanta, GA), Universidad de Rosario (Argentina), IBVF (Sevilla), and the University of Gèneve. Dr. Maria F. Fillat has published more than 100 research articles in international refereed journals and holds two patents. Her group has received several awards from the Economic and Social Council (Government of Aragón), the "Instituto de

Estudios Turolenses" and the Royal Academy of Sciences of Zaragoza. She led 13 research projects and six contracts with companies as a Principal Investigator (PI) and successfully organized nine international and three national scientific meetings. Dr. Fillat has also supervised 10 PhD theses (three more are in progress), nine Diploma theses, and 13 Master projects. She has also participated in four teaching innovation projects, two of them as PI.

Dr. Ajay Kumar

Dr. Ajay Kumar is a Visiting Scientist at the Agriculture Research Organization, Volcani Centre, Rishon-Lezion, Ministry of Agriculture and Rural Development, Israel. Dr. Kumar completed his doctoral research in Botany from the Institute of Science, Banaras Hindu University, Varanasi, India. He has published more than 150 research articles and book chapters in the leading international and national journals or books. He has a wide area of research experience, especially in plant–microbe interactions, microbial biocontrol, microbiology, molecular microbiology, and endophytes related to medicinal plants. He is an invited member of the editorial committee of the *International Journal of Current Microbiology and Applied Sciences* and *Journal of Plant Science and Agriculture Research*. He is also an active reviewer for journals including *PLOS ONE, Agriculture, Ecosystem and Environment, Biological Control, Plant Biochemistry and Biotechnology, 3Biotech*, etc.

Preface

It is our pleasure to introduce you to a new book that provides a perspective on the physiology of cyanobacteria in the context of the ongoing and potential biotechnological exploitation of these fascinating microorganisms. *Cyanobacterial Lifestyle and its Applications in Biotechnology* is the second volume of a series that covers selected aspects of the biology of cyanobacteria, highlighting the features which make them useful organisms for applications related to agriculture, nanotechnology or environmental microbiology, among other fields related to microbial biotechnology. Cyanobacteria have been used ancestrally to fertilize paddy fields and as a protein source for human consumption. However, the advances in the knowledge of cyanobacterial physiology and the development of novel techniques related to genetic engineering, synthetic biology, or nanotechnology widen the scope of applications derived from cyanobacteria and their metabolites. This book was motivated by the increasing interest of the scientific community for cyanobacteria in the last years and aims to attract specialized readers and undergraduate and PhD students interested in cyanobacteria-based microbial biotechnology. This volume's layout covers different cyanobacteria strategies for adaptation to hostile environments and how to improve its performance and photosynthetic metabolism under such conditions, including biotechnological perspectives in this field. The second series of chapters are focused on the use of cyanobacteria as fertilizers.

Furthermore, other applications in gene editing, bioremediation, biofuel production, or the use of cyanobacterial nanoparticles are revised. The book also includes original research about the use of cyanobacterial proteins to increase tolerance in yeasts. We hope that the experienced readers will enjoy the book, and also it will trigger the interest of novel cyanobacteriologists.

Finally, the editors would like to acknowledge all the authors who have generously contributed their time and expertise to write an excellent set of chapters.

Prashant Kumar Singh

María F. Fillat

Ajay Kumar

CHAPTER

1

Cyanobacterial diversity concerning the extreme environment and their bioprospecting

Arun Kumar Kashyap[a], *Sumit Kumar Dubey*[b], *Buddhi Prakash Jain*[c]

[a]Department of Biotechnology, Govt. E Raghavendra Rao Postgraduate Science College, Bilaspur, Chhattisgarh, India

[b]Department of Microbiology, D. L. S. Postgraduate College, Bilaspur, Chhattisgarh, India

[c]Department of Zoology, Mahatma Gandhi Central University, Motihari, Bihar, India

1.1 Introduction

An extreme environment is an environment in which it is difficult for a life form to survive. The harsh climate is natural (desert, permafrost, volcanoes, deep ocean, etc.) and artificial (spaceship, aircraft, etc.). Extreme environment prevails on most of the earth surface. All kind of organisms is known to survive, proliferate, and grow in such extreme environment. The organism found in the extreme environment is termed an extremophile. To survive the extreme environment, extremophiles adopt different strategies by synthesizing various biomolecules, because of which extremophiles open a way of finding new molecules with the potential biotechnological application.

Cyanobacteria are Gram-negative photosynthetic bacteria that could potentially meet the energy, food, medical, and industrial requirements due to their tolerance efficacy to extreme environmental conditions

Cyanobacterial Lifestyle and its Applications in Biotechnology.
DOI: https://doi.org/10.1016/B978-0-323-90634-0.00008-1

(Varshney et al., 2015). Among bacteria, the cyanobacterial strains are well known to acclimatize in the extreme environment. The macromolecular stability and improved genetic makeup help organisms to survive better in extreme conditions. These organisms have an improved or modified metabolic activity and physiological stability to colonize and survive in the harsh ecosystem. Extremophiles classified based on their extreme characters such as temperatures (i.e., psychrophiles, mesophiles, thermophiles, and hyper-thermophiles), pH (acidophiles, neutrophiles, and alkalophilic), salinity (halophiles), dryness (xerophiles), radiation-tolerant microbes, high concentration of heavy metals (bioleachates), and anaerobic environments (anaerobes) are shown in Fig. 1.1.

Moreover, they can grow in harsh environmental habitats with high toxicity, heavy metals concentration, and organic load (Rampelotto, 2013). Additionally, the cyanobacterial diversity in various environments could reveal the multiple origin and evolution of life on the earth's crust. The photosynthetic cyanobacteria are known for survival efficacy in extreme physiological conditions and played a significant role in agricultural and environmental sustainability. Cyanobacterial strains have been reported to survive in hot springs, higher salinity, low temperature, and deserts (Singh, 2014).

Cyanobacteria are rare in highly acidic conditions. Extremophiles need extreme conditions to grow, while extreme-tolerants can tolerate extreme physicochemical conditions to survive but grow efficiently at normal mesophilic niche. Extremophiles can survive in the following extreme environments (1) depths of around 6.0 km inside the earth's crust, (2) deep inside the ocean (around 10 km) at pressures of up to 110 MPa, (3) extreme acidic (up to pH 0.1) to basic (up to pH 12.0) conditions, (4) hydrothermal vents (up to 122 °C) to frozen up to −20 °C). The organisms that adapt to extreme environments offer a distinctive viewpoint on the elemental biological processes. For every extreme environmental condition investigated, a variety of organisms have shown that they can tolerate these conditions and often require those conditions for survival. Cyanobacteria can survive on a bare niche with the requirement of minimum light, carbon dioxide, and water for photosynthesis (Castenholz et al., 2001).

Due to the ability to grow under extreme habitats, cyanobacteria produce various biomolecules because it is an excellent source of biologically active constituents—the biologically active components produced as a primary and secondary metabolite (Drobac-Cik et al., 2007). Primary metabolites are produced during the active growth of cyanobacteria (i.e., log phase). The primary metabolites include amino acids, nucleotides, coenzyme, and other intermediates in the biosynthetic pathway. The secondary metabolites are not essential compounds for the growth of cyanobacteria and are produced during the stationary phase. Cyanobacteria are an excellent source of active biological secondary metabolite against various human diseases. Cyanobacteria produce secondary metabolites

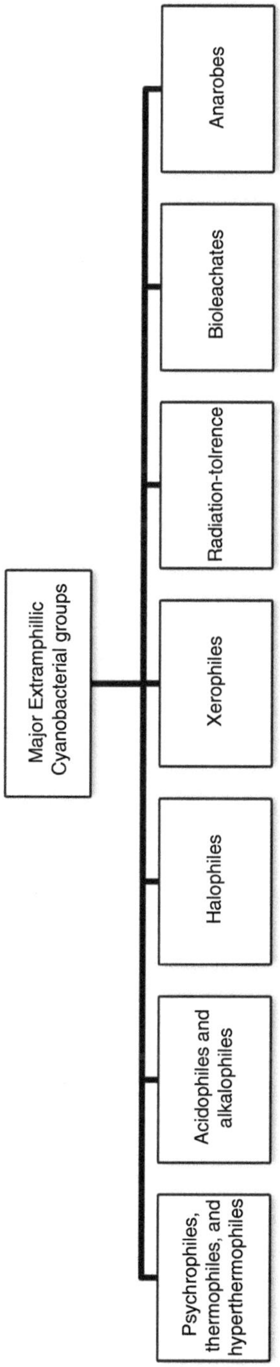

FIG. 1.1 Significant groups of extremophilic cyanobacteria.

with antibacterial, antifungal, anthelmintic, antiprotozoal, and cytotoxic properties (Drobac-Cik et al., 2007; Staley and Stanley, 1986).

Earlier research on cyanobacteria was concentrated on organisms found near aquatic habitats; in recent time, terrestrial microorganisms have also been studied. Terrestrial cyanobacteria are found in more extreme environments, because of which they possess a more robust mechanism of survival not found in aquatic cyanobacteria. Many studies suggest that harsh conditions require even more extreme metabolites in terrestrial cyanobacteria, providing a great range of metabolites with the potential biotechnological application (Drobac-Cik et al., 2007; Metting and Pyne, 1986).

1.2 Systematic study of extremophiles cyanobacteria

The advancement in understanding microbial habitats and microbial diversity and the technology is opening new avenues for bioactive metabolites. Cyanobacteria are most often cultivated in the Chu-10 and BG-11 medium (Chandra, 2019; Chu, 1942; Pathak et al., 2018) and grown at a temperature of 25 °C in the presence of 5% CO_2 and under the illumination by 40-W cool-white fluorescent lamps (Ferris and Hirsch, 1991). Cyanobacteria are closely related to bacteria and resemble algae due to chloroplast presence (Saini et al., 2018). Thus, the identification and classification of cyanobacteria are challenging because of versatile diversity in thallus (Komárek, 2016). The morphological identification of cyanobacterial species is most often studied by microscopic examination. Cyanobacteria such as *Synechococcus* and *Cyanothece* showed similar morphological characters that cause ambiguity during identification (Castenholz, 2004).

The development and availability of molecular characterization, that is, polymerase chain reaction (PCR) based *16S rRNA* gene sequencing with *rpoC1* genes, provides a competent identification strategy to the researchers' community for the last two decades (Lyra et al., 2001). The *16S rRNA* gene used as a marker gene due to its highly conserved nature during genetic inheritance. Along with *16S RNA* sequencing, restriction fragment length polymorphisms—PCR, random amplified polymorphic DNA—PCR, amplified fragment length polymorphism—PCR, and DNA Fingerprinting based on short tandemly repeated repetitive sequences and long tandemly repeated repetitive sequence have been used in different studies (Oinam et al., 2011).

The development of molecular biology techniques such as high-throughput DNA sequencing has revolutionized to expose extreme microbiology and investigate diversified microbial ecology with higher complexity. The understanding of the organism's physiology in culture is necessary for genomic and transcriptomic studies. Modern approaches cannot replace such, thus the combination of traditional and contemporary techniques has been applied to understand how the cyanobacterial strains can grow in extreme environmental conditions. The phylogenetic study of extremophiles provides a phylogenetic tree to predict the evolutionary relationship among different extremophiles. Studies suggest that extremophiles and hyperthermophiles share the same ancestor. This study could also be applicable to reveal the origins of life.

Cyanobacteria are naturally grown on average to the extreme ecosystem of terrestrial and aquatic habitats due to their morphological diversity (Lang-Yona et al., 2018). Thermophiles can grow at high temperature, which shows that their metabolic enzymes are thermostable. Thereby, such information can be considered as treasures in various research and industrial applications. The microbes grow at sea depths, feel high hydrostatic pressure, and have nominal light accessibility. Some environments comprise more than one extreme factor, and the microbes inhabiting those regions are considered polyextremophiles such as thermoacidophilic (high temperature and acidic condition) and xerophilic (hot climate and high UV irradiation). Thus, cyanobacteria have the potential to resolve several environmental issues. Some of them are listed next Table 1.1.

TABLE 1.1 List of extremophiles cyanobacteria and their applications.

S. No.	Function	Extremophile cyanobacteria	Reference
1	Biofertilizer and bioremediation	*Anabaena variabilis, Nostoc muscorum, Aulosira fertissliima,* and *Tolypothrix tenuis*	Singh and Singh (1987), Stewart et al., (1968)
2	Pesticide	*Nostoc* sp. (cryptophycin)	Biondi et al. (2004)
3	Antifungal against *S. sclerotiorum* and *Rhizoctonia solani*	*N. muscorum*	Kulik (1995)
4	Salt tolerance (7–15 g/L)	*Anabaena oscillarioides, A. aphanizomenoides,* and *Microcystis aeruginosa*	Moisander et al. (2002)
5	Insecticides (degradation of organophosphorus and organochlorine)	*Synechococcus elongatus, Anacystis nidulans,* and *Microcystis aeruginosa*	Vijayakumar (2012)
6	Degradation of lindane	*Oscillatoria, Synechococcus, Nodularia, Nostoc, Microcystis,* and Anabaena	El-Bestawy (2008)
7	Degradation of the organophosphorous herbicide glyphosate and mineralized glyphosate	*Anabaena* sp., *Lyngbya* sp., *Microcystis* sp., and *Nostoc* sp	Forlani et al. (2008)

(continued)

TABLE 1.1 (Cont'd)

S. No.	Function	Extremophile cyanobacteria	Reference
8	Degradation of glyphosate herbicide	*Spirulina* sp.	Lipok et al. (2009, 2007)
9	Degradation of crude oil	*Oscillatoria salina*, *Plectonema terebrans*, *Aphanocapsa* sp., and *Synechococcus* sp.	Raghukumar et al. (2001)Cohen (2002)
10	Degradation of n-alkanes	*Microcoleus chthonoplastes* and *Phormidium corium*	Zahra et al. (2020)
11	Oxidation of naphthalene to 1-naphthol	*Oscillatoria* sp. and *Agmenellum* sp.	Cerniglia et al. (1980, 1979)
12	Oxidize biphenyl to 4-hydroxybiphenyl	*Oscillatoria* sp.	Cerniglia et al. (1980)
13	The metabolization of phenanthrene into trans-9,10-dihydroxy-9,10-dihydrophenanthrene and 1-methoxy-phenanthrene.	*Agmenellum* sp.	Narro et al. (1992)
14	Hydrogen production	*Anabaena, Calothrix, Oscillatoria, Cyanothece, Nostoc, Synechococcus, Microcystis, Gloeobacter, Aphanocapsa, Chroococcidiopsis,* and *Microcoleus*	Masukawa et al. (2001), Nozzi et al. (2013)
15	Biomass production and CO_2	*Aphanothece microscopica*	Jacob-Lopes et al. (2008)
16	Thermophilic (CO_2 sequestration from the flue gas)	*Synechococcus aquatilis*, *Chlorogloeopsis* sp	Kumar et al. (2011), Miller et al. (2007)
17	Super food	*Arthrospira platensis*; *Spirulina platensis*	Plavšić et al. (2002), Prasanna et al. (2010)

1.2.1 Thermophiles

Thermophilic cyanobacteria have been explored for industrial bioprocess at high temperatures and documented for their application in bioremediation, CO_2 sequestration, and biofuel production (Patel et al., 2019). Amarouche-Yala et al. (2014) isolated cyanobacteria from nine hot springs (ranges from 39 to 93 °C), and the 16S rRNA sequences based molecular analysis showed 19 different identified cyanobacterial species on which the genera *Chroococcus, Cyanobacterium, Gloeocapsopsis Gloeocapsa, Geitlerinema, Fischerella, Synechocystis, Stigonema,* and *Microcoleus* have been observed. Cyanobacteria such as *Mastigocoleus laminosus* and *Phormidium tenue* have been documented to grow at 69 °C. Bacterial groups undergo genomic changes due to environmental changes such as high temperature. This genetic change provides thermal-tolerant abilities to the bacterial cells (Averhoff and Müller, 2010). The evolutionary changes are most often mediated by horizontal gene transfer (HGT) and gene mutations. The amino acid substitution takes place between lysine and arginine linked with thermophilic genetic adaptation (van Noort et al., 2013). Numerous genetic variations via mutations in the *glpF* and *fabA* genes have contributed to thermal tolerance in microbial cells (Wang et al., 2015).

Atomi et al. (2004) reported that when the reverse gyrase gene of *Thermococcus kodakarensis* was eliminated from the chromosome, the mutant *T. kodakarensis* can grow slowly under high temperatures and revealed the vital contribution of reverse gyrase to the adaptation of high temperature in microbial cells (Atomi et al., 2004). A schematic representation of genetic factors contributing to thermophilic adaptation in cyanobacterial cells is depicted in Fig. 1.2.

1.2.2 Psychrophiles

Psychrophiles are a group of organisms with the ability to grow and reproduce under low temperatures ranging from −20 °C to 10 °C (Sayed et al., 2020). *Synechococcales, Nostocales,* and *Oscillatoriales* are cyanobacteria orders predominantly found in the arctic region (Pushkareva et al., 2015; Steven et al., 2013). Few regional species of cyanobacteria can also be predominant in Antarctic soils (Namsaraev et al., 2010). Genetic modifications, acclimatization, and adaptations are the major survival strategies that could be implemented by cyanobacterial strains growing in the

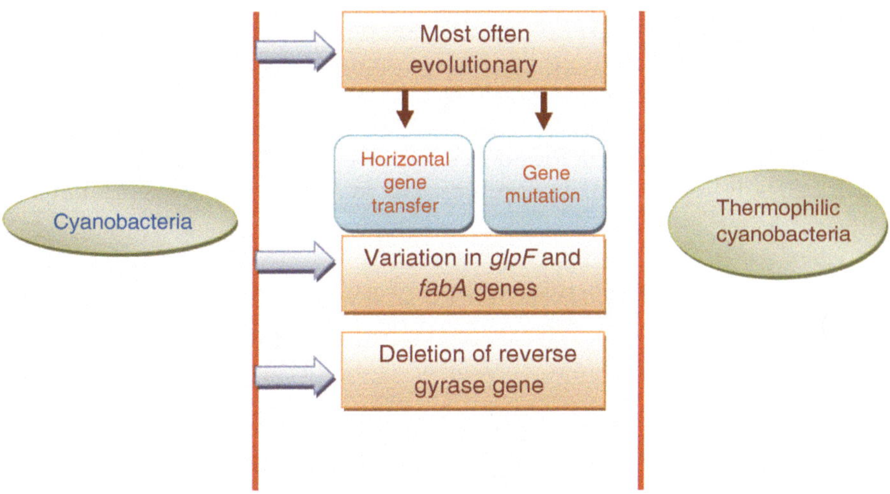

FIG. 1.2 Genetic factors contributing to thermophilic adaptation in cyanobacteria.

Polar region. HGT mechanisms have been reported for cold-adapted aromatic degrading bacteria (Panicker et al., 2006).

Moreover, exopolysaccharides synthesis stimulated dormancy, pigments production, sheaths, and specialized enzymes that played a critical role to prevent cell lysis during freeze-thaw cycles of the psychrophilic cells. The molecular analytical and gene mapping approach better perceives the microbial diversity and distribution pattern in polar environments (Komárek, 2016). The psychrophiles tolerate low temperatures because their cytosol conquers damage caused by freezing inside the cells. The psychrophiles cells produce antifreeze protein that permits the cells to stay alive below freezing point, and such antifreeze proteins are being used in the food technology as cryoprotectants. Deep investigation of psychrophiles could provide great insight into an environmental change due to global warming and astrobiology (Hoover et al., 2010). Psychrophilic bacteria have molecular adaptations in their membrane lipids to become accustomed to low temperature (Russell, 1997). The imperative parameters engaged in survival strategy and adaptation at low temperature in cyanobacterial cells involves modified lipid membrane fluidity, production of cold-active proteins, freeze tolerance adaptability, and cold-active enzymes (Jagannadham et al., 2000; Russell, 1997) (Fig. 1.3).

1.2.3 Halophiles

The identification and characterization of halophilic cyanobacteria have been done with an ideology that the mass production of a high-value cyanobacteria product would be potentially possible saline seawater (Pade and Hagemann, 2014). Unicellular, filamentous, and tightly coiled cyanobacteria are found in the hypersaline condition. *Aphanothece halophytica* and similar morphological cyanobacteria such as Cyanothece, Euhalothece, and Halothece are predominant in light-exposed hypersaline environments containing 15% salt (Oren, 2015). *Halospirulina tapeticola, Coleofasciculus chthonoplastes,* and *Halomicronema excentricum* are filamentous cyanobacteria found in a hypersaline solution containing salt up to 20% (Oren, 2015). Halophiles can synthesize and accumulate a

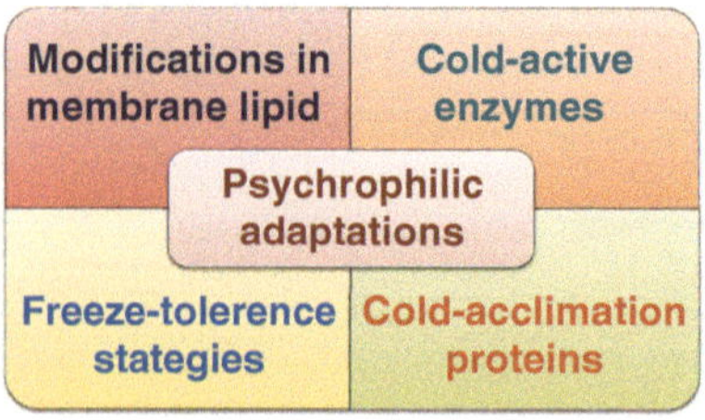

FIG. 1.3 Key factors involved in psychrophilic adaptation in cyanobacteria.

higher concentration of minerals in cytosol mediated by glycerol, which balances inner and outer osmotic pressure and prevents plasmolysis. Cytoplasmic osmotic pressure, and halophilic proteins played a key role to maintain cellular activities of halophiles in high salt concentration (Siglioccolo et al., 2011). The accumulation of high salt concentration in halophiles is driven by "salt-in" and "osmolyte" strategies.

1.2.4 Acidophiles

Acidophiles prokaryotes are known for their ability to use diversified energy sources (i.e., light, inorganic and organic) for cellular growth at both aerobic and anaerobic environments and at temperatures ranging between 4 °C and 96 °C (Rampelotto, 2013; Zahra et al., 2020). Nevertheless, no cyanobacterial strain has been found to be significant acidophilic nature, it is seldom (Gomez-Alvarez et al., 2012). However, few cyanobacteria are found that can survive at pH as low as 2.8 (West and Louda, 2011). The low pH assists metal solubility in water; Thus, acidic waters can retain higher metal concentrations than neutral and alkaline water (Pedrozo et al., 2001). The acidophilic cyanobacteria could significantly be applied in the removal of toxic heavy metals from soil and water.

1.2.5 Alkaliphiles

Photosynthetic prokaryotes uptake carbon dioxide during photosynthesis, which enhances pH in general (Boussiba and Gibson, 1991). Thus cyanobacteria most often tend to be exposed at higher pH during the peak time of photosynthesis. Some cyanobacterial genera, such as *Spirulina platensis*, *Synechococcus*, and *Anabaenopsis*, have been revealed to acclimatize in the high or extreme alkaline niche (Boussiba et al., 2000). *Plectonema nostocorum* has been shown to grow up to pH 13 (Boussiba et al., 2000; Boussiba and Gibson, 1991). Many alkaliphilic cyanobacteria (34 species) were discovered by Garasimenko et al. (1996).

1.2.6 Xerophilic

A diverse group of cyanobacteria are found in the desert crust (Garcia-Pichel et al., 2001). The cyanobacteria found in the arid and semiarid regions produce many bioactive compounds with antibacterial, antimicrobial, and antiinflammatory properties (Sayed et al., 2020). *Chroococcidiopsis* species, endolithic cyanobacteria, are commonly found in the desert because of the ability to tolerate solar radiation and aridity (Lacap-Bugler et al., 2017); they were first reported in the Negev desert (Potts and Friedmann, 1981). *Chroococcidiopsis* species were also reported from the hyperarid core of the Atacama desert (Wierzchos et al., 2006). The cyanobacteria have adapted to extreme conditions that prevail in the Atacama Desert (Gómez-Silva, 2018) and Namib Desert (Stomeo et al., 2013). Studies suggested that hyperarid regions of the Atacama Desert are dominated by *Chroococcidiopsis, Nostocales,* and *chloroflexi* (Lacap et al., 2011) species of cyanobacteria. In contrast, *Nostoc, Scytonema* and *synechococcophycideae* are present in comparatively low number (Vítek et al., 2014). In a study, *Xeronema,* especially Phormidium-like cyanobacteria and *Schizothrix* species, was found in Colorado Plateau (Garcia-Pichel et al., 2001). Cyanobacteria produce pigments to cope with the desert condition (Powell et al., 2015). The capability of masking excessive solar radiation (Photosynthetically Active Radiation (PAR) and UV-B) by the cyanobacterial cells (Sayed et al., 2020) to cope with arid and semiarid biomes could potentially be applicable in sunscreen research and development. The dry condition of the desert has posed constraints to cyanobacteria that it has diversified itself to cope with challenges such as desiccation, extreme temperature, and UV radiation (Garcia-Pichel et al., 2001).

1.3 Application of extremophile cyanobacteria

The molecular strategies are not fully explained for the study of organism's survival in extreme environments. It is believed that specific biomolecules and alternative biological pathways are involved in such survival capabilities developed in prokaryotes. Certain cyanobacterial species' stability and biological activity in extreme conditions make them the center of interest for biotechnological purposes. The enzymes are the fundamental requirement for the catalytic activity of biochemical processes carried out during cellular metabolism. The stability and catalytic efficacy of such enzymes under extreme environment attract the scientific communities to explore the industrial application of such microbial cells to improve the bioprospecting approach. Certain enzymes exhibit stability and activity in multiple extreme ecosystems (called poly-extremophilic), making them suitable for applied research in industrial sectors. The research on extremophiles would also be applicable in astrobiology. The investigation of extremophiles ecosystems stimulates the research possibility upon the hypotheses of "the life also presents elsewhere in the universe." Raman spectroscopy could be reported to be helpful in the study of extremophilic organisms (as a model organism) in the context of astrobiology. Nevertheless, physiology, metabolic engineering, and molecular biological studies are still necessary to explore the broader application of extremophiles cyanobacterial communities.

Extremophiles can survive in harsh natural conditions such as the desert, permafrost soil, deep sea, high temperature, etc. The extreme natural habitat selects microorganisms that can specifically grow on such harsh conditions to synthesize molecules with therapeutic potential. Several studies have shown that microorganisms growing in severe conditions are rich sources of diverse bioactive compounds (Sayed et al., 2020). Several bioactive compounds (i.e., alkaloids, amides, lipopeptides, indoles, and fatty acids) have been isolated from cyanobacteria and reported to have potential such as antiviral, antibacterial, antialgal antifungal (Mayer et al., 2009; Wase and Wright, 2008; Zahra et al., 2020). Apart from cyanobacteria's biocidal properties, extensive research is being carried out to explore the role of cyanobacteria to mitigate plant diseases (Biondi et al., 2004).

Cyanobacteria found in the arid and semiarid regions produce many bioactive compounds (Sayed et al., 2020), and cyanobacteria produce scyctonemin, porphyra-334 shinorine like pigment to cope with UV radiation; these compounds can be used to develop sunscreens (Powell et al., 2015). Syctonemin is produced by *Chroococcidiopsis* species and can work as a biomarker for cyanobacteria found in arid and semiarid regions (Vítek et al., 2014). Syctonemin can be used as a kinase inhibitor due to its property to inhibit *polo*-like kinase responsible for inflammation (D'Orazio et al., 2012). Inactivating NF-kB shinorine and Porphyra-334 can also work as antiinflammatory compounds (Becker et al., 2016).

Extremophile cyanobacteria found in extreme environments such as alkaline condition, desert, permafrost, hypersaline condition produce various metabolites in the form of alkaloids, cyclic peptide and amino acids, which can be classified as cytotoxins, neurotoxins, and hepatotoxins (Buratti et al., 2017; Metcalf et al., 2012). The biosynthesis of neurotoxins, such as 2,4,amino butyric acid, anatoxin-a (S), anatoxin-a, and beta methyl-amino-propionic acid is conducted in *Synechococcus, Arthrospira and Oscillatoria* cyanobacterial genera (Cirés et al., 2017). Cylindrospermopsin and nodularin, hepatotoxins, are also known to be biosynthesized in the same cyanobacterial genera. Apart from toxicological activity, anatoxin-a and cylindrospermopsin are proven to be estrogenic in both *in-vivo* and *in-vitro* conditions (Liu et al., 2018, 2019).

Halophilic cyanobacteria can survive in hypersaline condition with salt concentration up to 25%. Like other microorganism, cyanobacteria synthesize small zwitterionic or uncharged soluble organic solute to survive under high salt concentration (Oren, 2015). There is a clear relationship between the level of salinity and the type of solute synthesize in the cyanobacteria (Oren, 2015). Cyanobacteria grow in low salt concentration, synthesizing disaccharide or trehalose to cope with salinity (Oren, 2015). Cyanobacteria growing under moderate-to-high salinity biosynthesize glucosyl glycerol (2-O-a-D-glucopyranosyl-(1-2)-glycerol) (Klähn and Hagemann, 2011) and glycine betaine (N, N, N-trimethylglycine) (Goh et al., 2010) to tolerate salinity. Mycosporins, such as amino acids Mycosporine like amino acids (MAAs), glucosyl glycerate, and trimethylamine-N-oxide, are also found in salt-tolerant organisms and can help survive under high salt concentration (Oren, 1997).

Cyanobacteria have played a significant role in enhancing agricultural productivity and minimizing greenhouse gases (Singh et al., 2011). Cyanobacteria are known for nitrogen fixation. *Nodularia spumigena* and *Scytonema* found in the moderately saline lake are filamentous cyanobacteria that can fix the atmospheric nitrogen (Paerl et al., 2000; Roney et al., 2009). Many isolates of halophilic cyanobacteria have nitrogenase gene cluster and *nifH* gene, which are different from typical *nifH* gene found in cyanobacteria (Bolhuis et al., 2010). Many halophilic nonfilamentous, nonheterocystous cyanobacteria such as *Synechococcus, Coleofasciculus,* and *Aphanothece* can also fix nitrogen (Rothschild et al., 1994).

Cyanobacteria could also be used to refurbish degraded lands that lose natural productivity due to vigorous industrial pollution (Singh, 2014). Desertification is another concern in the agriculture domain. Cyanobacterial inoculums help eliminate desertification with a symbiotic association with other microbes, that is, bacteria, algae, fungi, and lichens. The biological soil crusts often consisted of diverse microbial strains as an alliance in distinct geographical regions (Rossi et al., 2017). Soil serves as an ecological niche for the primary colonization of microbes in deserted areas by supplementing the nutrient and moisture contents to microbial cells (Rossi et al., 2017). Cyanobacteria have been noted to persuade soil aggregation and water permeability in salty soil and help to improve the poor soil structure of desert lands. Cyanobacteria promote soil aggregation that leads to enhanced water holding capacity and aeration in soil, resulting in the rapid growth of soil microflora (Singh et al., 2016). A cyanobacterium assists in the reclamation of soil by binding the soil particles to improve soil permeability and aeration that results in higher agricultural productivity (Singh et al., 2011). Cyanobacteria secrete oxalic acid, which lowers soil's pH and electrical conductivity (Singh et al., 2016). The exopolysaccharides produced from cyanobacteria have been beneficial for the binding of soil particles to make soil aggregates in desert soils (Singh et al., 2016).

Cyanobacteria are also known for sustainable green energy production (i.e., methane, ethanol, hydrogen, butanol, and syngas (a mixture of hydrogen, carbon monoxide, and carbon dioxide)) during the last few decades (Koller et al., 2012). Many investigations have explored the applications of cyanobacterial strains toward the biofuels, superfoods, biofertilizers, biopesticides, and bioremediation (Benson and Orr, 2008; Yamaguchi, 1996). *Coleofasciculus,* a submerged Cyanobacteria and *Lyngbya,* an intertidal Cyanobacteria can ferment glycogen accumulated by photosynthesis in the day time and generate hydrogen gas and other products (Lee et al., 2014). The developed products are then used as an electron acceptor by sulfur bacteria. *Coleofasciculus chthonoplastes* generate acetate by fermentation of carbohydrates in the dark (Moezelaar et al., 1996), similarly *Oscillatoria limnetica* produces lactate in the dark (Oren and Shilo, 1979). Both the cyanobacteria can use sulfur as an electron acceptor and form sulfide at night (Moezelaar et al., 1996; Oren and Shilo, 1979).

Cyanobacteria are valuable for biological wastewater treatment to degrade or assimilate toxic substances present in the soil and water (Cohen, 2002). The diversification of cyanobacteria in ecology is retained due to their extensive metabolic activity. The significant role of extremophilic cyanobacteria has been documented for agriculture and environmental sustainability, bioenergy production, amelioration of degraded lands, and bioremediation (Singh et al., 2016; Singh and Strong, 2016). Psychrophilic cyanobacteria (genera *Oscillatoria*) have been noted for their ability to partially oxidize hydrocarbons (i.e., phenyl and naphthalene) (Cerniglia et al., 1979). Psychrotolerant cyanobacteria have been reported for significant bioremediation of hydrocarbon-contaminated soil of Antarctica because of their ability to maintain biochemical enzymatic activity at low temperature (Panicker et al., 2006). Plants in the arctic region cannot fix nitrogen and are helped by cyanobacteria for nitrogen and carbon dioxide uptake (Sayed et al., 2020).

The removal of Biological Oxygen Demand (BOD) per kg from activated sludge needs 1 kWh electric energy inputs for aeration. It generates 1 kg CO_2, while BOD per kg removal mediated by photosynthetic organism via oxygenation does not require energy. Byproduct generates sufficient algal biomass that can reduce methane to make 1 kWh of electric energy (Blok et al., 1985). The cyanobacterial-mediated wastewater treatment has created a center of attention worldwide since the 1960s due to its efficient and low-cost treatment efficiency (Rawat et al., 2011). The wastewater most often contains organic and inorganic content for cyanobacterial (such as *Anabaena oryzae*, *Anabaena variabilis*, and *Tolypothrix ceytonica*) growth and generates cyanobacterial biomass from which high-value compounds can be extracted (Chisti, 2020; El-Bestawy, 2008). The wastewater has high Chemical Oxygen Demand (COD) due to enough nitrogen and phosphorus content, which serves as a nutrient source for the optimal growth of cyanobacteria. By this means, the cyanobacterial cells decrease the nutrient load and make the wastewater usable for new purposes. *Oscillatoria* and *Phormidium* species have been documented for 75.22% and 81.69% nitrogen and 86% and 94% phosphorus removal efficiency (Kabariya and Ramani, 2018). (Fig. 1.4).

The survival capability of cyanobacteria in a high concentration of heavy metals could be possible due to the morphological plasticity, diverse metabolic route, absorption and adsorption efficacy (Sood et al., 2015). Thus the cyanobacterial growth in a polluted wastewater region improves water quality. Cyanobacterial could be significantly applicable in the remediation of wastewaters from contaminants , such as heavy metals. *Tolypothrix ceytonica* have been reported for the highest biosorption capacity, where 86.12 and 94.63% Removal Efficiencies (RE) were achieved for Zn and Cu, respectively (El-Bestawy, 2008)(Zahra et al., 2020).

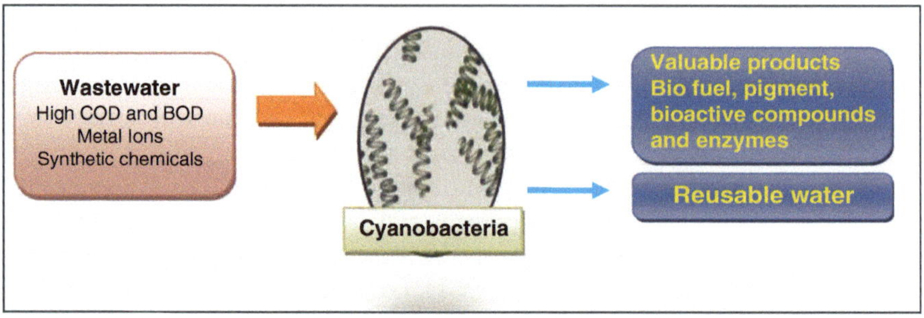

FIG. 1.4 Wastewater treatment and generation of valuable products by cyanobacteria.

The biodegradation of organic pollutants by cyanobacterial strains is also dependent upon mixotrophic growth. *Anabaena flosaquae* 4054 strain has been reported to degrade the endocrine-disrupting pollutants, that is phthalate esters (Babu and Wu, 2010), while another *Anabaena azotica* species significantly decompose pesticide hexachlorocyclohexane (also known as lindane) (Zhang et al., 2012). The cyanobacterial adaptability in a toxic niche makes it usable to remove the harmful components from the ecosystem by accumulation and degradation. The removal of toxic ingredients comprises pesticides (Megharaj et al., 1994), phenol (Ellis, 1977), and xenobiotics (Megharaj et al., 1987) from the wastewater. The low pH drives a higher rate of metal solubility in water. Thus, the acidophiles could be applied for remediation of heavy metals from wastewater.

Cyanobacterial species including *A. flosaquae* UTCC64, *Phormidium autumnale* UTEX1580, and *Synechococcus* sp. PCC7942 could potentially degrade textile dyes such as indigo and Remazol Brilliant Blue R. These cyanobacterial strains were studied in the tertiary treatment of wastewater effluent of textile industries. Among the *P. autumnale* UTEX1580 exhibited 100% degradation of the indigo dye (Dellamatrice et al., 2017). *Gloeocapsa pleurocapsoides* and *Phormidium ceylanicum* have been explored to decolorize cyclic azo dyes in different studies and showed approximately 80% decolorization of Acid Red 97 and FF Sky Blue dyes in 26 days. Moreover, 55% decolorization of Amido Black 10B dye was observed in *Chroococcus minutus* (Parikh and Madamwar, 2005). Cyanobacterial genus *Spirulina* has been reported to 100% remove nitrogen from wastewaters (Chevalier et al., 2000). *Phormidium tenue* has been documented to decolorize up to 60–90% wastewater (Klopsteg, 1966). Cyanobacteria have been explored for their essential role in the decontamination of oil derivatives from wastewaters and degradation of hydrocarbons (Cohen, 2002). Cyanobacteria hold an outstanding capability to become accustomed to diverse environments. Due to the augmented adaptive nature and faster rate of growing ability of cyanobacteria, it can significantly be applicable in the remediation of the contaminated ecosystem. The physiological and genetic adaptation helps the cyanobacteria cell to survive under a polluted ecosystem (Rampelotto, 2013). The high acclimatization and adaptation rate of cyanobacterial strains to diverse environmental conditions rely on mixotrophic growth. Cyanobacterial species have the potential to remove a broad range of contaminants from soil system and water bodies. In sustainable agricultural practice, the cyanobacterial species such as *Anabaena* sp., *Lyngya* sp., *Spirulina* sp., *Nostoc* sp. are well known to remove herbicide glyphosate from agricultural lands. Additionally, the genetic engineering tools could gain attention for the improvement of bioremediation practices to build up economic and eco-friendly remediation approach (Cuellar-Bermudez et al., 2017; Kuritz and Wolk, 1995).

1.4 Conclusions

Extremophilic bacteria, archaea, and other microbes enjoy cruel habitation as paradise due to their altered metabolic pathway and modified cellular activities. They are incredibly flexible to diverse niche and able to overcome severe ecological conditions and grow. Extremophiles can flourish in hot regions (as thermophiles), coldest province (as psychrophiles), salty area (as halophiles), acidic conditions (as acidophiles), alkaline niche (as alkaliphiles), desertic regions (as xerophiles), and anaerobic (absence of oxygen) based on their customized adaptation. The extremophile synthesizes various bioactive compounds used in medicine, biofuel production, the food industry, and agriculture by their altered metabolic activity. Extremophiles are the most prevalent source of diverse secondary metabolite. Our chemical libraries cannot match the diversity of bioactive compounds provided by the extremophile. The ability of cyanobacteria to survive in highly polluted water gives a new venue for wastewater treatment and suggests the possible application of cyanobacteria in waste treatments. Few species of cyanobacteria which can live in acidic condition provides new bioprospecting window in biomining. Recent findings in the bioprospecting of the extremophiles provide evidence that extreme situation naturally selects organism with modified biochemistry giving opportunity for new drug discoveries. Many drug candidates were identified from the organisms living in the extreme environment.

Further understanding in the identification of extremophile cyanobacteria by advanced genome sequencing method and bioinformatics tool can help discover new sources of therapeutic compounds. Further bioprospecting for bioactive compounds should look toward the underexplored biomes by advanced molecular techniques. These extremophiles can open up extreme ecological and biochemical aspects of living cells and show how to utilize such microbes for extensive industrial applications using biotechnological tools.

References

Amarouche-Yala, S., Benouadah, A., Bentabet, El Ouahab, A., López-García, 2014. Morphological and phylogenetic diversity of thermophilic cyanobacteria in Algerian hot springs. Extremophiles 18, 1035–1047. https://doi.org/10.1007/s00792-014-0680-7.

Atomi, H., Matsumi, R., Imanaka, T., 2004. Reverse gyrase is not a prerequisite for hyperthermophilic life. J. Bacteriol. 186, 4829–4833. https://doi.org/10.1128/JB.186.14.4829-4833.2004.

Averhoff, B., Müller, V., 2010. Exploring research frontiers in microbiology: recent advances in halophilic and thermophilic extremophiles. Res. Microbiol. 161, 506–514. https://doi.org/10.1016/j.resmic.2010.05.006.

Babu, B., Wu, J.-T., 2010. Biodegradation of phthalate esters by cyanobacteria1. J. Phycol. 46, 1106–1113. https://doi.org/10.1111/j.1529-8817.2010.00896.x.

Becker, K., Hartmann, A., Ganzera, M., Fuchs, D., Gostner, J.M., 2016. Immunomodulatory effects of the mycosporine-like amino acids *Shinorine* and *Porphyra*-334. Mar. Drugs 14. https://doi.org/10.3390/md14060119.

Benson, S.M., Orr, F.M., 2008. Carbon dioxide capture and storage. MRS Bull 33, 303–305. https://doi.org/DOI:10.1557/mrs2008.63.

Biondi, N., Piccardi, R., Margheri, M.C., Rodolfi, L., Smith, G.D., Tredici, M.R., 2004. Evaluation of Nostoc strain ATCC 53789 as a potential source of natural pesticides. Appl. Environ. Microbiol. 70, 3313–3320. https://doi.org/10.1128/AEM.70.6.3313-3320.2004.

Blok, J., de Morsier, A., Gerike, P., Reynolds, L., Wellens, H., 1985. Harmonization of ready biodegradability tests. Chemosphere 14, 1805–1820. https://doi.org/10.1016/0045-6535(85)90123-7.

Bolhuis, H., Severin, I., Confurius-Guns, V., Wollenzien, U.I.A., Stal, L.J., 2010. Horizontal transfer of the nitrogen fixation gene cluster in the cyanobacterium *Microcoleus chthonoplastes*. ISME J 4, 121–130. https://doi.org/10.1038/ismej.2009.99.

Boussiba, S., Gibson, J., 1991. Ammonia translocation in cyanobacteria. FEMS Microbiol. Rev. 8, 1–14. https://doi.org/10.1111/j.1574-6968.1991.tb04953.x.

Boussiba, S., Wu, X., Zarka, A., 2000. Alkaliphilic cyanobacteria in: Seckbach, J. (Ed.), Journey to Diverse Microbial Worlds: Adaptation to Exotic Environments,. Springer
Netherlands, Dordrecht, pp. 209–224. https://doi.org/10.1007/978-94-011-4269-4_15.

Buratti, F.M., Manganelli, M., Vichi, S., Stefanelli, M., Scardala, S., Testai, E., Funari, E., 2017. Cyanotoxins: producing organisms, occurrence, toxicity, mechanism of action and human health toxicological risk evaluation. Arch. Toxicol. 91, 1049–1130. https://doi.org/10.1007/s00204-016-1913-6.

Castenholz, R., 2004. Species usage, concept, and evolution in the cyanobacteria (blue-green algae). J. Phycol. 28, 737–745. https://doi.org/10.1111/j.0022-3646.1992.00737.x.

Castenholz, R.W., Wilmotte, A., Herdman, M., Rippka, R., Waterbury, J.B., Iteman, I., Hoffmann, L., 2001. Phylum BX. Cyanobacteria in Boone, D.R., Castenholz, R.W., Garrity, G.M. (Eds.), Bergey's Manual® of Systematic Bacteriology: Volume One : The Archaea and the Deeply Branching and Phototrophic Bacteria,. Springer New York, NY, pp. 473–599. https://doi.org/10.1007/978-0-387-21609-6_27.

Cerniglia, C.E., Gibson, D.T., Baalen, C.Van, 1979. Algal oxidation of aromatic hydrocarbons: formation of 1-naphthol from naphthalene by *Agmenellum quadruplicatum*, strain PR-6. Biochem. Biophys. Res. Commun. 88, 50–58. https://doi.org/10.1016/0006-291X(79)91695-4.

Cerniglia, C.E., Van Baalen, C., Gibson, D.T., 1980. Oxidation of biphenyl by the cyanobacterium, *Oscillatoria* sp., strain JCM. Arch. Microbiol. 125, 203–207. https://doi.org/10.1007/BF00446877.

Chevalier, P., Proulx, D., Lessard, P., (2000). Nitrogen and Phosphorous removal by high latitude mat-forming cyanobacteria for potential use in tertiary wastewater treatment. Journal of Applied Phycology 12, 105–112. https://doi.org/10.1023/A:1008168128654.

Chisti, Y., 2020. Chapter 1 - Microalgae biotechnology: A brief introduction. In: Jacob-Lopes, E., Maroneze, M.M., Queiroz, M.I., Zepka, L. (Eds.), Handbook of Microalgae-Based Processes and Products: Fundamentals and Advances in Energy, Food, Feed, Fertilizer, and Bioactive Compounds. Academic Press, Cambridge, MA, pp. 3–23. https://doi.org/10.1016/B978-0-12-818536-0.00001-4.

Chu, S.P., 1942. The influence of the mineral composition of the medium on the growth of planktonic algae: Part I. methods and culture media. J. Ecol. 30, 284–325. https://doi.org/10.2307/2256574.

Cirés, S., Casero, M.C., Quesada, A., 2017. Toxicity at the edge of life: a review on cyanobacterial toxins from extreme environments. Mar. Drugs 15. https://doi.org/10.3390/md15070233.

Cohen, Y., 2002. Bioremediation of oil by marine microbial mats. Int. Microbiol. 5, 189–193. https://doi.org/10.1007/s10123-002-0089-5.

Cuellar-Bermudez, S.P., Aleman-Nava, G.S., Chandra, R., Garcia-Perez, J.S., Contreras-Angulo, J.R., Markou, G., Muylaert, K., Rittmann, B.E., Parra-Saldivar, R., 2017. Nutrients utilization and contaminants removal. A review of two approaches of algae and cyanobacteria in wastewater. Algal Res 24, 438–449. https://doi.org/10.1016/j.algal.2016.08.018.

D'Orazio, N., Gammone, M.A., Gemello, E., De Girolamo, M., Cusenza, S., Riccioni, G., 2012. Marine bioactives: pharmacological properties and potential applications against inflammatory diseases. Mar. Drugs 10, 812–833. https://doi.org/10.3390/md10040812.

Dellamatrice, P.M., Silva-Stenico, M.E., Moraes, L.A.B.de, Fiore, M.F., Monteiro, R.T.R., 2017. Degradation of textile dyes by cyanobacteria. Braz. J. Microbiol. 48, 25–31. https://doi.org/10.1016/j.bjm.2016.09.012.

Drobac-Cik, A., Dulic, T., Stojanovic, D., Svircev, Z., 2007. The importance of extremophile cyanobacteria in the production of biologically active compounds. Zb. Matice Srp. Prir. Nauk., 57–66. https://doi.org/10.2298/zmspn0712057d.

El-Bestawy, E., 2008. Treatment of mixed domestic–industrial wastewater using cyanobacteria. J. Ind. Microbiol. Biotechnol. 35, 1503–1516. https://doi.org/10.1007/s10295-008-0452-4.

Ellis, B.E., 1977. Degradation of phenolic compounds by fresh-water algae. Plant Sci. Lett. 8, 213–216. https://doi.org/10.1016/0304-4211(77)90183-3.

Ferris, M.J., Hirsch, C.F., 1991. Method for isolation and purification of cyanobacteria. Appl. Environ. Microbiol. 57, 1448–1452. https://doi.org/10.1128/AEM.57.5.1448-1452.1991.

Forlani, G., Pavan, M., Gramek, M., Kafarski, P., Lipok, J., 2008. Biochemical bases for a widespread tolerance of cyanobacteria to the phosphonate herbicide glyphosate. Plant Cell Physiol 49, 443–456. https://doi.org/10.1093/pcp/pcn021.

Garcia-Pichel, F., López-Cortés, A., Nübel, U., 2001. Phylogenetic and morphological diversity of cyanobacteria in soil desert crusts from the Colorado Plateau. Appl. Environ. Microbiol. 67. 1902 –1910. https://doi.org/10.1128/AEM.67.4.1902-1910.2001.

Gerasimenko, L.M., Dubinin, A.V, Zavarzin, G.A., 1996. Alkaliphilic cyanobacteria from soda lakes of Tuva and their ecophysiology. Microbiology (New York, NY) 65, 736–740.

Goh, F., Barrow, K.D., Burns, B.P., Neilan, B.A., 2010. Identification and regulation of novel compatible solutes from hypersaline stromatolite-associated cyanobacteria. Arch. Microbiol. 192, 1031–1038. https://doi.org/10.1007/s00203-010-0634-0.

Gomez-Alvarez, V., Revetta, R.P., Santo Domingo, J.W., 2012. Metagenome analyses of corroded concrete wastewater pipe biofilms reveal a complex microbial system. BMC Microbiol 12 (122). https://doi.org/10.1186/1471-2180-12-122.

Gómez-Silva, B., 2018. Lithobiontic life: "Atacama rocks are well and alive. Antonie Van Leeuwenhoek 111, 1333–1343. https://doi.org/10.1007/s10482-018-1033-9.

Kabariya, H., Ramani, J., M.V., 2018. Dairy wastewater treatment by cyanobacteria for removal of nutrients with extraction of high value compounds from biomass. Int. J. Curr. Microbiol. Appl. Sci. 7, 1527–1538. https://doi.org/10.20546/ijcmas.2018.704.172.

Hoover, R., Pikuta, E., Hoover@nasa, R., Gov, 2010. Psychrophilic and psychrotolerant microbial extremophiles in polar environments. pp. 115–156. https://doi.org/10.1201/9781420083880-c5.

Jacob-Lopes, E., Scoparo, C.H.G., Franco, T.T., 2008. Rates of CO_2 removal by *Aphanothece microscopica Nägeli* in tubular photobioreactors. Chem. Eng. Process. Process Intensif. 47, 1365–1373. https://doi.org/10.1016/j.cep.2007.06.004.

Jagannadham, M.V, Chattopadhyay, M.K., Subbalakshmi, C., Vairamani, M., Narayanan, K., Rao, C.M., Shivaji, S., 2000. Carotenoids of an Antarctic psychrotolerant bacterium, *Sphingobacterium antarcticus*, and a mesophilic bacterium, *Sphingobacterium multivorum*. Arch. Microbiol. 173, 418–424. https://doi.org/10.1007/s002030000163.

Klähn, S., Hagemann, M., 2011. Compatible solute biosynthesis in cyanobacteria. Environ. Microbiol. 13, 551–562. https://doi.org/10.1111/j.1462-2920.2010.02366.x.

Klopsteg, P.E., 1966. Environmental sciences.Science80, 595. https://doi.org/10.1126/science.152.3722.595.

Koller, M., Salerno, A., Tuffner, P., Koinigg, M., Böchzelt, H., Schober, S., Pieber, S., Schnitzer, H., Mittelbach, M., Braunegg, G., 2012. Characteristics and potential of micro algal cultivation strategies: a review. J. Clean. Prod. 37, 377–388. https://doi.org/10.1016/j.jclepro.2012.07.044.

Komárek, J., 2016. A polyphasic approach for the taxonomy of cyanobacteria: principles and applications. Eur. J. Phycol. 51, 346–353. https://doi.org/10.1080/09670262.2016.1163738
.

Kulik, M.M., 1995. The potential for using cyanobacteria (blue-green algae) and algae in the biological control of plant pathogenic bacteria and fungi. Eur. J. Plant Pathol. 101, 585–599. https://doi.org/10.1007/BF01874863.

Kumar, K., Dasgupta, C.N., Nayak, B., Lindblad, P., Das, D., 2011. Development of suitable photobioreactors for CO_2 sequestration addressing global warming using green algae and cyanobacteria. Bioresour. Technol. 102, 4945–4953. https://doi.org/10.1016/j.biortech.2011.01.054.

Kuritz, T., Wolk, C.P., 1995. Use of filamentous cyanobacteria for biodegradation of organic pollutants. Appl. Environ. Microbiol. 61 234–238.

Lacap-Bugler, D.C., Lee, K.K., Archer, S., Gillman, L.N., Lau, M.C.Y., Leuzinger, S., Lee, C.K., Maki, T., McKay, C.P., Perrott, J.K., de los Rios-Murillo, A., Warren-Rhodes, K.A., Hopkins, D.W., Pointing, S.B., 2017. Global diversity of desert hypolithic cyanobacteria. . Front. Microbiol. 8 (867). https://doi.org/10.3389/fmicb.2017.00867.

Lacap, D.C., Warren-Rhodes, K.A., McKay, C.P., Pointing, S.B., 2011. Cyanobacteria and chloroflexi-dominated hypolithic colonization of quartz at the hyper-arid core of the Atacama Desert. Chile. Extremophiles 15, 31–38. https://doi.org/10.1007/s00792-010-0334-3.

Lang-Yona, N., Kunert, A.T., Vogel, L., Kampf, C.J., Bellinghausen, I., Saloga, J., Schink, A., Ziegler, K., Lucas, K., Schuppan, D., Pöschl, U., Weber, B., Fröhlich-Nowoisky, J., 2018. Fresh water, marine and terrestrial cyanobacteria display distinct allergen characteristics. Sci. Total Environ. 612, 767–774. https://doi.org/10.1016/j.scitotenv.2017.08.069.

Lee, J., Burow, L., Woebken, D., Everroad, R., Kubo, M., Spormann, A., Weber, P., Pett-Ridge, J., Bebout, B., Hoehler, T., 2014. Fermentation couples *Chloroflexi* and sulfate-reducing bacteria to cyanobacteria in hypersaline microbial mats. Front. Microbiol. 5, 61. https://doi.org/10.3389/fmicb.2014.00061.

Lipok, J., Owsiak, T., Młynarz, P., Forlani, G., Kafarski, P., 2007. Phosphorus NMR as a tool to study mineralization of organophosphonates—the ability of *Spirulina* spp. to degrade glyphosate. Enzyme Microb. Technol. 41, 286–291. https://doi.org/10.1016/j.enzmictec.2007.02.004.

Lipok, J., Wieczorek, D., Jewgiński, M., Kafarski, P., 2009. Prospects of in vivo 31P NMR method in glyphosate degradation studies in whole cell system. Enzyme Microb. Technol. 44, 11–16. https://doi.org/10.1016/j.enzmictec.2008.09.011.

Liu, J., Hernández, S.E., Swift, S., Singhal, N., 2018. Estrogenic activity of cylindrospermopsin and anatoxin-a and their oxidative products by Fe(III)-B*/H(2)O(2). Water Res 132, 309–319. https://doi.org/10.1016/j.watres.2018.01.018.

Liu, S.-W., Li, F.-N., Qi, X., Xie, Y.-Y., Sun, C.-H., 2019. *Nakamurella deserti* sp. nov., isolated from rhizosphere soil of Reaumuria in the Taklamakan desert. Int. J. Syst. Evol. Microbiol. 69, 214–219. https://doi.org/10.1099/ijsem.0.003132.

Lyra, C., Suomalainen, S., Gugger, M., Vezie, C., Sundman, P., Paulin, L., Sivonen, K., 2001. Molecular characterization of planktic cyanobacteria of Anabaena, *Aphanizomenon, Microcystis* and *Planktothrix* genera. Int. J. Syst. Evol. Microbiol. 51, 513–526. https://doi.org/10.1099/00207713-51-2-513.

Masukawa, H., Nakamura, K., Mochimaru, M., Sakurai, H., 2001. Photobiological hydrogen production and nitrogenase activity in some heterocystous cyanobacteria. In: J. Miyake, T. Matsunaga, and A.S. Pietro (Eds.). Biohydrogen II: An Approach to Environmentally Acceptable Technology, Elsevier, Amsterdam, pp. 63–66. https://doi.org/10.1016/B978-008043947-1/50005-3.

Mayer, A.M.S., Rodríguez, A.D., Berlinck, R.G.S., Hamann, M.T., 2009. Marine pharmacology in 2005-6: marine compounds with anthelmintic, antibacterial, anticoagulant, antifungal, anti-inflammatory, antimalarial, antiprotozoal, antituberculosis, and antiviral activities; affecting the cardiovascular, immune and nervous syst. Biochim. Biophys. Acta 1790, 283–308. https://doi.org/10.1016/j.bbagen.2009.03.011.

Megharaj, M., Madhavi, D.R., Sreenivasulu, C., Umamaheswari, A., Venkateswarlu, K., 1994. Biodegradation of methyl parathion by soil isolates of microalgae and cyanobacteria. Bull. Environ. Contam. Toxicol. 53, 292–297. https://doi.org/10.1007/BF00192047.

Megharaj, M., Venkateswarlu, K., Rao, A.S., 1987. Metabolism of monocrotophos and quinalphos by algae isolated from soil. Bull. Environ. Contam. Toxicol. 39, 251–256. https://doi.org/10.1007/BF01689414.

Metcalf, J.S., Richer, R., Cox, P.A., Codd, G.A., 2012. Cyanotoxins in desert environments may present a risk to human health. Sci. Total Environ. 421–422, 118–123. https://doi.org/10.1016/j.scitotenv.2012.01.053.

Metting, B., Pyne, J.W., 1986. Biologically active compounds from microalgae. Enzyme Microb. Technol. 8, 386–394. https://doi.org/10.1016/0141-0229(86)90144-4.

Miller, S.R., Castenholz, R.W., Pedersen, D., 2007. Phylogeography of the thermophilic cyanobacterium *Mastigocladus laminosus*. Appl. Environ. Microbiol. 73, 4751–4759. https://doi.org/10.1128/AEM.02945-06.

Moezelaar, R., Bijvank, S.M., Stal, L.J., 1996. Fermentation and sulfur reduction in the mat-building cyanobacterium *Microcoleus chthonoplastes*. Appl. Environ. Microbiol. 62, 1752–1758. https://doi.org/10.1128/AEM.62.5.1752-1758.1996.

Moisander, P.H., McClinton, E.3rd, Paerl, H.W., 2002. Salinity effects on growth, photosynthetic parameters, and nitrogenase activity in estuarine planktonic cyanobacteria. Microb. Ecol. 43, 432–442. https://doi.org/10.1007/s00248-001-1044-2.

Namsaraev, Z., Mano, M.J., Fernandez, R., Wilmotte, A., 2010. Biogeography of terrestrial cyanobacteria from Antarctic ice-free areas. Ann. Glaciol. 51, 171–177. https://doi.org/10.3189/172756411795931930.

Narro, M.L., Cerniglia, C.E., Van Baalen, C., Gibson, D.T., 1992. Metabolism of phenanthrene by the marine cyanobacterium *Agmenellum quadruplicatum* PR-6. Appl. Environ. Microbiol. 58, 1351–1359. https://doi.org/10.1128/AEM.58.4.1351-1359.1992.

Nozzi, N., Oliver, J., Atsumi, S., 2013. Cyanobacteria as a platform for biofuel production. Front. Bioeng. Biotechnol. 1, 7. https://doi.org/10.3389/fbioe.2013.00007.

Oinam, G., Tiwari, on, Sharma, G., 2011. PCR based molecular characterization of cyanobacteria with special emphasis on non-heterocystous filamentous cyanobacteria. Assam Univ. J. Sci Technol. 7, 101–113.

Oren, A., 2015. Cyanobacteria in hypersaline environments: biodiversity and physiological properties. Biodivers. Conserv. 24, 781–798. https://doi.org/10.1007/s10531-015-0882-z.

Oren, A., 1997. Mycosporine-like amino acids as osmotic solutes in a community of halophilic cyanobacteria. Geomicrobiol. J. 14, 231–240. https://doi.org/10.1080/01490459709378046.

Oren, A., Shilo, M., 1979. Anaerobic heterotrophic dark metabolism in the cyanobacterium *Oscillatoria limnetica*: sulfur respiration and lactate fermentation. Arch. Microbiol. 122, 77–84. https://doi.org/10.1007/BF00408049.

Pade, N., Hagemann, M., 2014. Salt acclimation of cyanobacteria and their application in biotechnology. Life 5, 25–49. (Basel, Switzerland). https://doi.org/10.3390/life5010025.

Paerl, H.W., Pinckney, J.L., Steppe, T.F., 2000. Cyanobacterial-bacterial mat consortia: examining the functional unit of microbial survival and growth in extreme environments. Environ. Microbiol. 2, 11–26. https://doi.org/10.1046/j.1462-2920.2000.00071.x.

Panicker, G., Aislabie, J., Bej, A.K., 2006. Analysis of aggregative behavior of *Pseudomonas* sp. 30-3 isolated from Antarctic soil. Soil Biol. Biochem. 38, 3152–3157. https://doi.org/10.1016/j.soilbio.2006.02.006.

Parikh, A., Madamwar, D., 2005. Textile dye decolorization using cyanobacteria. Biotechnol. Lett. 27, 323–326. https://doi.org/10.1007/s10529-005-0691-7.

Patel, A., Matsakas, L., Rova, U., Christakopoulos, P., 2019. A perspective on biotechnological applications of thermophilic microalgae and cyanobacteria. Bioresour. Technol. 278, 424–434. https://doi.org/10.1016/j.biortech.2019.01.063.

Pathak, J., Rajneesh, Maurya, P.K., Singh, S.P., Häder, D.-P., SinhaR.P, 2018. Cyanobacterial farming for environment friendly sustainable agriculture practices: innovations and perspectives. Front. Environ. Sci. 6 (7). https://doi.org/10.3389/fenvs.2018.00007.

Pedrozo, F., Kelly, L., Diaz, M., Temporetti, P., Baffico, G., Kringel, R., Friese, K., Mages, M., Geller, W., Woelfl, S., 2001. First results on the water chemistry, algae and trophic status of an Andean acidic lake system of volcanic origin in Patagonia (Lake Caviahue). Hydrobiologia 452, 129–137. https://doi.org/10.1023/A:1011984212798.

Plavšić, M., Terzic, S., Ahel, M., van den Berg, C.M.G., 2002. Folic acid in coastal waters of the Adriatic Sea. Mar. Freshw. Res. 53, 1245–1252. https://doi.org/10.1071/MF02044.

Potts, M., Friedmann, E.I., 1981. Effects of water stress on cryptoendolithic cyanobacteria from hot desert rocks. Arch. Microbiol. 130, 267–271. https://doi.org/10.1007/BF00425938.

Powell, J.T., Chatziefthimiou, A.D., Metcalf, J.S., Cox, P.A., Banack, S.A., 2015. Desert crust microorganisms, their environment, and human health. J. Arid Environ. 112, 127–133. https://doi.org/10.1016/j.jaridenv.2013.11.004.

Prasanna, R., Sood, A., Jaiswal, P., Nayak, S., Gupta, V., Chaudhary, V., Joshi, M., Natarajan, C., 2010. Rediscovering cyanobacteria as valuable sources of bioactive compounds. Prikl. Biokhim. Mikrobiol. 46, 133–147.

Pushkareva, E., Pessi, I.S., Wilmotte, A., Elster, J., 2015. Cyanobacterial community composition in Arctic soil crusts at different stages of development. FEMS Microbiol. Ecol. 91. https://doi.org/10.1093/femsec/fiv143.

Raghukumar, C., Vipparty, V., David, J.J., Chandramohan, D., 2001. Degradation of crude oil by marine cyanobacteria. Appl. Microbiol. Biotechnol. 57, 433–436. https://doi.org/10.1007/s002530100784.

Rampelotto, P.H., 2013. Extremophiles and extreme environments. Life 3, 482–485. (Basel, Switzerland). https://doi.org/10.3390/life3030482.

Rawat, I., Ranjith Kumar, R., Mutanda, T., Bux, F., 2011. Dual role of microalgae: phycoremediation of domestic wastewater and biomass production for sustainable biofuels production. Appl. Energy 88, 3411–3424. https://doi.org/10.1016/j.apenergy.2010.11.025.

Roney, H.C., Booth, G.M., Cox, P.A., 2009. Competitive exclusion of cyanobacterial species in the Great Salt Lake. Extremophiles 13, 355–361. https://doi.org/10.1007/s00792-008-0223-1.

Rossi, F., Li, H., Liu, Y., De Philippis, R., 2017. Cyanobacterial inoculation (cyanobacterisation): Perspectives for the development of a standardized multifunctional technology for soil fertilization and desertification reversal. Earth-Science Rev 171, 28–43. https://doi.org/10.1016/j.earscirev.2017.05.006.

Rothschild, L.J., Giver, L.J., White, M.R., Mancinelli, R.L., 1994. Metabolic activity of microorganisms in evaporites. J. Phycol. 30, 431–438. https://doi.org/10.1111/j.0022-3646.1994.00431.x.

Russell, N.J., 1997. Psychrophilic bacteria—molecular adaptations of membrane lipids. Comp. Biochem. Physiol. Part A Physiol. 118, 489–493. https://doi.org/10.1016/S0300-9629(97)87354-9.

Sood, A., Renuka, N., Prasanna, R., Ahluwalia, A.S., 2015. Cyanobacteria as potential options for wastewater treatment in: Ansari, A.A., Gill, S.S., Gill, R., Lanza, G.R., Newman, L. (Eds.)Phytoremediation: Management of Environmental Contaminants, Volume 2, . Springer International Publishing, Cham, pp. 83–93. https://doi.org/10.1007/978-3-319-10969-5_8.

Saini, D.K., Pabbi, S., Shukla, P., 2018. Cyanobacterial pigments: perspectives and biotechnological approaches. Food Chem. Toxicol. 120, 616–624. https://doi.org/10.1016/j.fct.2018.08.002.

Sayed, A.M., Hassan, M.H.A., Alhadrami, H.A., Hassan, H.M., Goodfellow, M., Rateb, M.E., 2020. Extreme environments: microbiology leading to specialized metabolites. J. Appl. Microbiol. 128, 630–657. https://doi.org/10.1111/jam.14386.

Shrivastava, D.K., Richa, M., Chandra, T.P., 2014. Evaluation of Cynobacterial toxicity against corresponding field- soil Bacteria. Int. J. of Pharm. Life Sci. 5(7), 3694–3700.

Siglioccolo, A., Paiardini, A., Piscitelli, M., Pascarella, S., 2011. Structural adaptation of extreme halophilic proteins through decrease of conserved hydrophobic contact surface. BMC Struct. Biol. 11, 50. https://doi.org/10.1186/1472-6807-11-50.

Singh, A.L., Singh, P.K., 1987. Influence of Azolla management on the growth, yield of rice and soil fertility. Plant Soil 102, 41–47. https://doi.org/10.1007/BF02370898.

Singh, J.S., Kumar, A., Rai, A.N., Singh, D.P., 2016. Cyanobacteria: a precious bio-resource in agriculture, ecosystem, and environmental sustainability. Front. Microbiol. 7, 529. https://doi.org/10.3389/fmicb.2016.00529.

Singh, J.S., Pandey, V.C., Singh, D.P., 2011. Efficient soil microorganisms: a new dimension for sustainable agriculture and environmental development. Agric. Ecosyst. Environ. 140, 339–353. https://doi.org/10.1016/j.agee.2011.01.017.

Singh, J.S., Strong, P.J., 2016. Biologically derived fertilizer: a multifaceted bio-tool in methane mitigation. Ecotoxicol. Environ. Saf. 124, 267–276. https://doi.org/10.1016/j.ecoenv.2015.10.018.

Singh, J., 2014. Cyanobacteria: a vital bio-agent in eco-restoration of degraded lands and sustainable agriculture. Climate Change Environ. Sustain. 2, 133–137.

Staley, J.T., Stanley, P.M., 1986. Potential commercial applications in aquatic microbiology. Microb. Ecol. 12, 79–100. https://doi.org/10.1007/BF02153224.

Steven, B., Lionard, M., Kuske, C.R., Vincent, W.F., 2013. High bacterial diversity of biological soil crusts in water tracks over permafrost in the high arctic polar desert. PLoS One 8, e71489. https://doi.org/10.1371/journal.pone.0071489.

Stewart, W.D.P., Fitzgerald, G.P., Burris, R.H., 1968. Acetylene reduction by nitrogen-fixing blue-green algae. Arch. Mikrobiol. 62, 336–348. https://doi.org/10.1007/BF00425639.

Stomeo, F., Valverde, A., Pointing, S.B., McKay, C.P., Warren-Rhodes, K.A., Tuffin, M.I., Seely, M., Cowan, D.A., 2013. Hypolithic and soil microbial community assembly along an aridity gradient in the Namib Desert. Extremophiles 17, 329–337. https://doi.org/10.1007/s00792-013-0519-7.

van Noort, V., Bradatsch, B., Arumugam, M., Amlacher, S., Bange, G., Creevey, C., Falk, S., Mende, D.R., Sinning, I., Hurt, E., Bork, P., 2013. Consistent mutational paths predict eukaryotic thermostability. BMC Evol. Biol. 13 (7). https://doi.org/10.1186/1471-2148-13-7.

Varshney, P., Mikulic, P., Vonshak, A., Beardall, J., Wangikar, P.P., 2015. Extremophilic microalgae and their potential contribution in biotechnology. Bioresour. Technol. 184, 363–372. https://doi.org/10.1016/j.biortech.2014.11.040.

Vijayakumar, S., 2012. Potential applications of cyanobacteria in industrial effluents-a review. J. Bioremediation Biodegrad. 03. https://doi.org/10.4172/2155-6199.1000154.

Vítek, P., Jehlička, J., Ascaso, C., Mašek, V., Gómez-Silva, B., Olivares, H., Wierzchos, J., 2014. Distribution of scytonemin in endolithic microbial communities from halite crusts in the hyperarid zone of the Atacama Desert. Chile. FEMS Microbiol. Ecol. 90, 351–366. https://doi.org/10.1111/1574-6941.12387.

Wang, Q., Cen, Z., Zhao, J., 2015. The Survival Mechanisms of thermophiles at high temperatures: an angle of omics. Physiology 30, 97–106. https://doi.org/10.1152/physiol.00066.2013.

Wase, N.V, Wright, P.C., 2008. Systems biology of cyanobacterial secondary metabolite production and its role in drug discovery. Expert Opin. Drug Discov. 3, 903–929. https://doi.org/10.1517/17460441.3.8.903.

West, M., Louda, J.W., 2011. Effect of external pH On Cyanobacterial Pigment Expression. Florida Sci 74, 181–186.

Wierzchos, J., Ascaso, C., McKay, C.P., 2006. Endolithic cyanobacteria in halite rocks from the hyperarid core of the Atacama Desert. Astrobiology 6, 415–422. https://doi.org/10.1089/ast.2006.6.415.

Yamaguchi, K., 1996. Recent advances in microalgal bioscience in Japan, with special reference to utilization of biomass and metabolites: a review. J. Appl. Phycol. 8, 487–502. https://doi.org/10.1007/BF02186327.

Zahra, Z., Choo, D.H., Lee, H., Parveen, A., 2020. Cyanobacteria: review of current potentials and applications. Environ. MDPI 7. https://doi.org/10.3390/environments7020013.

Zhang, H., Hu, C., Jia, X., Xu, Y., Wu, C., Chen, L., Wang, F., 2012. Characteristics of γ-hexachlorocyclohexane biodegradation by a nitrogen-fixing cyanobacterium, *Anabaena azotica*. J. Appl. Phycol. 24, 221–225. https://doi.org/10.1007/s10811-011-9670-7.

CHAPTER

2

Cyanobacterial nanoparticles: Application in agriculture and allied sectors

Vaishali Gupta[a], *Naveen Kumar Verma*[a], *Kaushal Kishore Choudhary*[b]

[a]Government College, Bichhua, Chhindwara, Madhya Pradesh, India
[b]Ram Jaipal College, Chapra, Bihar, India

2.1 Introduction

By 2050, the global population is estimated to reach nearly 9.6 billion. To meet food demand, agriculture production must increase by 70–100% (Rodrigues et al., 2017, Foley et al., 2011, Mueller et al., 2012). Agriculture is considered the backbone of developing countries, with more than 60% of the population dependent on it for their livelihood. Nanobiotechnology has the potential to revitalize and revolutionize the agriculture and food industry with novel tools for disease detection and management, plant nutrients absorption, slow-release fertilizers, and environmental protections.

2.2 Nanobiotechnology

Nanotechnology is an art and science where the matter or property of concern can be manipulated at the nano-level and can be implemented to improve life quality. Biotechnology is a technology that harnesses biological systems, living organisms, or part of it for improving the lives and health of planets. Nanobiotechnology is a dynamic and novel scientific approach blended with science, engineering, and technology at the nanoscale level.

Cyanobacterial Lifestyle and its Applications in Biotechnology.
DOI: https://doi.org/10.1016/B978-0-32-390634-0.00011-1

2.3 Nanoparticles

The term "Nano" is originated from the Greek word meaning dwarf. Nanoparticles (NPs) or ultrafine particles are usually defined based on the size at which basic properties discrete from the bulk material. NPs are lying in the range of 10^{-9} m and having dimensions of 1–100 nm (Banfield and HJRim, 2001, Abobatta, 2018). The physical properties of NPs are pretty different from the properties of the bulk material (Buffle, 2006). NPs may vary in their shape (Fig. 2.1).

2.3.1 Types of NPs

NPs are classified based on their dimension, composition, morphology, uniformity, and adhesion of particles; therefore categorized accordingly as natural NPs formed by the volcanic eruption, heavy metals fabrication through living microorganisms, forest fire, weathering of rocks, the explosion of clay minerals, and sandstorm (Baker et al., 2013). In contrast, incidental NPs are the byproducts of intentionally produced engineered NPs (Buzea et al., 2007). NPs are divided into the following types.

NPs based on shape and size are classified as follows:

1. Zero dimension
2. One dimension
3. Two-dimension.

NPs based on the nature of materials are classified as follows:

1. Organic NPs
   a. Liposomes
   b. Polymers
   c. Micelles.

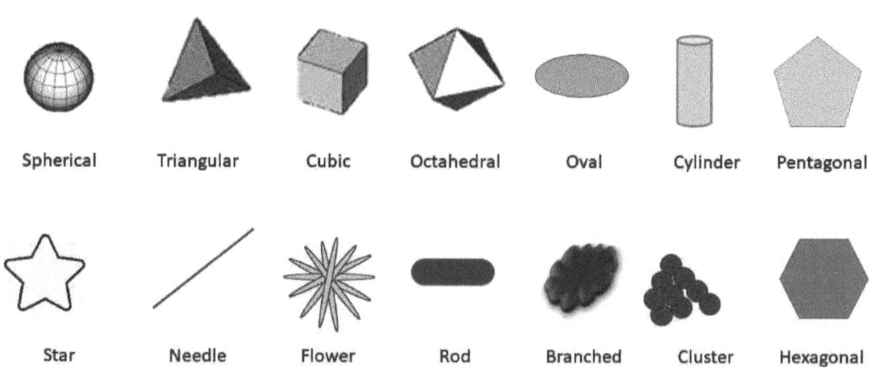

FIG.2.1 Various shape of nanoparticles.

2. Inorganic NPs
   a. Pure Metal
   b. Bimetallic
   c. Metallic oxides
   d. Metallic chalcogenides.

NPs based on phase composition are classified as follows:

1. Single-phase solids
2. Multi-phase solids.

NPs based on origin and source are classified as follows:

1. Natural
2. Human-made.

NPs based on crystallinity are classified as follows:

1. Amorphous
2. Crystalline
3. Polycrystalline.

2.4 Cyanobacterial NPs

Nanotechnology and NPs are widely used in the electronic, biomedical, cosmetic, and biotechnology field due to their unique characters such as affinity with other particles, large surface-to-volume ratio, and different shapes and sizes. The efficient crystallography, colloidal, optical polarizability, and electrical and thermal conductivity properties of NPs make nanotechnology a magnificent area to focus on. The synthesis of NPs is carried out using top-down and bottom-up systems, where the material is processed physically and chemically. Physical methods commonly used for the synthesis are attrition and pyrolysis, high with low production rates and high energy consumption. Traditional and widely used methods for chemical synthesis are based on different reducing agents such as sodium borohydride (Kim et al., 2007), methoxy polyethylene glycol, (Mallick et al., 2004) etc. Although chemical methods are cheaper for bulk production, a significant flaw of the synthesis is toxic substances that may be hazardous.

Moreover, toxic ingredients for chemicals synthesis may cause severe toxicity to humans and the environment. To overcome this problem, we need to replace toxic ingredients with environmentally safe biological materials. For decades, biological agents have been exploited in various fields, including NP production. Plants extracts (MJJoBs, 2010), algae (Jeyaraj et al., 2013), fungi (Balaji et al., 2009), lichen (Hamida et al., 2020), and bacteria (Saifuddin et al., 2009) have been shown to minimize metal precursor for their corresponding NPs (Asmathunisha et al., 2013). They are generally cost-effective, nontoxic and ecofriendly (Thakkar et al., 2010). In the recent years, the synthesis of NPs using cyanobacteria has become an exciting and valuable research field with promising applications (Lengke et al., 2006, 2007,Hamouda et al., 2019, MubarakAli et al., 2012).

Cyanobacteria are primitive and photosynthetic organisms that are ecologically and economically important. They are ubiquitous and found in different extreme environments (SJCJoS, 2011, Thajuddin and GJCs, 2005). Cyanobacteria have unique features of oxygenic photosynthesis, nitrogen fixation, high biomass yield, habitat diversity, and growth on nonarable lands, valuable byproducts, and biofuel generation, enhance soil fertility and reduced greenhouse gas emissions (Thajuddin and GJCs, 2005, Lee, 2018, Johansen, 2012, MAJJoap, 2013). Cyanobacteria-mediated synthesis of NPs is an ecofriendly, economic, expeditious, high yielding, and energy-efficient method. Cyanobacteria are considered biofactories that can synthesize NPs as extracellular and intracellular methods (Hamida et al., 2020).

Different cyanobacteria species such as *Lyngbya, Spirulina, Nostoc, Cylindrospermum, Oscillatoria, Synechococcus, Calothrix, Leptolyngbya, Aphanocapsa, Aphanothece, Phormidium,* and *Gloeocapsa* were used for the synthesis of NPs (Hamida et al., 2020). Among all the organisms, cyanobacteria's peculiar characters make it equally an important organism in the synthesis of NPs (Thakkar et al., 2010, Sharma et al., 2016, Rauwel et al., 2015). Hence, the current chapter explained green nanotechnology-based approaches, methods, and applications in the field of agriculture.

2.5 Synthesis of NPs

NPs can be synthesized in different shapes and sizes by physical, chemical, and biological methods (Iravani et al., 2014), used in diverse agriculture, medical, electronics, optics, and many others (Fig. 2.1). NPs prepared through physical, chemical, and biological methods were NPs synthesized through either atomic level (bottom-up system) or bulk material (top-down system).

2.5.1 Physical synthesis

Physical synthesis is a type of top-down fabrication method, where material such as metals are changed into their nanoforms using laser ablation (Amendola and MJPccp, 2009), laser pyrolysis techniques (Galvez et al., 2002), ball milling (De Carvalho et al., 2013), nucleation (Mackus et al., 2013), sonochemistry (Khandel et al., 2018), and sputtering (Mishra et al., 2007, Salem and Fouda, 2020). Evaporation and condensation of metals occur, where metallic NPs are formed by rearrangement and aggregation of metal atoms. Shape uniformity and purity of NPs due to the absence of solvents are the advantages of physical methods. Although the physical route of NP synthesis is unwieldy and expensive, highly sophisticated instruments are used in which high-energy consumption is required.

2.5.2 Chemical synthesis

Chemical synthesis strategies belong to the bottom-up route, in which the atom self-assembles or assembles nuclei and produces a uniform shape and size ranges of NPs (Salem and Fouda, 2020). Precursor metals, reducing and stabilizing agents, are the main components in chemical synthesis. Chemicals used as reducing agents are hydrazine, sodium borohydrate, sodium citrate, elemental hydrogen, and ascorbate. These agents reduce metal ions and led to metal atoms, followed by the accumulation into clusters. These clusters led to the formation of metallic colloidal particles. It is necessary to use stabilizing agents such as natural and synthetic polymers such as cellulose, natural rubber, and chitosan to prevent uncontrollable growth of particles and protect NPs that can be absorbed on or bind onto NPs' surface to avoid particle aggregation. These stabilizing chemicals are hydrophobic, so that they require some organic solvents such as ethane, dimethyl, formaldehyde, toluene, chloroform, and sodium carboxyl methylcellulose (Patel et al., 2017), which are toxic and nonbiodegradable. Toxic chemicals may contaminate the surface of NPs and make them inapt for specific allied applications. Moreover, the liberation of toxic byproducts in nature threatens the ecological system (Khanna et al., 2019).

In this context, to avoid all these drawbacks of physical and chemical methods, research emphasized the green synthesis of NPs using biological agents as an alternative NP synthesis process.

2.5.3 Biological synthesis of NPs

In the recent years, the green synthesis of NPs is an advanced version of the bottom-up route of fabrication of NPs. The objective of the biological synthesis of NPs is to reduce the use of toxic materials and to utilize biogenic sources to convert bulk material into nano-sized particles with unique properties (Bin-Meferij and RSJIJoN, 2019). Unicellular microorganisms include bacteria (Banerjee et al., 2020), cyanobacteria (Hamida et al., 2020), and diatoms (Chetia et al., 2017) and multicellular organisms such as algae and plants (Rajeshkumar et al., 2014, Seku et al., 2020) used for the biogenic synthesis of NPs. In contrast, the process is accomplished in the presence of proteins, enzymes, peptides, pigments, etc. (Asmathunisha et al., 2013). The methods for the biofabrication of NPs vary from unicellular to multicellular organisms. Different microbes follow different routes for NP production. However, the mode of synthesis of NPs for

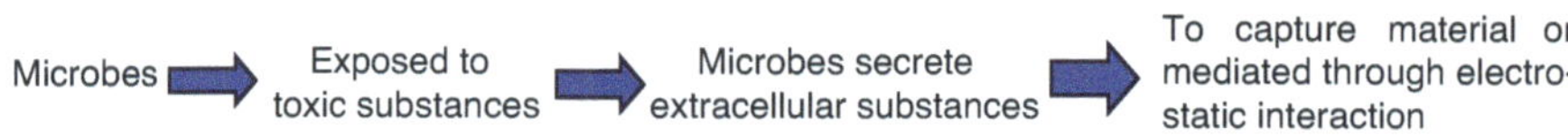

FIG. 2.2 Microbial route of synthesis of NPs

various microorganisms is quite similar, that is, intracellular or extracellular, while plants synthesize only by extracellular mode. Microbes synthesize NPs in nature, as shown in Fig. 2.2.

Microbes have their defense mechanism, where they produce different substances with various shapes and sizes to minimize toxification of target ions. Microbes follow a general pattern where they trap metal ions in the cell or on the microbial surface and then reduce it in NPs without additional stabilizing and reducing agents. The microbial reduction can be accomplished intracellularly or extracellularly through bioaccumulation, precipitation, biomineralization, or biosorption (Husen and KSJJon, 2014, Singh et al., 2015, Kanchi and Ahmed, 2018). Biotransformation of NPs by microbes can be classified into intracellular green synthesis and extracellular green synthesis.

2.5.3.1 *Intracellular green synthesis of NPs*

Biotransformation of NPs inside the host cell is performed by reducing bulk material into nano-size under controlled conditions such as temperature and pH (Sharma et al., 2016). The *in-vivo* biosynthesis of metal NPs inside the host cell involves the following five steps: (1) Culturing desired microorganism; (2) incubate living organisms with a definite precursor solution at desired temperature and pH; (3) separation of NPs from living cells; (4) purification of NPs with impurities; (5) characterization of NPs through physicochemical techniques (Lengke et al., 2007). For procuring NPs, the mixture of NPs and living microorganism is centrifuged at 12,000 *g* and pellets are used for direct examination after washing multiple times. NPs are obtained through lysis of cells using sonication (Kalabegishvili et al., 2012).

The yield of biosynthesized NPs is dependent on how microbial enzymes modify the composition of the reaction mixture or favor or inhibit nucleation of the initially formed particle (Zhang et al., 2011, Sastry et al., 2003, 2004).

2.5.3.2 *Extracellular green synthesis of NPs*

In extracellular synthesis, reduction occurs outside the cell or on the surface of the microbial cell in the presence of biological molecules such as exopolysaccharides, pigments, enzymes, peptide, proteins, hormones, and antioxidants (Hamida et al., 2020, Mata et al., 2009). These biomolecules play a role in a reducing and capping agent during the biofabrication process of NPs (Figs. 2.3 and 2.4).

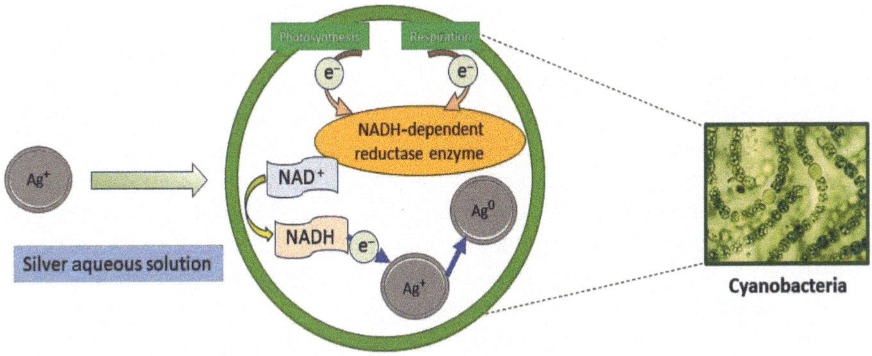

FIG. 2.3 Potential mechanism of green NPs synthesis.

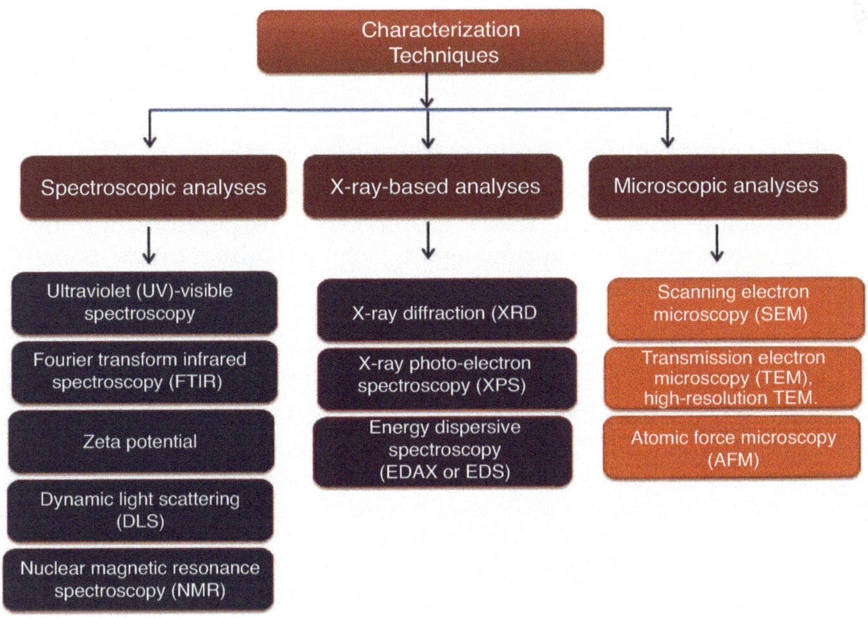

FIG. 2.4 Various tools for NPs characterization.

2.5.4 Cyanobacteria as a source of NPs synthesis

In the recent years, cyanobacteria have been used for the biosynthesis of NPs. Occurrence, abundance, fast growth rate, and adaptability in the extreme environment make them a good and valuable source of NPs synthesis. The synthesis of cyanobacteria-mediated NPs is carried by intracellular and extracellular syntheses.

2.5.4.1 *Cyanobacterial way of intracellular synthesis of NPs*

There are two different methods for NP synthesis: In the first process, the cellular biomass of the targeted cyanobacteria is cultured and collected in the logarithmic phase by centrifugation, and then it is washed to remove any traces of salts (Jena et al., 2013). A definite concentration of a precursor metal solution is mixed in cleaned biomass, dissolved in water, and incubated. In the second method, cyanobacteria are individually cultured and mixed with the bulk material solution and kept under culture condition (Merin et al., 2010). These methods are dependent on the dose and type of microorganism used in the synthesis.

During the intracellular synthesis of NPs, electrons produced via electron transport system, including energy-generating reaction in photosynthesis and to some extent respiratory electron transport system, will reduce metal ions. The biosynthesis of NPs takes place in thylakoid where $NADPH_2$ is reduced and oxidized in the presence of NADH-dependent reductase to fabricate NPs and obtained NPs get accumulated in the external membrane. A redox reaction occurs in the cytoplasm, thylakoid membrane, and cell membrane (Focsan et al., 2011).

2.5.4.2 *Cyanobacterial way of extracellular synthesis of NPs*

The biofabrication process of cyanobacterial NPs takes place outside the cell, and biological molecule pigments, proteins, ions, peptides, hormones, etc., act as reducing and capping agents. There are different methods for the extracellular synthesis of NPs.

2.5.4.2.1 Cell-free media

Extracellular culture is prepared by removing biomass by centrifugation and then mixed with a metal precursor solution under certain conditions (Hamida et al., 2020). It has been reported that biological compound protein, enzymes, antioxidants, and alkaloid compounds present in cell-free media control the biosynthesis process.

2.5.4.2.2 Cell biomass filtrate

It has been studied that cyanobacterial cell filtrates are also used for the biofabrication of NPs. In this process, cyanobacteria were harvested in the log phase, centrifuged at 4000 rpm for 20 min, and pellets were washed with distilled water twice to remove any trace elements. Freeze-dried pellets were crushed using a pestle and mortar filter with Whatman filter paper. The cyanobacterial extract is mixed with a metal precursor solution and incubated under a suitable condition. Biomolecule proteins and enzymes in the cell filtrate act as a stabilizing and capping agent (Bin-Meferij and RSJIJoN, 2019, Hamida et al., 2020, Al Rashed et al., 2018).

2.5.4.2.3 Biomolecule-based NP synthesis

- Pigments

Cyanobacteria. The oxygenic photosynthetic prokaryotes are distinguished by the presence of pigments such as chlorophyll, carotenoids and phycobiliprotein, and are capable to biosynthesis nanoparticles. Cyanobacterial species have a blue photosynthetic phycobiliprotein family accessory pigment C-phycocyanin attached to thylakoids (El-Naggar et al., 2017). Phycobilisomes are involved in light-harvesting and transferring electron toward photosystem II reaction center (MacColl RJJosb 1998). The ability to transfer electrons and bind with heavy metals by phycocyanin makes NPs' biofabrication worthwhile (Gelagutashvili, 2013). C-phycocyanin was isolated and purified from cyanobacteria and incubated under light and dark condition at 25 °C for 2 days after mixing with a metal solution (Rauwel et al., 2015, Gantar et al., 2012, Patel et al., 2015). Till now, pigment-mediated NPs have been reported in *Spirulina, Limnothrix, Nostoc,* and *Phormidium* species (Patel et al., 2015).

- Polysaccharides

NPs can be synthesized by cyanobacterial extracellular polysaccharides (Patel et al., 2015). Exopolysaccharides (EPS) is extracted from the potassium buffer solution of cyanobacteria through homogenization multiple times at ambient temperature. The aqueous solution is centrifuged at 6000 rpm for 10 min at 20 °C, and EPS-containing supernatant is used for NP synthesis, where EPS works as a potent reducing and capping agent.

2.6 Characterization of NPs

NPs are characterized by different techniques to assess their physical and chemical properties, including size, distribution, stability, charge, surface morphology, and dispersion (Menon et al., 2017, Shah et al., 2015). The spectroscopic techniques are used to determine wavelength ranges, confirm NP synthesis, their charges and hydrodynamic diameter, and the chemical composition of biogenic compounds. X-ray-based techniques such as X-ray diffraction, X-ray photoelectron spectroscopy, and energy-dispersive spectroscopy are implemented to assess the design, structure, and crystal phase. Microscopic techniques are also used to determine the size and morphological appearance of NPs. NPs exhibit magnetic properties that can be estimated through ferromagnetic resonance, vibrating sample magnetometer, etc.

2.7 Mode of action

Intracellular and extracellular biological syntheses of the NP are totally based on the oxidation-reduction process. These processes are dependent on enzymes; proteins act as a reducing, capping, and stabilizing agent. In cyanobacteria, reducing electrons are liberated during the electron transport system, electron-producing photosynthesis, and respiration process occurring in the cytoplasm, thylakoid membrane, and cell membrane. NADH-dependent reductase enzyme facilitates the process.

2.8 Applications of cyanobacterial-based NPs

Nanobiotechnology is a bright and innovative technique with significant roles in valuable products that enhance industrial productions, medicine, and agriculture. We need a rapid, secure, and cost-effective system to identify, manufacture, monitor, and solve different agriculture problems through nanobiotechnology in the agriculture field. Physiological and morphological changes occur in plants exposed to nanomaterials. Functional expressions of nanomaterials depend on several parameters such as host-plant and specific type of interaction, surface coating, size, dosage, time of exposure, and properties of NPs. NPs have been used in all stages of agriculture production, from seed production to yields, because of their unique feature. Nanobiotechnology has significant application in the agriculture field in the following various forms.

1. *Crop improvement and crop yield and growth.*
   a. *Nanofertilizer*: Growth and development of the plants depend on the nutrition they get from the soil, but these nutrients are not sufficiently available. Excessive use of chemical fertilizers has created many problems for the environment and human health. In this alarming situation, the use of cyanobacteria is a potential source of sustainable agriculture. It has been well studied that cyanobacteria fix nitrogen in the soil and make it available to plants. The capability of nitrogen fixation improves soil fertility through cyanobacteria. Different cyanobacterial species can be used as biofertilizer and maintain the decreasing level of nutrients in the soil and secrete plant growth hormones as secondary metabolites.
   Now, conventional biofertilizers have been overshaded by nanofertilizers. Nanofertilizers work as a vehicle to deliver nutrients to crop where nutrients are encapsulated within nanomaterial or coated under thin protective polymer layers or provided as an emulsion or as a particle (DeRosa et al., 2010). Nanofertilizers may be able to improve the quality of fertilizers through add-on antimicrobial additives. However, nanomaterials facilitate translocation of nutrients, storage of nutrients, and improved crop yield (Nair et al., 2010). It has been well studied that nanofertilizers have positive morphological and physiological effect on nitrogen assimilation, increment of photosynthetic rate, biomass enhancement, root and shoot length, and seed germination (Naderi and Abedi, 2012).

Nanotechnology is also applied for crop improvement through NP-mediated delivery of genetic material (Misra et al., 2016). Agronomic traits such as plant growth, biomass production, physiological parameters that influence yield and quality, stress tolerance, etc. can be altered through engineered nanomaterials and nanosensors (Rameshaiah et al., 2015, Qureshi et al., 2018). Nanoaptamers are used as an intelligent nanobiosensor, associated with biopolymer loaded with nanofertilizers to respond to the chemical signature of plant and deliver an optimized dose of nutrient deficient in soil (Hakeem and Pirzadah, 2020). Nano-biosensors (NBSs) are used to monitor plant health by early detection of disease, water deficiency, and soil availability. These NBSs are connected to internet for remote tracking and real-time data analysis.

b. *Seed germination*: Studies proved that NPs allow better penetration of water into the seeds (Khodakovskaya et al., 2009). The penetration of NPs into seeds enhances water and oxygen uptake efficiency, which fastens seed germination, accelerates nutrient absorption, plant growth, and increases resistance against stress.

2. *Crop disease management.*

Nanobiopesticides: Microbe, fungus, pests, and insects can be controlled through bionanopesticides that are safe, efficient, and ecofriendly. Nanoencapsulation, nanocages, and nanoemulsion are introduced to deliver nanopesticides to control pests under bionanotechnology (Bouwmeester et al., 2009, Bergeson, 2010). Biosubstances with pesticide effects or a combination with NPs are used as nanopesticides. Nanopesticides are better than the conventional pesticides because they reduce runoff of the organic solvents and minimize the hazardous effect of pesticides.

3. *Postharvest disease management.*

Nanobiotechnology is needed for efficient food production and decreased wastage of postharvest products due to resistant pathogen. Nanobiotechnology can be used to dry, store, and preserve products. Postharvesting phytopathogenic fungi that cause spoilage of fruits and vegetables after harvest can be controlled by nanofilms and nanocoatings. It also detects the quality of food and its spoilage rate.

4. *Nanobiosensor.*

Nanobiosensor are also able to check the crops in the field for precision farming for increasing the agriculture productivity with better management of time, fertilization and environment (Scott and Chen, 2013). Nanobiosensor can be used in the field of agriculture as given below:

a. The environmental factors (temperature, light, humidity, air quality) and stress (salt stress and drought) can be measured and managed through genetic modification by a nanosensor.

b. Nanobiosensor is also used to detect heavy metal in the soil.

c. Pathogen detection, monitoring, and eradication can be done by nanobiosensors.
d. Nanobiosensor can be implemented as an identification tool for pathogen detection in agriculture (Wong et al., 2016).
e. Nanobiosensor is employed for control management of nutrient in soil and crop.
Nanobiosensor can be used as real-time monitoring for cultivated field and crop. Wireless nanobiosensor are linked with a satellite system for remote sensing, data collection, monitoring, and geographic information (Brock et al., 2011, Javad, 2020).

5. *Bioremediation of heavy metals.*
Industrial, agricultural, and municipality wastewater has become a severe problem due to the accumulation of heavy metals. Various pollutants such as arsenic, cadmium, chromium, copper, lead, mercury, nickel, selenium, silver, and zinc are released. Nanotechnology provides a tool to identify and eliminate heavy metals from the environment. Nanobiofilm and nanocoating with ligands play an essential role to absorb heavy metals from the environment.
6. *Hydroponics.*
Hydroponics is a technique of agriculture where the plant is not grown in soil. It is a widely used technique used for the production of lettuce, tomatoes, cucumber, melon, broccoli, sweet pepper, chillies, eggplant, etc. Nutrient management is a crucial factor for hydroponics than soil-based production. Nanotechnology is used for the production of nanonutrients for the crop.

2.9 Future projections of cyanobacteria-based NPs

It has been assumed that NPs or nanofertilizers may prove to be a boon for quality food production in future to feed the growing population of human beings, which has been proved in medicine, materials, energy, etc. The end of NPs or nanofertilizers is promising, and scientists are excited about their significant contribution to food production with preliminary reports in certain parts of the world. A preliminary study on physicochemical properties suggests that high surface area, shape and size, and catalytic activity of NPs will certainly prove to be fruitful for sustainable agriculture practices in the future. European Commission has considered nanotechnology as one of its six "Key Enabling Technologies" that may significantly contribute to a greener economy (LJNvips, 2018). Future NP applications are directed toward controlled delivery of agrochemicals such as nutrients, fertilizers, pesticides, and herbicides, monitoring environmental stresses and crop conditions under field sensing system, and stress- and disease-tolerant traits in plants (LJNvips, 2018, Gogos et al., 2012, Chen and RJTiFS, 2011).

The literature on NPs suggests that NPs can boost up the rate of photosynthesis, thereby increasing the yield. It has been reported that the high photocatalytic activity of anatase crystal titanium oxide NPs ($nTiO_2$) has a significant role in the improvement of light absorbance during photosynthesis. It was assumed that $nTiO_2$ could protect chloroplast from ageing in the cell due to photochemical stress (LJNvips, 2018) (Hong et al., 2005, Hong et al., 2005). Further studies demonstrated that $nTiO_2$ could enhance the (1) carboxylation of RuBisCO, ultimately enhancing the photosynthesis rate (Gao et al., 2006, 2008, Linglan et al., 2008), (2) the activity of Chl-photophosphorylation (Qi et al., 2013), and (3) leaf absorbance and transpiration rate (Lei et al., 2007). Research work for the development of smart fertilizers and their delivery in the crop for replacing traditional fertilizers (Perrin et al., 1998) and encapsulation of NPs for controlled release to plants is in progress (Aouada and De Moura, 2015). A lot of studies on introducing NPs in plants are in progress. It has been assumed that it will create second green revolution in the agriculture sector for quantitative and qualitative food production in the coming future.

2.10 Limitations of NPs

Currently, the application of NPs in agriculture to enhance food production, pest control, etc. is in the infancy stage. Most importantly, information on the application of engineered NPs and Engineered nanomaterials (ENMs) in the field condition is either theoretically investigated or scarce. Technology-driven agriculture produces to feed the sizeable future population are facing two challenges: (1) increase in food production and (2) reduction in the impact of the environment on agriculture. As the application of ENMs is in the nascent stage, scientists will have to go the long run to assess the impact of ENMs on the metabolic ability of plants and cyanobacteria. This is posing tremendous pressure on scientists to investigate the primary observation concerning photosynthetic rate, nitrogen fixation, and other metabolic activity of the plants before recommending the application of ENMs in agriculture.

Furthermore, it is also necessary to investigate the impact of ENMs on soil microflora, which are essential biocomponents of soil. Soil microflora such as bacteria, cyanobacteria, Vesicular-arbuscular mycorrhiza (VAM) fungi, etc. help in the augmentation of soil with micro- and macronutrients by performing nutrient cycling, nitrogen fixation, and hydrolyzing complex organic phosphates, as well as increasing the surface area for much nutrient absorption. Hence, scientists need to summarize the impact on soil microflora before their application under field condition. It is noteworthy to mention here that scientists will have to address the effect of NPs in waste treatment to compete for green technology. These are some limitations in using NPs in agriculture for better qualitative and quantitative food production. Most importantly, handling of NPs in field condition, kind of precautions taken during their application in agro-environment, development of equipment for application of NPs in the agriculture field, and safety concerns for workers are major concerns and limitations for applying nanofertilizers under field conditions.

2.11 Conclusion

Undoubtedly, nanobiotechnology is a tremendous technique, which has the potential to solve agricultural and environmental problems. Nanobiotechnology is still in its developing phase, where it needs to be explored. Conventional agriculture techniques are not enough to feed an ever-increasing population with challenging environmental stress. Nanotechnology holds an essential position in reshaping agriculture and food production to fulfill the demands efficiently and economically. The tools and techniques of nanobiotechnology such as nanobiotechnology, nanobiofertilizer, nanobiosensor, and nanobiopesticides are boon for agricultural science; and agronomy fastens seed germination, slow release of nutrients, and real-time monitoring of data.

References

Abobatta. 2018. Nanotechnology application in agriculture. Acta Sci. Agric. 2(6).

Al Rashed, S., Al Shehri, S., Moubayed, N., 2018. Extracellular biosynthesis of silver nanoparticles from cyanobacteria. Biomed. Res. (0970-938X). 29 (13).

Amendola, V., Meneghetti, M.J., 2009. Laser ablation synthesis in solution and size manipulation of noble metal nanoparticles. Phys. Chem. Chem. Phys. 11 (20), 3805–3821.

Aouada, F.A., De Moura, M.R., 2015. Nanotechnology applied in agriculture: controlled release of agrochemicals. In: Nanotechnologies in Food and Agriculture. Springer, Cham, pp. 103–118.

Asmathunisha, N., Kathiresan, K.J.C., Biointerfaces, S.B., 2013. A review on the biosynthesis of nanoparticles by marine organisms. Colliod Surf. B. Interf. 103, 283–287.

Baker, S., Harini, B., Rakshith, D., Satish, S.J., 2013. Marine microbes: invisible nanofactories. J. Pharm. Res. 6 (3), 383–388.

Balaji, D., Basavaraja, S., Deshpande, R., Mahesh, D.B., Prabhakar, B., Venkataraman, A.J.C., et al., 2009. Extracellular biosynthesis of functionalized silver nanoparticles by strains of *Cladosporium cladosporioides* fungus. Colliod Surf. B. Interf 68 (1), 88–92.

Banerjee, A., Das, D., Rudra, S.G., Mazumder, K., Andler, R., Bandopadhyay, R.J., et al., 2020. Characterization of exopolysaccharide produced by *Pseudomonas* sp. PFAB4 for synthesis of EPS-Coated AgNPs with antimicrobial properties. J. Polym. Environ. 28 (1), 242–256.

Banfield, J.F., Zhang, H.J., 2001. Rim, geochemistry. Nanoparticles in the environment. Rev. Mineral. Geochem. 44 (1), 1–58.

Bergeson, L.L., 2010. Nanosilver: US EPA's pesticide office considers how best to proceed. 19 (3), 79–85.

Bin-Meferij, M.M., Hamida, R.S., 2019. Bio-fabrication and antitumor activity of silver nanoparticles utilizing novel *Nostoc* sp. Bahar M. Int. J. Med. 14, 9019.

Borowitzka, M.A., 2013. High-value products from microalgae—their development and commercialization. J. Appl. Phycol. 25 (3), 743–756.

Bouwmeester, H., Dekkers, S., Noordam, M.Y., Hagens, W.I., Bulder, A.S., De Heer, C., et al., 2009. Review of health safety aspects of nanotechnologies in food production. Regul. Toxicol. Pharmacol. 53 (1), 52–62.

Brock, D.A., Douglas, T.E., Queller, D.C., Strassmann, J.E., 2011. Primitive agriculture in a social amoeba. Nature. 469 (7330), 393–396.

Buffle, J., 2006. The key role of environmental colloids/nanoparticles for the sustainability of life. Environ. Chem. 3 (3), 155–158.

Buzea, C., Pacheco, II, Robbie, K.J.B., 2007. Nanomaterials and nanoparticles: sources and toxicity. Biointerphases 2 (4), MR17–MR71.

Chen, H., Yada, R., 2011. Nanotechnologies in agriculture: new tools for sustainable development. Trends Food Sci. Technol. 22 (11), 585–594.

Chetia, L., Kalita, D., Ahmed, G.A., 2017. JS, research b-s. Synthesis of Ag nanoparticles using diatom cells for ammonia sensing. Sens. Biosensing Res. 16, 55–61.

De Carvalho, J., De Medeiros, S., Morales, M., Dantas, A., Carriço, AJ., 2013. Synthesis of magnetite nanoparticles by high energy ball milling. Appl. Surf. Sci. 275, 84–87.

DeRosa, M.C., Monreal, C., Schnitzer, M, Walsh, R, Sultan, Y, 2010. Nanotechnology in fertilizers. Nat. Nanaotechnol. 5 (2), 91.

El-Naggar, NE-A., Hussein, M.H., El-Sawah, A.A., 2017. Bio-fabrication of silver nanoparticles by phycocyanin, characterization, in vitro anticancer activity against breast cancer cell line and in vivo cytotoxicity. Sci. Rep. 7 (1), 1–20.

Focsan, M., Ardelean, II, Craciun, C., Astilean, S., 2011. Interplay between gold nanoparticle biosynthesis and metabolic activity of cyanobacterium *Synechocystis* sp. PCC 6803. Nanotechnology 22 (48), 485101.

Foley, J.A., Ramankutty, N., Brauman, K.A., Cassidy, E.S., Gerber, J.S., Johnston, M., et al., 2011. Solutions for a cultivated planet. Nature 478 (7369), 337–342.

Galvez, A., Herlin-Boime, N., Reynaud, C., Clinard, C., Rouzaud, J.N., 2002. Carbon nanoparticles from laser pyrolysis. Carbon 40 (15), 2775–2789.

Gantar, M., Simović, D., Djilas, S., Gonzalez, W.W., Miksovska, J., 2012. Isolation, characterization and antioxidative activity of C-phycocyanin from *Limnothrix* sp. strain 37-2-1. J. Biotechnol. 159 (1-2), 21–26.

Gao, F., Hong, F., Liu, C, Zheng, L., Su, M., Wu, X., et al., 2006. Mechanism of nano-anatase TiO_2 on promoting photosynthetic carbon reaction of spinach. Biol. Trace Elem. Res. 111 (1), 239–253.

Gao, F., Liu, C., Qu, C., Zheng, L., Yang, F., Su, M., et al., 2008. Was improvement of spinach growth by nano-TiO_2 treatment related to the changes of Rubisco activase? Biometals. 21 (2), 211–217.

Gelagutashvili, E., 2013. Binding of heavy metals with C-phycocyanin: a comparison between equilibrium dialysis, fluorescence and absorption titration. J. Biomed. Life Sci. 1, 12–16.

Gilaki, M., 2010. Biosynthesis of silver nanoparticles using plant extracts. J. Biol. Sci. 10 (5), 465–467.

Gogos, A., Knauer, K., Bucheli, T.D., 2012. Nanomaterials in plant protection and fertilization: current state, foreseen applications, and research priorities. J. Agric. Food Chem. 60 (39), 9781–9792.

Hakeem, K.R., Pirzadah, T.B., 2020. Nanobiotechnology in Agriculture: An Approach Towards Sustainability. Springer.

Hamida, R.S., Abdelmeguid, N.E., Ali, M.A., Bin-Meferij, M.M., Khalil, M.I., 2020. Synthesis of silver nanoparticles using a novel cyanobacteria *Desertifilum* sp. extract: their antibacterial and cytotoxicity effects. Int. J. Nanomed. 15, 49.

Hamida, R.S., Ali, M.A., Redhwan, A., Bin-Meferij, M.M., 2020. Cyanobacteria–a promising platform in green nanotechnology: a review on nanoparticles fabrication and their prospective applications. Int. J. Nanomed. 15, 6033.

Hamouda, R.A, Hussein, M.H., Abo-elmagd, R.A., Bawazir, S.S., 2019. Synthesis and biological characterization of silver nanoparticles derived from the cyanobacterium *Oscillatoria limnetica*. Sci. Rep. 9 (1), 1–17.

Hong, F., Yang, F., Liu, C., Gao, Q., Wan, Z., Gu, F., et al., 2005. Influences of nano-TiO_2 on the chloroplast ageing of spinach under light. Biol. Trace Elem. Res. 104 (3), 249–260.

Hong, F., Zhou, J., Liu, C., Yang, F., Wu, C., Zheng, L., et al., 2005. Effect of nano-TiO_2 on photochemical reaction of chloroplasts of spinach. Biol. Trace Elem. Res. 105 (1), 269–279.

Husen, A., Siddiqi, K.S., 2014. Plants and microbes assisted selenium nanoparticles: characterization and application. J. Nanobiotechnol. 12 (1), 1–10.

Iravani, S., Korbekandi, H., Mirmohammadi, S.V., Zolfaghari, B.J., 2014. Synthesis of silver nanoparticles: chemical, physical and biological methods. Res. Pharm. Sci. 9 (6), 385.

Javad, S., 2020. Nanoagronomy. Springer; Nature.

Jena, J., Pradhan, N., Dash, B.P., Sukla, L.B., Panda, P.K., 2013. Biosynthesis and characterization of silver nanoparticles using microalga *Chlorococcum humicola* and its antibacterial activity. Int. J. Nanomater. Biostruct 3 (1), 1–8.

Jeyaraj, M., Sathishkumar, G., Sivanandhan, G., Mubarak Ali, D., Rajesh, M., Arun, R., et al., 2013. Biogenic silver nanoparticles for cancer treatment: an experimental report. Colliod Surf. B. Interf. 106, 86–92.

Johansen, M.N., 2012. Microalgae: Biotechnology, Microbiology, and Energy. Nova Science Publisher's.

Kalabegishvili, T., Kirkesali, E., Rcheulishvili, A., 2012. Synthesis of gold nanoparticles by blue-green algae *Spirulina platensis*. Frank Lab. of Neutron Physics.

Kanchi, S., Ahmed, S., 2018. Green metal nanoparticles: synthesis, characterization, and their applications. John Wiley & Sons.

Khandel, P., Yadaw, R.K., Soni, D.K., Kanwar, L., Shahi, S.K., 2018. Biogenesis of metal nanoparticles and their pharmacological applications: present status and application prospects. J. Nanostruct. Chem. 8 (3), 217–254.

Khanna, P., Kaur, A., Goyal, D., 2019. Algae-based metallic nanoparticles: Synthesis, characterization and applications. J. Microbiol. Methods 163, 105656. doi: 10.1016/j.mimet.2019.105656. Epub 2019 Jun 17. PMID: 31220512.

Khodakovskaya, M., Dervishi, E., Mahmood, M., Xu, Y., Li, Z., Watanabe, F, et al., 2009. Carbon nanotubes are able to penetrate plant seed coat and dramatically affect seed germination and plant growth. ACS Nano. 3 (10), 3221–3227.

Kim, J.S., Kuk, E., Yu, K.N., Kim, J-H, Park, S.J., Lee, H.J., et al., 2007. Antimicrobial effects of silver nanoparticles. Nanomed. Nanotechnol. Biol. Med. 3 (1), 95–101.

Kulasooriya S.A., 2011. Cyanobacteria: pioneers of planet earth. CJS (Bio. Sci.) 40 (2), 71–88.

Lee, R.E., 2018. Phycology. Cambridge university press.

Lei, Z., Mingyu, S., Chao, L., Liang, C., Hao, H., Xiao, W., et al., 2007. Effects of nanoanatase TiO_2 on photosynthesis of spinach chloroplasts under different light illumination. Biol. Trace Elem. Res. 119 (1), 68–76.

Lengke, M.F., Fleet, M.E., Southam, G.J., 2006. Morphology of gold nanoparticles synthesized by filamentous cyanobacteria from gold (I)– thiosulfate and gold (III)– chloride complexes. Langmuir. 22 (6), 2780–2787.

Lengke, M.F., Fleet, M.E., Southam, G.J., 2007. Biosynthesis of silver nanoparticles by filamentous cyanobacteria from a silver (I) nitrate complex. Langmuir. 23 (5), 2694–2699.

Linglan, M., Chao, L., Chunxiang, Q., Sitao, Y., Jie, L., Fengqing, G., et al., 2008. Rubisco activase mRNA expression in spinach: modulation by nanoanatase treatment. Biol. Trace Elem. Res. 122 (2), 168–178.

MacColl R., 1998. Cyanobacterial phycobilisomes. J. Struct. Biol. 124 (2-3), 311–334.

Mackus, A.J., Verheijen, M.A., Leick, Nm, Bol, A.A., Kessels, W.M., 2013. Influence of oxygen exposure on the nucleation of platinum atomic layer deposition: consequences for film growth, nanopatterning, and nanoparticle synthesis. Chem. Mater. 25 (9), 1905–1911.

Mallick, K., Witcomb, M., Scurrell, M.J., 2004. Polymer stabilized silver nanoparticles: a photochemical synthesis route. J. Mater. Sci. 39 (14), 4459–4463.

Marchiol, L., 2018. Nanotechnology in agriculture: new opportunities and perspectives. *New visions in plant science*, 9 (4), 161.

Mata, Y., Torres, E., Blazquez, M., Ballester, A., González, F., Munoz, J.A., 2009. Gold (III) biosorption and bioreduction with the brown alga *Fucus vesiculosus*. J. Hazard. Mater. 166 (2-3), 612–618.

Menon, S., Rajeshkumar, S., Kumar, V., 2017. A review on biogenic synthesis of gold nanoparticles, characterization, and its applications. Resource-Efficient Technol. 3 (4), 516–527.

Merin, D.D., Prakash, S., Bhimba, B.V., 2010. Antibacterial screening of silver nanoparticles synthesized by marine microalgae. Asian Pac. J. Trop. Med. 3 (10), 797–799.

Mishra, Y., Mohapatra, S., Kabiraj, D., Mohanta, B., Lalla, N., Pivin, J., et al., 2007. Synthesis and characterization of Ag nanoparticles in silica matrix by atom beam sputtering. Scripta Materialia. 56 (7), 629–632.

Misra, P., Shukla, P.K., Pramanik, K., Gautam, S., Kole, C., 2016. Nanotechnology for crop improvement. In: Plant Nanotechnology. Springer, Cham, pp. 219–256.

MubarakAli, D., Gopinath, V., Rameshbabu, N., Thajuddin, N., 2012. Synthesis and characterization of CdS nanoparticles using C-phycoerythrin from the marine cyanobacteria. Mater. Lett. 74, 8–11.

Mueller, N.D., Gerber, J.S., Johnston, M., Ray, D.K., Ramankutty, N., Foley, J.A., 2012. Closing yield gaps through nutrient and water management. Nature 490 (7419), 254–257.

Naderi, M.R., Abedi, A., 2012. Application of nanotechnology in agriculture and refinement of environmental pollutants. J. Nanotechnol., 11 (1), 18–26.

Nair, R., Varghese, S.H., Nair, B.G., Maekawa, T., Yoshida, Y., Kumar, D.S., 2010. Nanoparticulate material delivery to plants. Plant Sci. 179 (3), 154–163.

Patel, K., Bharatiya, B., Mukherjee, T., Soni, T., Shukla, A., Suhagia, B.J., et al., 2017. Role of stabilizing agents in the formation of stable silver nanoparticles in aqueous solution: characterization and stability study. J. Dispers. Sci. Technol. 38 (5), 626–631.

Patel, V., Berthold, D., Puranik, P., Gantar, M., 2015. Screening of cyanobacteria and microalgae for their ability to synthesize silver nanoparticles with antibacterial activity. Biotechnology Reports. 5, 112–119.

Perrin, T.S., Drost, D.T., Boettinger, J.L., Norton, J.M., 1998. Ammonium-loaded clinoptilolite: a slow-release nitrogen fertilizer for sweet corn. J. Plant Nutr. 21 (3), 515–530.

Qi, M., Liu, Y., T.J., Li, 2013. Nano-TiO_2 improve the photosynthesis of tomato leaves under mild heat stress. Biol. Trace Elem. Res. 156 (1), 323–338.

Qureshi, A., Singh, D., Dwivedi, S., 2018. Nano-fertilizers: a novel way for enhancing nutrient use efficiency and crop productivity. International Journal of Current Microbiology and Applied Sciences. 7 (2), 3325–3335.

Rajeshkumar, S., Malarkodi, C., Paulkumar, K., Vanaja, M., Gnanajobitha, G., Annadurai, G.J., 2014. Algae mediated green fabrication of silver nanoparticles and examination of its antifungal activity against clinical pathogens. Int. J. Metal 2014.

Rameshaiah, G.N., Pallavi, J., Shabnam, S., 2015. Nano fertilizers and nano sensors–an attempt for developing smart agriculture. Int. J. Eng. Res. General Sci. 3 (1), 314–320.

Rauwel, P., Küünal, S., Ferdov, S., Rauwel, E., 2015. A review on the green synthesis of silver nanoparticles and their morphologies studied via TEM. Adv. Mater. Sci. Eng. 2015.

Rodrigues, S.M., Demokritou, P., Dokoozlian, N., Hendren, C.O., Karn, B., Mauter, M.S., et al., 2017. Nanotechnology for sustainable food production: promising opportunities and scientific challenges. Environ. Sci.: Nano. 4 (4), 767–781.

Saifuddin, N., Wong, C., Yasumira, A.J., 2009. Rapid biosynthesis of silver nanoparticles using culture supernatant of bacteria with microwave irradiation. E-J. Chem. 6 (1), 61–70.

Salem, S.S., Fouda, A., 2021. Green synthesis of metallic nanoparticles and their prospective biotechnological applications: an overview. Biol. Trace Elem. Res. 199 (1), 344–370.

Sastry, M., Ahmad, A., Khan, M.I., Kumar R., 2004, Microbial nanoparticle production, Nanobiotechnology: Concepts, Applications and Perspectives 28, 126–135.

Sastry, M., Ahmad, A., Khan, M.I., Kumar, R., 2003. Biosynthesis of metal nanoparticles using fungi and actinomycete. Curr. Sci. 85 (5), 162–170.

Scott, N., Chen, H.J., 2013. Nanoscale science and engineering for agriculture and food systems. Ind. Biotechnol. 9 (1), 17–18.

Seku, K., Hussaini, S.S., Pejjai, B., Al Balushi, M.M.S., Dasari, R., Golla, N., et al., 2021. A rapid microwave-assisted synthesis of silver nanoparticles using *Ziziphus jujuba* Mill fruit extract and their catalytic and antimicrobial properties. Chem. Papers, 1341–1354.

Shah, M, Fawcett, D., Sharma, S., Tripathy, S.K., Poinern, G.E., 2015. Green synthesis of metallic nanoparticles via biological entities. Materials. 8 (11), 7278–7308.

Sharma, A., Sharma, S., Sharma, K., Chetri, S.P., Vashishtha, A., Singh, P., et al., 2016. Algae as crucial organisms in advancing nanotechnology: a systematic review. J. Appl. Phycol. 28 (3), 1759–1774.

Singh, R., Shedbalkar, U.U., Wadhwani, S.A., Chopade, B.A., 2015. Bacteriagenic silver nanoparticles: synthesis, mechanism, and applications. Appl Microbiol. Biotechnol. 99 (11), 4579–4593.

Thajuddin, N., Subramanian, G., 2005. Cyanobacterial biodiversity and potential applications in biotechnology. Curr. Sci., 47–57.

Thakkar, K.N., Mhatre, S.S., Parikh, R.Y., 2010. Biological synthesis of metallic nanoparticles. Nanomed. (Lond). 6 (2), 257–262.

Wong, M.H., Misra, R.P., Giraldo, J.P., Kwak, S-Y., Son, Y., Landry, M.P., et al., 2016. Lipid exchange envelope penetration (LEEP) of nanoparticles for plant engineering: a universal localization mechanism. Nano Lett. 16 (2), 1161–1172.

Zhang, X., Yan, S., Tyagi, R., Surampalli, R.J., 2011. Synthesis of nanoparticles by microorganisms and their application in enhancing microbiological reaction rates. Chemosphere 82 (4), 489–494.

CHAPTER

3

Cyanobacterial photosynthetic reaction center in wobbly light: Modulation of light energy by orange carotenoid proteins (OCPs)

Rahul Prasad Singh[a], *Sandeep Kumar Singh*[a], *Ajay Kumar*[b], *Arpan Modi*[b], *Mukesh Kumar Yadav*[c], *Punuri Jayasekhar Babu*[c], *Garima Singh*[d], *Kaushalendra*[e], *Naveen Chandra Joshi*[f], *Avinash Chandra Rai*[b], *Sandip Ghuge*[b], *Anil Kumar*[b], *Avinash Singh*[g], *Prashant Kumar Singh*[c], *Wenjing Wang*[h]

[a]Department of Botany, Institute of Science, Banaras Hindu University, Varanasi, Uttar Pradesh, India

[b]Agriculture Research Organization (ARO), The Volcani Center, Rishon-Lezion, Israel

[c]Department of Biotechnology, Mizoram University (A Central University), Pachhunga University College Campus, Aizawl, Mizoram, India

[d]Department of Botany, Mizoram University (A Central University), Pachhunga University College Campus, Aizawl, Mizoram, India

[e]Department of Zoology, Mizoram University (A Central University), Pachhunga University College Campus, Aizawl, Mizoram, India

[f]Amity Institute of Microbial Technology (AIMT), Amity University, Noida, Uttar Pradesh, India

[g]Department of Biochemistry and Molecular Pharmacology, New York University, New York, USA

[h]Department of Biology and Food Science, Shangqiu Normal University, Shangqiu, The People Republic of China

Cyanobacterial Lifestyle and its Applications in Biotechnology.
DOI: https://doi.org/10.1016/B978-0-323-90634-0.00007-X

3.1 Introduction

Oxygenic photoautotrophs such as plants, green algae, and cyanobacteria are blessed with the capacity to convert solar energy into chemical energy and releases oxygen as a byproduct; both are required for the fundamental growth of life on the earth (Kirilovsky and Kerfeld, 2012). The light quality and intensity vary and largely depend on the day length, climate, season, etc. The light that is fundamental for life can get deadly when its energy surpasses the limit of photosynthetic life forms to utilize it for carbon fixation. Under these conditions, unsafe reactive oxygen species (ROS) is produced. Hence, although light is essential for photosynthesis, it can be wrecking if photon energy caught by the light-harvesting antenna complex proceeds even after the photosynthetic electron transport saturation. For instance, if the energy of singlet excited particles of antenna chlorophyll cannot be exploited totally by the reaction center, at that point, this singlet state will be converted into long-lived triplet states. Excited chlorophyll triplets may associate with oxygen atoms and prompt the formation of conceivably ruinous singlet oxygen species (Morosinotto and Bassi, 2014) and are profoundly reactive (oxidizing agents) that can react with various biomolecules such as protein, lipid, and DNA (Cogdell and Gardiner, 2015) as well as lead to photosynthetic electron transport chain annihilation (Karapetyan, 2007). Photosystem II (PSII) is the most light-sensitive complex in the photosynthetic apparatus (Aro and Virgin, 1993; Prasil et al., 1992; Tyystjärvi, 2008; Vass and Aro, 2007; Singh et al., 2018). Temperature and water accessibility also vary daily; in like manner, and become growth-limiting factors, for example, a low concentration of iron in the oceans. In this way, photosynthetic life forms have created physiological mechanisms that permit them to adapt and make due to a broad scope of natural conditions. These components reacting to varieties of the light and/or their intensities include quick and moderate changes in the photosynthetic apparatus, going from seconds to hours. The fast cycles include conformational changes or a rearrangement of the photosynthetic complexes.

In contrast, the synthesis or debasement of proteins and protein complexes are slower processes. Two of the fast cycles include changes are: (1) the compelling size of the PSII antenna and state interchanges (van Thor et al., 1998; Williams and Allen, 1987; Wollman, 2001) and (2) non-photochemical-quenching (NPQ) (Horton, 1996; Niyogi, 1999). In green algae and plants, the light-harvesting complex (LHCII), and the PSII antenna, is made of six hydrophobic layers of natural polypeptides of the LHC family noncovalently tied to chlorophyll and carotenoids. In cyanobacteria and red-green algae, this membrane-bound chlorophyll antenna is missing, and sun-oriented energy is consumed by a substantial extramembrane (soluble) complex known as the phycobilisome (PB). The

Phycobilisome (PB) is made of a different kind of chromophorylated phycobiliproteins and linker peptides (The linker peptides are required for the PBs structural organization and functioning) (Adir, 2005; Glazer, 1984; Grossman et al., 1993; MacColl, 1998; Tandeau and de Marsac, 2003). Phycobilisome (PBS) are attached with the thylakoid membrane outer surface and radiate like rods (Gantt and Conti, 1966a, b). In most freshwater cyanobacterial species, these rod-like structures are made up of entire phycocyanin (PC). In contrast, in large numbers of marine cyanobacterial species, phycoerythrin (PE) or phycoerythrocyanin (PEC) is also present on the distal ends of these rod-like structures. Allophycocyanin (APC) is the primary central phycobiliprotein present in PBS. Besides this, three other chromophorylated proteins namely; ApcD, ApcF, and ApcE or Lcm (linker center layer) fill in as terminal energy acceptors that trap the light energy and transfer the chlorophylls of photosystems-I (Ashby and Mullineaux, 1999; Glindt et al., 1994; Dong et al., 2009). Likewise, the Lcm connects the PB to the thylakoids.

In green algae and plants, under the saturating light, a decline in the pH of the chloroplast lumen triggers a mechanism that changes LHCII into a proficient energy dissipator (Hortan et al., 1996; Niyogi, 1999; Pascal et al., 2005; Ruban et al., 2007). This component includes the xanthophyll cycle initiation, PsbS protein protonation (an Lhc family protein), and LHC (and/or potentially minor antenna) conformational changes, which changes the association among chlorophylls and carotenoids. A lessening of PSII-related fluorescence discharge joins energy scattering, otherwise called nonphotochemical-quenching fast (NPQf) or qE, which usually fills in as a proportion of the dispersal interaction. This NPQ process is quickly reversible (in no time) in the dark and absence of protein synthesis.

The identical mechanism in cyanobacteria is not initiated by a drop in pH of thylakoid lumen under high irradiance. However, in the presence of high blue-green (or white) light, cyanobacterial cells accumulate a soluble carotenoid protein, the orange carotenoid protein (OCP), which, by interfacing with the phycobilisome, builds energy dispersal in the form of heat and hence reducing the energy of light hitting to the reaction (Wilson et al., 2006). The fluorescence decline related to this energy dispersal in the antenna complex is denoted as non-photochemical quenching in cyanobacteria (qEcya) to separate this system from that existing in plants. The qEcya-mediated mechanisms not by any mean the only one that ensures cyanobacteria survival under high irradiance. Zhu et al. (2010) advocated that mutants lacking zeaxanthin are susceptible to high light than the qEcya mutants and hence demonstrated the role of carotenoid in photoprotection.

Moreover, under high light conditions and nutrient deprivation, temperature stress, heavy metal stress, UV-B stress, and other unfavorable environments, single-helix chlorophyll-binding proteins such as high light inducible proteins/early light inducible proteins(HLIPs/ELIPs), of the LHC protein family, are produced and known to involve in photoprotection in cyanobacteria (Funk and Vermaas, 1999; He et al., 2001; Singh et al., 2015). Kilian et al. (2008) have advocated a novel two-domain fusion protein, Chlorophyll a-b-Binding proteins from the cyanobacteria, which have similarity with ELIP/HLIP superfamily proteins, expresses under multiple stress conditions such as including high light. Iron starvation–induced-A protein (IsiA) expressed under iron stress also demonstrated as another Lhc family protein binding with chlorophyll/carotenoid offers cell endurance. This protein either attached with PSI antenna complex or in an unattached form dissipates energy or involves quenching of fluorescence (Ihalainen et al., 2005; Wilson et al., 2007; Yeremenko et al., 2004). In this way, cyanobacteria have built up a vast assortment of machinery to shield themselves from various stresses and associated life forms. For instance, OCP-related energy-dispersal increments under conditions where there is an overabundance of practically separated phycobilisomes (e.g., iron starvation, PSI mutants) (Wilson et al., 2007). This chapter discusses mainly the phycobilisome-related light-instigated photoprotective component intervened by the OCP. Later primary and research outcomes will be portrayed and discussed briefly to other blue-light responsive proteins; significant past consequences will be referenced.

3.2 Fates of light energy absorbed by pigments

On absorption of light energy, the chlorophyll or its accessory pigments get excited and promoted to higher energy levels or its first singlet excited state from its ground state. This excited state then has three primary fates (as shown in Fig. 3.1). Either the energy is: (1) by Förster resonance energy transfer passed to another chlorophyll molecule and gradually passed to the photochemical reaction centers (photosystem I and photosystem II) where energy is used in photosynthesis (called photochemical quenching)), or (2) the excited state by emitting the energy as heat (known as nonphotochemical quenching, NPQ) can return to the ground state, or (3) the excited state can return to the ground state by emitting a photon (fluorescence).

Suppose the light absorption continues beyond the saturation of the pigments. In that case, molecules lead to enhancement in the lifetime of singlet excited chlorophyll molecules, which leads to the formation of long-lived chlorophyll triplet states. The triplet chlorophyll serves as a potent molecular oxygen photosensitizer and farming singlet oxygen that can cause oxidative damage to the proteins, pigments, lipids, and the photosynthetic thylakoid membrane. To counter this problem, one photoprotective mechanism is the so-called NPQ, which relies upon the conversion and dissipation of the excess excitation energy into heat.

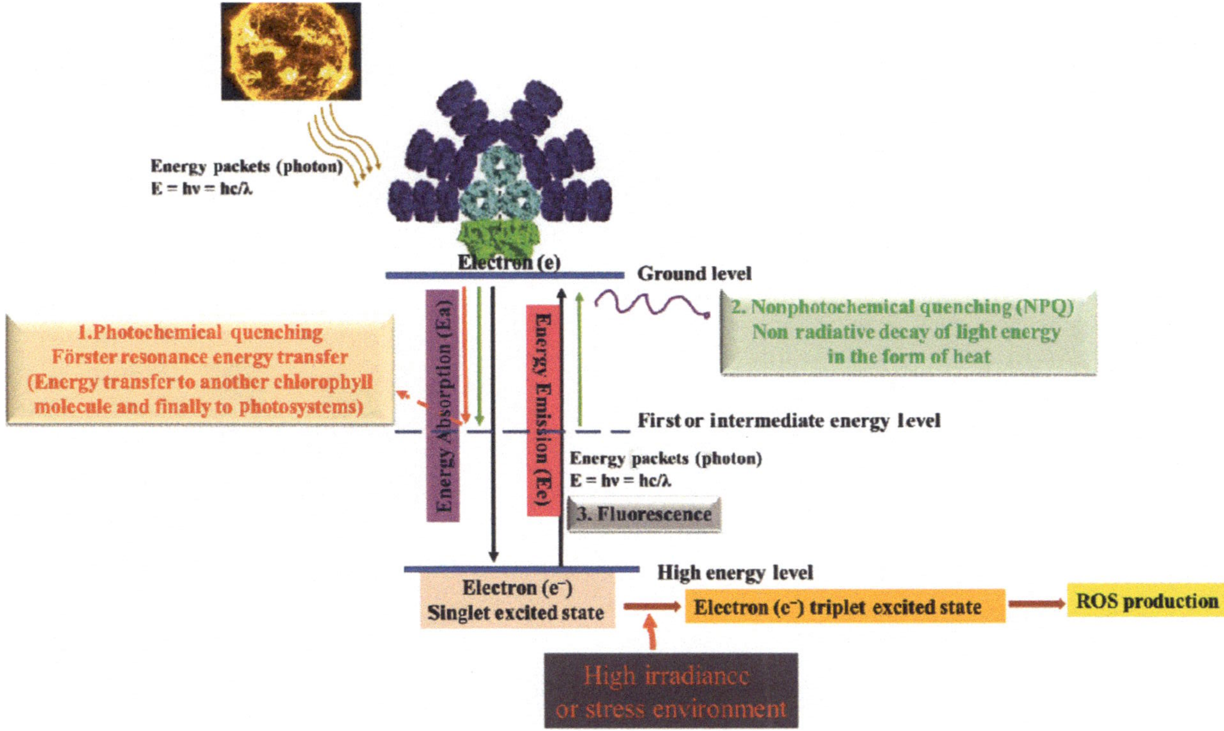

FIG. 3.1 The fate of energy absorbed by photosynthetic as well as accessory pigments in cyanobacteria. Following the absorption of light, the pigment molecules get excited. Electrons from the ground state (zero oxidation state) get energized and promoted to a highly excited state, which is very short-lived (10^{-15} s). Soon the excited molecule returns to the ground energy level. This excited state of the molecule then has three primary fates. Either the energy is (1) transferred to another chlorophyll molecule by Förster Resonance Energy Transfer, where energy is exploited for photosynthesis (also known as photochemical quenching); or (2) the excited state by emitting energy in the form of heat returns back to ground level (called nonphotochemical quenching); or (3) the excited state can return to the ground state by emitting a photon (fluorescence). In case of high radiance or stress conditions, the enthusiastic (singlet) state of molecule converted into a triplet excited state, leading to ROS production. In excess light or stresses, the lifetime of singlet excited chlorophyll increases, leading to its conversion into long-lived chlorophyll triplet states by intersystem crossing. Triplet chlorophyll molecules are highly reactive to the molecular oxygen and farming singlet oxygen which can cause oxidative damage to the pigments, lipids, and proteins of the photosynthetic thylakoid membrane.

3.3 Light-harvesting complex organization in cyanobacteria

Photosynthetic organisms are challenged primarily to speedy and consistent light intensities fluctuations. Due to light-restricting conditions, photosynthetic life forms have developed cautious light-gathering receiving antenna complexes that ideally can help retain excitation energy. Notwithstanding, high light energies can be highly poisonous to these life forms where the light-harvesting antenna complex catches energy even after the immersion of photochemistry. To stay away from the conceivably phototoxic impact of overabundance absorbed light energy, cyanobacteria have developed a technique that directs the progression of excitation energy to reaction centers (Sluchanko et al., 2017). The cyanobacterial oxygenic photosynthesis machinery is similar to eukaryotic algae and higher plants. However, they differ in position and organization of light-harvesting antenna complexes (which fuels the reaction centers). Plants and eukaryotic green algae utilize integral membrane light-harvesting systems that are fit for diffusing along the lateral side across the thylakoid membrane. At the same time, the leading light-harvesting antenna complex in cyanobacteria is PBS, a supramolecular complex of water-soluble accessory pigment found in cyanobacteria. In connection with the membrane-associated reaction center, PBS is enclosed on the thylakoid membrane surface in the form of arrays, as shown in Fig. 3.2 (Bryant et al., 1979; Karapetyan, 2007). Gantt and Conti (1966) were the pioneered workers on PBS and identified it first from employing red alga *Porphyridium cruentum*. Later PBS has been advocated from many cyanobacterial species (Glazer et al., 1976).

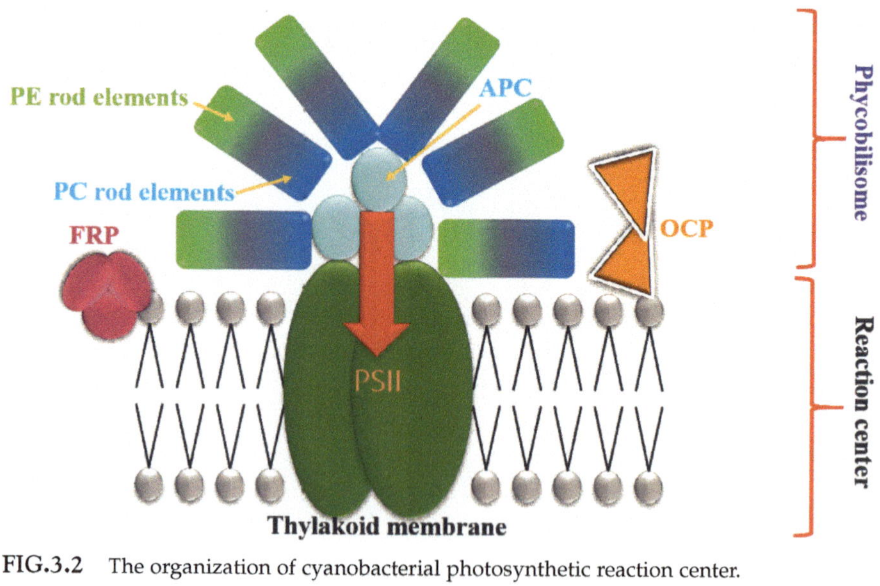

FIG.3.2 The organization of cyanobacterial photosynthetic reaction center.

PBS has two substructures, namely the rod and core. These structures have trimers and hexamers of pigment phycobiliproteins (PBPs) accompanied by nonpigmented linker polypeptides, which stabilize the entire complex. PBPs are made of α and β subunits (in some phycoerythrin, a remarkable γ subunit can moreover be found) that bear covalently joined open-chain tetrapyrroles named as phycobilins. Based on the composition of pigment and their spectral properties, PBPs have been classified into the following four groups:

1. APC, ~ 650 nm—absorption maxima,
2. PC, ~ 620 nm—absorption maxima,
3. PEC, ~ 595 nm—absorption maxima, and
4. PE, ~ 560 nm—absorption maxima.

The rod comprises PC and PE, although few cyanobacteria only include just PCs, such as *Synechocystis* sp. PCC 6803, while the core explicitly contains APC. The nonpigmented linker polypeptides (not seen in the figure) are one of the substantial parts of the PBS that are elementary for the PBPs organization into rod and core and contribute to building up the exceptionally robust cyanobacterial light-harvesting complex. The PE and PC of the rod assimilate excitation energy effectively and transfer them to the APC core. This collected excitation energy finally relocates to the photosynthesis reaction center, PS II (Bailey and Grossman, 2008; Liu et al., 2005; Saer and Blankenship, 2017). As indicated by a new underlying model proposed by Liu et al. (2017), PBS moves excitation energy to both PS I and PS II by securing a megacomplex-like structure with both reaction centers (Liu et al., 2017) (Fig. 3.3).

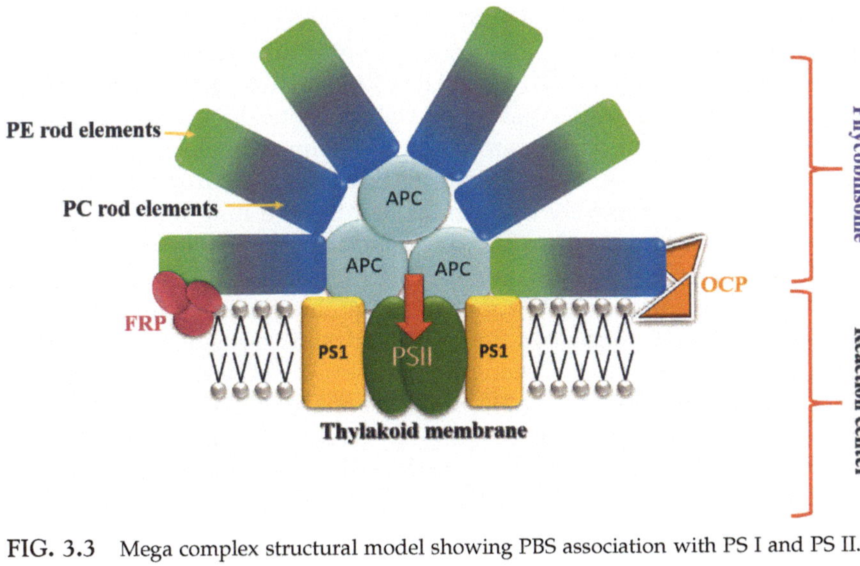

FIG. 3.3 Mega complex structural model showing PBS association with PS I and PS II.

3.4 Modulation of light energy in cyanobacteria (photoprotective mechanism)

To oblige under high light condition, photoautotrophs have built up a systems/mechanism for modulation of light. The regulation/modulation of these systems is essential to keep up the photosynthetic productivity by limiting photodamage. Under light overabundance, NPQ works variously among the more significant part of the photosynthetic life forms. In plants and green algae, NPQ embraces membrane-bound chlorophyll-containing light reaping complex (LHC II) and start by the drop in thylakoid lumen pH (Muller et al., 2001). Thylakoid lumen acidification led to xanthophyll cycle components deepoxidation, which involves violaxanthin oxidation to zeaxanthin through antheraxanthin intermediate (Demmig-Adams and Adams, 1996; Ort, 2001) and initiates the PsbS (a subunit of photosystem II that has a place with LHC superfamily) protonation (Li et al., 2000, 2004). PsbS is firmly discerning to low pH (Morosinotto and Bassi, 2014), and it lacks chlorophyll and xanthophyll. Each of these cycles induces conformational changes in LHC II that alter the connection between chlorophyll and carotenoids and improves the heat dissemination of excess assimilated energy to safeguard the PS II integrity. Comparable photoprotective systems also happen in numerous types of *Bacillariophyceae, Chloromonadophyceae, Chrysophyceae, Euglenophyceae, and Xanthophyceae Dinophyceae* that depends on the light-animated change of diadinoxanthin to diatoxanthin (Gorbunov et al., 2011).

In contrast to plants and eukaryotic algae, cyanobacteria follow a different mechanism of NPQ, which involves a sizeable water-soluble light-harvesting complex, PBS, bounded on the surface of the thylakoid membrane and a carotenoid as chromophore, which is activated by intense light. Despite differences in the organization of light-harvesting antenna complexes, carotenoids play a crucial role in plants and cyanobacteria. Considering their light-harvesting and photoprotective capability under fluctuating light condition, carotenoids are imperative for the proper functioning of photosynthetic apparatus (Zakar et al., 2016). NPQ mechanism is studied exclusively in vascular plants for a long time. However, a photoprotective mechanism in cyanobacteria was uncovered in the mid-late 1980s (Holt and Krogmann, 1981). Cyanobacteria were considered incapable of performing NPQ for protecting their photosynthetic machinery under light-saturating condition. Instead, state transition involves the redistribution of energy between photosystem II (PS II) and photosystem I (PS I), considered central photoprotective mechanisms to divert excess excitation flux away. Several pieces of research have proved that cyanobacteria can also operate NPQ as one of the essential photoprotective strategies to deal with high light. NPQ in cyanobacteria essentially requires the following three essential components:

(i) A distinct, large water-soluble light-harvesting complex, phycobilisome (PBS).
(ii) An OCP with a sensor that senses elevated light level and an effector that quenches excess fluorescence after interacting with PBS core.
(iii) Fluorescence recovery protein (FRP) that helps in dissociation of OCP from PBS core (Shirshin et al., 2017).

Fluorescence NPQ is one of the effective procedures to quickly disperse the abundance of excitation energy in the form of heat (Fig. 3.1). The underlying mechanism of NPQ varies in plants, eukaryotic green algae, and cyanobacteria. Unlikely pH-adjustment in higher plants and green algae, as mentioned earlier, cyanobacteria require OCP and FRP, which serve as a photoprotective component.

3.4.1 Photoprotective mechanism mediated through phycobilisome

For the first time, El Bissati et al. (2000) portrayed the proof for the presence of a blue-green light-activated phycobilisome-mediated photoprotective mechanism. They observed that on exposure to intense blue-green light incited PSII fluorescence quenching in *Synechocystis* under conditions in which the plastoquinone (PQ) pool was oxidized to a great extent oxygen-evolving activity not saturated. Besides, the blue-green light-mediated fluorescence quenching unhindered in the presence of 3-(3,4-dichlorophenyl)-1,1-dimethylurea (DCMU) (an inhibitor of electron transport between a primary quinone electron acceptor (QA) and secondary quinone electron acceptor (QB) quinones) and, in this way, forestalling the PQ pool decrease. These outcomes demonstrated that the fluorescence reduction was not identified with state transitions, which are actuated by removing the PQ pool (Van Thor et al., 1998; Williams and Allen, 1987; Wollman, 2001). When the "quenched" cells were moved to low light intensities, recuperation of fluorescence occurred even within sight of interpretation of inhibitors (El Bissati et al., 2000), showing that the blue-green light-incited quenching is unrelated to photoinhibition (Aro and Virgin, 1993; Prasil et al., 1992; Tyystjärvi, 2008; Vass and Aro, 2007). On account of photodamage, the fluorescence expansion related to the PSII-impaired D1 polypeptide; this interaction requires D1 synthesis.

Furthermore, El Bissati et al. (2000) recommended that high light energy instigate phycobilisome fluorescence quenching associatively with a reduction of energy move from the phycobilisome to the photosystems. Like this, unearthly and energy information affirmed the presence of this phycobilisome-related fluorescence quenching (Rakhimberdieva et al., 2004). Blue-light-mediated reversible quenching of the phycobilisome emission has been demonstrated in a *Synechocystis* mutant lacking reaction center II and the Chl antenna complexes (CP43 and CP47), later these findings have been asserted by two different labs (Ruban et al., 2007; Scott et al., 2006). Conversely, blue-green light could not activate fluorescence quenching in a phycobilisome-deficient mutant or a mutant containing just the PC poles of the phycobilisome. In contrast, the presence of just the center of the phycobilisome is adequate to instigate the quenching (Wilson et al., 2006). In Wild Type (WT) *Synechocystis* cells, a reduction in phycobilisome-related fluorescence was seen without change in the chlorophyll-related fluorescence emission (Wilson et al., 2006; Scott et al., 2006), trapping phycobilisomes as fundamental parts of the blue-light-actuated qEcya system. Picosecond time-settled fluorescence decay information was predictable with a fluorescence quenching at the phycobilisome center (Scott et al., 2006). The phycobilisome center contains APC producing at 660 nm (at room temperature and 77 K) and terminal emitters (ApcD, ApcF, and Lcm (ApcE)), transmitting at around 680 nm (room temperature; at 77 K: 683 nm). Recently, Rakhimberdieva et al. (2010) proposed that fluorescence quenching begins from the short-frequency APC (660 nm outflow), yet this stay to be demonstrated tentatively. Under conditions in which all PSII centers are shut, within sight of DCMU+light, chlorophyll excitation could also be quenched by the phycobilisome-mediated quenching system. In these conditions, energized chlorophyll molecules could be adequately extensive to permit the exciton to be moved from PSII chlorophyll to the center of the phycobilisome and their quenching (Rakhimberdieva et al., 2007). The blue-green light-initiated fluorescence quenching connects with a decline of the antenna complex's viable size, leading to reduced energy arrival at the photochemical center (Ruban et al., 2007). A quantitative estimation of light reaching to reaction center by utilizing *Synechocystis* mutants lacking PSI or PSII showed that in the extinguished cells only 60%–70% of the energy phycobilisomes absorbed energy reaches the PSII or the PSI (Rakhimberdieva et al., 2010).

qEcya mechanism largely relies on the nature of light; just blue-green light can trigger it. The action spectrum for phycobilisome emission quenching in the *Synechocystis* PS II–deficient mutants identified that a carotenoid molecule could be associated with this interaction (Rakhimberdieva et al., 2004). Wilson et al. (2006) illustrated that OCP, a soluble 35 kD protein containing a solitary, noncovalently bound keto-carotenoid (Holt and Krogmann, 1981; Kerfeld, 2004a, 2004b), is explicitly associated with the qEcya system. Without the OCP (in a ΔOCP *Synechocystis* mutant or then again in cyanobacterial strains coming up short on the OCP quality), the fluorescence quenching prompted by high blue-green light is repressed. Furthermore, the cells are more susceptible to high light intensities; this is shown as a quicker diminishing PSII movement under high irradiance (Wilson et al., 2006). Wilson et al. (2006) employed immunogold marking and electron microscopy to identify the position of OCP and advocated that the OCPs are present in the interthylakoid cytoplasmic region on the membrane phycobilisome side. The coisolation of the OCP upheld the presence of an interaction between the OCP and the phycobilisomes and thylakoids with the phycobilisome-associated membrane fraction (Wilson et al., 2006; Wilson et al., 2007). Blue-green light incites an exceptionally huge fluorescence quenching under the iron-limiting conditions, a lot bigger than standard medium (Wilson et al., 2007; Rakhimberdieva et al., 2007; Cadoret et al., 2004; Joshua et al., 2005). This enormous light-instigated phycobilisome-related fluorescence quenching corresponds with a high OCP concentration, compared to chlorophyll and phycobiliproteins (Rakhimberdieva et al., 2007; Cadoret et al., 2004). Subsequently, the OCP-mediated photoprotection enhancements under a stress environment highlight their significance for the cyanobacterial acclimation to ecosystem perturbations. The connection between the OCP convergence and excitation energy scattering was later affirmed utilizing a *Synechocystis* mutant OCP overexpressing. Large amounts of OCP (8–10 times more than the WT) presenting an enormous fluorescence quenching was noticed (Wilson et al., 2008). The quenching of maximal fluorescence was in increments from 25% to 30% in WT cells and 60%-70% in overexpressing OCP cells (Kirilovsky and Kerfeld, 2012). The evaluation of OCP and phycobilisomes recommends that in WT *Synechocystis* cells, there is one OCP atom for each two to three phycobilisomes; in the overexpressing OCP strain, there are three to four OCP particles per phycobilisome (Kirilovsky and Kerfeld, 2012).

3.4.2 The orange carotenoid proteins (OCP)

Holt and Krogman (1981) advocated a soluble protein from three different cyanobacterial strains: *Arthrospira maxima, Microcystis aeruginosa,* and *Aphanizomenon flosaquae* keto-carotenoid pigment, that is, 3-hydroxyechinenone (3-hECN). Wu and Krogmann (1997), for the first time, identified *slr1963* gene encodes for OCP in *Synechocystis* sp PCC 6803 by taking the first sequenced cyanobacterial genome as a reference. Kerfeld et al. (2003) extracted OCP from *A. maxima* and resolved the crystal structure, and Wilson et al. (2010) resolved the crystal structure of OCP from WT from *Synechocystis*. OCP (~35 kDa) is the only known protein that harbors carotenoid as a chromophore and water-soluble protein. As shown by the proposed crystal structure, OCP includes two discrete primary domains (Kirlovsky and Kerfeld, 2016; Bao et al., 2017; Kerfeld et al., 2017; Sluchanko et al., 2017); (i1) N-terminal space (NTD; 18 kDa all helices) and (2) a C-terminal domain (CTD; having mixed α-helical/β-sheet). NTD is explicit to just cyanobacteria, while the CTD, a member of nuclear transport factor-2 (NTF-2), is found in both prokaryotes and eukaryotes (Kirlovsky and Kerfeld, 2016; Kerfeld et al., 2017). Furthermore, an adaptable

25 amino acids long linker molecule links these two domains, recently it has been advocated that this linker is not required to connect two domains of OCP (Moldenhauer et al., 2018; Sluchanko et al., 2017). OCP gains a compact globular structure, well-adjusted by interdomain and interaction at the primary interface with a highly conserved R155 (NTD)/E244 (CTD) salt bridge. The second minor interface between the two domains is settled by a hydrophobic connection between α-A helix inside N-terminal expansion (NTE) of NTD and a few β-sheets of CTD (Sluchanko et al., 2017). OCP two domains (CTD and NTD) epitomize a solitary keto-carotenoid (Lu et al., 2016) that extends into NTD and CTD totally, simply 4% presented to the hydrophobic environment (Sluchanko et al., 2017) (Fig. 3.4). The purified OCP from various cyanobacterial species shows that it can bind to multiple sorts of keto-carotenoids (such as echineone,3′- hydroxy-echineone, myxoxanthin, canthaxanthin, etc.). The 4-keto oxygen assumes to play a very crucial role in the formation of carotenoid-protein interactions (H-bond) with two conserved amino acid residues (tyrosine-Y201 and tryptophan-288) of CTD. This hydrogen bonding is very crucial for OCP photoactivity. It has been demonstrated essentially that replacement with zeaxanthin (lacking carbonyl group) can lessen OCP photoactivity (Punginelli et al., 2009). Under light-curbing condition, OCP stays in their orange structure (OCP^O), and the phycobilisome complex (PBS) works effectively to increase light absorption. When excitation energy (blue-green

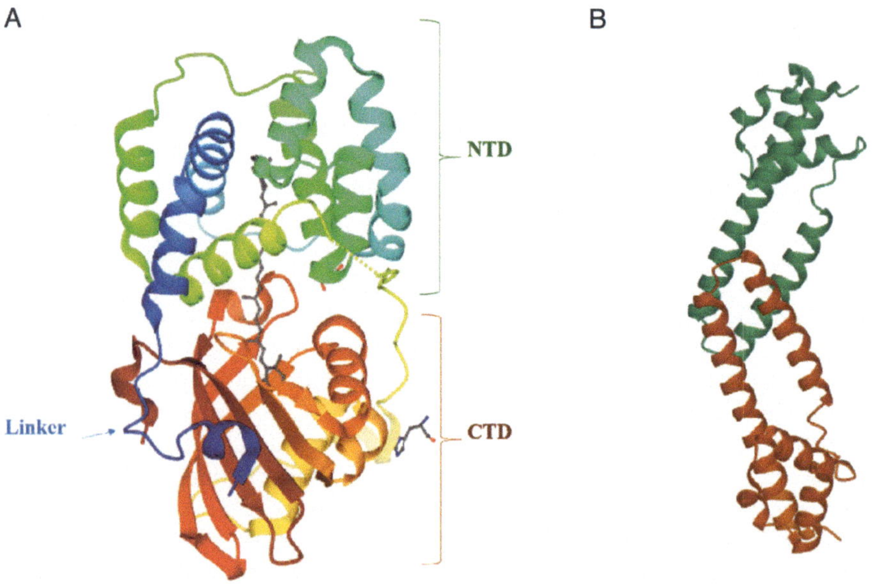

FIG. 3.4 The OCP and FRP structures: (A) OCP dimer: NTD colored in green, CTD in orange, the linker in blue, and carotenoid is shown as gray sticks. (B) FRP: the dimeric form.

light) surpasses the ideal, it plunges the reaction center. Associatively, abundance energy switches the dormant orange type of OCP (OCP^O) to the dynamic red form (OCP^R), which is a more limited OCP variation (Cogdell and Gardiner, 2015). In 2003, Kerfeld appeared, although OCP^O and OCP^R can quench singlet oxygen, extinguishing OCP^R relatively higher than the OCP^O. Holt and Krogmann (1981) identified OCP^R from some cyanobacterial species. Subsequently, Wu and Krogmann (1997) identified OCP^R as an OCP proteolytic component, conveying that most of the NTD domain alongside total carotenoid lacks CTD.

In this manner, OCP endures blue-green light-mediated dissociation of the domain, which plays a very crucial role in the conversion of OCP^O (inactive orange OCP) to the OCP^R (active red OCP) and sets the cyanobacterial NPQ platform.

3.4.3 Full antenna capacity recovery: the role of FRP

Subsequently, after the OCP (*slr1963* gene product in *Synechocystis)* unearthing, it has been advocated that the *slr1963* associated consistently with another Open Reading Frame (ORF) (Boulay et al., 2010), named as FRP, which is critical for regulation of OCP-intervened cyanobacterial fluorescence quenching. Two oligomeric and conformational FRP states from *Synechocystis* haves been crystallized. One is an active dimer (having compact fold, which interacts with CTD of OCP (as shown in Fig. 3.3B), and another one, a dormant tetramer (having an extended α-helix with a bit of spiral cap), is a storage protein type. Both these structures of FRP are interconvertible. However, although the FRP design is presently known, how they associate with OCP is as yet questionable.

3.5 Mechanism of OCP-mediated light modulation in cyanobacteria

In plants and eukaryotic green algae, NPQ operates at a light-harvesting chlorophyll antenna complex implanted in the thylakoid layer. In light starvation, these edifices upgrade the light-catching effectiveness to use every photon. Upon exposure to high light, pH of lumen diminishes, which changes these antenna complexes from light receiving state to light defensive state by xanthophylls cycle (cycle incorporates change of violaxanthin to zeaxanthin under high light intensity) initiation. Zeaxanthin accumulation is essential as it goes about as an effective quencher of the energized chlorophyll molecule. However, contrary to plant and algal NPQ mechanism, which relies on the alteration in pH, the cyanobacterial NPQ is governed mainly by OCP, located on the water-soluble light-harvesting antenna complex present fringe to thylakoid membrane (Kirlovsky, 2015).

3.5.1 Cyanobacterial reaction center in high light: OCP is a significant game-changer

Upon enlistment by blue-green light or high irradiance, the OCP undergoes conformational changes, leading to conversion of dormant OCP^O to a functional OCP^R structure by separating two OCP domains, that is, NTD and CTD (Wilson et al., 2008) (Fig. 3.5). This change is due to the salt-bridge breakage at R155-E244. The significant interface between two domains and NTE separated from CTD at minor interface permits the separation of two OCP domains (Sluchanko et al., 2017). When OCP^R formed, it obtains the capacity to binds with PBS after the interaction between positively charged R155 on the NTD surface of OCP^R and OCP^R gets stabilized by negative surface charges on PBS core (Bao et al., 2017). Subsequently, proficient fluorescence quenching occurs from the excited PBS core. It has been recommended that overabundance excitation energy moves from bilin shades of PBS to OCP^R carotenoid chromophore for the antenna complex de-excitation (Tian et al., 2012). This cycle appears as though OCP^R shuts the tap toward the finish of the light-harvesting antenna complex to limit the PS II (reaction center) energy progression,

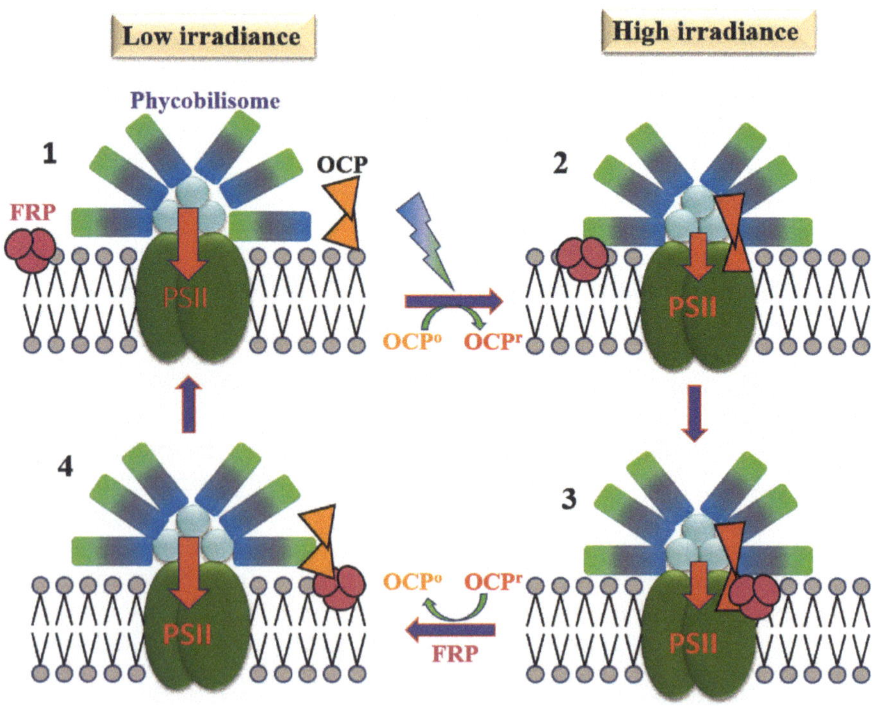

FIG. 3.5 Modulation of light energy in cyanobacteria: a proposed working model adopted and modified from Kirilovsky and Kerfeld (2012).

which shields photosynthetic apparatus from photograph harm. Next to these progressions, light additionally incites the movement of carotenoid 12 Å more profound from CTD to NTD. Moving of carotenoid is basic as in it brings the carotenoid conclusion to bilin shade of PBS center (Leverenz et al., 2012).

1. Under dark and low irradiance, the OCP primarily in the orange form (OCP^{o}) and free and/or weakly attached with PBS.
2. The blue-green light absorption tempts the changes in the carotenoid and the OCP, converted from OCP^{o} (orange; inactive) to the OCP^{R} (red form; active). The OCP^{R} binds with the APC trimers of the PBS core by its C-terminal region and induces fluorescence quenching. The phycobilisome protein and chromophore that interact with the OCP^{R} remains to be identified.
3. As a trimer, the FRP binds with the OCP^{R} N-terminal.
4. The OCP^{R} detachment from the PBS medicated by FRP. Leading to the accelerated conversion of the OCP^{R} to OCP^{O}. In darkness (or blur light), conditions in which the seperated OCP is not converted to the reduced form, hence the FRP action induces the fluorescence recovery.

3.5.2 Cyanobacterial reaction center under low light/dark: FRP is a significant game-changer

OCP^{R} remains metastable until and unless high light environment operates. Soon after this, with the emergence of dark and/or low irradiance, this condition can revert to inactive OCP° form. FRP, a second protein found as a subsequent perusing outline reliably present with the quality that codes for OCP, helps this inversion. FRP stays dynamic in dimer structure and gets capacity to tie with CTD of the occupied OCP territory (OCP^{R}) (Lu et al., 2016). In active OCP state, NTD and CTD, the two domains get separated, which permits NTD to tie with PBS, while CTD hangs openly employing between domain linker. FRP binds with openly portable CTD and turns off the fluorescence quenching by decoupling OCP^{R} from PBS core (Sutter et al., 2013). Site-directed mutagenesis and native mass spectrometry have demonstrated that the head area of FRP can bind with CTD. This limiting at CTD causes changes in FRP adaptation that helps in spanning the two OCP domains again (Lu et al., 2016). Preceding keep away from inefficient fluorescence quenching under light immersing condition, FRP assumes a significant part by aiding the aggregation of the absorbed photon at the reaction center antenna complex. Like this, it is named FRP (Fig. 3.4). Although it is anticipated that FRP capacities would fluctuate in response to changing light conditions, it is unknown if natural circumstances will direct FRP capabilities. Besides how FRP cooperates with PBS, the molecular mechanisms that work with their limiting are scrutinized.

3.6 Conclusions

We presume that OCP is fundamental for effectuating cyanobacterial NPQ. By and large, OCP, which includes two emphatically united domains: NTD and CTD, happens in two distinct states: dormant orange state (OCP^O) and dynamic red structure (OCP^R). NTD of OCP is normal in OCP^O and OCP^R. Under dull and light restricting conditions, the two forms of OCP stay unblemished and do not include an energy-scattering mechanism. It is being viewed that blue-green or high white light is important to break the hydrogen bonds and hydrophobic, combining these two domains. When OCP prompted by light, the two domains get separated, and OCP turns out in structure OCP^R.

Further OCPR communicates with the PBS center and helps in dispersing satiate energy. To recuperate the limit of the antenna complex for augmenting light assimilation under light-restricting condition, the FRP decouples OCP^R from PBS and returns OCP^R to the OCP^O structure. However, a great deal of exploration has been done on the primary and utilitarian association of PBS, OCP, and FRP. Yet, the whole cyanobacterial NPQ component is equivocal; besides, how FRP ties to CTD and recovery of fluorescence will be future exploration.

Acknowledgment

Wenjing Wang is thankful to National China Science Foundation, China, for financial support. Prashant Kumar Singh is grateful to University Grant Commission, New Delhi, India, for the UGC start-up research grant.

References

Adir, N., 2005. Elucidation of the molecular structures of components of the phycobilisome: reconstructing a giant. Photosynth. Res. 85, 15–32.

Aro, E.M., Virgin, I., Andersson, B., 1993. Photoinhibition of photosystem II, inactivation, protein damage and turnover. Biochim. Biophys. Acta 1143, 113–134.

Ashby, M.K., Mullineaux, M.K., 1999. The role of ApcD and ApcF in energy transfer from phycobilisomes to PSI and PSII in a cyanobacterium. Photosynth. Res. 61, 169–179.

Bailey, S., Grossman, A., 2008. Photoprotection in cyanobacteria: regulation of light harvesting. Photochem. Photobiol. 84, 1410–1420.

Bao, H., Melnicki, M.R., Kerfeld, C.A., 2017. Structure and functions of orange carotenoid protein homologs in cyanobacteria. Curr. Opin. Plant Biol. 37, 1–9.

Boulay, C., Wilson, A., D' Haene, S., Kirilovsky, D., 2010. Identification of a protein required for recovery of full antenna capacity in OCP-related photoprotective mechanism in cyanobacteria. Proc. Natl. Acad. Sci. 107, 11620–11625.

Bryant, D.A., Guglielmi, G., de Marsac, N.T., Castets, A.M., Cohen-Bazire, G., 1979. The structure of cyanobacterial phycobilisomes: a model. Arch. Microbiol. 123, 113–127.

Cadoret, J.C., Demouliere, R., Lavaud, J., van Gorkom, H.J., Houmard, J., Etienne, A.L., 2004. Dissipation of excess energy triggered by blue light in cyanobacteria with CP43′(IsiA). Biochim. Biophys. Acta 1659, 100–104.

Cogdell, R.J., Gardiner, A.T., 2015. Activated OCP unlocks nonphotochemical quenching in cyanobacteria. Proc. Natl. Acad. Sci. 112, 12547–12548.

Demmig-Adams, B., Adams, W.W., 1996. The role of xanthophyll cycle carotenoids in the protection of photosynthesis. Trends Plant Sci. 1, 21–26.

Dong, C., Tang, A., Zhao, J., Mullineaux, C.W., Shen, G., Bryant, D.A., 2009. ApcD is necessary for efficient energy transfer from phycobilisomes to photosystem I and helps to prevent photoinhibition in the cyanobacterium *Synechococcus* sp. PCC 7002. Biochim. Biophys. Acta 1787, 1122–1128.

El Bissati, K., Delphin, E., Murata, N., Etienne, A., Kirilovsky, D., 2000. Photosystem II fluorescence quenching in the cyanobacterium *Synechocystis* PCC 6803: involvement of two different mechanisms. Biochim. Biophys. Acta 1457, 229–242.

Funk, C., Vermaas, W., 1999. A cyanobacterial gene family coding for single-helix proteins resembling part of the light-harvesting proteins from higher plants. Biochemistry 38, 9397–9404.

Gantt, E., Conti, Conti S.F., 1966a. Granules associated with chloroplast lamellae of *Porphyridium cruentum*. J. Cell Biol. 29, 423–434.

Gantt, E, Conti, S.F., 1966b. Phycobiliprotein localization in algae. Brookhaven Symp Biol 19, 393–405.

Gindt, Y.M., Zhou, J., Bryant, D.A., Sauer, K., 1994. Spectroscopic studies of phycobilisome subcore preparations lacking key core chromophores: assignment of excited state energies to the Lcm, β 18 and α AP-B chromophores. Biochim. Biophys. Acta 1186, 153–162.

Glazer, A.N., Apell, G.S., Hixson, C.S., Bryant, D.A., Rimon, S., Brown, D.M., 1976. Biliproteins of cyanobacteria and Rhodophyta: homologous family of photosynthetic accessory pigments. Proc. Natl. Acad. Sci. USA 73 (2), 428–431 Feb.

Glazer, A.N., 1984. Phycobilisome-a macromolecular complex optimized for light energy-transfer. Biochim. Biophys. Acta 768, 29–51.

Gorbunov, M.Y., Kuzminov, F.I., Fadeev, V.V., Kim, J.D., Falkowski, P.G., 2011. A kinetic model of non-photochemical quenching in cyanobacteria. Biochim. Biophys. Acta 1807, 1591–1599.

Grossman, A.R., Schaefer, M.R., Chiang, G.G., Collier, J.L., 1993. The phycobilisome, a light harvesting complex responsive to environmental-conditions. Microbiol. Rev. 57, 725–749.

He, Q., Dolganov, N., Bjorkman, O., Grossman, A.R., 2001. The high light-inducible polypeptides in *Synechocystis* PCC6803. Expression and function in high light. J. Biol. Chem. 276, 306–314.

Holt, T.K., Krogmann, D.W., 1981. A carotenoid-protein from cyanobacteria. Biochim. Biophys. Acta 637, 408–414.

Horton, P., Ruban, A.V., Walters, R.G., 1996. Regulation of light harvesting in green plants. Annu. Rev. Plant Physiol. Plant Mol. Biol. 47, 655–684.

Ihalainen, J.A., D'Haene, S., Yeremenko, N., van Roon, H., Arteni, A.A., Boekema, E.J., van Grondelle, R., Matthijs, H.C., Dekker, J.P., 2005. Aggregates of the chlorophyll-binding protein IsiA (CP43) dissipate energy in cyanobacteria. Biochemistry 44, 10846–10853.

Joshua, S., Bailey, S., Mann, N.H., Mullineaux, C.W., 2005. Involvement of phycobilisome diffusion in energy quenching in cyanobacteria. Plant Physiol 138, 1577–1585.

Karapetyan, N.V., 2007. Non-photochemical quenching of fluorescence in cyanobacteria. Biochemistry (Moscow) 72, 1127–1135.

Kerfeld, C.A., Melnicki, M.R., Sutter, M., Dominguez-Martin, M.A., 2017. Structure, function and evolution of the cyanobacterial orange carotenoid protein and its homologs. New Phytol. 215, 937–951.

Kerfeld, C.A., Sawaya, M.R., Brahmandam, V., Cascio, D., Ho, K.K., Trevithick-Sutton, C.C., Krogmann, D.W., Yeates, T.O., 2003. The crystal structure of a cyanobacterial water-soluble carotenoid binding protein. Structure 11, 55–65.

Kerfeld, C.A., 2004a. Structure and function of the water-soluble carotenoid-binding proteins of cyanobacteria. Photosynth. Res. 81, 215–225.

Kerfeld, C.A., 2004b. Water-soluble carotenoid proteins of cyanobacteria. Arch. Biochem. Biophys. 430, 2–9.

Kilian, O., Steunou, A.S., Grossman, A.R., Bhaya, D., 2008. A novel two domain-fusion protein in cyanobacteria with similarity to the CAB/ELIP/HLIP superfamily: evolutionary implications and regulation. Mol. Plant 1 (1), 155–166. doi:10.1093/mp/ssm019.

Kirilovsky, D., Kerfeld, C.A., 2016. Cyanobacterial photoprotection by the orange carotenoid protein. Nat. Plants 2, 16180.

Kirilovsky, D., Kerfeld, C.A., Jan 2012. The orange carotenoid protein in photoprotection of photosystem II in cyanobacteria. Biochim. Biophys. Acta 1817 (1), 158–166. doi:10.1016/j.bbabio.2011.04.013.

Kirilovsky, D., 2015. Modulating energy arriving at photochemical reaction centers: orange carotenoid protein-related photoprotection and state transitions. Photosynth. Res. 126, 3–17.

Leverenz, R.L., Sutter, M., Wilson, A., Gupta, S., Thurotte, A., de Carbon, C.B., Kerfeld, C.A., 2012. A 12 Å carotenoid translocation in a photo switch associated with cyanobacterial photoprotection. Science 348, 1463–1466.

Li, X.P., Bjorkman, O., Shih, C., Grossman, A.R., Rosenquist, M., Jansson, S., Niyogi, K.K., 2000. A pigment-binding protein essential for regulation of photosynthetic light harvesting. Nature 403, 391–395.

Li, X.P., Gilmore, A.M., Caffari, S., Bassi, R., Golan, T., Kramer, D., Niyogi, K.K., 2004. Regulation of photosynthetic light harvesting involves intra-thylakoid lumen pH sensing by the PsbS protein. J. Biol. Chem. 279, 22866–22874.

Liu, H., Zhang, H., Niedzwiedzki, D.M., Prado, M., He, G., Gross, M.L., Blankenship, R.E., 2017. Phycobilisomes supply excitations to both photosystems in a megacomplex in cyanobacteria. Science 342, 1104–1107.

Liu, L.N., Chen, X.L., Zhang, Y.Z., Zhou, B.C., 2005. Characterization, structure and function of linker polypeptides in phycobilisomes of cyanobacteria and red algae: an overview. Biochim. Biophys. Acta -Bioenergetics 1708, 133–142.

Lu, Y., Liu, H., Saer, R.G., Zhang, H., Meyer, C.M., Li, V.L., Blankenship, R.E., 2016. Native mass spectrometry analysis of oligomerization states of fluorescence recovery protein and orange carotenoid protein: two proteins involved in the cyanobacterial photoprotection cycle. Biochemistry 56, 160–166.

MacColl, R., 1998. Cyanobacterial phycobilisomes. J. Struct. Biol. 124, 311–334.

Moldenhauer, M., Sluchanko, N.N., Tavraz, N.N., Junghans, C., Buhrke, D., Willoweit, M., Chiappisi, L., Schmitt, F.J., Vukojević, V., Shirshin, E.A., Ponomarev, V.Y., Paschenko, V.Z., Gradzielski, M., Maksimov, E.G., Friedrich, T., 2018. Interaction of the signaling state analog and the apoprotein form of the orange carotenoid protein with the fluorescence recovery protein. Photosynth. Res. 135, 125–139.

Morosinotto, T., Bassi, R., 2014. Molecular mechanisms for activation of non-photochemical fluorescence quenching: from unicellular algae to mosses and higher plants. In: Demmig-Adams, B., Garab, G., Adams, W.W. III, Govindjee (Eds.), Non-Photochemical Quenching and Energy Dissipation in Plants, Algae and Cyanobacteria. Springer, Netherlands, pp. 315–331.

Müller, P., Li, X.P., Niyogi, K.K., 2001. Non-photochemical quenching. A response to excess light energy. Plant Physiol. 125, 1558–1566.

Niyogi, K.K., 1999. Photoprotection revisited: genetic and molecular approaches. Annu. Rev. Plant Physiol. Plant Mol. Biol. 50, 333–359.

Ort, D.R., 2001. When there is too much light. Plant Physiol. 125, 29–32.

Pascal, A.A., Liu, Z.F., Broess, K., van Oort, B., van Amerongen, H., Wang, C., Horton, P., Robert, B., Chang, W.R., Ruban, A., 2005. Molecular basis of photoprotection and control of photosynthetic light-harvesting. Nature 436, 134–137.

Prasil, O., Adir, N., Ohad, I., 1992. Dynamics of photosystem II: mechanism of photoinhibition and recovery processes. In: Barber, J. (Ed.), The Photosystems: Structure, Function and Molecular Biology. Elsevier Science Publishers, Amsterdam, pp. 295–348.

Punginelli, C., Wilson, A., Routaboul, J.M., Kirilovsky, D., 2009. Influence of zeaxanthin and echinenone binding on the activity of the orange carotenoid protein. Biochim. Biophys. Acta-Bioenergetics 1787, 280–288.

Rakhimb, erdieva, M.G., Elanskaya, I.V., Vermaas, W.F.J., Karapetyan, N.V., 2010. Carotenoid triggered energy dissipation in phycobilisomes of *Synechocystis* sp. PCC 6803 diverts excitation away from reaction centers of both photosystems. Biochim. Biophys. Acta 1797, 241–249.

Rakhimberdieva, M.G., Stadnichuk, I.N., Elanskaya, I.V., Karapetyan, N.V., 2017. Carotenoid-induced quenching of the phycobilisome fluorescence in photosystem II-deficient mutant of Synechocystis sp. FEBS Lett. 574 (1-3), 85–88.

Rakhimberdieva, M.G., Vavilin, D.V., Vermaas, W.F., Elanskaya, I.V., Karapetyan, N.V., 2007. Phycobilin/chlorophyll excitation equilibration upon carotenoid-induced nonphotochemical fluorescence quenching in phycobilisomes of the cyanobacterium *Synechocystis* sp. PCC 6803. Biochim. Biophys. Acta 1767, 757–765.

Ruban, A.V., Berera, R., Ilioaia, C., van Stokkum, I.H.M., Kennis, J.T.M., Pascal, A., van Amerongen, H., Robert, B., Horton, P., van Grondelle, R., 2007. Identification of a mechanism of photoprotective energy dissipation in higher plants. Nature 450, U522–U575.

Saer, R.G., Blankenship, R.E., 2017. Light harvesting in phototrophic bacteria: structure and function. Biochem. J. 474, 2107–2131.

Scott, M., McCollum, C., Vasil'ev, S., Crozier, C., Espie, G.S., Krol, M., Huner, N.P., Bruce, D., 2006. Mechanism of the down regulation of photosynthesis by blue light in the Cyanobacterium *Synechocystis* sp. PCC 6803. Biochemistry 45, 8952–8958.

Shirshin, E.A., Nikonova, E.E., Kuzminov, F.I., Sluchanko, N.N., Elanskaya, I.V., Gorbunov, M.Y., Fadeev, V.V., Friedrich, T., Maksimov, E.G., 2017. Biophysical modelling of in vitro and in vivo processes underlying regulated photoprotective mechanism in cyanobacteria. Photosynth. Res. 133, 261–271.

Singh, P.K., Shrivastava, A.K., Chatterjee, A., Pandey, S., Rai, S., Singh, S., Rai, L.C., 2015. Cadmium toxicity in diazotrophic *Anabaena* spp. adjudged by hasty up-accumulation of transporter and signaling and severe down-accumulation of nitrogen metabolism proteins. J Proteom 127 (Pt A), 134–146. doi:10.1016/j.jprot.2015.05.019 Sep 8.

Singh, P.K., Wang, W., Shrivastava, A.K., 2018. Cadmium-mediated morphological, biochemical and physiological tuning in three different *Anabaena* species. Aquat. Toxicol 202, 36–45. doi:10.1016/j.aquatox.2018.06.011.

Sluchanko, N.N., Slonimskiy, Y.B., Moldenhauer, M., Friedrich, T., Maksimov, E.G., 2017. Deletion of the short N-terminal extension in OCP reveals the main site for FRP binding. FEBS Lett. 591, 1667–1676.

Sutter, M., Wilson, A., Leverenz, R.L., Lopez-Igual, R., Thurotte, A., Salmeen, A.E., Kirilovsky, D., Kerfeld, C.A., 2013. Crystal structure of the FRP and identification of the active site for modulation of OCP-mediated photoprotection in cyanobacteria, Proc. Natl. Acad. Sci., 110, 10022–10027.

Tandeau, N., de, Marsac., 2003. Phycobiliproteins and phycobilisomes: the early observations. Photosynth. Res. 76, 197–205.

Tian, L., van Stokkum, I.H., Koehorst, R.B., van Amerongen, H., 2012. Light harvesting and blue-green light induced non-photochemical quenching in two different C-phycocyanin mutants of *Synechocystis* PCC 6803. J. Phys. Chem. B117, 11000–11006.

Tyystjärvi, E., 2008. Photoinhibition of photosystem II and photodamage of the oxygen evolving manganese cluster. Coord. Chem. Rev. 252, 361–376.

van Thor, J.J., Mullineaux, C.W., Matthijs, H.C.P., Hellingwert, K.J., 1998. Light harvesting and state transitions in cyanobacteria. Bot. Acta 111, 430–443.

Vass, I., Aro, E.M., 2007. Photoinhibition of Photosystem II electron transport. In: Renger, G. (Ed.), Primary Processes of Photosynthesis: Basic Principles and Apparatus. Royal Society of Chemistry, Cambridge, pp. 393–411.

Williams, W.P., Allen, J.F., 1987. State-1/State-2 changes in higher-plants and algae. Photosynth. Res. 13, 19–45.

Wilson, A., Ajlani, G., Verbavatz, J.M., Vass, I., Kerfeld, C.A., Kirilovsky, D., 2006. A soluble carotenoid protein involved in phycobilisome-related energy dissipation in cyanobacteria. Plant Cell 18, 992–1007.

Wilson, A., Boulay, C., Wilde, A., Kerfeld, C.A., Kirilovsky, D., 2007. Light-induced energy dissipation in iron-starved cyanobacteria: roles of OCP and IsiA proteins. Plant Cell 19, 656–672.

Wilson, A., Kinney, J.N., Zwart, P.H., Punginelli, C., D'Haene, S., Perreau, F., Klein, M.G., Kirilovsky, D., Kerfeld, C.A., 2010. Structural determinants underlying photoprotection in the photoactive orange carotenoid protein of cyanobacteria. J. Biol. Chem. 285, 18364–18375.

Wilson, A., Punginelli, C., Gall, A., Bonetti, C., Alexandre, M., Routaboul, J.M., Kerfeld, C.A., van Grondelle, R., Robert, B., Kennis, J.T., Kirilovsky, D., 2008. A photoactive carotenoid protein acting as light intensity sensor. Proc. Natl. Acad. Sci. USA 105, 12075–12080.

Wollman, F.A., 2001. State transitions reveal the dynamics and flexibility of the photosynthetic apparatus. EMBO J 20, 3623–3630.

Wu, Y.P., Krogmann, D.W., 1997. The orange carotenoid protein of *Synechocystis* PCC 6803. Biochim. Biophys. Acta 1322, 1–7.

Yeremenko, N., Kouril, R., Ihalainen, J.A., D'Haene, S., van Oosterwijk, N., Andrizhiyevskaya, E.J., Keegstra, W., Dekker, H.L., Hagemann, M., Boekema, E.J., Matthijs, H.C., Dekker, J.P., 2004. Supramolecular organization and dual function of the IsiA chlorophyll-binding protein in cyanobacteria. Biochemistry 43, 10308–10313.

Zakar, T., Laczko-Dobos, H., Toth, T.N., Gombos, Z., 2016. Carotenoids assist in cyanobacterial photosystem II assembly and function. Front. Plant Sci. 7, 295.

Zhu, Y., Graham, J.E., Ludwig, M., Xiong, W., Alvey, R.M., Shen, G., Bryant, D.A., 2010. Roles of xanthophyll carotenoids in protection against photoinhibition and oxidative stress in the cyanobacterium *Synechococcus* sp. strain PCC 7002. Arch. Biochem. Biophys. 504, 86–99.

CHAPTER

4

Back to the past: Improving photosynthesis with cyanobacterial genes

Rocío C. Arce, Paula Vicino*, Mariana Demarchi, Anabella F. Lodeyro*

Instituto de Biología Molecular y Celular de Rosario (IBR-UNR/ CONICET), Facultad de Ciencias Bioquímicas y Farmacéuticas, Universidad Nacional de Rosario (UNR), Rosario, Argentina. *These authors contributed equally to this work

4.1 Introduction

Agricultural productivity must be increased by more than 50% to supply human's necessities by 2050, as the population is estimated to grow by about 1.2% per year (Dann and Leister, 2017). Similar to the situation with food, a much larger energy supply will also be required in the future. Crops are required for producing sustainable biofuels and bioproducts to reduce the dependence on fossil fuels. It will be a difficult challenge to meet this demand without significant increases in water, mineral nutrients, or fertilizer use, resources that are becoming limited (Tilman et al., 2011). In addition, to mitigate environmental damage caused by extensive agriculture it will be necessary to satisfy global food demands without increasing the amount of cultivable land, which emphasizes the necessity to improve crop yields. One of the available approaches to achieve this goal is genetic engineering (Ort et al., 2015); however, it is limited by the genetic stock of extant plant species. Promising sources for potential beneficial crop traits are bacteria and archaea, especially cyanobacteria because of their evolutionary relationship to chloroplasts. A cyanobacterium was the likely ancestor of chloroplasts (Jensen and Leister, 2014); an endosymbiosis

Cyanobacterial Lifestyle and its Applications in Biotechnology.
DOI: https://doi.org/10.1016/B978-0-323-90634-0.00013-5

that was assumed crucial for land colonization by plants (Cavalier-Smith, 2010). Cyanobacteria represent ideal model organisms for photosynthesis research, offering a rich source of genes for plant genetic engineering, due to similarities and differences from the plant genetic systems. In particular, their responses differ substantially from plants as they are unique, rapid and to survive, they have developed several global adaptive responses (Rachedi et al., 2020). Many plant metabolic and regulatory pathways were evolved from cyanobacterial ancestors and several of these routes have not diverged much, thus allowing productive interactions of the transgenic products with the corresponding endogenous systems. On the other hand, the sequence divergence between plant proteins and their cyanobacterial counterparts also prevents in most cases the undesired consequences of gene silencing and cosuppression, evading the plant regulatory networks. During evolution, some genes of cyanobacterial origin have disappeared from the plant genome, therefore their introduction in higher plants open possibilities to regain some of the adaptive advantages that allowed cyanobacteria to thrive in extreme environments. Genetic modification using microbial genes has the potential to generate crops with enhanced photosynthesis and abiotic stress tolerance, key factors in crop yield (Orr et al., 2017; Rae et al., 2017).

Although it was generally believed that plant photosynthesis had already been optimized during evolution to perform at its maximum efficiency (Leister, 2012), further observations have demonstrated that photosynthesis in terrestrial plants can be considered remarkably inefficient (Éva et al., 2019; Ort et al., 2015). In that way, most attention in research that could result in an improvement of the productivity of crops has been directed to alter photosynthetic processes (Nowicka et al., 2018). A range of avenues designed at enhancing the photosynthetic capacity has been proposed. Most attractive intervention points include; (1) improving the efficiency of the primary CO_2-fixing enzyme Rubisco, (2) optimizing elements of the Calvin–Benson–Bassham (CBB) cycle, (3) introducing the carboxysome-based CO_2-concentrating mechanism (CCM) from cyanobacteria, and (4) improving the photochemical response of photosynthesis to rapid changes in light conditions (Kubis and Bar-Even, 2019; Orr et al., 2017).

In addition, plants in nature are exposed to a variety of suboptimal conditions, culminating in abiotic stress, such as drought, extreme temperature, salinity, high light intensity, among others; and the majority of crops are not growing under optimal conditions in the field. Crop yield can thus be reduced by over 50% due to environmental hardships (Atkinson and Urwin, 2012). The development of plants capable of coping with environmental stress is critical. The current global goal is to develop crops with higher yields or tolerant to unfavorable stress situations. The use of cyanobacterial genes to fulfill this purpose has still been relatively limited, despite the many potential advantages of their use (Hennacy and Jonikas, 2020). We discuss herein cyanobacterial genes and pathways, which have been introduced in plants to enhance their stress tolerance or improve photosynthetic processes using this strategy (Fig. 4.1). We aim to provide an overview of current attempts to improve yield, especially under abiotic stress using cyanobacterial genes. In the last section, we explore future opportunities including cyanobacterial genes lost in plant genomes.

4.2 Engineering cyanobacterial genes not related to photosynthesis

Environmental abiotic stresses, such as drought, salinity, and extreme temperatures, are detrimental factors limiting the development and yield of crops. Improving agriculture is therefore vital to fill the gap between population growth and food production, which is widening due to the severity of adverse conditions impact (Yadav et al., 2020). In this context, cyanobacteria inhabit environments ranging from deserts to the cryosphere, and they have to cope with various stresses and kinds of deprivation that threaten their growth and viability. To adapt to these adverse conditions and survive, they have developed several global adaptive responses that modulate the patterns of gene expression and cellular functions (Rachedi et al., 2020). Genes involved in these responses are of particular interest because they are a promising strategy to create stress-tolerant crops. In this section, we present some cyanobacterial interventions not related to photosynthesis which improved abiotic stress tolerance in plants. The summary of these discoveries could be found in Table 4.1.

4.2.1 Stress response

As mentioned before, cyanobacteria have the plasticity to adapt to extreme environments to survive. There are several examples in which the expression of cyanobacterial genes in higher plants led to improved stress tolerance. Particularly, overexpression of DnaK1 protein from the halotolerant cyanobacterium *Aphanothece halophytica* conferred tolerance to high temperature in tobacco plants, showing higher germination rates and better growth under stress in early development (Ono et al., 2001). Protein denaturation, instabilities in nucleic acids and cytoskeletal structure, inactivation of enzymes, and decrease in membrane integrity are the main damages under heat stress, defined as a rise in normal temperature sufficient to cause permanent damage to plant growth and development (Lamaoui et al., 2018). DnaK1 belongs to the heat-shock protein 70 family,

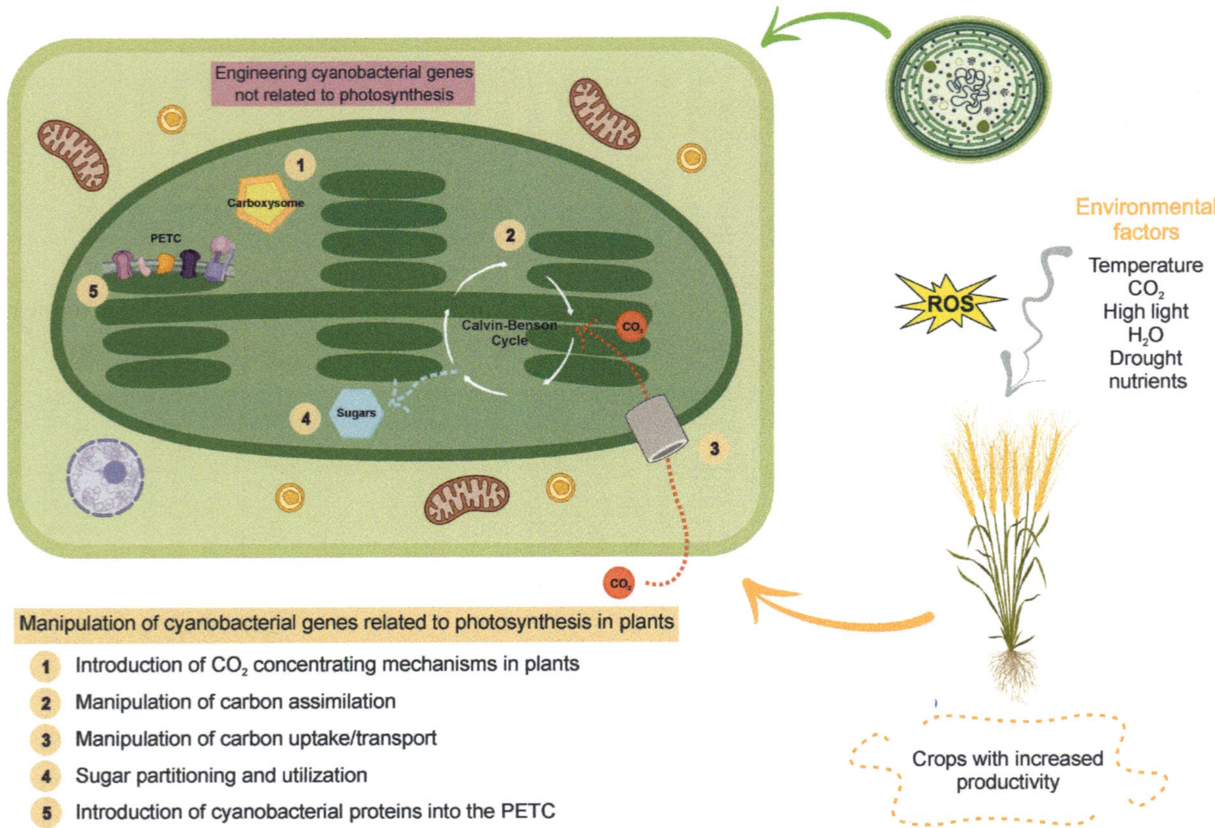

FIG. 4.1 **Engineered higher plants pathways using cyanobacterial genes.** Crops are subject to multiple environmental factors that can negatively affect them and cause significant losses. Most of the strategies designed to enhance adverse environmental conditions or boost the productivity involve the modification of different photosynthetic processes, such as: 1, the introduction of CO_2 concentrating mechanisms in carboxysome-like structures; 2, manipulation of carbon assimilation (Calvin-Benson-Bassham cycle); 3, manipulation of carbon uptake/transport through the membranes; 4, modification of sugar partitioning and utilization; and 5, introduction of external acceptors into the photosynthetic electron transport chain (PETC). Image created using BioRender.

TABLE 4.1 Cyanobacterial genes not related to photosynthesis which improve the growth of plants properties under stress.

Pathway	Protein(s)	Donor organism	Plant	Phenotype	References
Stress response	DnaK1	*Aphanothece halophytica*	*Nicotiana tabacum* cv Petit Havana	High-temperature tolerance (greater germination rate, plant elongation and quantum yields of electron transport).	Ono et al. (2001)
	DnaK2	*Synechocystis* spp. PCC6803 *Synechocystis* spp. PCC6906	*Arabidopsis thaliana*	Salt tolerance (greater root length, chlorophyll content, fresh weight, and survival rate; reduced H_2O_2 accumulation).	Kim et al. (2014)
	Drnf1	*Nostoc flagelliforme*	*Arabidopsis thaliana*	Salt tolerance (greater germination rate and shoot growth).	Cui et al. (2018)
	Wspa1	*Nostoc commune*	*Arabidopsis thaliana*	Osmotic stress tolerance (lower relative electrolyte leakage, higher proline accumulation, and ΦPSII).	Ai et al. (2014)
Amino acid metabolism	ApGSMT and ApDMT	*Aphanothece halophytica*	*Arabidopsis thaliana* *Nicotiana tabacum* cv. Wisconsin 38 *Gossypium hirsutum*	Salt, drought, and cold tolerance (greater germination rate, fresh weight, survival rate, yield and ΦPSII).	He et al. (2011)., Song et al. (2018), Waditee et al. (2005)
	PGDH	*Aphanothece halophytica*	*Arabidopsis thaliana*	Salt and cold tolerance (greater germination rate and fresh weight).	Waditee et al. (2007)
Lipid metabolism	DesC	*Synechococcus vulcanus*	*Nicotiana tabacum* cv. Samsun	Cold stress tolerance (greater germination rate, less tissue damage).	Orlova et al. (2003)
	HPT, TCY, and TMT	*Synechocystis* sp. PCC6803	*Nicotiana tabacum* cv. Petit Havana	Cold stress tolerance (less tissue damage).	Lu et al. (2013)

ApDMT: *A. halophytica* dimethylglycine methyltransferase; ApGSMT: *A. halophytica* glycine sarcosine methyltransferase; DesC: Acyl-lipid Δ9-desaturase; Drnf1: drought-resistant genes of *Nostoc flagelliforme* 1; HPT: homogentisate phytyltransferase; PGDH: 3-phosphoglycerate dehydrogenase; TCY: tocopherol cyclase; TMT: γ-tocopherol methyltransferase; Wspa1: acidic water stress protein 1; ΦPSII: quantum yield of PSII.

which is involved in stress response and functions as a molecular chaperone that binds non-native states of other proteins and assists them to reach a functional conformation (Bukau and Horwich, 1998). Focus on DnaKs gained momentum as they were shown to be up-regulated under abiotic stress (Kanesaki et al., 2002; Rupprecht et al., 2007). Other proteins from the same family such as DnaK2s from freshwater *Synechocystis* spp. PCC6803 and marine *Synechocystis* spp. PCC6906 conferred salt tolerance to *Arabidopsis* plants. Transgenic plants from both cyanobacterial strains displayed longer roots, higher pigment content, and enhanced fresh weight among other advantages under high concentrations of NaCl (Kim et al., 2014).

The drought-adapted cyanobacteria *Nostoc flagelliforme* and *Nostoc commune* are arising as a valuable resource for gene discovery and potential applications in plant breeding (Ye and Gao, 2015). Among them, the drought-resistant gene 1 (*drnf1*) of *N. flagelliforme* and the acidic water stress protein 1 (*wspa1*) from *N. commune* were expressed in *Arabidopsis* and the resulting lines exhibited abiotic stress tolerance. In the case of *drnf1*, transgenic plants had a better growth performance, in addition to higher germination rate and net photosynthesis under salt stress (Cui et al., 2018). Plants expressing *wspa1* showed increased tolerance to osmotic stress. Physiological measurements such as relative electrolyte leakage, malondialdehyde (a proxy for lipid peroxidation), proline accumulation, and the maximal quantum efficiency of photosystem (PS) II were improved in transgenic plants upon osmotic stress compared to wild-type (WT) siblings (Ai et al., 2014).

4.2.2 Amino acid metabolism

Manipulations on plant amino acid metabolism by expressing cyanobacterial enzymes led to successful results in terms of facing environmental adverse situations. Glycine betaine (*N,N,N*-trimethylglycine) is of particular interest because it functions as an important osmoprotectant, and it is synthesized by many crops in response to abiotic stresses. In plants, glycine betaine is synthesized from choline by a two-step oxidation reaction via betaine aldehyde catalyzed by a ferredoxin-dependent choline monooxygenase and NAD^+-dependent betaine aldehyde dehydrogenase, respectively (Annunziata et al., 2019). In *A. halophytica*, an alternative biosynthesis pathway from glycine was found catalyzed by two *N*-methyltransferase enzymes. One enzyme, glycine sarcosine methyltransferase (ApGSMT), catalyzes the methylation reactions of glycine to sarcosine and of sarcosine to dimethylglycine; and the other, dimethylglycine methyltransferase (ApDMT), catalyzed the specific methylation of dimethylglycine to glycine betaine (Waditee et al., 2003).

Coexpression of these proteins in *Arabidopsis*, tobacco, and cotton conferred ample abiotic stress tolerance (He et al., 2011; Song et al., 2018; Waditee et al., 2005). Transgenic *Arabidopsis* plants subjected to salt, drought, and low-temperature stresses showed higher germination rates and a more stable quantum yield of PSII over days (Waditee et al., 2005). In tobacco plants, their performance under drought was also characterized. When plants coexpressed ApGSMT and ApDMT, they exhibited higher germination rates and relative water content, less cell membrane damage, and better photosynthetic capacity measured as photosynthetic gas exchange and change of chlorophyll fluorescence under the adverse condition (He et al., 2011). Finally, transgenic cotton plants showed increased salt tolerance both in the greenhouse conditions and in saline fields. They had an increased seed yield in saline fields and showed better performance of PSII and greater photosynthetic capacity than WT siblings (Song et al., 2018).

Another cyanobacterial gene related to amino acid metabolism whose expression improved plant performance under stress was the 3-phosphoglycerate dehydrogenase (PGDH). This enzyme catalyzes the first step of the phosphorylated pathway of serine biosynthesis and was isolated from *A. halophytica*. Similar to the approach described above, as serine and glycine are interconvertible by serine hydroxymethyltransferase, enhancement of serine biosynthesis would eventually lead to an increase in glycine levels. Effectively, the expression of PGDH increased glycine betaine levels. Transgenic *Arabidopsis* plants expressing *A. halophytica* PGDH displayed salt and cold tolerance evidenced by their better growth under stress (Waditee et al., 2007).

4.2.3 Lipid metabolism

Low-temperature photoinhibition is one of the major factors that limit plant productivity. As the main components of cell membranes, lipids have been recognized as having a significant role in cold stress, both as a mechanical defense through leaf surface protection and plasma membrane remodeling, and as signal transduction molecules (Barrero-Sicilia et al., 2017). Due to this, modifications in genes related to lipid biosynthesis are a viable strategy to improve extreme low-temperature tolerance in crops. One approach based on cyanobacterial genes involved the expression of acyl-lipid Δ9-desaturase (*desC*) from thermophilic *Synechococcus vulcanus* in tobacco plants. This gene was characterized in *Escherichia coli* to potentially mediate the conversion of stearic acid into oleic acid. Moreover, DesC protein levels increased about 10-fold during the shift in temperature from 55 to 45 °C (Kiseleva et al., 2000). In tobacco lines, the expression of this gene conferred chilling tolerance, which was evidenced by higher germination rates and less injury of leaf tissues in comparison to WT counterparts (Orlova et al., 2003).

Another genetic intervention in lipid metabolism was the coexpression of three *Synechocystis* sp. PCC6803 genes involved in the tocochromanol pathway. In plants, these lipid compounds play an important role in defense and physiological processes. Tocochromanols are highly potent antioxidants and efficient quenchers of reactive oxygen species (ROS) as they are able to donate a hydrogen atom from the hydroxyl group in the chromanol ring to a free radical, followed by mesomeric stabilization of the molecule's own unpaired electron. Besides, their role in controlling the extent of lipid peroxidation has also been confirmed in higher plants (Falk and Munné-Bosch, 2010). Because of these observations, and to avoid limitations from plant-specific regulation of enzyme activities, three *Synechocystis* proteins needed in tocopherol biosynthetic pathway were expressed in transplastomic tobacco plants: homogentisate phytyltransferase, tocopherol cyclase, and γ-tocopherol methyltransferase (TMT). Transgenic plants were shown to contained higher levels of tocochromanol and photosynthetic pigments, and under cold stress, they showed less symptoms of photooxidative damage in comparison to WT siblings (Lu et al., 2013).

4.3 Manipulation of cyanobacterial genes related to photosynthesis in plants

The main driving force supporting growth and biomass production in plants is photosynthesis, which supplies the energy and carbon required for the biosynthesis of organic compounds necessary for development (Nowicka et al., 2018). As mentioned before, photosynthesis is considered an inefficient process. The maximum theoretical amount of photosynthetically active light that plants can successfully convert into biomass is 9.4% for C3 plants and 12.3% for C4 plants (Zhu et al., 2010). Despite great strides in increasing yield during the "Green Revolution," conversion of light energy to plant biomass is far from the theoretical maximum (Zhu et al., 2010), and it has not been directly increased by traditional breeding (Koester et al., 2016). Conversion of CO_2 to biomass is a particular problem for C3 plants that are even less efficient due to photorespiration. Photosynthetic inefficiencies exist in both the oxygenic photosynthetic reactions and in the CBB cycle. A lot of effort had been focused on over-expressing endogenous proteins or introducing extrinsic genes involved in such processes to optimize them. Particularly, cyanobacterial photosynthetic genes contribute to a new genetic source for plant breeding as they were the ancestors of chloroplasts but had evolved differently according to their needs (Leister, 2019). In this section, we describe several examples in which different stages of the photosynthetic process have been modified by introducing cyanobacterial genes and the obtained outcome, highlighting the successful cases. These findings are summarized in Table 4.2.

4.3.1 Introduction of CCMs in plants

One of the major determinants for plant productivity is photosynthetic carbon metabolism. Endogenous genes had been used with the aim to modify carbon assimilation and, in consequence, improve plant growth and yield. However, most of these studies have failed to show an improvement in plant biomass production, as endogenous genes are subjected to regulation *in vivo*. In this context, the use of cyanobacterial genes opens a window of new possibilities as they are under different regulatory mechanisms.

The CBB cycle, the primary route for carbon assimilation in the chloroplasts of C3 plants (Sharkey, 1985), is divided into three phases. The first one involves carboxylation of the CO_2 acceptor molecule, ribulose-1,5-bisphosphate (RuBP) catalyzed by ribulose-1,5-bisphosphate-carboxylase/oxygenase (Rubisco), generating 3-phosphoglycerate (3PGA). In the second phase, the reduction step, 3PGA, is converted to triose phosphate, and finally, the regeneration phase produces the acceptor molecule RuBP for CO_2 assimilation. Assimilates formed are used either to synthesize transitory starch in the plastids or to produce sucrose in the cytosol (Quick and Neuhaus, 1997). In addition to carboxylation, Rubisco can catalyze the oxygenation of RuBP in a pathway known as photorespiration, leading to a net loss of fixed carbon. Despite being the most abundant enzyme on the planet and responsible for 90% of inorganic carbon fixation, Rubisco is inefficient and slow (turnover numbers of 1–10 CO_2 s^{-1}) (Erb and Zarzycki, 2018). All of these factors make Rubisco an attractive target for carbon assimilation improvement.

Cyanobacteria and C4 plants have evolved forms of CCM that allow them to increase the availability of CO_2 in the surrounding of Rubisco, which, in the case of cyanobacteria, is an enzyme with high catalytic rates but low affinity for CO_2. One of the forms of CCM present in cyanobacteria and several autotrophic prokaryotes is a polyhedral microcompartment known as carboxysome, where Rubisco and carbonic anhydrase are enclosed. This mechanism maintains elevated CO_2 levels in the vicinity of Rubisco (Price et al., 2013; Zarzycki et al., 2013). On the other hand, C3 plants, which lack a CCM, are constrained to express forms of Rubisco with higher CO_2 affinity but a relatively low rate of turnover (Whitney et al., 2011).

Two carboxysome types have arisen through convergent evolution: α-carboxysomes, which encapsulate Form-1A Rubisco in proteobacteria

TABLE 4.2 Cyanobacterial genes related to photosynthesis which improve growth properties.

Pathway	Protein(s)	Donor organism	Plant	Phenotype	References
CO_2-concentrating mechanism (CCM)	RbcL and RbcS with RbcX or CcmM35	*Synechococcus elongatus* PCC7942	*Nicotiana tabacum*	Improved rates of CO_2 fixation (higher Rubisco catalytic activity) and slower plant growth	Lin et al. (2014)
	RbcL and RbcS	*Synechococcus elongatus* PCC6301	*Nicotiana tabacum* cv. Samsun	Higher CO_2 fixation (enhanced Rubisco expression, improved growth rates)	Occhialini et al. (2016)
	CbbL, CbbS, CsoS1A, and CsoS2	*Cyanobium marinum* PCC7001	*Nicotiana tabacum*	α-carboxysomes production and autotrophic growth at elevated CO_2	Long et al. (2018)
Carbon assimilation (CBB cycle)	FBP/SBPase	*Synechococcus* sp. PCC7942	*Nicotiana tabacum* cv. Xanthi	Increased photosynthetic capacity, carbohydrate accumulation and accelerated growth rate.	Miyagawa et al. (2001)
	FBPase-II	*Synechococcus* sp. PCC7942	*Nicotiana tabacum* cv. Xanthi	Greater photosynthetic activities (higher activation of Rubisco, growth rate, height, size and dry weight)	Tamoi et al. (2006)
	FBP/SBPase	*Synechococcus* sp. PCC7942	*Lactuca sativa* cv. Green wave	Greater photosynthetic capacity and accelerated growth (increased leaf size and number)	Ichikawa et al. (2010)
	FBP/SBPase	*Synechococcus* sp. PCC7942	*Glycine max* cv. Thorne	Higher carbon assimilation under heat and elevated CO_2 treatment (higher seed yield, number of pods, seed weight and harvest index)	Köhler et al. (2017)
Carbon assimilation (CBB cycle)	FBP/SBPase and CcaA	*Synechococcus* PCC7942	*Oryza sativa*	Increased stomatal and mesophyll conductance to CO_2, net photosyn- thetic rate, and carboxylation efficiency	Gong et al. (2017)
	FBP/SBPase, (cyanobacteria) Cyt C_6 (algae), SBPase (plant)	*Synechocystis* sp. PCC 7942	*N. tabacum* cv. Petit Havana/ Samsun	Higher CO_2 assimilation (increased photosynthetic capacity, growth, and biomass) and higher iWUE	López-Calcagno et al. (2020)

Carbon uptake/ transport	IctB	*Synechococcus* PCC7942	*N. tabacum* and *A. thaliana*	Higher photosynthetic rates under CO_2 limiting conditions, relative growth rates and dry weight	Lieman-Hurwitz et al. (2003)
	IctB	*Synechococcus* PCC7942	*Oryza sativa*	Higher photosynthetic rates and better growth (more tillers or panicles and grains per plant)	Yang et al. (2008)
	IctB and FBP/ SBPase	*Synechococcus elongatus* PCC7942	*Oryza sativa*	Higher mesophyll conductance, net photosynthetic rate, Rubisco activity, and height.	Gong et al. (2015)
	SBPase, FBPA, and IctB	*Synechococcus* PCC7942	*N. tabacum* cv. Samsun	Higher rates of photosynthetic carbon assimilation, leaf area, and biomass yield under the high light regime.	Simkin et al. (2015)
Carbon uptake/ transport	IctB	*Synechococcus* PCC7942	*Glycine max* cv. Thorne	Higher photosynthetic CO_2 uptake and dry weight. Increased photosynthetic rates and biomass production under drought stress.	Hay et al. (2017)
Sugar partitioning	FBPase-II cytosolic	*Synechococcus elongatus* PCC7942	*Arabidopsis thaliana*	Higher sugar levels, increased number of lateral shoots (at elevated CO_2 levels) and lower expression of genes related to hormone metabolism	Otori et al. (2017)
	FBP/SBPase cytosolic	*Synechococcus* sp. PCC7942	*Nicotiana tabacum* cv. Xanthi	Increased lateral shoot, leaf number, and fresh weight increased CO_2 levels, higher photosynthetic activity.	Tamoi et al. (2011)
Sugar utilization	PEPC	*Synechococcus vulcanus*	*Arabidopsis thaliana* Heynh	Changes in the proportion of aromatic/ nonaromatic amino acids	Chen et al. (2004)

(continued)

TABLE 4.2 (Cont'd)

Pathway	Protein(s)	Donor organism	Plant	Phenotype	References
Photosynthetic electron transport chain (PETC)	CrtO	*Synechocystis* sp. PCC6803	*Nicotiana glauca*	Higher carotenoids content, not improved photosynthesis	Zhu et al. (2007)
Photosynthetic electron transport chain (PETC)	Fld	*Anabaena* PCC7119	*Nicotiana tabacum* *Arabidopsis thaliana* *Agrostis stolonifera* *Solanum tuberosum*	Increased both biotic and abiotic stress tolerance (higher pigment content, higher photosynthetic parameters, less ROS production, less leaf damage, higher survival under stress).	Li et al. (2017), Pierella Karlusich et al. (2020), Rossi et al. (2017), Su et al. (2018), Tognetti et al. (2006), Tognetti et al. (2007b), Zurbriggen et al. (2009)
	sFNR and Fld	*Anabaena* PCC7119	*Nicotiana tabacum*	Tolerance to redox-cycling oxidants, lower damage, and decreased ROS accumulation.	Giró et al. (2011)
	Flv1-3	*Synechocystis* sp. PCC6803	*Nicotiana tabacum*	Faster recovery of ΦPSII and NPQ during dark-light transitions. More oxidized PETC and enhanced PMF of dark-adapted leaves.	Gómez et al. (2018)
	Flv1-3	*Synechocystis* sp. PCC6803	*Arabidopsis thaliana*	Faster recovery of ΦPSII and NPQ during dark-light transitions. Enhanced biomass under high light.	Tula et al. (2020)
	Flv1-3	*Synechocystis* sp. PCC6803	*Hordeum vulgare*	Earlier heading, increased biomass, more spikes and grains per plant, higher total grain weight per plant under drought.	Shahinnia et al. (2021)

Photosynthetic electron transport chain (PETC)	Flv2-4	*Synechocystis* sp. PCC6803	*Nicotiana tabacum* *Arabidopsis thaliana*	High irradiation, salinity, oxidants, and drought tolerance (better growth, preservation of photosynthetic activity, and membrane integrity). Enhanced accumulation of soluble sugars and amino acids under drought.	Vicino et al. (2021)

CbbL: Cyanobium Form-1A Rubisco large subunit; CbbS: Cyanobium Form-1A Rubisco small subunit; CcaA: formerly icfA, inorganic carbon fixation A which encodes a β-CA (carbonic anhydrase) enzyme; CcmM35: Internal carboxysomal protein; CrtO: β-carotene ketolase; CsoS1A and CsoS2: Structural proteins from the shell of the carboxysomes; FBP/SBPase: Fructose-1,6/sedoheptulose-1,7-bisphosphatase; FBPA: fructose-1,6-bisphosphate aldolase; FBPase: Fructose-1,6-bisphosphatase; Fd: Ferredoxin; Fld: Flavodoxin; Flv: flavodiiron protein; IctB: inorganic carbon transporter B; iWUE: intrinsic water-use efficiency; PMF: proton motive force; RbcL: Ribulose bisphosphate carboxylase large chain; RbcS: Ribulose bisphosphate carboxylase small chain; RbcX: Assembly chaperone; SBPase: Sedoheptulose-1,7-bisphosphatase; sFNR: Soluble cyanobacterial ferredoxin–$NADP^+$ reductase; SvPEPC: Phosphoenolpyruvate carboxylase; ΦPSII: quantum yield of PSII.

and some unicellular cyanobacteria, and β-carboxysomes, which encapsulate the plant-like Form-1B Rubisco in the remaining cyanobacteria (Badger et al., 2002; Rae et al., 2013). The assembly of some types of carboxysomes requires coordinated expression of 13 genes (Long et al., 2016), with the biogenesis and composition of each carboxysome type being unique (Hennacy and Jonikas, 2020).

First attempts to replace the different plant Rubisco subunits for cyanobacterial ones failed in creating functional complexes (Bracher et al., 2017). Lin et al. (2014) assembled both Rubisco subunits *rbcL* (LSU) and *rbcS* (SSU) as a functional enzyme within *Nicotiana tabacum* chloroplasts. They co-expressed them along with either RbcX or CcmM35 from *Synechococcus elongatus*, which are a chaperone involved in the assembly of the LSU core complex and an internal carboxysomal protein, respectively. They were able to observe structures with high similarity to pro-carboxysomes, an important early stage in the carboxysome assembly, and could assemble an active cyanobacterial Rubisco, even though transgenic plants exhibited slower growth but normal appearance. They concluded that RbcX is not essential for the assembly of functional cyanobacterial Rubisco in chloroplasts but CcmM35 might assist in the process (Lin et al., 2014). Two years later, Occhialini et al. (2016) demonstrated that neither RbcX nor CcmM35 was needed for assembly of active cyanobacterial Rubisco in tobacco chloroplasts. Through the alteration of the gene regulatory elements on the Rubisco transgenes, cyanobacterial Rubisco expression was enhanced and the transgenic plants grew at near WT growth rates, although still requiring elevated CO_2. This evidence confirmed that it was possible to introduce a Rubisco with higher catalytic rates into vascular plants and improve plant growth if adequate CO_2 is available near the enzyme. This precedent was an important step toward the introduction of a complete and functional CCM into the chloroplasts of higher plants.

A different strategy was pursued by Long et al. (2018) who engineered simplified carboxysomes in tobacco plants. *Cyanobium marinum* Form-1A Rubisco large and small subunit genes (*cbbL* and *cbbS)* were expressed in chloroplasts, along with genes for two key α-carboxysome structural proteins (CsoS1A and CsoS2). This minimal gene set produced simplified carboxysomes, which encapsulated the introduced Rubisco and enabled autotrophic growth at elevated CO_2. Even though the expression of structurally intact carboxysomes within a C3 plant chloroplast is still a critical and complex engineering milestone toward the longer term goal of attaining a functional chloroplastic CCM in C3 plants, they demonstrated a way forward to functional carboxysome construction in C3 plants (Long et al., 2018).

4.3.2 Manipulation of carbon assimilation

Sedoheptulose-1,7-bisphosphatase (SBPase) and fructose-1,6-bisphosphatase (FBPase) exist at the branching point of the CBB cycle where metabolites either participate in the regeneration of RuBP to continue the cycle or diverge off into starch synthesis (Miyagawa et al., 2001). Mathematical modeling has suggested that the amount of SBPase and FBPase is not optimal for photosynthesis at current atmospheric CO_2 levels and the problem will be amplified at the elevated CO_2 levels predicted for the future (Zhu, X.-G., et al., 2007). While it has been shown that decreases in FBPase do not lead to significant changes in CO_2 assimilation, small decreases in SBPase resulted in lower photosynthetic activities and impaired growth, highlighting its importance (Harrison et al., 2001). The relative contribution of SBPase and FBPase was further evaluated through the creation of transgenic plants expressing these enzymes (Miyagawa et al., 2001; Simkin et al., 2015; Tamoi et al., 2006). It was shown that SBPase-overexpressing plants had increased carbon fixation and growth (Simkin et al., 2015), whereas the FBPase-overexpressing plants had elevated starch content, but no effects on growth or carbon fixation (Tamoi et al., 2006).

Cyanobacteria has an enzyme which is a combined SBPase/FBPase performing both activities at nearly equal rates (Miyagawa et al., 2001). It is worth noting that this enzyme exhibits particular characteristics such as (1) a primary structure different from that of typical chloroplastic and cytosolic FBPases from eukaryotic cells, (2) activity to hydrolyze FBP and sedoheptulose 1,7-bisphosphate with almost equal specific activities, and (3) no inhibition by fructose- 2,6-bisphosphate. The use of a cyanobacterial SBPase/FBPase is particularly useful as the plant SBPase and FBPase are subject to strict redox regulation by the thioredoxin system, while the cyanobacterial versions are not (Tamoi et al., 2006). Transgenic tobacco expressing a *Synechococcus* SBPase/FBPase in chloroplasts had increased Rubisco activity, CO_2 fixation, and biomass under atmospheric conditions (360 ppm CO_2). Also, transformed plants exhibited higher levels of CBB cycle intermediates and carbohydrates than those in WT plants (Miyagawa et al., 2001). The expression of the same SBPase/FBPase has also been examined in transgenic rice and soybean and both cases resulted in improvement of photosynthesis. In the case of rice, it was evidenced by higher mesophyll conductance and net photosynthetic rate (Gong et al., 2015; Köhler et al., 2017). Transgenic soybean was tested in field trials, obtaining better seed yield for plants exposed to higher temperatures and increased CO_2 assimilation rates compared to the WT control lines (Köhler et al., 2017). Ichikawa et al. (2010) have obtained similar results with lettuce producing a greater number of leaves and a higher size. Additionally, Gong et al., 2017 have coexpressed cyanobacterial SBPase/FBPase and CcaA which encodes a carbonic anhydrase (β-CA) located in the cytoplasm. The outcome of this strategy resulted in plants with better photosynthetic capacity, and a synergistic effect was observed when both proteins were expressed under field conditions (Gong et al., 2017). Recently, an approach consisting of the coordinated expression of genes from the electron transport chain (red algal protein cyt c6) and RuBP regeneration (*Synechocystis* SBPase/FBPase and plant SBPase) produced even higher increases in biomass accumulation under both glasshouse and field conditions. These studies suggest that the use of cyanobacterial SBPase/FBPase is a promising strategy to enhance plant biomass through acceleration of the CBB cycle flux and a possibility to improve intrinsic water-use efficiency under field conditions (López-Calcagno et al., 2020).

4.3.3 Manipulation of carbon uptake/transport

Cyanobacterial CCMs enable relatively rapid CO_2 fixation by elevating intracellular inorganic carbon as bicarbonate and then concentrating it as CO_2 around the enzyme Rubisco in the carboxysomes. Related to that, there are genes associated with carbon accumulation, but without a known function. The cyanobacterial *ictB* gene was initially identified from a high CO_2 requiring mutant (Bonfil et al., 1998). This gene was originally hypothesized to be a bicarbonate transporter, but that possibility has been already ruled out (Bonfil et al., 1998; Price et al., 2011). The function of IctB is still unknown. The *ictB* gene from *Synechococcus* was first transformed into *Arabidopsis* and tobacco, and proved to increase CO_2 uptake, and enhanced plant growth under low humidity conditions (Lieman-Hurwitz et al., 2003). This gene was also expressed in rice and soybean (Gong et al., 2015; Hay et al., 2017; Yang et al., 2008). Increases in CO_2 uptake were observed in both species, and increases in plant biomass were obtained in soybean (Gong et al., 2015; Hay et al., 2017). Transgenic soybean plants expressing *ictB* were also used to perform field trials, which showed increased CO_2 uptake and biomass (Hay et al., 2017). The mechanism underlying the ability of this single cyanobacterial gene to boost photosynthesis is currently uncertain. Additionally, IctB has been transformed as part of a multigene stacking approach in tobacco with SBPase and fructose-1,6-bisphosphate aldolase (FBPA) (Simkin et al., 2015). Plants expressing these genes individually displayed increases in both photosynthesis and biomass, while plants expressing the three proteins showed the best performance compared to WT siblings (Simkin et al., 2015).

4.3.4 Sugar partitioning and utilization

As it was mentioned before, sucrose is the final product of the CBB cycle and is the major translocated form of carbon toward sink organs in plants. Due to that function, its synthesis might be a rate-limiting step to establish a balanced partitioning between sucrose and starch biosynthesis. Sucrose is primarily formed from Uridine diphosphate (UDP)-glucose and fructose-6-phosphate followed by dephosphorylation of the resulting sucrose-6-phosphate to yield sucrose. This process takes place in the cytosol, and FBPase and sucrose-phosphate synthase are the key enzymes for sucrose synthesis (Serrato et al., 2009). It was evidenced that *Arabidopsis* plants with reduced endogenous FBPase expression levels led to inhibition of sucrose synthesis, accumulation of phosphorylated intermediates, and stimulation of starch synthesis (Rojas-González et al., 2015; Strand et al., 2000). In addition, this line resulted in a pronounced inhibition of plant growth (Strand et al., 2000). The same behavior was described in mutant rice plants lacking FBPase, which showed markedly decreased photosynthetic rates and severe growth retardation with reduced chlorophyll content, resulting in plant death (Lee et al., 2008).

A cyanobacterial FBPase-II from *S. elongatus* PCC7942 was overexpressed in the cytosol of *Arabidopsis* plants. The resultant transgenic lines presented better growth at elevated CO_2 levels (1000 ppm) evidenced by more secondary cauline and primary rosette branches, higher photosynthetic rates and sugar contents, and surprisingly decreased levels in the expression of genes related to hormone metabolism and signaling (Otori et al., 2017). These results set a direct relationship between alterations in sugar partitioning with increases in the development of shoot branching. However, the correlation with hormone response was unclear. A similar approach was based on the expression of FBP/SBPase in the cytosol of tobacco plants (Tamoi et al., 2011). Transgenic plants grown under elevated CO_2 levels (1200 ppm) exhibited higher shoot and root dry weight in comparison with WT counterparts. In addition, the rate of photosynthesis increased significantly under high light intensity in FBP/SBPase expressing plants and they did not accumulate hexose, an intermediate of sucrose synthesis, under elevated levels of CO_2. Therefore, these changes might lead to improvement of translocation efficiency in the source and sink leaves of transgenic plants (Tamoi et al., 2011).

Among the variety of strategies designed to improve photosynthesis, introducing the C4 pathway into C3 plants was considered. The main difference between them is that C4 plants have an additional cycle responsible for fixing inorganic carbon, which does not replace the CBB cycle but enables effective carbon assimilation followed by a release of CO_2 under conditions where it can be refixed by Rubisco with high efficiency (Nowicka et al., 2018). Between the enzymes in the C4 photosynthetic pathway, phosphoenolpyruvate carboxylase (PEPC) is one of the most important because it performs primary fixation of atmospheric CO_2. PEPC is a ubiquitous enzyme found in all photosynthetic organisms and some non-photosynthetic ones that catalyze the addition of CO_2 to phosphoenolpyruvate to produce oxaloacetate. The latter compound is a direct precursor for the synthesis of a number of amino acids such as aspartate, asparagine, threonine, methionine, and lysine; also it works as an intermediate in the citric acid cycle in mitochondria (O'Leary et al., 2011). The PEPC from *S. vulcanus* was extensively characterized in *E. coli*, where it remained active even after incubation at 50 °C for 60 min or exposure to chaotropic agents at 30 °C for 20 h, being more stable than the C4-type PEPC from *Zea mays* (Chen et al., 2002). Because of these encouraging results, this gene was expressed in *Arabidopsis* plants to evaluate its viability and effect in transformant lines. Resultant transgenic plants presented higher amino acid contents per fresh weight (Chen et al., 2004). This intervention appears to be a promising biotechnological application, but other components may be required to avoid aberrant phenotypes.

4.3.5 Introduction of cyanobacterial proteins into the photosynthetic electron transport chain

The light-dependent reactions use light energy to make two molecules needed for the next stage of photosynthesis: the energy storage molecule ATP and the reduced electron carrier NADPH. They take place through the PETC in the thylakoid membranes. There are several genes in common in addition to some substantial differences among plants and cyanobacteria. Because they are derived from the same common ancestor, the introduction and/or replacement of some components opened a window to examine new crop improvements (Leister, 2019).

The driving energy of photosynthesis, the sunlight, is absorbed by a variety of pigments associated with photosystems. These complexes consist of the photochemical reaction center which is the site of primary charge separation, surrounded by light-harvesting (or antenna) complexes (LHCs). The optimal absorption wavelength range for LHCs is in the red region (680–690 nm), where the energy is utilized by chlorophyll to split water and reduce ferredoxin (Fd). Different LHC complexes differ in the number of pigments and their composition and structure in a way that they are an optimized energy collector system (Guidi et al., 2017). Increased contents of carotenoids could be a way to improve light capture and energy transfer to chlorophylls. Moreover, hydroxylated and ketolated carotenoids are high-value compounds with industrial, food, and feed applications. In this way, it was overexpressed by β-carotene ketolase *crtO* gene from *Synechocystis* PCC6803 in *Nicotiana glauca* plants to improve ketocarotenoid production. The resulting plants had a high content of carotenoids in both old and young leaves, as well as in some flower tissues. Although they did not show improvements in photosynthetic efficiency, these results showed promising advances in this field and the synthesis of products of industrial interest by plants (Zhu, C., et al., 2007).

4.3.5.1 *The cyanobacterial lost genes: flavodoxin as a source for photosynthesis stress protection*

All environmental adversities generate secondary oxidative stress, which negatively affects key cellular components and functions. A common feature of abiotic stress situations is the generation of ROS such as singlet oxygen ($^1O_2$), hydrogen peroxide (H_2O_2) and the superoxide ($O2^{\cdot-}$), and hydroxyl ($HO^{\cdot}$) radicals. They react with different biomolecules, inactivating them and causing severe damage (Czarnocka and Karpiński, 2018). However, it is increasingly apparent that a major role played by ROS during plant responses to both developmental and environmental stimuli is to act as signal transducers rather than toxic compounds (Foyer et al., 2017). Indeed, it has been demonstrated that some processes that involve cell death such as natural senescence or biotic stress responses are genetically controlled and ROS play a critical signaling role (Rossi et al., 2017; Wu et al., 2012).

Chloroplasts are the main site of ROS propagation in illuminated leaves and most of them are produced by adventitious electron transfer from components of the PETC to molecular oxygen (Czarnocka and Karpiński, 2018). Stress conditions inhibit CO_2 assimilation and hence NADPH oxidation in the regenerative stages of the CBB cycle (Apel and Hirt, 2004). Besides, many environmental hardships cause down-regulation of Fd expression in plants and cyanobacteria, which is the final electron acceptor of the PETC (Mazouni et al., 2003; Tognetti et al., 2006; Zimmermann et al., 2004). These conditions thus result in a decrease in the levels of the main physiological PETC acceptors, oxidized Fd, and $NADP^+$, contributing to PETC over-reduction and ROS propagation (Apel and Hirt, 2004). Fds are small, mobile metalloproteins containing a [2Fe–2S] cluster as a prosthetic group (Hase et al., 2006) and collect reducing equivalents generated in the PETC, delivering them to a plethora of metabolic, regulatory, dissipative, and developmental processes. The largest fraction of photoreduced Fd is used for $NADP^+$ reduction in the reaction catalyzed by the Fd-$NADP^+$-oxidoreductase (FNR) (Ceccarelli et al., 2004). Reduced Fd also delivers electrons for nitrogen and sulfur assimilation; amino acid, fatty acid, and secondary metabolism; antioxidant defense; reductive activation of enzymes, among others (Zurbriggen et al., 2008). Knocked-down plants in which the levels of either FNR or Fd had been decreased by RNA antisense or RNA interference approaches display enhanced susceptibility to several sources of stress, in addition to growth and reproductive penalties (Hajirezaei et al., 2002; Hald et al., 2008; Holtgrefe et al., 2003; Palatnik et al., 2003). However, the transformation of tobacco plants with a cyanobacterial plastid-targeted Fd failed to improve stress tolerance, due to down-regulation of the transgene under oxidative conditions (Ceccoli et al., 2011). Fds are extremely sensitive to environmental stress and their expression is transcriptionally and post-transcriptionally down-regulated by numerous adverse environmental conditions (Holtgrefe et al., 2003; Petracek et al., 1998; Tognetti et al., 2006).

Many prokaryotes, including cyanobacteria, and some algae express a flavoprotein, flavodoxin (Fld), an isofunctional electron carrier that contains flavin mononucleotide (FMN) as a prosthetic group instead of an [2Fe–2S] center (Sancho, 2006). Fld properties as redox shuttle largely match those of Fd, being able to replace the metalloprotein in many reactions, such as in $NADP^+$ reduction, H_2 formation, and N_2 fixation (see table 1 in Lodeyro et al., 2012). When both carriers are present in the same genome, the Fld gene is typically induced as an adaptive resource under environmental or nutritional hardships that compromise Fd expression or activity (i.e., iron limitation). Fld induction has been used for many years as a reliable marker of iron deficiency in the oceans and constitutes a key selective advantage for colonization of iron-poor waters by phytoplankton (Erdner and Anderson, 1999). Given the key position occupied by Fd in the network of electron distribution, its decline affects vast portions of central metabolism, as well as defensive, regulatory, and developmental processes (Zurbriggen et al., 2008). Fld induction mitigates these effects, allowing survival and reproduction under adverse conditions, although the relevance of Fld replacement varies among species. Fld is absent in plant genomes; the distribution of the flavoprotein in nature suggests that the gene was lost somewhere in the evolutionary transition between green algae and terrestrial plants (Pierella Karlusich et al., 2014).

In the case of Fd and Fld, the constraints for binding of both to cognate partners are less stringent than those of typical enzyme-substrate interactions. Purified Fld could productively interact *in vitro* with plant enzymes whose prokaryotic ancestors used this flavoprotein as their common or occasional substrate. They include FNR, PSI, and thioredoxin reductase for $NADP^+$ and thioredoxin reduction (Tognetti et al., 2006), as well as cyclic electron transport by isolated plant thylakoids (Scheller, 1996).

These observations strengthened the possibility that Fld could still function *in planta*, and that its introduction into the plant genome could improve stress tolerance in the same way as in microorganisms. The hypothesis was proven by introducing the *Anabaena* Fld gene into tobacco plants, in which the transgenic product was targeted to chloroplasts (Tognetti et al., 2006). Plants expressing Fld in plastids showed enhanced tolerance to environmental stresses, including high temperature, chilling, drought, high light intensities, UV radiation, and exposure to the redox-cycling herbicide methyl viologen, all having as a common feature the establishment of an oxidative stress condition (Tognetti et al., 2006). The accumulation of ROS such as hydrogen or organic peroxides and $O_2^{\cdot -}$, which was prominent in stressed WT plants, was mitigated in the Fld transformants. Tolerance was illustrated in the transformants by the preservation of leaf turgor, cellular and membrane integrity, and plastid ultrastructure. Fld expression also preserved photosynthetic capacity, maintaining high levels of CO_2 fixation. Fld involvement in $NADP^+$ photoreduction and thioredoxin reduction as a substrate of Fd-thioredoxin reductase helped to relieve the electron pressure imposed onto the PETC by the stress condition (Tognetti et al., 2006). These transformants, expressing a similar level of Fld to endogenous Fd, were able to survive and even to increase in height, fresh weight, and dry weight when they we're subjected to iron-deficiency treatments; whereas WT specimens exhibited the typical symptoms of this nutrient scarcity such as interveinal chlorosis, growth retardation, and high lethality rates. Furthermore, these plants displayed increased tolerance to nitrotoluene derivatives and were able to significantly eliminate these widespread industrial pollutants from contaminated media (Tognetti et al., 2007a).

As mentioned before, plastid-targeted Fld has been previously shown to specifically block the accumulation of ROS in the chloroplasts of transgenic plants exposed to various sources of abiotic stress (Tognetti et al., 2006); and it is generally accepted that ROS are involved in the orchestration of the response to pathogen attack. ROS play fundamental roles in plant responses to pathogen infection, including modulation of cell death processes and defense-related gene expression. Fld-expressing plants are particularly useful to evaluate the role of ROS in biotic interactions. WT tobacco plants inoculated with the nonhost pathogen *Xanthomonas campestris* pv. vesicatoria accumulated high levels of ROS in chloroplasts, followed by the appearance of typically localized lesions. In contrast, plants expressing plastid-targeted Fld exhibited lower ROS levels and significantly reduced lesions. Metabolic routes normally inhibited by pathogens were protected in the transformants (Zurbriggen et al., 2009). On the other hand, the possible function of Fld as a cell death inhibitor during infection by necrotrophic pathogens also deserves attention. For this purpose, Fld-expressing tobacco lines were inoculated with the typical necrotrophic fungus, *Botrytis cinerea*. Tissue damage and fungal growth were significantly inhibited in infected leaves of the transformants compared with infected WT counterparts. ROS build-up triggered by *Botrytis* infection and associated with chloroplasts was significantly decreased (70–80%) in transgenic leaves relative to the WT plants. In addition, the expression of pathogenesis-related genes was induced to a lower degree in Fld plants. The impact of fungal infection on photosynthetic activity was also lower in transgenic plants. The results indicate that chloroplast-generated ROS play a major role in lesion development during *Botrytis* infection (Rossi et al., 2017). Transgenic lines expressing Fld in chloroplasts of *Arabidopsis* were used to demonstrate the importance of photosynthetic inhibition during effector-triggered immunity. The expression of Fld-alleviated photosynthetic inhibition and decreased

ROS accumulation in chloroplasts were triggered by conditional expression of avirulence effectors (Su et al., 2018). Taken together, the results indicate that Fld expression could compensate successfully for Fd decline occurring upon iron deficiency or by its down-regulation under stress, engaging in at least some of the Fd-dependent pathways of plant chloroplasts, improving the delivery of reducing equivalents to productive pathways of the chloroplast, which resulted in down-regulation of ROS levels and increased tolerance to a wide range of abiotic, biotic, and xenobiotic stresses (Zurbriggen et al., 2009).

For all types of stresses assayed, the effect of Fld was dose dependent, and targeting the ectopic product to chloroplasts was mandatory. Transgenic plants that expressed high amounts of Fld in the cytosol displayed WT phenotype under stress conditions (Tognetti et al., 2006; Tognetti et al., 2007b; Zurbriggen et al., 2009). The ability of the foreign Fld to be reduced by Fd-dependent chloroplast systems (e.g., the PETC) appeared essential to its protective effect (Tognetti et al., 2006).

As NADPH build-up is another unwanted consequence of stress episodes and detrimental to the photosynthetic processes, the overexpression of FNR seems to be the logical choice to use the excess of NADPH as an electron source for Fd/Fld reduction. Soluble FNR (sFNR) catalyzes Fld reduction by NADPH with high efficiency (Carrillo and Ceccarelli, 2003), but FNR normally binds to the thylakoid membrane, and when membrane-bound, the preferred direction of the reaction is still under debate (Benz et al., 2010). A gene encoding a plastid-directed sFNR was obtained by replacing the N-terminal domain of *Anabaena* FNR with a chloroplast-targeting transit peptide. The expression of sFNR in tobacco chloroplasts led to transgenic plants whose growth phenotypes and stress tolerance were similar to those of their WT siblings (Giró et al., 2011). However, the design of an *in vivo* system for Fld reduction and NADPH oxidation, involving coexpression of Fld and sFNR in tobacco chloroplasts, resulted in transgenic lines displayed higher tolerance to oxidative stress and lower ROS build-up compared to WT plants (Giró et al., 2011).

The expression of a cyanobacterial Fld in plants has been proposed as a potentially powerful biotechnological tool to generate multiple-tolerant crops. The *Anabaena* Fld gene was introduced into the model legume *Medicago truncatula,* and the transgenic plants obtained were subjected to salt stress. However, Fld-expressing *M. truncatula* plants did not display enhanced tolerance to saline stress under the experimental conditions, as they did not display significant differences in morphometric parameters or photochemical efficiency when compared to WT plants. Nevertheless, nitrogen fixation was less affected by salt stress in nodules of Fld-expressing transgenic lines than in WT plants (Coba de la Peña et al., 2010). When Fld technology was transferred to creeping bentgrass (*Agrostis stolonifera*), an important perennial crop, the transgenic turfgrass exhibited enhanced performance under oxidative condition, drought, and heat stress than WT controls, modified expression of heat-shock protein genes, production of more reduced thioredoxin, elevated N accumulation and total chlorophyll content as well as upregulated expression of nitrite reductase and N transporter genes (Li et al., 2017).

The third most important food crop in the world, potato (FAO, 2014; Visser et al., 2009), is vulnerable to drought, salinity, and other environmental stresses, which affect tuber yield and quality (Muñiz García et al., 2018; Vasquez-Robinet et al., 2008). The need for drought-tolerant potato lines has prompted to evaluate the Fld approach in this crop. Manipulation of chloroplast electron distribution by introducing the alternative electron carrier in potato chloroplasts protected photosynthetic activities even at a pre-symptomatic stage of drought. Transcriptional and metabolic profiling revealed an attenuated response to the adverse condition in Fld-expressing plants, correlating with their increased stress tolerance. Interestingly, tuber yield losses under chronic water limitation were mitigated in Fld-expressing lines, indicating that the flavoprotein has the potential to improve major agronomic traits in potato (Pierella Karlusich et al., 2020).

These examples establish a key role of cyanobacterial Fld modulating plant responses to multiple sources of adverse environmental conditions in crop species and demonstrate the feasibility of manipulating a single gene in crop species for genetic engineering of plant stress tolerance, to improve plant productivity, and to utilize otherwise vast nonproductive agricultural areas. Fig. 4.2 shows the intervention point tackled by Fld in PETC from transgenic plants.

4.3.5.2 The cyanobacterial lost genes: flavo-diiron proteins, boosting higher plants photosynthetic processes

In algae, cyanobacteria, and plants, sunlight fuels the linear electron transport (LET) from water to $NADP^+$, committing on two photosystems, PSI and PSII (Jagannathan et al., 2012). LET is associated with the generation of a proton motive force, including both a proton concentration component (ΔpH) and an electric field component ($\Delta\psi$) that drives ATP synthesis (Kramer and Evans, 2011). The activation of the CBB cycle is slower than the onset of photosynthetic electron transport and, when light intensity changes, this causes a temporary unbalance between LET and carbon fixation rates. In consequence, the electron carriers become over-reduced, O_2 can replace CO_2 as a terminal electron acceptor driving the harmful formation of ROS, and potentially damaging photosynthetic apparatus (Foyer, 2018; Janasch et al., 2019). Alternative electron transports (AET) are used to adjust the ATP/NADPH ratio and the ΔpH magnitude to the requirements of assimilatory routes. The most widely employed is cyclic

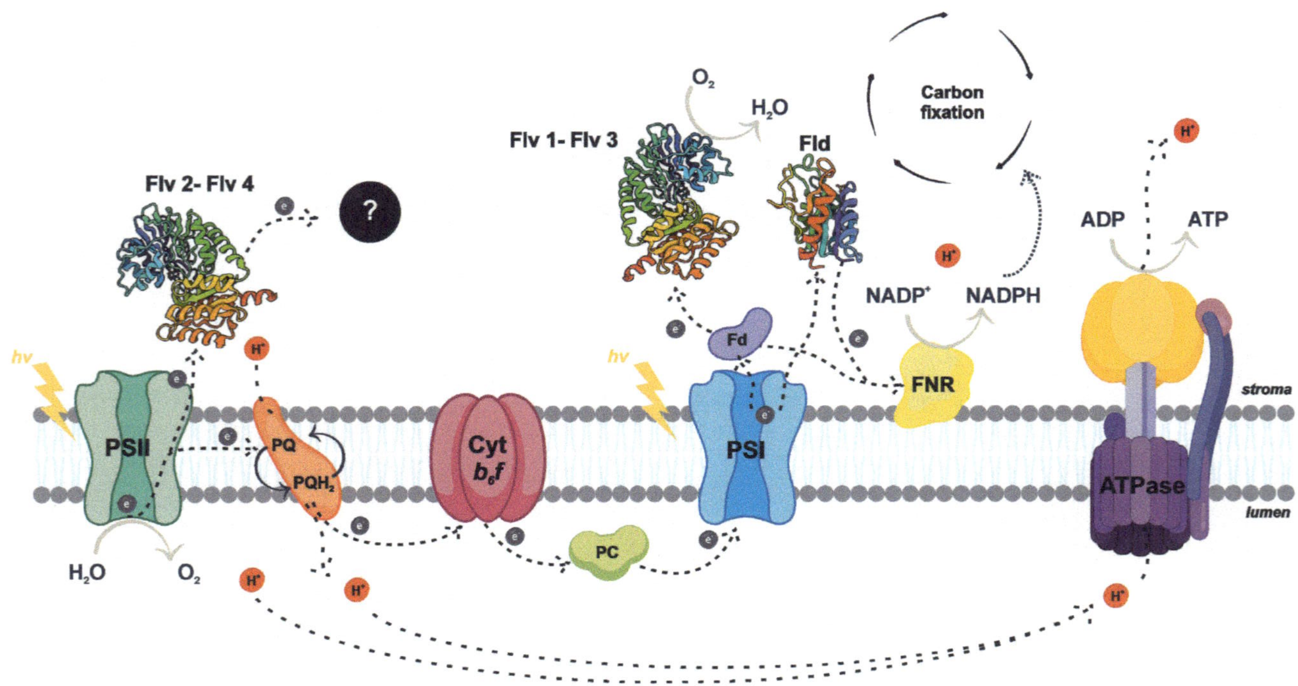

FIG. 4.2 **Schematic representation of the photosynthetic electron transport chain (PETC) in plants. Only the linear electron transport from H_2O to NADP+ is shown. Intervention points of Fld and Flvs in chloroplasts of transgenic plants.** While Fld accepts electrons directly from PSI compensating the decline of Fd levels under stress conditions, Flv1-Flv3 mediates the direct reduction of O_2 to H_2O using Fd as electron donor without the production of ROS. Flv2-Flv4 provides an additional electron sink accepting electrons from PSII to an unknown final electron acceptor. Abbreviations: PSI: photosystem I, PSII: photosystem II, PQ: plastoquinone, PQH_2: plastoquinol, Cytb_6f: cytochrome b_6f, PC: plastocyanin, FNR: ferredoxin-NADP(H) oxidoreductase, Fd: ferredoxin, Fld: flavodoxin, Flv: flavo-diiron protein. Image created using BioRender.

electron transport (CET) around PSI, which is mediated by Fd via PGR5/PGRL1 and/or NAD(P)H dehydrogenase-like pathways (Yamori and Shikanai, 2016) producing ATP without $NADP^+$ reduction, relieving the reducing pressure at the PSI acceptor side by providing the CBB cycle with ATP to re-oxidized NADPH and limiting electron transfer upstream of PSI (Wang et al., 2015). Another AET to dissipate the excess of excitation energy over the PETC is through the Mehler reaction and the water-water cycle (Asada, 1999), also termed pseudocyclic electron transport (PCET). In this additional electron sink, H_2O is split by PSII and then resynthesized when O_2 replaces $NADP^+$ as the final electron acceptor downstream of PSI, detoxifying $O_2^{\cdot -}$ produced by this PS (Allahverdiyeva et al., 2015a; Yamamoto et al., 2016). As mentioned before in this chapter, photorespiration is still another dissipative route available in C3 plants (Hagemann et al., 2013). Finally, photosynthetic microorganisms, nonvascular plants, and gymnosperms harbor a different type of AET based on the activity of a family of enzymes called flavo-diiron proteins (Flvs) that have been shown to contribute to PCET (Allahverdiyeva et al., 2013). These enzymes have a modular organization and they all share a common core composed of three functional domains: a metallo β-lactamase-like domain harboring a nonheme diiron catalytic site, an Fld-like fold with bound FMN, and a C-terminal NAD(P)H-flavin oxidoreductase-like domain (Vicente et al., 2008, 2002). This domain could be capable of accepting electrons directly from NADH/NADPH and data from two-hybrid system showed potential interactions between Flv and Fd isoforms in both *Synechocystis* PCC6803 (Cassier-Chauvat and Chauvat, 2014) and *Chlamydomonas reinhardtii* (Peden et al., 2013), suggesting that Fd might also be involved in electron transport from PSI to Flv (Alboresi et al., 2019). Some α-cyanobacteria and algae possess only two Flvs (Flv1 and Flv3), whereas β-cyanobacteria harbor four isoforms (Flv1–Flv4). They have been reported to act as heterodimers, with the pairs Flv1-Flv3 and Flv2-Flv4 being the functional species *in vivo* (Allahverdiyeva et al., 2015a). Heterocyst-forming cyanobacteria still contain two other isoforms (Flv1B and Flv3B) which accumulate specifically in that compartment and participate in O_2 elimination, allowing N_2 fixation by oxygen-sensitive nitrogenase (Ermakova et al., 2013).

In cyanobacteria, Flv1 and Flv3 are responsible for safeguarding PSI under rapidly fluctuating light intensities; this system can alleviate photodamage by transferring reducing equivalents from NADPH to oxygen avoiding over-reduction of the PETC (Allahverdiyeva et al., 2015b). Although the PETC has remained relatively unchanged in both plant chloroplasts and their cyanobacterial ancestors, angiosperms do not have Flvs to protect their photosystems. Instead, they have developed different mechanisms to cope with abruptly changing light intensities in the environment.

Illumination of plant canopies in the field varies widely during the day, with dynamic light transitions in various time-scales caused by clouds, other leaves, etc. Model simulations indicate that plants grown under fluctuating light could decrease CO_2 assimilation up to 20% compared to steady-state light conditions (Zhu et al., 2010). Higher plants such as *Arabidopsis* have evolved different protective mechanisms (Takahashi and Badger, 2011). In particular, the PGR5 protein controls electron flow during sudden changes in light intensity by allowing the regulation mostly via the cytochrome *b6f* complex. In *Arabidopsis*, PGR5 interacts with the PGRL1 protein and this heterodimer is proposed to mediate CET (Dal Corso et al., 2008). The expression of Flv1 and Flv3 homologs from the moss *Physcomitrella patens* in a *pgr5* mutant of *Arabidopsis* partially restored the WT phenotype (Yamamoto et al., 2016). In *pgr5* mutants, under high light conditions, electrons are trapped in the PSI reaction center (P700), resulting in PSI photoinhibition (Munekage et al., 2002). When expressed in a WT background; however, Flvs did not compete with CET during steady-state photosynthesis, but contributed to PS protection under fluctuating light and allowed for faster recovery from dark adaptation (Yamamoto et al., 2016). Besides the function of PGR5, plants have also acquired other dynamic mechanisms, among them non-photochemical quenching (NPQ) is one of the most important, allowing dissipation of the excess of excitation energy at PSII as heat (Duffy and Ruban, 2015). Photosynthetic activity is down-regulated during this process and requires some time to recover after the light stress has abated. In a first attempt to determine if cyanobacterial Flvs could mediate electron transfer from the PETC to oxygen when expressed in plants, Flv1 and Flv3 from *Synechocystis*, the best studied of all Flvs, were introduced in tobacco chloroplasts and their contribution to photosynthetic dark recovery was characterized (Gómez et al., 2018). During photosynthesis induction, prior to the complete activation of the NADPH-consuming CBB cycle, the expression of Flv proteins in tobacco chloroplasts provided an additional electron sink that kept the PETC in a more oxidized state. This led to an enhancement of electron transfer efficiency, as well as a faster generation of proton motive force across the thylakoid membrane. As a result, energy-dependent NPQ processes were activated sooner, preventing over-reduction of the PETC and maintaining higher photochemical performance (Gómez et al., 2018). Thus, avoiding PSI electron acceptor limitation by the introduction of additional electron sinks highlights the potential to improve photosynthetic efficiency and plant productivity.

Flvs could enhance the photosynthetic efficiency of plants exposed to sudden changes in light intensity even in the presence of other dissipative systems (Gómez et al., 2018). This effect might improve growth and yield in the field, where changing environmental conditions are the rule. This hypothesis was tested first in *Arabidopsis* (Tula et al., 2020).

Transgenic plants co-expressing *Synechocystis* Flv1-Flv3 in chloroplasts showed enhanced efficiency of light utilization, boosting the plant's capacity to accumulate biomass as the growth light intensity was raised. The Flv1-Flv3 lines displayed an increased production of ATP and an acceleration of carbohydrate metabolism. The results indicate that plants harboring Flv1-Flv3 grew significantly better under a range of moderate to high light intensities and displayed increased carbohydrate and ATP levels (Tula et al., 2020). Cyanobacterial Flvs may act as regulators of photosynthesis when expressed in angiosperms, specifically avoiding over-reduction of the PETC under conditions of electron pressure (Fig. 4.2).

Plants expressing Flv1 and Flv3 have been assayed for their photosynthetic performance and light responses (Gómez et al., 2018; Tula et al., 2020; Wada et al., 2018; Yamamoto et al., 2016). More recently, the effects of this additional chloroplast electron sink on stress tolerance and yield were reported. This strategy has been employed in barley, the fourth most important cereal as a source for food and fodder. The heterologous expression of *Synechocystis* Flv1-Flv3 in barley chloroplasts accelerated days to heading, increased biomass, promoted the number of spikes and grains per plant, and improved the total grain weight per plant of transgenic lines exposed to drought. Improved growth correlated with several metabolic changes, enhanced availability of soluble sugars, and a higher turnover of amino acids. In Flv1-Flv3 lines, the energy status of the leaves was maintained in the stressed plants by converting sucrose to glucose and fructose, immediate precursors for energy production to support plant growth under drought (Shahinnia et al., 2021). Common features of drought-stressed barley plants are diminished grain yield (evidenced by a reduced number of tillers, grains, and spikes) and individual grain weight (Farooq et al., 2009). When barley plants were exposed to drought at the seedling stage, the heterologous expression of Flv1-Flv3 resulted in the acceleration of heading time and flowering. These plants, under stress conditions, beat their nontransgenic controls in the accumulation of aboveground biomass, the number of grains set, and the total grain weight per plant (Shahinnia et al., 2021), showing that the dimer Flv1-Flv3 is also functional in a monocotyledonous species, maintaining growth in a situation where an excess of electrons are produced.

Flv2 and Flv4 are restricted to β-cyanobacteria. Studies on *Synechocystis* have shown that Flv2-Flv4 can accept electrons from PSII (Bersanini et al., 2014; Zhang et al., 2009) and catalyze O_2 photoreduction under atmospheric (low) CO_2 levels (Santana-Sánchez et al., 2019). However, the two activities apparently result from different electron transfer pathways mediated by the Flv2-Flv4 complex (Santana-Sanchez et al., 2019). Oxygen consumption occurs at the reducing side of PSI, similar to Flv1-Flv3, and the excess of reducing power at the level of PSII is relieved by an Flv2-Flv4-driven AET from PSII to an unknown final electron acceptor different from O_2 (Bersanini et al., 2014; Santana-Sánchez et al., 2019). In cyanobacteria, the expression of the operon is induced by high irradiation and low CO_2 levels (Zhang et al., 2009), and overexpressing cells were more resistant to photoinhibition of PSII and grew significantly faster than their WT counterparts under high light intensities (Bersanini et al., 2014). This evidence encourages the hypothesis that the activity of Flv2-Flv4 as an electron sink might be operative in other photosynthetic organisms, such as plants. Therefore, tobacco and *Arabidopsis* plants expressing *Synechocystis* Flv2-Flv4 in chloroplasts were prepared to investigate whether they could functionally engage in electron exchange reactions with the plant PETC. Transgenic plants from the two species displayed increased tolerance to high irradiation, evidenced by protection to photosynthetic components and activities, and the prevention of ROS build-up (Vicino et al., 2021). These observations support the assumption that the Flv2-Flv4 dimer is able to interact with the PETC and drive novel AET pathways in the chloroplasts of transformed plants. Flv2-Flv4 expressing lines showed enhanced tolerance to other sources of stress such as salinity, redox-cycling oxidants, and water deprivation, reflected by differential preservation of pigments and cell membrane integrity, photosynthetic activity, biomass production, and metabolic performance (Vicino et al., 2021). Introduction of the Flv2-Flv4 system in chloroplasts of the two species conferred significant stress protection for all these parameters in the stressed plants, indicating that the Flv2-Flv4 function as electron sink is conserved across species. This was the first report describing the transformation of two species of flowering plants, tobacco, and *Arabidopsis*, with cyanobacterial genes encoding the Flvs most restricted in their distribution. The results showed that they can functionally interact with the chloroplast PETC despite the evolutionary divergence between cyanobacteria and angiosperms (Fig. 4.2).

The broad range advantages conferred by lost cyanobacterial flavoproteins when introduced in plants indicate that their use might have biotechnological applications in the aim to improve the field performance of crops grown in suboptimal environments.

4.4 Concluding remarks

As chloroplasts originated from ancestral cyanobacteria, both are able to perform oxygenic photosynthesis and share many biochemical and physiological pathways. However, as these organisms face different environments and challenges, they have differentiated during evolution according to their needs. Cyanobacteria inhabiting very hostile environments have acquired different strategies to survive, especially regarding protection involved in photosynthesis, which is the main process that supplies the energy and carbon required for the biosynthesis of organic compounds necessary for development. Because of that, introducing some of these genes to the plant genome offers a novel genetic source to improve crop tolerance to environmental stresses, a critical challenge in the context of food necessary to supply the increase of population with regard to climate change. Several examples described in this chapter illustrate the potential of this approach as a biotechnological tool for the generation of crops with increased yield and performance in the field. However, looking forward to the future, deeper research is needed to exploit promising cyanobacterial interventions in plants.

Acknowledgments

PV and RA are Fellows and AFL is Staff Researcher from CONICET, Argentina. AFL is Faculty member of the Facultad de Ciencias Bioquímicas y Farmacéuticas, Universidad Nacional de Rosario, Argentina. This work was supported by Grant from the Agencia Nacional de Promoción Científica y Tecnológica, Argentina PICT 2017-3080 to AFL, and MD is Fellow from this Institution. We wish to thank N. Carrillo for his critical reading of the draft.

References

Ai, Y., Yang, Y., Qiu, B., Gao, X., 2014. Unique WSPA protein from terrestrial macroscopic cyanobacteria can confer resistance to osmotic stress in transgenic plants. World J. Microbiol. Biotechnol. 30, 2361–2369.

Alboresi, A., Storti, M., Cendron, L., Morosinotto, T., 2019. Role and regulation of class-C flavodiiron proteins in photosynthetic organisms. Biochem. J. 476, 2487–2498.

Allahverdiyeva, Y., Isojärvi, J., Zhang, P., Aro, E.-M., 2015a. Cyanobacterial oxygenic photosynthesis is protected by flavodiiron proteins. Life 5 (1), 716–743.

Allahverdiyeva, Y., Mustila, H., Ermakova, M., Bersanini, L., Richaud, P., Ajlani, G., Battchikova, N., Cournac, L., Aro, E.-M., 2013. Flavodiiron proteins Flv1 and Flv3 enable cyanobacterial growth and photosynthesis under fluctuating light. Proc. Natl. Acad. Sci. USA 110, 4111–4116.

Allahverdiyeva, Y., Suorsa, M., Tikkanen, M., Aro, E.-M., 2015b. Photoprotection of photosystems in fluctuating light intensities. J. Exp. Bot. 66, 2427–2436.

Annunziata, M.G., Ciarmiello, L.F., Woodrow, P., Dell'aversana, E., Carillo, P., 2019. Spatial and temporal profile of glycine betaine accumulation in plants under abiotic stresses. Front. Plant. Sci. 10, 230.

Apel, K., Hirt, H., 2004. REACTIVE OXYGEN SPECIES: metabolism, oxidative stress, and signal transduction. Annu. Rev. Plant. Biol. 55, 373–399.

Asada, K., 1999. THE WATER-WATER CYCLE IN CHLOROPLASTS: scavenging of active oxygens and dissipation of excess photons. Annu. Rev. Plant. Physiol. Plant. Mol. Biol. 50, 601–639.

Atkinson, N.J., Urwin, P.E., 2012. The interaction of plant biotic and abiotic stresses: from genes to the field. J. Exp. Bot. 63, 3523–3544.

Badger, M.R., Hanson, D., Price, G.D., 2002. Evolution and diversity of CO_2 concentrating mechanisms in cyanobacteria. Funct. Plant Biol. 29, 161–173.

Barrero-Sicilia, C., Silvestre, S., Haslam, R.P., Michaelson, L.V., 2017. Lipid remodelling: unravelling the response to cold stress in *Arabidopsis* and its extremophile relative *Eutrema salsugineum*. Plant Sci 263, 194–200.

Benz, J.P., Lintala, M., Soll, J., Mulo, P., Bölter, B., 2010. A new concept for ferredoxin–NADP(H) oxidoreductase binding to plant thylakoids. Trends Plant Sci 15, 608–613.

Bersanini, L., Battchikova, N., Jokel, M., Rehman, A., Vass, I., Allahverdiyeva, Y., Aro, E.-M., 2014. Flavodiiron protein Flv2/Flv4-related photoprotective mechanism dissipates excitation pressure of PSII in cooperation with phycobilisomes in cyanobacteria. Plant Physiol 164, 805–818.

Bonfil, D.J., Ronen-Tarazi, M., Sültemeyer, D., Lieman-Hurwitz, J., Schatz, D., Kaplan, A., 1998. A putative HCO_3^- transporter in the cyanobacterium *Synechococcus* sp. strain PCC 7942. FEBS Lett 430, 236–240.

Bracher, A., Whitney, S.M., Hartl, F.U., Hayer-Hartl, M., 2017. Biogenesis and metabolic maintenance of rubisco. Annu. Rev. Plant. Biol. 68, 29–60.

Bukau, B., Horwich, A.L., 1998. The Hsp70 and Hsp60 chaperone machines. Cell 92, 351–366.

Carrillo, N., Ceccarelli, E.A., 2003. Open questions in ferredoxin-$NADP^+$ reductase catalytic mechanism. Eur. J. Biochem. 270, 1900–1915.

Cassier-Chauvat, C., Chauvat, F., 2014. Function and regulation of ferredoxins in the cyanobacterium, *Synechocystis* PCC6803: recent advances. Life 4.

Cavalier-Smith, T., 2010. Deep phylogeny, ancestral groups and the four ages of life. Philos. Trans. R. Soc. B Biol. Sci. 365, 111–132.

Ceccarelli, E.A., Arakaki, A.K., Cortez, N., Carrillo, N., 2004. Functional plasticity and catalytic efficiency in plant and bacterial ferredoxin-NADP(H) reductases. Biochim. Biophys. Acta. 1698, 155–165.

Ceccoli, R.D., Blanco, N.E., Medina, M., Carrillo, N., 2011. Stress response of transgenic tobacco plants expressing a cyanobacterial ferredoxin in chloroplasts. Plant. Mol. Biol. 76, 535–544.

Chen, L.M., Li, K.Z., Miwa, T., Izui, K., 2004. Overexpression of a cyanobacterial phosphoenolpyruvate carboxylase with diminished sensitivity to feedback inhibition in *Arabidopsis* changes amino acid metabolism. Planta 219, 440–449.

Chen, L.M., Omiya, T., Hata, S., Izui, K., 2002. Molecular characterization of a phosphoenolpyruvate carboxylase from a thermophilic cyanobacterium, *Synechococcus vulcanus* with unusual allosteric properties. Plant. Cell. Physiol 43, 159–169.

Coba de la Peña, T., Redondo, F.J., Manrique, E., Lucas, M.M., Pueyo, J.J., 2010. Nitrogen fixation persists under conditions of salt stress in transgenic *Medicago truncatula* plants expressing a cyanobacterial flavodoxin. Plant Biotechnol. J. 8, 954–965.

Cui, L., Liu, Y., Yang, Y., Ye, S., Luo, H., Qiu, B., Gao, X., 2018. The *drnf1* gene from the drought-adapted cyanobacterium *Nostoc flagelliforme* improved salt tolerance in transgenic *Synechocystis* and *Arabidopsis* plant. Genes. (Basel) 9, 441.

Czarnocka, W., Karpiński, S., 2018. Friend or foe? Reactive oxygen species production, scavenging and signaling in plant response to environmental stresses. Free Radic. Biol. Med. 122, 4–20.

DalCorso, G., Pesaresi, P., Masiero, S., Aseeva, E., Schünemann, D., Finazzi, G., Joliot, P., Barbato, R., Leister, D., 2008. A complex containing PGRL1 and PGR5 is involved in the switch between linear and cyclic electron flow in *Arabidopsis*. Cell 132, 273–285.

Dann, M., Leister, D., 2017. Enhancing (crop) plant photosynthesis by introducing novel genetic diversity. Philos. Trans. R. Soc. B. Biol. Sci. 372 (1730) Article ID:20160380.

Duffy, C.D.P., Ruban, A.V, 2015. Dissipative pathways in the photosystem-II antenna in plants. J. Photochem. Photobiol. B. Biol. 152, 215–226.

Erb, T.J., Zarzycki, J., 2018. A short history of RubisCO: the rise and fall (?) of nature's predominant CO_2 fixing enzyme. Curr. Opin. Biotechnol. 49, 100–107.

Erdner, D.L., Anderson, D.M., 1999. Ferredoxin and flavodoxin as biochemical indicators of iron limitation during open-ocean iron enrichment. Limnol. Oceanogr. 44, 1609–1615.

Ermakova, M., Battchikova, N., Allahverdiyeva, Y., Aro, E.-M., 2013. Novel heterocyst-specific flavodiiron proteins in *Anabaena* sp. PCC 7120. FEBS Lett 587, 82–87.

Éva, C., Oszvald, M., Tamás, L., 2019. Current and possible approaches for improving photosynthetic efficiency. Plant. Sci 280, 433–440.

Falk, J., Munné-Bosch, S., 2010. Tocochromanol functions in plants: antioxidation and beyond. J. Exp. Bot. 61, 1549–1566.

FAO, 2014. Food and Agricultural Organization of the United Nations Database. Rome.

Farooq, M., Wahid, A., Kobayashi, N., Fujita, D., Basra, S.M., 2009. Plant drought stress: effects, mechanisms and management. Agron. Sustain. Dev. 29, 185–212.

Foyer, C.H., 2018. Reactive oxygen species, oxidative signaling and the regulation of photosynthesis. Environ. Exp. Bot. 154, 134–142.

Foyer, C.H., Ruban, A.V, Noctor, G., 2017. Viewing oxidative stress through the lens of oxidative signalling rather than damage. Biochem. J. 474, 877–883.

Giró, M., Ceccoli, R.D., Poli, H.O., Carrillo, N., Lodeyro, A.F., 2011. An *in vivo* system involving co-expression of cyanobacterial flavodoxin and ferredoxin–$NADP^+$ reductase confers increased tolerance to oxidative stress in plants. FEBS Open Bio 1, 7–13.

Gómez, R., Carrillo, N., Morelli, M.P., Tula, S., Shahinnia, F., Hajirezaei, M.-R., Lodeyro, A.F., 2018. Faster photosynthetic induction in tobacco by expressing cyanobacterial flavodiiron proteins in chloroplasts. Photosynth. Res. 136, 129–138.

Gong, H.Y., Li, Y., Fang, G., Hu, D.H., Jin, W.Bin, Wang, Z.H., Li, Y.S., 2015. Transgenic rice expressing Ictb and FBP/Sbpase derived from cyanobacteria exhibits enhanced photosynthesis and mesophyll conductance to CO_2. PLoS One 10, e0140928.

Gong, H.Y., Li, Y., Fang, G., Hu, D.H., Wang, Z.H., Yu, G.H., Li, Y.S., 2017. *ccaA* and FBP/SBPase derived From cyanobacteria improved photosynthetic capacity in rice. Plant. Mol. Biol. Rep. 35, 619–633.

Guidi, L., Tattini, M., Landi, M., 2017. How does chloroplast protect chlorophyll against excessive light? In: Jacob-Lopes, E., Zepka, L.Q., Queiroz, M.I. (Eds.). InTechopen, London, pp. 22–26.

Hagemann, M., Fernie, A.R., Espie, G.S., Kern, R., Eisenhut, M., Reumann, S., Bauwe, H., Weber, A.P.M., 2013. Evolution of the biochemistry of the photorespiratory C2 cycle. Plant Biol 15, 639–647.

Hajirezaei, M.-R., Peisker, M., Tschiersch, H., Palatnik, J.F., Valle, E.M., Carrillo, N., Sonnewald, U., 2002. Small changes in the activity of chloroplastic $NADP^+$-dependent ferredoxin oxidoreductase lead to impaired plant growth and restrict photosynthetic activity of transgenic tobacco plants. Plant J 29, 281–293.

Hald, S., Nandha, B., Gallois, P., Johnson, G.N., 2008. Feedback regulation of photosynthetic electron transport by NADP(H) redox poise. Biochim. Biophys. Acta 1777, 433–440.

Harrison, E.P., Olcer, H., Lloyd, J.C., Long, S.P., Raines, C.A., 2001. Small decreases in SBPase cause a linear decline in the apparent RuBP regeneration rate, but do not affect rubisco carboxylation capacity. J. Exp. Bot. 52, 1779–1784.

Hase, T., Schürmann, P., Knaff, D.B., 2006. The interaction of ferredoxin with ferredoxin-dependent enzymes. In: Golbeck, J.H. (Ed.), Photosystem I: The Light-Driven Plastocyanin:Ferredoxin Oxidoreductase. Springer, Dordrecht, pp. 477–498.

Hay, W.T., Bihmidine, S., Mutlu, N., Hoang, K.Le, Awada, T., Weeks, D.P., Clemente, T.E., Long, S.P., 2017. Enhancing soybean photosynthetic CO_2 assimilation using a cyanobacterial membrane protein, ictB. J. Plant. Physiol. 212, 58–68.

He, Y., He, C., Li, L., Liu, Z., Yang, A., Zhang, J., 2011. Heterologous expression of ApGSMT2 and ApDMT2 genes from *Aphanothece halophytica* enhanced drought tolerance in transgenic tobacco. Mol. Biol. Rep. 38, 657–666.

Hennacy, J.H., Jonikas, M.C., 2020. Prospects for engineering biophysical CO_2 concentrating mechanisms into land plants to enhance yields. Annu. Rev. Plant Biol. 71, 461–485.

Holtgrefe, S., Bader, K.P., Horton, P., Scheibe, R., von Schaewen, A., Backhausen, J.E., 2003. Decreased content of leaf ferredoxin changes electron distribution and limits photosynthesis in transgenic potato plants. Plant Physiol 133, 1768–1778.

Ichikawa, Y., Tamoi, M., Sakuyama, H., Maruta, T., Ashida, H., Yokota, A., Shigeoka, S., 2010. Generation of transplastomic lettuce with enhanced growth and high yield. GM Crops 1, 322–326.

Jagannathan, B., Shen, G., Golbeck, J.H., 2012. The evolution of type I reaction centers: the response to oxygenic photosynthesis. In: Burnap, R., Vermaas, W. (Eds.), Functional Genomics and Evolution of Photosynthetic Systems. Springer, Dordrecht, pp. 285–316.

Janasch, M., Asplund-Samuelsson, J., Steuer, R., Hudson, E.P., 2019. Kinetic modeling of the Calvin cycle identifies flux control and stable metabolomes in *Synechocystis* carbon fixation. J. Exp. Bot. 70, 973–983.

Jensen, P.E., Leister, D., 2014. Chloroplast evolution, structure and functions. F1000Prime Rep 6, 40.

Kanesaki, Y., Suzuki, I., Allakhverdiev, S.I., Mikami, K., Murata, N., 2002. Salt stress and hyperosmotic stress regulate the expression of different sets of genes in *Synechocystis* sp. PCC 6803. Biochem. Biophys. Res. Commun. 290, 339–348.

Kim, J., Ahn, M.S., Park, Y.M., Kim, S.W., Min, S.R., Jeong, W.J., Liu, J.R., 2014. *Synechocystis* PCC6803 and PCC6906 dnaK2 expression confers salt and oxidative stress tolerance in *Arabidopsis* via reduction of hydrogen peroxide accumulation. Mol. Biol. Rep. 41, 1091–1101.

Kiseleva, L.L., Serebriiskaya, T.S., Horvàth, I., Vigh, L., Lyukevich, A.A., Los, D.A., 2000. Expression of the gene for the Δ9 acyl-lipid desaturase in the thermophilic cyanobacterium. J. Mol. Microbiol. Biotechnol. 2, 331–338.

Koester, R.P., Nohl, B.M., Diers, B.W., Ainsworth, E.A., 2016. Has photosynthetic capacity increased with 80 years of soybean breeding? An examination of historical soybean cultivars. Plant Cell Environ 39, 1058–1067.

Köhler, I.H., Ruiz-Vera, U.M., VanLoocke, A., Thomey, M.L., Clemente, T., Long, S.P., Ort, D.R., Bernacchi, C.J., 2017. Expression of cyanobacterial FBP/SBPase in soybean prevents yield depression under future climate conditions. J. Exp. Bot. 68, 715–726.

Kramer, D.M., Evans, J.R., 2011. The importance of energy balance in improving photosynthetic productivity. Plant. Physiol 155, 70–78.

Kubis, A., Bar-Even, A., 2019. Synthetic biology approaches for improving photosynthesis. J. Exp. Bot. 70, 1425–1433.

Lamaoui, M., Jemo, M., Datla, R., Bekkaoui, F., 2018. Heat and drought stresses in crops and approaches for their mitigation. Front. Chem. 6, 26.

Lee, S.K., Jeon, J.S., Börnke, F., Voll, L., Cho, J.Il, Goh, C.H., Jeong, S.W., Park, Y.Il, Kim, S.J., Choi, S.B., Miyao, A., Hirochika, H., An, G., Cho, M.H., Bhoo, S.H., Sonnewald, U., Hahn, T.R., 2008. Loss of cytosolic fructose-1,6-bisphosphatase limits photosynthetic sucrose synthesis and causes severe growth retardations in rice (*Oryza sativa*). Plant Cell Environ. 31, 1851–1863.

Leister, D., 2012. How can the light reactions of photosynthesis be improved in plants? Front. Plant. Sci. 3, 199.

Leister, D., 2019. Genetic engineering, synthetic biology and the light reactions of photosynthesis. Plant Physiol 179, 778–793.

Li, Z., Yuan, S., Jia, H., Gao, F., Zhou, M., Yuan, N., Wu, P., Hu, Q., Sun, D., Luo, H., 2017. Ectopic expression of a cyanobacterial flavodoxin in creeping bentgrass impacts plant development and confers broad abiotic stress tolerance. Plant Biotechnol. J. 15, 433–446.

Lieman-Hurwitz, J., Rachmilevitch, S., Mittler, R., Marcus, Y., Kaplan, A., 2003. Enhanced photosynthesis and growth of transgenic plants that express ictB, a gene involved in HCO_3^- accumulation in cyanobacteria. Plant Biotechnol. J. 1, 43–50.

Lin, M.T., Occhialini, A., Andralojc, P.J., Parry, M.A.J., Hanson, M.R., 2014. A faster rubisco with potential to increase photosynthesis in crops. Nature 513, 547–550.

Lodeyro, A.F., Ceccoli, R.D., Pierella Karlusich, J.J., Carrillo, N., 2012. The importance of flavodoxin for environmental stress tolerance in photosynthetic microorganisms and transgenic plants. Mechanism, evolution and biotechnological potential. FEBS Lett 586, 2917–2924.

Long, B.M., Hee, W.Y., Sharwood, R.E., Rae, B.D., Kaines, S., Lim, Y.-L., Nguyen, N.D., Massey, B., Bala, S., von Caemmerer, S., Badger, M.R., Price, G.D., 2018. Carboxysome encapsulation of the CO_2-fixing enzyme Rubisco in tobacco chloroplasts. Nat. Commun. 9, 3570.

Long, B.M., Rae, B.D., Rolland, V., Förster, B., Price, G.D., 2016. Cyanobacterial CO_2-concentrating mechanism components: function and prospects for plant metabolic engineering. Curr. Opin. Plant Biol. 31, 1–8.

López-Calcagno, P.E., Brown, K.L., Simkin, A.J., Fisk, S.J., Vialet-Chabrand, S., Lawson, T., Raines, C.A., 2020. Stimulating photosynthetic processes increases productivity and water-use efficiency in the field. Nat. Plants 6, 1054–1063.

Lu, Y., Rijzaani, H., Karcher, D., Ruf, S., Bock, R., 2013. Efficient metabolic pathway engineering in transgenic tobacco and tomato plastids with synthetic multigene operons. Proc. Natl. Acad. Sci. USA. 110, 623–632.

Mazouni, K., Domain, F., Chauvat, F., Cassier-Chauvat, C., 2003. Expression and regulation of the crucial plant-like ferredoxin of cyanobacteria. Mol. Microbiol. 49, 1019–1029.

Miyagawa, Y., Tamoi, M., Shigeoka, S., 2001. Overexpression of a cyanobacterial fructose-1,6-/sedoheptulose-1, 7-bisphosphatase in tobacco enhances photosynthesis and growth. Nat. Biotechnol. 19, 965–969.

Munekage, Y., Hojo, M., Meurer, J., Endo, T., Tasaka, M., Shikanai, T., 2002. PGR5 is involved in cyclic electron flow around photosystem I and is essential for photoprotection in *Arabidopsis*. Cell 110, 361–371.

Muñiz García, M.N., Cortelezzi, J.I., Fumagalli, M., Capiati, D.A., 2018. Expression of the *Arabidopsis ABF4* gene in potato increases tuber yield, improves tuber quality and enhances salt and drought tolerance. Plant. Mol. Biol. 98, 137–152.

Nowicka, B., Ciura, J., Szymańska, R., Kruk, J., 2018. Improving photosynthesis, plant productivity and abiotic stress tolerance—current trends and future perspectives. J. Plant. Physiol. 231, 415–433.

O'Leary, B., Park, J., Plaxton, W.C., 2011. The remarkable diversity of plant PEPC (phosphoenolpyruvate carboxylase): recent insights into the physiological functions and post-translational controls of non-photosynthetic PEPCs. Biochem. J. 436, 15–34.

Occhialini, A., Lin, M.T., Andralojc, P.J., Hanson, M.R., Parry, M.A.J., 2016. Transgenic tobacco plants with improved cyanobacterial Rubisco expression but no extra assembly factors grow at near wild-type rates if provided with elevated CO_2. Plant J 85, 148–160.

Ono, K., Hibino, T., Kohinata, T., Suzuki, S., Tanaka, Y., Nakamura, T., Takabe, T., Takabe, T., 2001. Overexpression of DnaK from a halotolerant cyanobacterium *Aphanothece halophytica* enhances the high-temperature tolerance of tobacco during germination and early growth. Plant Sci 160, 455–461.

Orlova, I.V., Serebriiskaya, T.S., Popov, V., Merkulova, N., Nosov, A.M., Trunova, T.I., Tsydendambaev, V.D., Los, D.A., 2003. Transformation of tobacco with a gene for the thermophilic acyl-lipid desaturase enhances the chilling tolerance of plants. Plant. Cell. Physiol 44, 447–450.

Orr, D.J., Pereira, A.M., da Fonseca Pereira, P., Pereira-Lima, Í.A., Zsögön, A., Araújo, W.L., 2017. Engineering photosynthesis: progress and perspectivesF1000Research6, 1891.

Ort, D.R., Merchant, S.S., Alric, J., Barkan, A., Blankenship, R.E., Bock, R., Croce, R., Hanson, M.R., Hibberd, J.M., Long, S.P., Moore, T.A., Moroney, J., Niyogi, K.K., Parry, M.A.J., Peralta-Yahya, P.P., Prince, R.C., Redding, K.E., Spalding, M.H., Van Wijk, K.J., Vermaas, W.F.J., Von Caemmerer, S., Weber, A.P.M., Yeates, T.O., Yuan, J.S., Zhu, X.G., 2015. Redesigning photosynthesis to sustainably meet global food and bioenergy demand. Proc. Natl. Acad. Sci. USA. 112, 8529–8536.

Otori, K., Tamoi, M., Tanabe, N., Shigeoka, S., 2017. Enhancements in sucrose biosynthesis capacity affect shoot branching in *Arabidopsis*. Biosci. Biotechnol. Biochem. 81, 1470–1477.

Palatnik, J.F., Tognetti, V.B., Poli, H.O., Rodríguez, R.E., Blanco, N., Gattuso, M., Hajirezaei, M.-R., Sonnewald, U., Valle, E.M., Carrillo, N., 2003. Transgenic tobacco plants expressing antisense ferredoxin-NADP(H) reductase transcripts display increased susceptibility to photo-oxidative damage. Plant J 35, 332–341.

Peden, E.A., Boehm, M., Mulder, D.W., Davis, R., Old, W.M., King, P.W., Ghirardi, M.L., Dubini, A., 2013. Identification of global ferredoxin interaction networks in *Chlamydomonas reinhardtii*. J. Biol. Chem. 288, 35192–35209.

Petracek, M.E., Dickey, L.F., Nguyen, T.T., Gatz, C., Sowinski, D.A., Allen, G.C., Thompson, W.F., 1998. Ferredoxin-1 mRNA is destabilized by changes in photosynthetic electron transport. Proc. Natl. Acad. Sci. USA 95, 9009–9013.

Pierella Karlusich, J.J., Arce, R.C., Shahinnia, F., Sonnewald, S., Sonnewald, U., Zurbriggen, M.D., Hajirezaei, M.R., Carrillo, N., 2020. Transcriptional and metabolic profiling of potato plants expressing a plastid-targeted electron shuttle reveal modulation of genes associated to drought tolerance by chloroplast redox poise. Int. J. Mol. Sci. 21, 1–22.

Pierella Karlusich, J.J., Lodeyro, A.F., Carrillo, N., 2014. The long goodbye: the rise and fall of flavodoxin during plant evolution. J. Exp. Bot. 65, 5161–5178.

Price, G.D., Badger, M.R., von Caemmerer, S., 2011. The prospect of using cyanobacterial bicarbonate transporters to improve leaf photosynthesis in crop plants. Plant Physiol 155, 20–26.

Price, G.D., Pengelly, J.J.L., Forster, B., Du, J., Whitney, S.M., Von Caemmerer, S., Badger, M.R., Howitt, S.M., Evans, J.R., 2013. The cyanobacterial CCM as a source of genes for improving photosynthetic CO_2 fixation in crop species. J. Exp. Bot. 64, 753–768.

Quick, W.P., Neuhaus, H.E., 1997. A molecular approach to primary metabolism in higher plants. In: Foyer, C.H., Quick, W.P. (Eds.), The Regulation and Control of Photosynthetic Carbon Assimilation1st edn. Taylor and Francis Ltd, London, pp. 41–62.

Rachedi, R., Foglino, M., Latifi, A., 2020. Stress signaling in cyanobacteria: a mechanistic overview. Life 10, 312.

Rae, B.D., Long, B.M., Badger, M.R., Price, G.D., 2013. Functions, compositions, and evolution of the two types of carboxysomes: polyhedral microcompartments that facilitate CO_2 fixation in cyanobacteria and some proteobacteria. Microbiol. Mol. Biol. Rev. 77, 357–379.

Rae, B.D., Long, B.M., Förster, B., Nguyen, N.D., Velanis, C.N., Atkinson, N., Hee, W.Y., Mukherjee, B., Price, G.D., McCormick, A.J., 2017. Progress and challenges of engineering a biophysical CO_2-concentrating mechanism into higher plants. J. Exp. Bot. 68, 3717–3737.

Rojas-González, J.A., Soto-Súarez, M., García-Díaz, Á., Romero-Puertas, M.C., Sandalio, L.M., Mérida, Á., Thormählen, I., Geigenberger, P., Serrato, A.J., Sahrawy, M., 2015. Disruption of both chloroplastic and cytosolic FBPase genes results in a dwarf phenotype and important starch and metabolite changes in *Arabidopsis thaliana*. J. Exp. Bot. 66, 2673–2689.

Rossi, F.R., Krapp, A.R., Bisaro, F., Maiale, S.J., Pieckenstain, F.L., Carrillo, N., 2017. Reactive oxygen species generated in chloroplasts contribute to tobacco leaf infection by the necrotrophic fungus *Botrytis cinerea*. Plant. J 92, 761–773.

Rupprecht, E., Gathmann, S., Fuhrmann, E., Schneider, D., 2007. Three different DnaK proteins are functionally expressed in the cyanobacterium *Synechocystis* sp. PCC. 6803. Microbiology 153, 1828–1841.

Sancho, J., 2006. Flavodoxins: sequence, folding, binding, function and beyond. Cell Mol. Life. Sci. C. 63, 855–864.

Santana-Sánchez, A., Solymosi, D., Mustila, H., Bersanini, L., Aro, E.-M., Allahverdiyeva, Y., 2019. Flavodiiron proteins 1–to-4 function in versatile combinations in O_2 photoreduction in cyanobacteria. Elife 8, e45766.

Scheller, H.V., 1996. In vitro cyclic electron transport in barley thylakoids follows two independent pathways. Plant Physiol 110, 187–194.

Serrato, A.J., De Dios Barajas-López, J., Chueca, A., Sahrawy, M., 2009. Changing sugar partitioning in FBPase-manipulated plants. J. Exp. Bot. 60, 2923–2931.

Shahinnia, F., Tula, S., Hensel, G., Reiahisamani, N., Nasr, N., Kumlehn, J., Gómez, R., Lodeyro, A.F., Carrillo, N., Hajirezaei, M.R., 2021. Plastid-targeted cyanobacterial flavodiiron proteins maintain carbohydrate turnover and enhance drought stress tolerance in barley. Front. Plant. Sci. 11, 613731.

Sharkey, T.D., 1985. Photosynthesis in intact leaves of C3 plants: physics, physiology and rate limitations. Bot. Rev. 51, 53–105.

Simkin, A.J., McAusland, L., Headland, L.R., Lawson, T., Raines, C.A., 2015. Multigene manipulation of photosynthetic carbon assimilation increases CO_2 fixation and biomass yield in tobacco. J. Exp. Bot. 66, 4075–4090.

Song, J., Zhang, R., Yue, D., Chen, X., Guo, Z., Cheng, C., Hu, M., Zhang, J., Zhang, K., 2018. Co-expression of ApGSMT2g and ApDMT2g in cotton enhances salt tolerance and increases seed cotton yield in saline fields. Plant. Sci 274, 369–382.

Strand, Å., Zrenner, R., Trevanion, S., Stitt, M., Gustafsson, P., Gardeström, P., 2000. Decreased expression of two key enzymes in the sucrose biosynthesis pathway, cytosolic fructose-1,6-bisphosphatase and sucrose phosphate synthase, has remarkably different consequences for photosynthetic carbon metabolism in transgenic *Arabidopsis thaliana*. Plant. J 23, 759–770.

Su, J., Yang, L., Zhu, Q., Wu, H., He, Y., Liu, Y., Xu, J., Jiang, D., Zhang, S., 2018. Active photosynthetic inhibition mediated by MPK3/MPK6 is critical to effector-triggered immunity. PLoS Biol 16, e2004122.

Takahashi, S., Badger, M.R., 2011. Photoprotection in plants: a new light on photosystem II damage. Trends Plant Sci 16, 53–60.

Tamoi, M., Hiramatsu, Y., Nedachi, S., Otori, K., Tanabe, N., Maruta, T., Shigeoka, S., 2011. Increase in the activity of fructose-1,6-bisphosphatase in cytosol affects sugar partitioning and increases the lateral shoots in tobacco plants at elevated CO_2 levels. Photosynth. Res. 108, 15–23.

Tamoi, M., Nagaoka, M., Miyagawa, Y., Shigeoka, S., 2006. Contribution of Fructose-1,6-bisphosphatase and Sedoheptulose-1,7-bisphosphatase to the photosynthetic rate and carbon flow in the Calvin cycle in transgenic plants. Plant Cell Physiol 47, 380–390.

Tilman, D., Balzer, C., Hill, J., Befort, B.L., 2011. Global food demand and the sustainable intensification of agriculture. Proc. Natl. Acad. Sci. USA. 108, 20260–20264.

Tognetti, V.B, Monti, M.R., Valle, E.M., Carrillo, N., Smania, A.M., 2007a. Detoxification of 2,4-dinitrotoluene by transgenic tobacco plants expressing a bacterial flavodoxin. Environ. Sci. Technol. 41, 4071–4076.

Tognetti, V.B., Palatnik, J.F., Fillat, M.F., Melzer, M., Hajirezaei, M.R., Valle, E.M., Carrillo, N., 2006. Functional replacement of ferredoxin by a cyanobacterial flavodoxin in tobacco confers broad-range stress tolerance. Plant Cell 18, 2035–2050.

Tognetti, V.B., Zurbriggen, M.D., Morandi, E.N., Fillat, M.F., Valle, E.M., Hajirezaei, M.-R., Carrillo, N., 2007b. Enhanced plant tolerance to iron starvation by functional substitution of chloroplast ferredoxin with a bacterial flavodoxin. Proc. Natl. Acad. Sci. USA 104, 11495–11500.

Tula, S., Shahinnia, F., Melzer, M., Rutten, T., Gómez, R., Lodeyro, A.F., von Wirén, N., Carrillo, N., Hajirezaei, M.R., 2020. Providing an additional electron sink by the introduction of cyanobacterial flavodiirons enhances growth of *A. thaliana* under various light intensities. Front. Plant Sci. 11, 902.

Vasquez-Robinet, C., Mane, S.P., Ulanov, A.V, Watkinson, J.I., Stromberg, V.K., De Koeyer, D., Schafleitner, R., Willmot, D.B., Bonierbale, M., Bohnert, H.J., Grene, R., 2008. Physiological and molecular adaptations to drought in Andean potato genotypes. J. Exp. Bot. 59, 2109–2123.

Vicente, J.B., Gomes, C.M., Wasserfallen, A., Teixeira, M., 2002. Module fusion in an A-type flavoprotein from the cyanobacterium *Synechocystis* condenses a multiple-component pathway in a single polypeptide chain. Biochem. Biophys. Res. Commun. 294, 82–87.

Vicente, J.B., Justino, M.C., Gonçalves, V.L., Saraiva, L.M., Teixeira, M., 2008. Chapter 2 - Biochemical, spectroscopic, and thermodynamic properties of flavodiiron proteins. In: Poole, R.K.B.T.-M. (Ed.), Globins and Other Nitric Oxide-Reactive Proteins. Academic Press, Cambridge, MA, pp. 21–45 Part B.

Vicino, P., Carrillo, J., Gómez, R., Shahinnia, F., Tula, S., Melzer, M., Rutten, T., Carrillo, N., Hajirezaei, M.-R., Lodeyro, A.F., 2021. Expression of flavodiiron proteins Flv2-Flv4 in chloroplasts of *Arabidopsis* and tobacco plants provides multiple stress tolerance. Int. J. Mol. Sci. 22, 1178.

Visser, R.G.F., Bachem, C.W.B., de Boer, J.M., Bryan, G.J., Chakrabati, S.K., Feingold, S., Gromadka, R., van Ham, R.C.H.J., Huang, S., Jacobs, J.M.E., Kuznetsov, B., de Melo, P.E., Milbourne, D., Orjeda, G., Sagredo, B., Tang, X., 2009. Sequencing the potato genome: outline and first results to come from the elucidation of the sequence of the world's third most important food crop. Am. J. Potato Res. 86, 417.

Wada, S., Yamamoto, H., Suzuki, Y., Yamori, W., Shikanai, T., Makino, A., 2018. Flavodiiron protein substitutes for cyclic electron flow without competing for CO_2 assimilation in rice. Plant. Physiol 176, 1509–1518.

Waditee, R., Bhuiyan, M.N.H., Rai, V., Aoki, K., Tanaka, Y., Hibino, T., Suzuki, S., Takano, J., Jagendorf, A.T., Takabe, Tetsuko, Takabe, Teruhiro, 2005. Genes for direct methylation of glycine provide high levels of glycine betaine and abiotic-stress tolerance in *Synechococcus* and *Arabidopsis*. Proc. Natl. Acad. Sci. USA., 102, 1318–1323.

Waditee, R., Bhuiyan, N.H., Hirata, E., Hibino, T., Tanaka, Y., Shikata, M., Takabe, T., 2007. Metabolic engineering for betaine accumulation in microbes and plants. J. Biol. Chem. 282, 34185–34193.

Waditee, R., Tanaka, Y., Aoki, K., Hibino, T., Jikuya, H., Takano, J., Takabe, Tetsuko, Takabe, Teruhiro, 2003. Isolation and functional characterization of N-methyltransferases that catalyze betaine synthesis from glycine in a halotolerant photosynthetic organism *Aphanothece halophytica*. J. Biol. Chem. 278, 4932–4942.

Wang, C., Yamamoto, H., Shikanai, T., 2015. Role of cyclic electron transport around photosystem I in regulating proton motive force. Biochim. Biophys. Acta., 1847, 931–938.

Whitney, S.M., Houtz, R.L., Alonso, H., 2011. Advancing our understanding and capacity to engineer nature's CO_2-sequestering enzyme. Rubisco. Plant Physiol. 155, 27–35.

Wu, A., Allu, A.D., Garapati, P., Siddiqui, H., Dortay, H., Zanor, M.-I., Asensi-Fabado, M.A., Munné-Bosch, S., Antonio, C., Tohge, T., Fernie, A.R., Kaufmann, K., Xue, G.-P.,

Mueller-Roeber, B., Balazadeh, S., 2012. JUNGBRUNNEN1, a reactive oxygen species–responsive NAC transcription factor, regulates longevity in *Arabidopsis*. Plant Cell 24, 482–506.

Yadav, S., Modi, P., Dave, A., Vijapura, A., Patel, D., Patel, M., 2020. Effect of abiotic stress on crops. In: Hasanuzzaman, M., Fujita, M., Filho, M.C.M.T., Nogueira, T.A.R. (Eds.), Sustainable Crop Production. IntechOpen, London 3–24.

Yamamoto, H., Takahashi, S., Badger, M.R., Shikanai, T., 2016. Artificial remodelling of alternative electron flow by flavodiiron proteins in *Arabidopsis*. Nat. Plants 2, 16012.

Yamori, W., Shikanai, T., 2016. Physiological functions of cyclic electron transport around photosystem I in sustaining photosynthesis and plant growth. Annu. Rev. Plant. Biol. 67, 81–106.

Yang, S.-M., Chang, C.-Y., Yanagisawa, M., Park, I., Tseng, T.-H., Ku, M.S.B., 2008. Transgenic rice expressing cyanobacterial bicarbonate transporter exhibited enhanced photosynthesis, growth and grain yield. In: Allen, J.F., Gantt, E., Golbeck, J.H., Osmond, B. (Eds.), Photosynthesis. Energy from the Sun. Springer, Cham, pp. 1243–1246.

Ye, S., Gao, X., 2015. Excavating abiotic stress-related gene resources of terrestrial macroscopic cyanobacteria for crop genetic engineering: dawn and challenge. Bioengineered 6, 313–315.

Zarzycki, J., Axen, S.D., Kinney, J.N., Kerfeld, C.A., 2013. Cyanobacterial-based approaches to improving photosynthesis in plants. J. Exp. Bot. 64, 787–798.

Zhang, P., Allahverdiyeva, Y., Eisenhut, M., Aro, E.-M., 2009. Flavodiiron proteins in oxygenic photosynthetic organisms: photoprotection of photosystem II by Flv2 and Flv4 in *Synechocystis* sp. PCC 6803. PLoS One 4, e5331.

Zhu, C., Gerjets, T., Sandmann, G., 2007. *Nicotiana glauca* engineered for the production of ketocarotenoids in flowers and leaves by expressing the cyanobacterial *crtO* ketolase gene. Transgenic Res 16, 813–821.

Zhu, X.-G., de Sturler, E., Long, S.P., 2007. Optimizing the distribution of resources between enzymes of carbon metabolism can dramatically increase photosynthetic rate: a numerical simulation using an evolutionary algorithm. Plant Physiol 145, 513–526.

Zhu, X.-G., Long, S.P., Ort, D.R., 2010. Improving photosynthetic efficiency for greater yield. Annu. Rev. Plant. Biol. 61, 235–261.

Zimmermann, P., Hirsch-Hoffmann, M., Hennig, L., Gruissem, W., 2004. GENEVESTIGATOR. *Arabidopsis* microarray database and analysis toolbox. Plant Physiol 136, 2621–2632.

Zurbriggen, M.D., Carrillo, N., Tognetti, V.B., Melzer, M., Peisker, M., Hause, B., Hajirezaei, M.-R., 2009. Chloroplast-generated reactive oxygen species play a major role in localized cell death during the non-host interaction between tobacco and *Xanthomonas campestris* pv. vesicatoria. Plant J 60, 962–973.

Zurbriggen, M.D., Tognetti, V.B., Fillat, M.F., Hajirezaei, M.-R., Valle, E.M., Carrillo, N., 2008. Combating stress with flavodoxin: a promising route for crop improvement. Trends Biotechnol 26, 531–537.

CHAPTER

5

Promises and challenges for expanding the use of N_2-fixing cyanobacteria as a fertilizer for sustainable agriculture

Rafael Ambrosio[a,b], *Lara Sanchez Rizza*[a,b], *Mauro Do Nascimento*[a,b], *Hugo Guillermo Jimenez Pacheco*[c], *Lilia Mary Miranda Ramos*[d], *Jose A. Hernandez*[e,f], *Leonardo Curatti*[a,b]

[a]Instituto de Investigaciones en Biodiversidad y Biotecnología (INBIOTEC-CONICET), Mar del Plata, Argentina
[b]Fundación para Investigaciones Biológicas Aplicadas, Argentina
[c]Universidad Católica de Santa María, Arequipa, Perú
[d]Universidad Nacional de San Agustín de Arequipa, Arequipa, Perú
[e]Department of Biochemistry and Molecular Genetics, College of Graduate Studies, Midwestern University, Glendale, AZ, United States
[f]Arizona College of Osteopathic Medicine, Midwestern University, Glendale, AZ, United States

5.1 Food security, sustainable agriculture, and N-fertilizers

The Second Sustainable Development Goal (SDG) of United Nations "End hunger, achieve food security and improved nutrition and promote sustainable agriculture" by 2030 represents one of the most challenging targets for Humanity of the present decade (Griggs, 2013; UN SDGs, 2019). Although hunger declined steadily during the last decades, the global undernourishment trend reverted in 2015 and remained globally at a level slightly below 11%. Dangerously, last year's assessments showed

Cyanobacterial Lifestyle and its Applications in Biotechnology.
DOI: https://doi.org/10.1016/B978-0-32-390634-0.00002-0

a significant rise in the prevalence of undernourishment in almost all subregions of Africa and, to a lesser extent, in Western Asia and Latin America, and the Caribbean, with a prevalence of undernourishment close to 20%, 12%, and 7%, respectively. In 2018, more than 820 million people in the world suffered from hunger and about 2 billion people in the world experienced moderate or severe food insecurity (FAO, 2019). This situation might be severely worsened by demographic projections toward the next decades. The world's population is projected to grow from 7.7 billion in 2019 to 8.5 billion in 2030 (10% increase), and further to 9.7 billion in 2050 (26%) and 10.9 billion in 2100 (42%). The population of sub-Saharan Africa is projected to double by 2050 (99%) (FAO, IFAD, UNICEF, WFP and WHO, 2019). Africa alone will contribute to ~58% of the world's population growth by 2050, representing roughly one quarter of the world population (Ciceri and Allanore, 2019).

Thus, calculations anticipated that to feed the world in 2050, global food production may have to increase by 60–110%. This will require a very large boost in food productivity mostly by the 2.5 billion smallholders that control 500 million small farms and which provide up to 80% of the food supply in Asia and sub-Saharan Africa (Rockström *et al.*, 2017).

Promotion of sustainable agriculture is an integral part of the second SDG of the United Nations. Slow or inexistent economic growth, social and political conflicts and instability, and climate change led to major shifts in the food production systems, especially in agriculture, posing new challenges for the next decades (FAO, 2019).

One of the most damaging consequences of climate change is the increase in frequency and intensity of climatic events, such as waves of intense heat or cold, droughts, heavy rains and floods, tropical cyclones and storms, and forest fires (WMO, 2020). Food insecurity is closely associated with drought. The number of undernourished people in drought-sensitive countries has increased by 45.6% since 2012 (FAO, 2019).

Both sustainable intensification and extensification of agriculture will be needed to fill the projected gap of agriculture output toward future food security. Approximately 40% of the world's terrestrial surface has already been transformed into agricultural land. There is a small margin for expanding agriculture toward fertile/humid areas, which currently provide critical ecosystem services related to the preservation of biodiversity, being prominent carbon sinks. These spaces contribute actively to lessen the trend of climate change, maintaining these and other planetary livelihood conditions within the "safe operating space" of a stable and the resilient earth system (Steffen *et al.*, 2015). On the other hand, expanding crop-production toward arid lands is more appealing but extremely challenging (Kleijn *et al.*, 2019). This is mostly because water and nutrients colimit plants growth and crop yields. While poor soil fertility limits the ability of plants to efficiently use water by enabling the development of a healthy root system, soil water content is the single most important factor controlling nutrient availability to plants (Drechsel *et al.*, 2014). While the world's cultivated area has grown by 12% over the last 50 years, the global irrigated area has doubled, and fertilizer use has increased more than fivefold. This, together with the adoption of improved seeds and best management practices, enabled a 2.5- to 3-fold increase in the yields of major crops (Drechsel *et al.*, 2014). Irrigated agriculture accounts for 20% of the total cultivated land and contributes 40% of the total food produced worldwide, consuming 70% of all water withdrawn from aquifers, streams, and lakes (FAO, 2019).

It is expected that fertilizers will be a critical player to fill the productivity gaps across regions of the world (Ciceri and Allanore, 2019). The world average application of N, P_2O_5, and K_2O in 2015 was of about 70, 30, and 24 kg/ha, respectively. In contrast, in Africa these fertilizers were much less used at an average rate of about 15, 5, and 2.5 kg/ha, respectively, while in some regions were not applied at all (Drechsel *et al.*, 2015). While the world average rate of fertilizer application raised almost steadily during the last decades, it remained almost the same in Africa. On the other hand, the average rate of application in Asia is currently by far the largest in the world and increased markedly above the world's average since the last decades (Drechsel *et al.*, 2015; Ciceri and Allanore, 2019). A recent forecast toward 2022 suggested a potential negative balance (the difference between supply and demand) for all the three main fertilizers in South Asia and Latin America and the Caribbean, and a negative balance for at least one of the fertilizers in most regions of the world (FAO, 2020).

Taking advantage of fertilizers application to boost crop yields is currently problematic in most poor regions of the world. Dependence on imported fertilizers at international prices, lack of credit to finance fertilizer purchases, inadequate infrastructure, high cost of transportation from production sites to farms, insufficient knowledge of proper fertilizer management, among other reasons, broaden the gap between current and potential demand of fertilizer in these regions. All these factors contribute to making commercial fertilizers pay minimal dividends for most subsistence farmers (Ciceri and Allanore, 2019).

Fertilizers are produced from natural resources, which are slowly recycled in the biogeochemical cycles of the planet, making its availability highly susceptible to depletion. P and K are mined from rock or sediments, whose reserves are unevenly distributed across regions (Manning, 2010; Pufahl and Groat, 2017). Thus, demand impacts the price of these commodities (Ciceri and Allanore, 2019).

Synthetic N fertilizer is mostly obtained using the Haber–Bosch process, in which atmospheric N_2 is reacted with H_2 obtained by steam reformation of natural gas (CH_4) at very high pressure and temperature over an iron catalyst to produce ammonia. The carbon footprint of this industrial process represents around 1.5% of all greenhouse gas emissions and 3–5% of the world's natural gas production, which corresponds to about 1–2% of the world's annual energy supply. Cost of N-fertilizer is strongly influenced by the price of oil and natural gas. Globally, the industry produces around 500 Mt/year of N-containing fertilizer with a value of about $250 billion (Wang *et al.*, 2018).

Conversely to the impact of low fertilization on food security, in other regions of the world such as Asia, overuse of N and P fertilizers, and/or improper management related to the timing of application and/or wrong balance between fertilizers and water, constitutes one of most impacted parameters of environmental deterioration that already surpassed the boundary of a stable and the resilient earth system (Steffen *et al.*, 2015). Global data indicated current fertilizer efficiencies for major cereal crops in the first year of application of 40–65% for N, 30–50% for K, and 15–25% for P. However, while applied P remains available to crops over long periods, often for a decade or longer, up to 70–80% of the added N can be lost in rain-fed conditions when not properly managed, and up to 60–70% in irrigated fields (Drechsel *et al.*, 2015). Approximately 40% of the N-fertilizer that is lost to the environment is denitrified back to unreactive atmospheric N_2, with the sole damage of a large futile waste of the energy use for its synthesis from N_2 and H_2. In addition, partial denitrification produces N_2O, a greenhouse gas with warming potential 296 times higher than an equal mass of CO_2 for a large impact on climate change. Other species of reactive N reach water reservoirs and aquifers, reducing the quality of drinking water and promoting eutrophication and changes in biodiversity. Paradoxically, most of these unintended effects of over-fertilization contribute largely to endanger food security (Sutton *et al.*, 2011; Steffen *et al.*, 2015; Rockström *et al.*, 2020).

Balanced fertilization is an essential practice for improving nutrient and water use efficiency. Conversely, in situations when this is not attained, due to lack of knowledge, financial capacity, or logistic constraints, fertilizers are not only wasted but converted into an environmental hazard (Drechsel *et al.*, 2015).

Many alternatives for sustainable fertilization, both affordable and non- or less-contaminant, are currently in use or under accelerated research and development. Some of these alternatives represent a renewed interest in the promotion of ancestral practices that are more recently conceptualized as "circular economy" (Schroeder *et al.*, 2019), such as recycling of nutrients in manure (Awasthi *et al.*, 2019) and other solid organic waste (Chen *et al.*, 2020), in domestic (Chojnacka *et al.*, 2020) or industrial wastewater, for example, fermentation vinasse (Prado *et al.*, 2013). These alternatives would comprise direct application to more or less sophisticatedly treated residues to enhance its fertilizing and soil conditioner performances. Excellent reviews of these alternatives are already available and will not be further described in this chapter.

5.2 Biological nitrogen fixation

There is a rising interest in finding alternative ways of reducing atmospheric N_2 to ammonia, mainly for its use as fertilizer, but also for other purposes. Progress was made to substitute H_2O electrolysis for CH_4 steam-reformation as the source of H_2 for ammonia synthesis. This promising technology would produce ammonia directly from N_2 and H_2O powered by sunlight or any other source of clean and renewable energy (Wang *et al.*, 2018; MacFarlane *et al.*, 2020).

Atmospheric N_2 is naturally reduced to ammonia during biological N_2 fixation (BNF). Global BNF contributes about 351 Tg of reactive N to terrestrial and marine ecosystems annually. About 17% of the global BNF corresponds to an anthropogenic component promoted for agricultural purposes and about 120 Tg corresponds to the Haber–Bosch process (Zhang *et al.*, 2020).

All known N_2-fixing organisms (diazotrophs) correspond to the bacterial or archaeal domains of life. Nonetheless, some eukaryotes (including plants and algae) engage in different kinds of associations with N_2-fixing bacteria, from commensalistic to symbiotic, that enable indirect access to atmospheric N_2 (Zehr and Capone, 2020). Last years' multidisciplinary analysis uncovered a much more prominent role of marine N_2-fixing cyanobacteria in the global N-cycle for an almost even N-balance between N_2-fixation and denitrification in marine environments (Zehr and Capone, 2020; Zhang *et al.*, 2020).

The global BNF correlated with agriculture represents about 60 Tg N/year, most of it corresponding to symbiotic BNF between legume crops and rhizobial bacteria (Herridge *et al.*, 2008). On one hand, the most bulk staple food for humans does not correspond to this group of plants, but cereals. On the other hand, specific legumes and rhizobia only engage in a mutualistic cross-feeding relationship after a sophisticated and specialized cross-talk of molecular signals, preventing a more generalized use of symbiotic BNF in agriculture (Mus *et al.*, 2016). In nonlegume crops, BNF occurs through less stable interactions with a variety of diazotrophic bacteria for a significant contribution to the geochemical cycle of N in natural environments (Zhang *et al.*, 2020), but a minor N-input to intensive agriculture (Herridge *et al.*, 2008; Mus *et al.*, 2016).

There is currently a variety of nonmutually exclusively approaches to increase the share of BNF in the agriculture of nonlegume crops: (1) engineering artificial symbioses between cereals and N_2-fixing bacteria, which would mimic the legume-rhizobium symbiosis (Mus *et al.*, 2016; Bailey-Serres *et al.*, 2019); (2) enhancement of N_2-fixing bacterial endophytes that are naturally associated to cereals (Mus *et al.*, 2016; Bloch *et al.*, 2020); (3) direct transfer of bacterial *nif* genes into cereals and other crop plants (Curatti and Rubio, 2014; Mus *et al.*, 2016; Burén and Rubio, 2018). Much progress has been made in the last few years in all of these challenging alternatives, which would entail more or less extreme genetic engineering of either plants or bacteria. These are very promising strategies for which the reader is referred to those comprehensive reviews cited above, and are no longer going to be analyzed further in this chapter.

In this chapter, we will analyze the prospects of a controlled eutrophication approach, originally proposed more than three decades ago (Benemann, 1979), in view of the latest findings. This strategy would concentrate nutrients from industrial or domestic wastewater into the biomass of microalgae or cyanobacteria, which might be directly used as an organic fertilizer and soil conditioner. Use of N_2-fixing cyanobacteria would further increase the input of N in a process mostly supported by sunlight. This "circular economy" strategy would integrate and improve wastewater reclamation and nutrients recycling as fertilizer and soil conditioner.

5.3 Cyanobacterial BNF

The major biological building blocks of all living beings on the earth consist of the elements carbon, hydrogen, nitrogen (N_2), oxygen, phosphorus, and sulfur. A set of chemical reactions ensure the movement of those elements within complex biogeochemical cycles (Ciais *et al.*, 2013). Several pathways of assimilation by cells have been described for most of them. For example, six alternative autotrophic pathways exist in prokaryotes for the fixation of inorganic carbon into organic material (Fuchs, 2011; Berg, 2011). However, there is only one entry point for the capture of N_2 (g): the reduction of N_2 to ammonia catalyzed by the nitrogenase enzyme. Hence, nitrogenase represents a limiting step, as it is the only evolutionary solution of nature for the conversion of N_2 to NH_3 (Allen *et al.*, 2019).

BNF is the conversion of N_2 into NH_3, which can be subsequently utilized by other organisms, including fungi, plants, and animals (Hernandez *et al.* 2009; Curatti *et al.*, 2014). Even though N_2 (g) represents 78% (w/v) of the earth's atmosphere, its low reactivity limits its transformation into fixed N_2 molecules. Hence, the global balance of N_2 on the earth's biosphere depends on nitrogenase activity to supply fixed N_2 into the ecosystems (Falkowski, 1997).

The ability to fix N_2 at moderate temperature and atmospheric pressure is limited to a special group of prokaryotic organisms (bacteria and archaea) termed diazotrophs. There are no N_2-fixing eukaryotes, although remarkable progress to create diazotrophic eukaryotes of important agricultural relevance has been carried out in the last decade (Curatti and Rubio, 2014; Burén and Rubio, 2018; Burén *et al.*, 2018).

5.3.1 Evolutionary origins

Cyanobacteria resemble a vast, heterogeneous group that can perform wide, complex, and diverse metabolic activities that diverged from a common ancestor shared with Firmicutes and Proteobacteria (Soo *et al.*, 2014) and provided an evolutionary context in which cyanobacteria evolved as a distinct group. This transition marked the development of oxygenic photosynthesis, allowing cyanobacteria to use water as an unrestricted source of electrons for the reduction of CO_2 and radically changed the composition of the earth surface (Garcia-Pichel *et al.*, 2020).

Cyanobacteria have been intensively studied for their ability to fix atmospheric N_2 while also performing oxygenic photosynthesis, and for their contribution to the nitrogen cycle on the earth. It has been proposed that nitrogenases arose in this group prior to oxygenic photosynthesis and provided a source of available N_2 for life to proliferate (Poudel *et al.*, 2018; Mus *et al.*, 2019).

Cyanobacteria are widely accepted as the photosynthetic ancestors of chloroplasts (Margulis, 1981). As no N_2-fixing chloroplasts exist, this suggests that either genes for N_2 fixation were acquired by horizontal gene transfer early during evolution, or N_2 fixation was already present before the divergence of cyanobacteria (Thompson *et al.*, 2013). Phylogenetic studies suggest that BNF emerged in an anaerobic, thermophilic ancestor of hydrogenotrophic methanogens and later diversified via lateral gene transfer into anaerobic bacteria, and eventually aerobic bacteria including cyanobacteria. Isotopic evidence indicates that nitrogenase activity existed at 3.2 Ga, prior to the advent of oxygenic photosynthesis and the rise of oxygen in the atmosphere, implying the presence of favorable environmental conditions for oxygen-sensitive nitrogenase to evolve. The diversification of organisms able to carry out N_2 fixation also increased the number of *nif* encoded genes. There is a tendency to the increased complexity of *nif* genes in recent aerobic lineages such as cyanobacteria and proteobacteria. The majority of them appear to be associated with metal cluster biosynthesis and nitrogenase regulation (Mus *et al.*, 2019).

5.3.2 Ecological implications of N_2-fixing cyanobacteria

Cyanobacteria are one of the major primary producers. Estimated global cyanobacterial biomass in the main reservoirs of cyanobacteria on the earth, including marine and freshwater plankton, arid land soil crusts, and endoliths, show typical population density values of 120×10^9 kg of carbon (kg C) in marine plankton global biomass of *Prochlorococcus*, or 43×10^9 kg of C of *Synechococcus*. This makes these cyanobacterial species the most abundant cyanobacteria on the earth. Tropical marine blooms of *Trichodesmium* account for an additional 10×10^9 kg C worldwide. In terrestrial environments, the mass of cyanobacteria in arid land soil crusts is estimated to reach 54×10^9 kg C and that of arid land endolithic communities an additional 14×10^9 kg C. The global biomass of planktic cyanobacteria in lakes is estimated to be around 3×10^9 kg C. These estimates account for a total global cyanobacterial biomass in the order of 3×10^{11} kg C. Moreover, these estimates did not account for some potentially significant biomass reservoirs such as polar and subarctic areas, top soils in subhumid climates, and shallow marine and freshwater benthos, so total global amounts could be larger (Garcia-Pichel *et al.*, 2003).

Diazotrophic activity accounts for around 66% of total fixed N_2 into the biosphere (Rubio and Ludden, 2008) despite this group represents only approximately 7.8% of all archaeal and bacterial cultivars with complete genomic sequences available (Poudel *et al.*, 2018). N_2 fixation rates estimates by nonheterocystous forming cyanobacteria argue that around 170–270 Tg N/year are fixed on the earth (Paul, 1978). N_2 fixation ranges from 0.002 to 0.7 g N/m^2 year (Howarth *et al.*, 1988; Karl *et al.*, 1992) and can account for 5.4 Tg N/year for blooms or 4.7 Tg N/year only from *Trichodesmium* spp (Capone and carpenter, 1982, Carpenter and Capone, 1992; Bergman *et al.*, 1997) contributing as much as 6% of global N_2 fixation. This is only one genus of many diazotrophs and is contrasting with around 20% for N_2 fixation carried by legumes. Cyanobacteria, including heterocyst-forming and nonforming cyanobacteria, and cyanobacteria that cannot be cultured such as symbionts may be a major contributor to BNF and N cycle on the earth (Bergman *et al.*, 1997).

This makes cyanobacteria one of the major drivers of primary productivity in the oceans, but also in arid lands, with global biomass surpassing a thousand million metric tons (Garcia-Pichel *et al.*, 2003) proving the ecological importance as primary producers and its potential for agricultural use and establishment in low fertility soils.

5.3.3 Nitrogenase enzyme

There are three established classes of nitrogenase enzymes, with a similar amino acid sequence and metal cofactor structure: Mo-nitrogenase, V-nitrogenase, and Fe-only nitrogenase (Rubio and Ludden, 2008; Hernandez *et al.*, 2009). The most ubiquitous, best characterized, and most efficient nitrogenase is the molybdenum-dependent nitrogenase, which contains the iron-molybdenum cofactor FeMo-co in its active site, where the binding of N_2 and reduction into NH_3 takes place. All diazotrophs reported so-far contain a Mo-nitrogenase, and some diazotrophs, in addition, also produce the alternative V-nitrogenase and/or Fe-only nitrogenase, containing a FeMo-co-like metallocluster termed FeV-co or FeFe-co, respectively. FeMo-co ([7Fe-9S-C-Mo-R-homocitrate) is arguably the most complex metallocluster in nature, composed of an inorganic [6Fe-9S] core that coordinates a central carbon atom, a capping iron on one end, and a terminal R-homocitrate coordinated to a molybdenum atom on the other end of the cofactor (Rubio and Ludden, 2008).

The synthesis of active nitrogenase with a full complement of metal clusters requires the orchestrated participation of a set of N_2 fixation *nif* genes (Mo-containing nitrogenase), *vnf* genes (V-containing nitrogenase), or *anf* genes (Fe-containing nitrogenase). The primary nitrogenase reported in cyanobacteria is the molybdenum nitrogenase, identified in >50% of sequenced cyanobacterial genomes, including terrestrial and marine strains (Poudel, 2018). Vanadium nitrogenase has been also found in several *Nostoc* and *Anabaena* isolates (Bothe *et al.*, 2010). Recently, genes encoding for a vanadium nitrogenase (*vnf*) were reported in four plants *Nostoc* cyanobionts (Nelson *et al.*, 2019). A notable exception is *Anabaena variabilis*, which expresses three different nitrogenases: a heterocyst-specific Mo-nitrogenase, a vegetative cell-specific Mo-nitrogenase, and a heterocyst-specific V-nitrogenase (Thiel and Pratte, 2014). No other well-characterized cyanobacterial strain expresses two or three nitrogenases. Genes coding for the Fe-only nitrogenase have not been described in cyanobacteria. It has been suggested that typical conditions found in waters, with abundant Mo and V but limiting Fe, may have favored the expression of Mo and V nitrogenases (Emerson and Huested, 1991).

5.3.3.1 *nif genes*

Some of the best characterized cyanobacterial *nif* gene clusters (Thiel, 2019) include *A. variabilis* ATCC 29413 (Thiel and Pratte, 2014), *Cyanothece* sp. ATCC 51142 (Stockel *et al.*, 2011; Liu *et al.*, 2018), *Leptolyngbya* sp. (Tsujimoto *et al.*, 2018), and *Thermoleptolyngbya* sp. (Yoon *et al.*, 2017). N_2 fixation or N_2 fixation–related genes, including the structural genes for nitrogenase, *nifHDK*, are typically clustered together as follows: *nifB-fdxN-nifS-nifU-nifH-nifD-nifK-nifE-nifN-nifX-orf2-nifW-hesA-hesB-fdxH*. These genes are organized in at least six transcriptional units (Esteves-Ferreira *et al.*, 2017): *nifB-fdxN-nifS-nifU*, *nifHDK*, *nifEN*, *nifX-orf2*, *nifW-hesA-hesB*, and *fdxH*. *nifP* genes have been found in *Synechococcus* strain RF-1 upstream from *nifB*, in the opposite orientation (Haselkorn and Buikema, 1992; Flores *et al.*, 1999; Huang, *et al.*, 1999). Downstream from *nifB*, *fdxH* has been identified in heterocyst formers (Böhme, 1998). On the other hand, the *nifJ* gene and, the *nifV-nifZ-nifT* operon are found in other locations in the chromosome of *Anabaena* sp. strain PCC 7120 (*Nostoc* sp. 7120) (Böhme, 1998). A monomeric hemoglobin gene (*glbN*), has also been found in *Nostoc* next to *nifU* and *nifH* (Potts *et al.*, 1992; Hill *et al.*, 1996). The *nif* and N_2 fixation–related genes are not expressed in cultures supplemented with combined N_2 (Flores *et al.*, 1999; Huang *et al.*, 1999). Regulation of the expression in some cyanobacteria appears to be subject to a circadian rhythm (Colón-López *et al.*, 1997; Chen *et al.*, 1998; Huang *et al.*, 1999) and can be confined to the heterocysts in the heterocyst formers (Elhai and Wolk, 1990). As mentioned above, some heterocyst-forming cyanobacteria, such as *A. variabilis* strain ATCC 29413, bear a second *nif* gene cluster that is expressed in the vegetative cells under microaerobic or anaerobic conditions, suggesting possible nitrogenase activity in a special condition (Schrautemeier *et al.*, 1995; Thiel *et al.*, 1995).

Nitrogenase activity requires the assembly of an active nitrogenase enzyme through a complex process requiring (1) the synthesis of the P-cluster on the nitrogenase component, (2) the synthesis of the FeMo-cofactor elsewhere, and (3) the insertion of FeMo-co into apo-nitrogenase. In the current model for FeMo-co biosynthesis (Burén *et al.*, 2020), NifU serves as a molecular scaffold for the assembly of [4Fe-4S] clusters, with NifS participating as cysteine desulfurase involved in the biosynthesis of the metalloclusters. The Fe-S clusters are transferred to NifB, which uses them for the synthesis of NifB-co, a [8Fe-9S-C] precursor, in an S-adenosylmethionine (SAM)-dependent reaction. NifB-co is transferred to NifEN, which acts as a scaffold protein to synthesize FeMo-co in a NifH-dependent reaction requiring molybdate and homocitrate, the metabolic product of NifV. FeMo-co is finally inserted into apo-dinitrogenase (apo-NifDK) to produce the mature holoenzyme NifDK. Dinitrogenase is the catalytic component, containing the active site cofactor. It is a dimer of two αβ dimers. Each αβ dinitrogenase dimer contains the FeMo-cofactor and a P-cluster [8Fe-7S]. Dinitrogenase reductase serves as the specific electron donor to dinitrogenase. It is an α2 dimer containing a single [4Fe-4S] cluster coordinated in between the two subunits. Additional players in this intricate biosynthetic pathway include NifX (metallochaperone of FeMo-co precursors), NifW (involved in apo-NifDK binding prior to P-cluster maturation), NifZ (involved in P-cluster maturation), NifT (unknown role), FdxN (involved in the formation of NifB-co), NifJ (pyruvate-flavodoxin (Fld) oxidoreductase required for nitrogenase activity in iron deficiency in *Anabaena* (Bauer *et al.*, 1993), and NifP (serine acetyltransferase involved in the expression of nitrogenase activity). The genes for NifQ (involved in the processing of molybdenum for incorporation into FeMo-co), NifM (involved in the maturation of NifH), and NafY (involved in stabilization of apo-NifDK and binding of FeMo-co) are not reported in cyanobacterial genomes (Thiel and Pratte, 2014).

5.3.3.2 *Nitrogenase sensitivity to oxygen*

Nitrogenase is highly susceptible to O_2 inactivation and cyanobacteria have developed several strategies to prevent O_2-dependent inactivation of nitrogenase (Weare and Benemann, 1974). Some cyanobacteria species (e.g., *Plectonema boryanum*) can only fix N_2 either anaerobically or microaerobically. Aerobic N_2 fixation required one of following evolutionary strategies (Issa *et al.*, 2014): (1) spatial separation from oxygenic photosynthesis in heterocystous cyanobacteria (HC) (e.g., *Anabaena*), or (2) temporal separation from oxygenic photosynthesis in nonheterocystous cyanobacteria (NHC) (e.g., *Gloeothece, Cyanothece, Lyngbya*). Within NHC, a special case includes *Trichodesmium* and *Katagnymene*, which are able to perform both a temporal and spatial separation from oxygenic photosynthesis.

Due to the high energy required for BNF, at least 16 ATP and 8 electrons, equivalents to fix 1 N_2 into 2 NH_3 (Inomura *et al.*, 2017), the evolution of mechanisms that allow for N_2 fixation during O_2-dependent respiration was an impactful innovation. In obligate aerobes, mechanisms have evolved to exploit respiration itself as a mechanism to prevent O_2 inactivation of nitrogenase (Soto-Urzúa and Baca, 2001) but other mechanisms could be operating as well (Oelze, 2000). The physiological electron donors to cyanobacterial nitrogenase are ferredoxin (Fd) (Martin *et al.*, 1989; Sandmann *et al.*, 1990; Yakunin and Hallenbeck, 1998; Egener *et al.*, 2001) and Fld (Gennaro *et al.*, 1996; Yang *et al.*, 2016;). Reduction of Fd or Fld in anaerobic diazotrophs can occur via the oxidation of pyruvate via the activity of pyruvate Fld oxidoreductase (Nieva-Gomez, *et al.*, 1980; Wahl and Orme-Johnson, 1987; Yakunin and Hallenbeck, 1998; Yakunin and Hallenbeck, 1998; Saeki, 2004) or the oxidation of hydrogen through the activity of one of several [FeFe]- or [NiFe]-hydrogenases (Bothe, 1980; Schmitz *et al.*, 1995; Rey *et al.*, 2006; Gutekunst *et al.*, 2014; Khanna and Lindblad, 2015; Therien *et al.*, 2017). Oxygenic phototrophic cyanobacteria can utilize photosystem I to energize electrons to low enough potentials to drive the reduction of Fd (Dutton, 1986; Jagannathan and Golbeck, 2009). However, this Fd is not available to N_2 fixation as it must be temporally or spatially separated from oxygenic photosynthesis. Cyanobacteria encode Fd-$NADP^+$ oxidoreductase (FNR) that can function in reverse to reduce Fd or Fld with NADPH generated by carbohydrate oxidation in heterocysts or when O_2 tensions are low (Schrautemeier and Böhme, 1985: Apte *et al.*, 1978; Aliverti *et al.*, 2001; Seki, 2004). Some cyanobacteria also encode pyruvate Fd oxidoreductase (Poudel *et al.*, 2018).

5.3.4 N_2 fixation in HC: spatial separation

Cyanobacteria have diversified through evolution, producing both unicellular and multicellular forms. Most cyanobacteria divide by binary fission, but some exhibit alternatives to the standard mode of bacterial division, reproducing by budding or by multiple fission of enlarged cells (Rippka and Herdman, 1985; Angert, 2005). In some multicellular, filamentous cyanobacteria, vegetative cells can differentiate into heterocysts, which are specialized for the fixation of N_2 gas (Haselkorn, 1978). While vegetative cells are photosynthetically active, in heterocyst this process is lost, relying on spatial separation of N_2-fixing modules from oxygenic photosynthetic ones. This differentiation is permanent and nonreversible, and heterocyst cells cannot replicate themselves (Fay, 1992).

The number of heterocysts varies in proportion between species, but is about 5–10%, and is often used as a morphological trait for classification purposes (Bergman *et al.*, 1997, Böhme, 1998).

All cyanobacteria contain a Gram-negative type of cell envelope with an outer membrane external to the cytoplasmic membrane and to the murein sacculus, which is thicker than in most other well-characterized Gram-negative bacteria (Drews and Weckesser, 1982; Hoiczyk and Hansel, 2000). The outer membrane of heterocyst-forming cyanobacteria is composed of lipopolysaccharides and proteins in a complex matrix that does not enter the septum between adjacent cells in the filament. These cells are surrounded by their cytoplasmic membrane and peptidoglycan layers (Ris and Singh, 1961; Wildon and Mercer, 1963; Nierzwicki-Bauer *et al.*, 1984; Flores *et al.*, 2006; Flores *et al.*, 2010; Herrero *et al.*, 2016), and display a continuous periplasmic space between the cytoplasmic membrane and outer membrane, (Flores *et al.*, 2006) which allows molecules movement along the filament (Mariscal *et al.*, 2007; Herrero *et al.*, 2016). In the heterocyst-specific cell, this envelope is composed of two layers deposited outside of the outer membrane, one made of glycolipids (heterocyst glycolipid or Hgl) and the other made of polysaccharide (heterocyst envelope polysaccharide or Hep) (Herrero, *et al.*, 2013). The Hgl layer appears to have a reduced permeability to gases, whereas the Hep layer has been suggested to have a protective role for the Hgl layer (Xu *et al.*, 2008). The outer membrane also contains proteins including porins, which usually take the form of β-barrels with a central pore that allows the passage of small ions and molecules, making the outer membrane permeable to small molecules (Nikaido, 2003). Cell to cell joints, involved in cell to cell adhesion, have been observed in the cell poles in the intercellular septa, presumably on the cytoplasmatic face with suggested plasmodesmata functions (Buikema and Haselkorn, 1991; Ernst *et al.*, 1992; Bauer *et al.*, 1995; Flores *et al.*, 2007, Omairi-Nasser *et al.*, 2014). However, there is no clear evidence of true plasmodesmata transport. Cytoplasmic-membrane continuity is now termed "septal junctions." Cytoplasmic communication has been reported between adjacent cells in a gradient manner, summarizing that both continuous periplasmic space and cell-to-cell joining complexes are important in molecular exchange and cell communication (Giddings and Staehelin, 1978; Mariscal *et al.*, 2007; Flores and Herrero, 2010; Mariscal, 2014; Mullineaux and Nurnberg, 2014; Flores *et al.*, 2016; Herrero *et al.*, 2016).

Nitrogenase, the N_2 fixing enzyme, is encoded by the *nifHDK* genes and surrounded by several accessory genes in six transcriptional operons that are highly regulated by a transcriptional regulator (NtcA) and a master regulator (HetR) relying on N_2 content and 2-oxoglutarate levels (Herrero *et al.*, 2001; Bothe *et al.*, 2010). These signals also enhance Fe uptake by increasing expression *furA* (Lopez-Gomollon *et al.*, 2007), and lead to *nifD* rearrangement by a XisA-mediated deletion of a genetic spacer between the two fragments in *nifD* gene, regrouping *nifD* into a single transcriptional unit. The latter has been reported in HC but not in NHC (Brusca *et al.*, 1989; Haselkorn, 1992; Carrasco *et al.*, 2005).

Due to nitrogenase sensitivity to O_2, HC mutants that lack heterocyst retain the ability to fix N_2 only when incubated under anoxic or microoxic conditions (Rippka and Stanier, 1978; Spence and Stewart, 1987). This microoxic environment is provided by the heterocyst, spatially separating the N_2 fixation from the photosynthetic and oxygen-evolving cyanobacterial vegetative cells in the filament (Flores and Herrero, 2010). HC constitutes a monophyletic group, separated from NHC, in an oxic environment, that could have acted as a selection pressure for the evolution of heterocysts around 2.1 billion years ago, suggesting that this cell form only evolved from another cell form, such as akinetes resistance structures (Tomitani *et al.*, 2006; Flores and Herrero, 2010). The concept of oxygen-inhibition of nitrogenase has relevant evolutionary implications, and it has been suggested as a key factor to limit O_2 accumulation in the atmosphere and delay its rise for over two billion years (Allen *et al.*, 2019).

Besides differences in the outer layer, internally heterocysts also undergo intracytoplasmic membranes (such as thylakoids) differentiation from that observed in vegetative cells, probably reflecting the particular bioenergetics of the heterocysts. Ribulose 1,5 bisphosphate carboxylase–oxygenase (Rubisco) and water splitting of photosystem II) activity is lost, and heterocysts lack photosynthetic CO_2 fixation, thus avoiding intracellular O_2 generation and contributing to the formation of a microoxic environment (Wolk *et al.*, 1994). Gases such as O_2 and N_2 probably could move between poles of adjacent cells (Walsby, 2007). Additionally, respiration rates and probably specific respiratory oxidases contribute to oxygen consumption in heterocysts and bioenergetics for N_2 fixation (Wolk *et al.*, 1994; Valladares *et al.*, 2003). Furthermore, photosystem I–dependent photophosphorylation provides additional energetic and reductant power (Almon and Böhme, 1982; Wolk *et al.*, 1994; Magnuson, 2019). Due to the autotrophy lost, heterocysts depend on fixed carbon from neighboring cells, probably in the form of reduced compounds such as sucrose and other sugars (Wolk, 1968; Wolk *et al.*, 1994; Curatti *et al.*, 2002) with an important catabolic contribution to the bioenergetics from the oxidative pentose pathway (Böhme and Schrautemeier, 1987; Summers *et al.*, 1995).

Due to the lack of glutamine-2-oxoglutarate amido-transferase (GOGAT), ammonium fixed by nitrogenase is incorporated into amino acids only in the form of glutamine by ATP-dependent amidation of glutamate catalyzed by glutamine synthetase (GS), which is in high levels in heterocyst cells (Wolk *et al.*, 1976; Wolk *et al.*, 1994; Martín-Figueroa *et al.*, 2000). Evidence of N_2 compounds transport between cells has been observed, suggesting a net export of N_2 from heterocyst (Wolk *et al.*, 1976; Thomas *et al.*, 1977). Additionally, cyanophycin (multi-L-arginyl-poly-(L-aspartic acid)) accumulates at heterocyst pole regions and could serve as a N_2 buffer and store of fixed N_2 to be shared with vegetative cells (Carr, 1983; Sherman *et al.*, 2000; Ziegler *et al.*, 2001; Picossi *et al.*, 2004; Burnat *et al.*, 2014).

5.3.4.1 N_2 fixation in NHC: temporal separation only, or temporal plus spatial separation (*Trichodesmium*)

Unicellular cyanobacteria are not able to differentiate in specialized cells, and some can perform both N_2 fixation and oxygenic photosynthesis in the same cell. Despite being able to fix atmospheric N_2 in theoretically all cells, they are considered less efficient than heterocyst-forming cyanobacteria because N_2 fixation rates in both populations of cyanobacteria are similar. Strategies for avoiding oxygen inactivation of nitrogenase rely on temporal separation of these two processes, where photosynthesis occurs during daytime and N_2 fixation during the night (Bergman *et al.*, 1997). Cyanobacteria in these communities conduct oxygenic photosynthesis during the day resulting in CO_2 fixation and the storage of carbon in the form of polysaccharides. At night, the respiratory capacity of the concentrated mat biomass can quickly consume the available O_2 resulting in an anoxic environment. Cyanobacteria, in turn, transition to an anaerobic mode of metabolism and fix N_2 with energy provided through the fermentation of stored carbon (Steunou *et al.*, 2006). While this process is common in most unicellular and NHC, there are some exceptions, such as *Trichodesmium*, that show patterns of cell specialization to N_2 fixation in some regions of the colony, and show daytime N_2 fixation (Bergman *et al.*, 1997; Bergman *et al.*, 2013).

NHC are a heterogeneous group with a wide range of metabolic adaptations. Many of them can fix N_2, but few of them can do it aerobically and are often observed forming mats into a dense polysaccharide matrix that prevents O_2 diffusion and nitrogenase inactivation (Potts *et al.*, 1978; Pearson *et al.*, 1979). This group of cyanobacteria contributes significantly to the global N cycle. *Trichodesmium* is extremely abundant in tropical and subtropical oceans and can form massive blooms covering very large areas. It is presumably one of the major contributors to N_2 fixation in oceans together with *Synechococcus*. Both *Synechococcus* and *Aphanothece* are also capable of BNF in aerobic conditions in soils and freshwater ponds (Huang and Chow, 1986; Roger *et al.*, 1986; Roger *et al.*, 1986b; Huang and Chow, 1988; Bergman *et al.*, 1997).

As opposed to HC, where its capacity for photosynthesis, CO_2 fixation, and O_2 production is lost, there are no specialized cells for BNF in NHC (Bergan *et al.*, 1997), and NHC can perform N_2 fixation and photosynthetic CO_2 fixation in the same cell. An exception should be made for *Trichodesmium*, an NHC that produces diazocytes, a differentiated cell specialized in N_2 fixation (Sandh *et al.*, 2012; Bergman *et al.*, 2013).

NHC includes more than 70 strains able to perform BNF from marine, freshwater, and terrestrial environments, and even symbiotic associations with diatoms and lichens have been reported (Bergman *et al.*, 1997). Most of these strains only can fix atmospheric N_2 in microaerobic or anoxic conditions, but interestingly few are capable to fix N_2 in oxic conditions and also sustain oxygenic photosynthesis in the same cell. They do so by protecting nitrogenase from O_2 inactivation separating temporarily this process and taking advantage of the reductant and ATP generated in photosynthesis (Bergman *et al.*, 1997).

Molybdenum nitrogenase is present in NHC. It is formed by the Fe-protein (dinitrogenase reductase) and MoFe-protein (dinitrogenase) encoded by *nifH* and *nifDK* genes, respectively. Other accessory genes also participate in regulation and protein maturation. Despite being present in some HC, no alternative nitrogenases (iron and vanadium nitrogenases encoded by *anf* and *vnf* genes, respectively) are found in NHC. A comparison of the molybdenum nitrogenase nucleotide sequences shows a high similarity between diazotrophic bacteria and NHC *nif* genes. Even arrangements of *nifHDK* are contiguous, showing that *nif* structural genes are highly conserved (Bergman *et al.*, 1997) and is posttranslationally regulated by ammonium and other N_2 compounds, O_2 content, and light/dark conditions (Reich and Böger, 1989; Ernst *et al.*, 1990; Böhm *et al.*, 1992; Ohki *et al.*, 1992; Durner *et al.*, 1994; Zhang *et al.*, 1995).

NHC that can fix N_2 aerobically synthesize nitrogenase under N_2 limitation conditions in the presence of air. On the other hand, anoxic or microoxic conditions are required for nitrogenase derepression and activity (Stewart and Lex, 1970; Weare and Benemann, 1974; Rippka and Waterbury, 1977; Weisshaar and Böger, 1983; Giani and Krumbein, 1986; Almon and Böger, 1988; Maryan *et al.*, 1986; Rai *et al.*, 1992) in cyanobacteria that cannot fix N_2 aerobically, probably due to lack of mechanisms to prevent nitrogenase inactivation by oxygen (Bergman *et al.*, 1997).

Two contrasting situations are described within NHC, with some organisms capable of N_2 fixation at night and others during the daytime. The first is supported by the idea that photosynthetic activity is neglectable at night and there is no O_2 formation, so with an additional high metabolic O_2 consumption rate, the conditions for N-fixation could be performed. On the other hand, BNF during the daytime could be supported by the idea that ATP and reductant power required for N_2 conversion to ammonia is generated by photosynthetic activity (Gallon 1980; Gallon and Hamadi, 1984; Giani and Krumbein, 1986; Maryan *et al.*, 1986).

In NHC, N_2 fixation provides ammonium and is assimilated by GS and glutamate synthase or GOGAT pathway. In aerobic NHC, N_2 fixation could rely on illumination conditions and have a cyclic rhythm that goes along with light or darkness conditions. N_2 fixation and photosynthesis often show opposite activities in NHC, performing N_2 fixation at night and photosynthesis with daylight, which achieves temporal separation of N_2 fixation from oxygenic photosynthesis. In sea mats, nitrogenase activity was neglectable during the daytime, with O_2 levels peaking during the day and dropping at night. This condition is not favorable for N_2 fixation during the day but could provide a suitable environment at night coupled with aerobic respiration (Villbrandt *et al.*, 1990; Bergman *et al.*, 1997). Also, nitrogenase activity shows peaks of activity during the transition from dark to light (sunrise). This can be explained by metabolic energy at night not being enough to sustain stable rates of N_2 fixation, and requiring to be coupled with low light intensity for photosynthetic activity with still low O_2 content and production (Heyer *et al.*, 1989; Bergman *et al.*, 1997).

In *Trichodesmium* colonies there is a spatial separation of N-fixation in the central region, in which cells are differentiated and specialized. This region contains lightly pigmented filaments that do not fix CO_2, limit O_2 production (diazocytes), and bear a separated CO_2-fixation specialized region (Siddiqui *et al*, 1992; Prufert-Bebout *et al.*, 1993; Janson *et al.*, 1994; Fredriksson and Bergman, 1995; Paerl *et al.*, 1995; Bergman *et al.*, 1997; Bergman *et al.*, 2013). *Trichodesmium* fixes N_2 exclusively during the daytime and is accompanied by a circadian rhythm with synthesis and degradation during morning and evening, respectively (Capone *et al.*, 1990; Fredriksson and Bergman, 1995;). A similar pattern is also evidenced in *Lyngbya aestuarii*, with terminal cells photosynthetically inactive, confining nitrogenase (Paerl *et al.*, 1991). Considerable nitrogenase activity during daytime is also reported in other NHC and complex mats from natural environments. (Bebout *et al.*, 1987; Paerl *et al.*, 1989; Capone *et al.*, 1990; Villbrandt *et al.*, 1990; Jones, 1992). NHC are predominant N_2 fixers in mats, although it is also possible that nitrogenase activity in mats could be due to association with other microorganisms (Stal *et al.*, 1984; Paerl and Prufert, 1987; Villbrandt *et al.*, 1990; Bergman *et al.*, 1997).

Little is known regarding the changes that permit adaption from a N_2-sufficient condition to N_2 fixation in N_2-limiting conditions in NHC, as opposed to HC. Similar to HC, a decline in photosynthetic activity, an increase in respiratory activity, and repression of nitrate reductase expression is observed in NHC. In some cases, such as *Trichodesmium* and

P. boryanum, there is also an increase in GS activity. Conversely to HC, where there is a loss CO_2 fixation and a permanent morphological differentiation into heterocyst, NHC retain Rubisco and phycobiliproteins, and changes are reversible (Bergman *et al.*, 1997).

5.4 Cyanobacteria as biofertilizers

The inoculation of cyanobacteria very often enhances several parameters of plant growth, such as shoot/root length, fresh/dry weight, the weight of seeds, chlorophyll content, and yield. Thus, its prospects as a natural fertilizer have been repeatedly proposed and quite extensively reviewed in the literature in the last decades. We suggest some recent reviews for the last updates in the subject (Rai *et al.*, 2019; Renuka *et al.*, 2018; Win *et al.*, 2018; Alobwede *et al.*, 2019; Do Nascimento *et al.*, 2019; Garlapati *et al.*, 2019; Chittora *et al.*, 2020; Joshi *et al.*, 2020; Kuraganti *et al.*, 2020).

There is a long and quite successful history of using cyanobacterial BNF to restore soil fertility. N-fertilization of rice paddy-fields is a paradigmatic case of study (Venkataraman, 1975). As such, taking advantage of naturally occurring cyanobacteria, or on purpose inoculated into rice paddy-fields, has been largely covered in the literature (Choudhury and Kennedy, 2004). Rice is the major food crop of nearly half of the world's population and represents a high N-demanding crop removing around 16–17 kg N from the soil for the production of each ton of rough rice including straw. Although urea is applied to cover this large demand, its efficiency of use is very low (30–40%) due to the general issues stated before (Drechsel *et al.*, 2015). Rice crops are mostly grown in wetlands (about 85% of the total rice-cropped area) (Choudhury and Kennedy, 2004). Under these conditions, both aerobic and anaerobic, free-living and symbiotic diazotrophic bacteria can thrive. Cyanobacteria often form N_2-fixing symbiotic association with flowering plants, algae, fungi, liverworts, and ferns (Foster and Zehr, 2019). In rice paddy-fields, it is especially worth mentioning the small free-floating freshwater fern *Azolla pinnata* (and other related species) because it engages in a symbiotic association with the N_2-fixing cyanobacterium, *Anabaena azolla* (Mahanty *et al.*, 2017). The fern forms a green mat over water which holds the cyanobacterial symbiont in the dorsal fleshy lobe of the leaf, within a central cavity (Mishra and Dash, 2014). In submerged paddy fields, about 50–60% of N requirement by rice is met through the combination of mineralization of soil organic N and BNF by free-living and *Azolla-Anabaena* symbionts, which can fix atmospheric N_2 at around 20–30 kg N/ha (Mishra and Dash, 2014). It has been calculated that N_2 fixation by free-living cyanobacteria and cyanobacteria in a symbiosis with *Azolla* in paddy fields might globally contribute around 5 Tg N/year (Herridge *et al.*, 2008).

Thus, free-living and symbiotic cyanobacteria provide N to crops by two co-occurring pathways at the field level: one that depends on compounds continuously released by living bacteria and another one that relies on the biomass as it decomposes leading to mineralization (necromass).

5.4.1 Living cyanobacteria–dependent traits

Cyanobacteria have long been considered as part of the large group of plant growth–promoting rhizobacteria (Rai *et al.*, 2019). These bacteria exert three main activities in the soil, namely biofertilization, biostimulation, and/or biocontrol. Biofertilizers are usually defined as organic compounds from living microorganisms that enhance soil fertility by adding essential macronutrients such as N, P, K, and others micronutrients. Biostimulants are bacteria (or other microorganisms) that produce organic molecules, such as plant hormones, vitamins, siderophores, exopolysaccharides, etc. able to promote plant growth by improving seed germination, plant development, or improve soil characteristics. Biocontrollers are beneficial microorganisms that assist plants in their defense against pathogens by different mechanisms (competition, production of antimicrobials, parasitism, etc.) (Singh *et al.*, 2016; Tiwari, 2016; Mahapatra *et al.*, 2018; Win *et al.*, 2018; Chittora *et al.*, 2020; Joshi *et al.*, 2020; Kuraganti *et al.*, 2020). The main roles in the soil and paddy-fields of free-living cyanobacteria are CO_2 and N_2-fixation, enhancement of the availability of P, the release of growth-promoting hormones, secretion of mucilage/polysaccharides that improve soil structure/porosity, and reduction of the rate of water loss and soil erosion (Prasanna *et al.*, 2013).

5.4.1.1 *Release of fixed N_2*

Rice paddies have abundant free-living and symbiotic N_2-fixing cyanobacteria among other diazotrophic bacteria. In this environment, most free-living microorganisms assimilate the NH_4^+ produced by nitrogenase for their growth and, apparently, do not excrete it into the environment so that it can be used by crops. On the other hand, symbiotic N_2-fixing cyanobacteria, as those engaged in a symbiotic relationship with the fern *Azolla*, lack an effective NH_4^+ assimilation pathway, releasing some of the fixed N as NH_4^+, which is mostly taken up by the mutualistic partner. As a result, similar to free-living diazotrophic cyanobacteria, only a very limited amount of fixed-N remains immediately available for rice plants until mineralization of organic N in the sediment (Latorre *et al.*, 1986) (see below).

Since the first isolation of a *Klebsiella pneumoniae* mutant strain deficient in N-assimilation and derepressed for N_2-fixation, which resulted in NH_4^+ released into the culture medium (Shanmugam and Valentine, 1975), many other mutants have been isolated in cyanobacteria (Spiller *et al.*, 1986) and other free-living diazotrophs (Barney, 2020).

Pioneering experiments evaluated the inoculation of rice plants with an *A. variabilis* mutant strain (SA-1), which bears a mutation in GS, and as a result, presents deficient N-assimilation and excretes ammonium. This resulted in a substantially increased dry weight and N content compared to non-N-fertilized plants or plants inoculated with the parental strain (SA-0) (Latorre *et al.*, 1986). A subsequent study used a variety of ammonium-excreting strains such as *A. variabilis* SA-1 or *A. azolla* derivatives, immobilized in polyurethane foam or sugarcane waste (bagasse). When used as an inoculant in both laboratory and rice field experiments, data showed increased chlorophyll content of the plants and increased rice grain and straw yields (Kannaiyan *et al.*, 1997). Characteristically, most of these mutant strains impaired in NH_4^+ assimilation presented slower growth than its parental strains (Latorre *et al.*, 1986).

More contemporary research conducted in a natural environment showed that the colony-forming cyanobacterium *Aphanizomenon* sp. releases about half of its fixed N_2 as NH_4^+, which is transferred to the prokaryotic and eukaryotic plankton with a short turnover time of ~ 5 h. It was proposed that transference of newly fixed N_2 into fast-sinking biomass can have a profound impact on ecosystem productivity and biogeochemical processes at shorter time scales than previously thought (Adam *et al.*, 2016).

5.4.1.2 Release of fixed carbon

Cyanobacteria, as well as microalgae, tend to release part or their fixed C into their surroundings, which can represent up to 80% of photosynthetically assimilated C (Lakaniemi *et al.*, 2012). In some cases, such as cyanobacteria of the genus *Anabaena*, the release of hydrated exopolymeric substances (EPS) could be very conspicuous (González López *et al.*, 2009). The main constituents of EPS are exopolysaccharides, and also lower quantities of proteins, nucleic acids, and lipids. High water retention capacity of the EPS, especially by the exopolysaccharides fraction, helps to protect cyanobacteria from desiccation, UV radiation, heavy metals, predation, and enzymatic degradation. Exopolysaccharides can be attached to the cell surface, in which case they form the cyanobacterial capsule, or be released into the environment. As the sugar compositions of these two fractions of polysaccharides are markedly different, it was suggested that they are produced by different biosynthetic pathways (Di Pippo *et al.*, 2013). Cyanobacterial exopolysaccharides are composed of a few different monosaccharides, deoxyhexoses (fucose and rhamnose), acidic hexoses (glucuronic and galacturonic acids), peptidic and acetylated amino sugars, and other noncarbohydrate constituents. The uronic acids and sulfate groups confer an anionic nature, sticky behavior, and high affinity toward metal cations (Colica and Philippis, 2013).

Notably, mostly due to these hygroscopic exopolysaccharides, cyanobacteria can absorb many times their dry weight in water (Mager and Thomas, 2011). Filamentous forms of N_2-fixing cyanobacteria with the capability of producing exopolysaccharides can develop a superficial network of filaments on the soil, with adhesive properties that bind the soil particles, contributing to the improvement of soil moisture and the topsoil aggregation. This soil aggregates or biological soil crusts (BSCs) are more characteristic of arid and semiarid environments where they contribute to soil formation and fertility and prevent soil erosion (Singh *et al.*, 2016; Renuka *et al.*, 2018; Chamizo *et al.*, 2020).

The EPS matrix in BSCs is capable of both physical and chemical immobilization of nutrients acting as a reservoir (Colica and Philippis, 2013). Thus, it was proposed that inoculation with cyanobacteria could prevent some runoff of fertilizers, keeping soil fertility and ameliorating percolation into groundwater and eutrophication of water bodies (Renuka *et al.*, 2018).

Other studies showed that, in addition to maintaining a high matric potential (Ψm) in native BSCs, cyanobacterial EPS enhanced the fitness and activity of the microbial community in the soil crusts under water-limited conditions, which plays a key function in soil nutrients mineralization (Adessi *et al.*, 2018) (see next).

In saline soils, the inoculation of EPS-excreting cyanobacteria can trap excess Na^+ ions, which can prevent some detrimental effect on crops growth (Renuka *et al.*, 2018).

5.4.1.3 Enhancement of phosphorus availability

It has also been suggested that cyanobacteria can improve the bioavailability of P to the plants by solubilizing and mobilizing the insoluble organic PO_4^{3-} present in the soil organic matter (including added manures) with the help of phosphatase enzymes. Cyanobacteria, as well as other soil bacteria, present different combinations of inducible cell-bound or extracellular phosphomonoesterase, phosphodiesterase, and phytase activities, which can be further classified based on their optimum pH for maximum activity in acid or alkaline phosphatases (Whitton *et al.*, 1991). Cyanobacteria also share with other microbes the ability to solubilize insoluble forms of P, such as $(Ca)_3(PO_4)_2$, $FePO_4$, $AlPO_4$, and $Ca_5(PO_4)_3OH$ in soils and sediments by means of one or more different mechanisms. Some cyanobacteria synthesize organic acids, such as phthalic acid, which can solubilize P, without changing the pH, by chelating polycations, and/or competing with PO_4^{3-} for adsorption sites in the soil (Yandigeri *et al.*, 2011). Alternatively, or additionally, acidification of the medium by the release of organic acids also results in some solubilization of P (Singh *et al.*, 2016).

5.4.1.4 Release of phytohormones

Cyanobacteria from the most genus are able to produce one or more of the plant growth–promoting phytohormones, gibberellins (GAs), cytokinin (CK), auxin; indole-3-acetic acid (IAA), indole-3-butyric acid (IBA), indole acetamide, indole carboxylic acid), and other growth regulators such as abscisic acid (ABA), ethylene (ET), and salicylic acid (SA) (Hashtroudi *et al.*, 2013; Singh, 2014; Singh *et al.*, 2016; Singh *et al.*, 2017; Renuka *et al.*, 2018; Wenz *et al.*, 2019).

Hussain and Hasnain (2011) demonstrated a positive correlation between cyanobacterial CK and auxin and plant growth parameters. A significant enhancement of the plant's endogenous phytohormones pool was observed as a result of the plant–cyanobacteria association in the rhizosphere (Hussain and Hasnain, 2011). This and other studies showed the predominant accumulation of IBA in several filamentous cyanobacteria, which is noteworthy considering that IBA is considerably more effective than IAA at promoting adventitious plants root formation. This may be explained by (1) a lower susceptibility of IBA to the degradation enzymes compared to IAA, and (2) the steady supply of free IAA produced by the slow conversion of IBA to IAA, which exerts auxin activity on the plant (Woodward and Bartel, 2005).

In another study, a rise in levels of endogenous and exogenous auxin in cyanobacteria was observed as a result of the presence of wheat plants (Mazhar *et al.*, 2013). These studies indicated active phytohormones-mediated crosstalk between plants and cyanobacteria.

A definitive proof for the involvement of cyanobacterial CK-type phytohormones in plant growth stimulation was provided by a study in which the gene encoding an isopentenyl transferase enzyme, which catalyzes the first step of CKs biosynthesis, was inactivated in an endophytic *Nostoc* sp. strain. The mutation resulted in a sharp decline (up to 80%) in the zeatin-type CK released into the culture medium, failure to colonize rice and wheat roots, and failure to promote plants growth compared to the parental strain (Hussain *et al.*, 2013).

Although less frequently, gibberellic acid-like substances were also demonstrated in cyanobacterial cell-extracts or growth medium (Haroun and Hussein, 2003; Osman *et al.*, 2010). Nevertheless, several studies in which the effect of cyanobacterial extracts on plant growth was compared to that of purified GAs were interpreted as an indication of cyanobacterial gibberellic acid–like substances being involved (Singh, 2014).

ABA (Finkelstein *et al.*, 2002) and ET (Husain *et al.*, 2020) production and signaling are generally associated with stress tolerance in plants. Both additionally promoted many agronomically important aspects. For example, ABA regulates the synthesis of seed storage proteins and lipids, the promotion of seed desiccation tolerance and dormancy, and the inhibition of the phase transitions from embryonic to germinative growth and from vegetative to reproductive growth (Finkelstein *et al.*, 2002). Similarly, ET is also involved in adventitious root formation, and stem and petiole elongation (Husain *et al.*, 2020).

ABA production and release into the medium was demonstrated in many cyanobacterial species (Hartung, 2010). Maršálek and Šimek showed that external application of ABA significantly stimulated nitrogenase activity in the cyanobacterium *Nostoc muscorum* and another N_2-fixing rhizobacterium, such as *Azotobacter chroococcum* (Maršálek and Šimek, 1992), although at an ABA concentration which could be not physiologically relevant (Hartung, 2010). ABA is produced by cyanobacteria and/or other known prolific producers in the rhizosphere, such as fungi, in close proximity to N_2-fixers. Whether ABA would contribute to plant nutrition and growth in the environment is mostly speculative at this time.

Conversely, an increase of ABA biosynthesis under salt stress in some cyanobacterial strains (Maršálek and Šimek, 1992) may represent one of the mechanisms by which cyanobacteria promote their associated plant's tolerance to stress (Hartung, 2010).

Strains from some cyanobacterial genera, especially from Nostocales were reported to naturally produce ET (Musatenko, 2001). Overproduction of ET has been accomplished in many different studies by introducing a *Pseudomonas syringae* gene responsible for the conversion of 2-oxoglutarate and arginine into ET in the non-N_2-fixing strains *Synechocystis* sp PCC 6803 and *Synechococcus elongatus* PCC 7942. It was demonstrated more recently that depletion of 2-oxoglutarate would be a limiting factor for ET production, which can be overcome by additional mutations to enhance C-flux toward the tricarboxylic acid (TCA) cycle (Carbonell *et al.*, 2019). Apparently, neither effect of these ET overproducing-strains was tested for its effect on plants, nor similar genetic modifications were introduced into N_2-fixing Nostocales, in which N_2-fixation could be enhanced according to the expected lower levels of 2-oxoglutarate.

Besides phytohormones, several other metabolites can be found in cyanobacteria: phenolics, phytoene/terpenoids, phytols, sterols, free fatty acids, photo-protective compounds (MAAs, scytonemin, carotenoids, polysaccharides, halogenated compounds, etc.), cyanotoxins, biocides (algaecides, herbicides and insecticides), and other plant growth regulators such as amino acids, sugars, and vitamins (Singh *et al.*, 2017). These effectors can promote plant growth and/or defense either directly or through more complex interactions with other microbes in the rhizosphere (Singh, 2014).

5.4.2 Cyanobacterial biomass–dependent traits: mineralization pathway

The beneficial effects of adding organic matter–rich manures and composts to soils fertility have been known since very long ago (Sikora and Yakovchenko, 1996). N is mostly present as organic N in the soil, and especially in manures. However, plants can only use inorganic forms of N, taking only a little fraction of small organic forms under extreme conditions (Li *et al.*, 2019). Soil mineralization is the conversion of organic into inorganic nutrients. This process is essential for nutrient cycling and productivity in both natural and agricultural environments. Besides physicochemical conditions such as pH, humidity and temperature, soil microbiota play a key role in organic matter breakdown and depolymerization, toward nutrient mineralization and soil health (Li *et al.*, 2016; Espenberg *et al.*, 2018; Li *et al.*, 2019; Risch et at., 2019; Wade *et al.*, 2020).

Coming back to the paradigmatic example of fertilization of rice paddyfields, *Azolla-Anabaena* biomass contains 4–5% N on a dry basis. Because it is decomposed efficiently and quickly in the soil, it represents a useful organic "green-manure" contributing a good balance of C, N, P, K, S, Zn, Fe, Mo, etc. for rice (Al Abboud *et al.*, 2014). Thus, *Azolla-Anabaena* was cultivated by vegetative means at a commercial scale in specific facilities in India, with a mean yield of 1.5 kg per square meter in a week, and then incorporated in the rice field as "green manure" prior to rice planting (Mazid and Khan, 2015; Mahanty *et al.*, 2017).

Direct application of microalgal or cyanobacterial biomass as "green manure" into the soil has a similar impact on soil formation and fertility. In fact, large-scale cultivation of microalgae/cyanobacteria at the expense of agricultural runoff, or other wastewater, according to a controlled eutrophication approach for capturing N and P in the form of algal biomass, was envisioned more than three decades ago (Benemann, 1979). The use of this biomass as "green manure" would largely promote recycling of these nutrients and sustainability in agriculture (Benemann *et al.*, 2003). Furthermore, if this process relied on N_2-fixing cyanobacteria, it might involve upgrading of the wastewater as a fertilizer, by attaining a better balance among the nutrients. Besides this sustainable increase in the N-content by direct capture from the atmosphere, other bioactive compounds in the biomass are likely to improve the overall performance (Benemann, 1979; Do Nascimento *et al.*, 2019).

The application of cyanobacteria-based "green manure" would be advantageous in comparison with the application of relatively small doses of live cyanobacteria that need to proliferate and compete successfully with native soil microbes mostly under adverse and changing environments. It was suggested that separation of biological fertilizer production (large-scale cultivation inappropriate facilities, see below) from fertilizer application and crop cultivation would allow specific optimization of both processes, bypassing the odds of cyanobacterial acclimation to the rhizosphere and better adjustment of the application of "green manure" according to specific crop needs for precision agriculture (Benemann, 1979; Bongiovanni and Lowenberg-Deboer, 2004; Do Nascimento *et al.*, 2019).

Recently, Win and colleagues (Win *et al.*, 2018) reviewed most of those reports which demonstrated that the use of microalgal or cyanobacterial biomass as an organic fertilizer on impoverished soils sustained growth of wheat (Alobwede *et al.*, 2019), wheat and millet (Nisha *et al.*, 2007), tomatoes (Coppens *et al.*, 2016; Garcia-Gonzalez and Sommerfeld, 2016), and corn and cucumber (Mulbry *et al.*, 2005). One of these studies was based on a mixture of potentially N_2-fixing strains, which were grown using a defined mineral medium containing an N-fertilizer (Nisha *et al.*, 2007). A more recent report showed strong fertilizing activity of diazotrophically cultivated *Nostoc* sp. M2 biomass at the expense of N-poor wastewater (ethanol fermentation vinasse) on wheat, maize, and common beans (Do Nascimento *et al.*, 2019). Importantly, these two reports evidenced soil conditioner properties of the cyanobacterial biomass by raising wheat productivity (Nisha *et al.*, 2007; Do Nascimento *et al.*, 2019) and millet (Nisha *et al.*, 2007) plants under conditions of water scarcity.

Natural arid and semi-arid soils are generally N-limited and BSCs) usually represent the major input of N to sustain trophic chains in those environments. BSCs closely associate with mineral grains and organic matter containing varying proportions of cyanobacteria and algae forming lichens and mosses (Belnap, 2001). BSCs can be permanent at later stages or successional during initial stages of soil formation. Successional stages of crust communities are often characterized by the first stage of substrate colonization by large cyanobacteria of the order Nostocales, comprising filamentous and heterocyst-forming N_2-fixing cyanobacteria, as one of the main contributors of BNF in BSCs of arid and semi-arid soils (Belnap *et al.*, 2001). The next stage is frequently dominated by smaller cyanobacteria and eukaryotic algae, and then by bryophytes and lichens, at the later stages of the succession (Langhans *et al.*, 2009). BSCs influence soil erodibility, availability of organic C and N, soil structure and moisture, soil respiration, UV reflection, and soil temperature regulation. Most of these aspects work concertedly to allow mineralization of biomass to incorporate or to replenish soil nutrients and the evolution of the soil that would allow the development of plants (Abinandan *et al.*, 2019; Crowther *et al.*, 2019; Joshi *et al.*, 2020).

Cyanobacteria and microalgae are also abundant in agroecosystems, accounting together for about 27% of total biomass in agricultural lands (Abinandan *et al.*, 2019). The most predominant cyanobacterial groups in these artificial environments are *Oscillatoriales, Nostocales, Chroococcales,*

Synechococcales, Chroococcidiopsidales, Pleurocapsales, Microcoleaceae, and *Chlorellales* (Triverdi *et al.*, 2016).

Similar to manure from other sources, microalgal and/or cyanobacterial biomass can still exert its soil fertilizing properties even as necromass (nonliving and decomposing biomass) as most of its organic matter constituents, and some bioactive compounds, remain stable and active for some time after cell death and lysis (Abinandan *et al.*, 2019).

Soil N mineralization is usually more prominent in well-aerated soils from more humid climates at lower latitudes reflecting the controls of soil temperature, moisture, and oxygen content over microbial activity (Risch *et al.*, 2019). This N cycling is driven by abiotic decomposition, and biotic assimilative and dissimilative processes, including microorganisms-mediated pathways such as N_2-fixation, nitrification, denitrification, dissimilatory nitrate reduction to ammonium, anaerobic ammonium oxidation, and complete oxidation of ammonium to nitrate (Espenberg *et al.*, 2018; Li *et al.*, 2019). A great variety of microorganisms participate in the process: fungi, bacteria, archaea, animals, protists (Crowther *et al.*, 2019), and also cyanobacteria and microalgae (Trivedi, 2016; Abinandan *et al.*, 2019). Cyanobacteria and microalgae participation in soil mineralization has been comparatively less studied (Abinandan *et al.*, 2019).

A variety of studies identified soil organic C as one of the main drivers of soil net N mineralization across different types of soil ecosystems, which correlated positively with soil microbial biomass and activity (Wang *et al.*, 2017; Espenberg *et al.*, 2018; Crowther *et al.*, 2019; Risch *et al.*, 2019). Some cyanobacteria, as photosynthetic and N_2-fixing primary producers participate largely in soil formation, determination of soil structure, fertility and plant growth, given their roles in regulating the exchanges of elements (mainly O_2, CO_2, N_2) between the atmospheric and soil pools (Crowther *et al.*, 2019). In addition to providing a significant amount of organic C and N into the soil, microalgae and cyanobacteria produce and release other nutritive and/or stimulating products (see above) that enhance proliferation and activity of other soil bacteria, fungi, and invertebrates, which concertedly contribute to organic matter mineralization (Trivedi *et al.*, 2016; Abinandan *et al.*, 2019).

It was reported that alteration of this natural process and shifts in the relative abundance of these elements in the soil in agroecosystems, mainly due by improper management of fertilization, caused losses in microbial biomass and diversity, general soil health, nutrient availability, and fertility (Sahu *et al.*, 2018; Abinandan *et al.*, 2019; Li *et al.*, 2019). Conversely, better crop management promotes increased microbial biomass, C and N mineralization rates, and soil fertility (Wang *et al.*, 2019). Additionally, healthier soils enabled higher crop yields per unit of applied fertilizer, and prevented some N loss into the environment, maintaining soil fertility, and ameliorating the negative environmental impact of fertilization (Wade *et al.*, 2020).

Inorganic phosphate, solubilized by the activity of soil cyanobacteria and /or other microorganisms, gets incorporated by the growing population of the microbial biomass, including cyanobacteria, which temporarily prevents P accessibility to plants. However, upon microbial death, the PO_4^{3-} locked in the biomass initiates a mineralization pathway that again can be incorporated into plants or a new geochemical cycle of microbial biomass (Singh *et al.*, 2016). Soil microalgae and cyanobacteria accumulate in their biomass inorganic phosphate and polyphosphates, which represent easily available forms of phosphate to plants (Mukherjee *et al.*, 2015; Abinandan *et al.*, 2019). Pioneering studies showed that plants uptake more efficiently P from algal organic matter than inorganic PO_4^{3-}, when both were provided in equal amounts over a long period of time (Fuller and Rogers, 1952).

5.5 Cyanobacteria and microalgae mass culture technology

Relying primarily on a cyanobacterial biomass mineralization pathway largely bypasses the odds of bacterial survival and maintenance in the agricultural field. It would also facilitate adjusting the fertilizer application dose and the timing according to the crop's demand.

However, this technology would reduce the production of massive amounts of cyanobacterial biomass. A key aspect for such an endeavor is the production of large amounts of cyanobacterial biomass at the lowest possible cost, ensuring the economic sustainability of the process (Fasahati *et al.*, 2019).

Large-scale cultivation of cyanobacteria has been investigated since the late 1940s (Borowitzka and Borowitzka, 1988).

Much scientific, technological, and economic knowledge has been accumulated since then. This subject has been extensively covered in the literature, mostly related to the production of microalgae biomass as a feedstock for biofuels, but also the production of feed supplements, and wastewater reclamation, among other applications (Brennan and Owende, 2010; Mata *et al.*, 2010; Williams and Laurens, 2010; Novoveská *et al.*, 2016; Laurens *et al.*, 2017; Borowitzka and Vonshak, 2017; Acién Fernández *et al.*, 2020).

Currently, the two main technologies for algal biomass production are classified into open-to-the-environment or closed systems (Qin *et al.*, 2019). The efficiency of these two concepts is frequently assessed using indicators such as their efficiency of light utilization, control of temperature, evaporation, hydrodynamic stress to cells, susceptibility to contamination with other algae or predators/grazers, and the possibility of escalation. These factors dominate the decision on whether to use closed or open ponds, indoors or outdoors, to cultivate cyanobacteria (Pathak *et al.*, 2018). Additionally, the choice of the cultivation systems depends on the quality needed for the final product, and on biological factors of the desired algae in terms of the biochemical composition of the biomass, growth speed and resilience to competitors, specific nutritional, physical and environmental requirements, etc. (Chisti, 2020).

5.5.1 Open systems

5.5.1.1 Raceway ponds

Open ponds (OP) can be grouped into natural systems (e.g., shallow lakes and lagoons), artificial ponds, or containers (Costa *et al.*, 2019). They usually consist of shallow open tanks (typically 0.2–0.5 m deep) exposed to the environment and lined on concrete, clay, or plastic materials. They have a variety of shapes and sizes with different configurations (shallow big ponds, circular ponds, tanks, and raceway ponds) (Davis *et al.*, 2016). Due to their relative low-cost and simple scaling-up, most of the world's microalgae biomass is produced using these systems, especially open raceway ponds (Acién Fernández *et al.*, 2019).

High rate algal ponds (HRAPs) are widely used for algal biomass production (e.g., mostly for *Chlorella* and *Spirulina*), coupled to wastewater treatment (Chisti, 2016). In these systems, culture mixing to optimize cells access to nutrients and light is achieved by the action of paddle wheels, propellers, or pumps which aided by the geometry of the ponds creates a water current at a speed of 0.2–0.5 m/s. As these systems normally involve a minimum power expense for operation, are easier to maintain, and are cleaner than other culture systems, they are the most frequent choice for rural, peri-urban, and remote communities (Acién Fernández *et al.*, 2019).

This is especially important for applications such as biomass for "green manure" fertilizers for which a certain level of biomass contamination or overall quality biomass of the algal/cyanobacterial biomass could be tolerated (Cohen, 1986; Borowitzka and Vonshak, 2017; Young *et al.*, 2017). Nevertheless, the risks of predation and/or contamination by other robust growing competitors and local fauna could be ameliorated by targeting to microalgae or cyanobacteria requiring or tolerating harsh growth conditions. For example, *Spirulina*, able to grow at high pH up to 10, and *Dunaliella*, able to grow at high salt concentrations, can succeed in conditions not permissible for growth of most potential competitors or predators (Varshney *et al.*, 2015; Chisti, 2016).

The main disadvantages of OPs are lower productivity and final biomass density than other systems. This is due to poorer light utilization by the cells, evaporative losses of water, and less efficient utilization of concentrated sources of CO_2 caused by higher diffusion into the atmosphere. Productivity is mostly dependent on the season, which largely limits the geographical locations at which these systems can be operated for high productivity. Maximum biomass productivities up to 20 gm^2/d were consistently observed in temperate climates, especially in the summer season. However, large variability from 5 to 40 gm^2/d was most frequent registered. Only in tropical and subtropical regions, these high daily productivities can be achieved year-round for high annual mean productivities in the range of 90 MT dry biomass/ha/year (Williams and Laurens, 2010; Laurens, 2017).

Demonstrations of large-scale cultivation of N_2-fixing cyanobacteria using OP systems are relatively less. N_2-fixing strains of *Anabaena* and *Nostoc* frequently trap a high content of N within their biomass (8–13% (w/w) of their dry biomass) (Vargas *et al.*, 1998). Moreno and colleagues demonstrated a productivity of 56.6 MT dry biomass/ha/year for *Anabaena* sp. strain ATCC 33047 cultivated in raceway ponds in the south of Spain (Moreno *et al.*, 2003). Thus, according to this productivity, the atmospheric N_2 trapped in a 1-ha facility for *Anabaena* cultivation in raceway ponds in appropriate weather might produce enough fertilizer for about 100–150 ha for crops production area at a rate of 50 kg N/ha/year (Do Nascimento *et al.*, 2019). As a reference, this fertilizer dose would surpass the target set by the Abuja Declaration on Fertilizer for an African Green Revolution at least 50 kg NPK/ha by 2015 (Summit, 2006) which has not been met yet (Njoroge *et al.*, 2018).

A recent study showed the optimization/improvement of ABA, IAA, and SA concentrations in an *Anabaena* sp. cultured diazotrophically in open raceway by manipulating the photosynthetically active radiation (PAR) availability according to culture parameters and control of biomass density. It was proposed that this strategy might offer N fertilizer, upgraded with phytohormone concentrations that could be regularly adjusted to address changing crop needs and crop stressors throughout the season (Wenz *et al.*, 2019).

5.5.1.2 Thin-layer cascades

Thin-layer cascades (TLCs) are another cultivation system that has been developed in the early 1960s and has evolved greatly since then. These systems consist of shallow sloping tanks or channels (made of stainless steel, polyvinyl chloride, polyethylene and zinc-galvanized parts), through which a high concentration of cells is constantly recirculating (Acién *et al.* 2017. Photobioreactors (PBRs) for the production of microalgae. In Microalgae-based biofuels and bioproducts (pp. 1–44). Woodhead Publishing; Grivalský *et al.*, 2019). The operation is divided into two stages: a circulation stage during the day, taking advantage of sunlight, and a storage stage in a retention tank during the dark period. The cells are impelled by a pump to the top of the cascade and then slowly fall by gravity. The current TLC systems allow to regulate the slope of the cascades (0–3%), the thickness of the culture layer (< 50 mm), and the regulation of the flow speed (0.2–0.5 m/s), optimizing irradiation to the culture at different cell densities.

In comparison with OP systems, TLC allows the cultivation of algae at higher cell concentrations (> 10 g/L of dry biomass), much higher productivities (> 30 g m^2/d), and also allows the storage of cells in retention tanks during the night or in case of adverse weather conditions. Also, they are more effective in the self-regulation of temperature, as the culture can be heated up quickly in the morning and/or decrease the temperature by evaporation in case of excessive heat (Grivalský *et al.*, 2019). A recent report using this system demonstrated a biomass productivity of 14.9–21.3 gm^2/d of the microalga *Microchloropsis salina* at a very high density of 30 g cell dry weight per liter (Schädler *et al.*, 2020). Much less information is available for the cultivation of cyanobacteria in TLC systems. In a comparative study, a 30% higher biomass productivity of *Arthrospira maxima* (*Spirulina*) was shown for a TLC in comparison with an OP under both optimal (33 °C) and suboptimal (25 °C) temperatures (Benavides *et al.*, 2017).

The major disadvantages of TLC in comparison to a raceway or circular ponds are higher costs of installation and operation, and an even higher propensity for water evaporation (Grivalský *et al.*, 2019). The evaporation rate could be as high as 5–7 Lm^2/d in shallow ponds and cascade systems operated under warm and dry conditions. Thus, Apel and coworkers proposed an ingenious approach for collecting rainwater using the channel of thin-layer systems while storing the algae suspension in the tank (Apel *et al.*, 2017).

5.5.1.3 *Turf scrubbers*

Algal turf systems (ATS) share some characteristics with the open systems mentioned above (OP and TLC). They are shallow open systems, through which nutrient-rich water flows on a sloping surface where one or more species of algae are attached (Kesaano and Sims, 2014; Siville and Boeing, 2020). The ATS technology was originally introduced in the 1980s mainly for the removal of the N and P during wastewater treatment (e.g., agricultural runoff, aquaculture wastewater, dairy and swine manure effluent, primarily domestic wastewater and secondary effluent) (Adey *et al.*, 1993; Hoffmann, 1998).

ATS are mostly intended to sustain the growth of complex communities of microalgae, cyanobacteria, and other bacteria that assemble spontaneously from microbes in the wastewater. However, species/strain dominance could be modulated by specific inoculations and/or cultivation parameters (Kesaano and Sims, 2014).

Recent reports demonstrated that ATS are capable of producing high yields of biomass up to 35–60 g m^2/d (Liu *et al.*, 2020).

These periphytic or biofilm-forming systems have several advantages over other systems where the cells are suspended. Besides moderate implementation costs, they are easier to operate and provide a simpler and more efficient method of harvesting biomass, which is obtained at very high density. Together, these aspects contribute to reducing biomass production costs (Park *et al.*, 2018).

A disadvantage of ATS is that they tend to be more susceptible to extreme shifts in climate conditions than other submerged systems. This could limit the geographical locations at which ATS can be established for high and consistent productivity. It has been reported to be extremely susceptible to weather conditions, and the time it takes to reestablish the algal turf considerably affects temporal productivity. Hence, it must operate continuously, avoiding interruptions. Large terrestrial grazers such as snails and other organisms often become attracted by this massive concentration of immobilized biomass (Nwoba *et al.*, 2020). However, these two disadvantages can be mitigated by the use of greenhouses to protect the crop. One study demonstrated a tenfold increase in productivity by optimal design of an algae greenhouse with a controlled environment in comparison to an outdoor culture system (Soni *et al.*, 2017).

Examples of on purpose cultivation of N_2-fixing cyanobacteria in ATS are scarce. Nevertheless, the study by Velu and colleagues compared the productivity of an N_2-fixing *Tolypothrix* sp, in vertical bags (suspension cultures) and ATS run side by side. Although productivity largely favored the submerged culture (45–49 g dry weight/m^2/d vs 1.0–1.2 g dry weight g/m^2/d), both biomass density and total water expenditure favored the ATS system. A high enrichment in the *Tolypothrix* sp. biomass produced (94–98% of the photosynthetic community) could have been favored by the use of a defined culture medium lacking any other source of N than atmospheric N_2 (Velu *et al.*, 2015). Overall, this research suggested that *Tolypothrix* sp. (or related cyanobacteria) cultivated in ATS would represent a promising platform for low-cost production of cyanobacterial biomass using low N-containing wastewater for green manure and other applications.

5.5.2 Closed systems

Closed systems or PBRs, are mainly designed to overcome the problems associated with open-pond production systems (Narala *et al.*, 2016).

PBRs vary widely in dimensions and construction materials. They can be classified into several types, mainly dominated by tubular and flat panel reactors (e.g., bag systems, alveolar panels and other flat plate reactors, vertical bubbled columns, and airlift reactors) (Yen *et al.*, 2019).

This system makes it possible to maintain the right growth conditions for a specific strain, not necessarily adapted to growth in extreme conditions, avoiding or at least reducing the risk of contamination by other organisms that may affect the culture. In contrast, to OPs, PBRs allow better control over water evaporation, pH, temperature, reduce carbon dioxide loss, and provide uniform illumination of the culture surface (Narala *et al.*, 2016). PBRs allow a more efficient mass transfer of CO_2 and O_2 which is required for higher biomass productivity and density. However, these systems are difficult to build and operate and require a high initial investment which represents the main limiting factor for their commercialization for low-price commodities, such as fertilizers (Clippinger and Davis, 2019).

Plastic bag PBRs have attracted great interest for large-scale production of the phototrophic organism due to their simplicity and low cost. They consist of large plastic bags of up to 2000 L and approximately 0.5 m in diameter. Currently, plastic bags systems are principally used in aquafeed production. Only a few pilot-scale projects for biodiesel and biofertilizer production from algae have been reported (De Pauw and Persoone G, 1988; Molina-Grima *et al.*, 2009; Velu *et al.*, 2015; Qin *et al.*, 2019).

Flat-plate PBRs usually have a rectangular shape and are built of transparent materials (glass, plexiglass, polycarbonate, etc.) for maximum utilization of solar light energy, with a light path of 1–10 cm and a height of up to 2 m. Cells mixing is often achieved by bubbling air (or CO_2-enriched air) from one side of the chamber to impulse a constant circulation of water (it could be a formal "air-lift" system or not). Although less common, mixing can be done mechanically (Singh and Sharma, 2012). Flat-panel PBRs have potential productivity between of 10.4 and 35.6 g dry weight/m^2/d (Banerjee and Ramaswamy, 2019).

The main disadvantages of PBRs are high costs of construction and operation (including cleaning) and the need to control culture temperature frequently contributes largely to the cost (Tan *et al.*, 2020).

Due to the benefits related to higher operational control for producing high quality and high density microalgal/cyanobacterial slurries, PBRs represent a suitable choice for the preparation of starters to inoculate larger open systems.

5.6 Use of wastewater for cyanobacteria culture

Irrigation is the most important factor in increasing crop production. Recycling of water and nutrients from both industrial and domestic wastewaters for agricultural purposes is conceptually very appealing for overall sustainability, especially in the aridest and poorer regions of the world. The main benefits of wastewater reuse include protection of water resources, prevention of coastal pollution, recovery of water and nutrients for agriculture, savings in clear water use and wastewater treatment costs, etc. (Capra and Scicolone, 2007). The volume of sewage water generated by domestic, industrial, and commercial sources has increased along with the increasing population, urbanization, improved living conditions, and economic development (Qadir *et al.*, 2010). Up to 92% of wastewater in developing low-income countries is discharged untreated and is most frequently used directly or indirectly for food production (Medlicott and Bos, 2014; United Nations World Water Assessment Programme (WWAP), 2016).

Popular practical knowledge in these regions tends to sustain the belief that, at least in the near-term, wastewater, especially untreated one, favors agricultural production and hence food security (Medlicott and Bos, 2014). However, this alternative is not free of technological and/or public health problems. Technological issues are primarily due to lack or poor urbanization and basic infrastructures such as roads, rural electrification, systems for water supply and irrigation in cropping areas (Spencer, 1996), and infrastructure for wastewater treatment and transportation (Medlicott and Bos, 2014). Public health concerns include farmers' and consumers' exposure to pathogens including parasitic helminth worms, protozoa, viruses, and bacteria, and also to organic and inorganic pollutants, such as heavy metals (Qadir *et al.*, 2010; World Health Organization (WHO), 2010).

Farmers and their families are frequently life-long exposed to the infectious and/or toxic effects of untreated wastewater. As a result of mostly cultural reasons in some regions of the world, women are most vulnerable when they are in charge of farming, including watering with wastewater. They also become a vector of these infectious diseases to the rest of the family members, as they are also primarily responsible for preparing the meals. Lack of access to proper treatments tends to exacerbate this public health problem (Qadir *et al.*, 2010; World Health Organization (WHO), 2010).

Numerous studies documented the transfer of toxic heavy metals to vegetables (World Health Organization (WHO), 2010). For example, a study showed that the Ni, Pb, Cd, and Cr content of edible portions of vegetables irrigated with wastewater were 90%, 28%, 83%, and 63% above the safe limits, respectively (Balkhair and Ashraf, 2016).

Besides water itself, the most important constituents that make wastewater attractive as fertilizer are C, N, and P. These nutrients in wastewater originate from human feces and urine, food, detergents, pesticides, pharmaceuticals, industrial inputs, and agricultural drainage. Each cubic meter of domestic wastewater contains 300–600 g organic-C, 40–60 g N (mostly as NH_4^+ and organic-N), and 5–20 g P (PO_4^{3-} and organic-P) (Li *et al.*, 2015).

Similar to the agricultural run-off from heavily fertilized farms (see the introductory section of this chapter), these constituents are the main cause of eutrophication. This process comprises massive proliferation of microalgae and cyanobacteria, oxygen depletion due to oxidation of algal biomass by heterotrophic bacteria, and mayor shifts in the biodiversity of water bodies (Smith *et al.*, 2006; Steffen *et al.*, 2015; Rockström *et al.*, 2020).

More than three decades ago, John Benemann, when at the University of California-Berkeley, proposed a controlled eutrophication approach for capturing N and P from industrial or domestic wastewater in the form of microalgal/cyanobacterial biomass in HRAPs. Separation of the algal biomass by decantation or other means would render treated water together with immobilized and concentrated N and P (and C), to be used mostly as a green manure–type fertilizer (Benemann, 1979; Benemann *et al.*, 2003).

HRAPs are shallow, open raceway ponds, which have been used for treatment of municipal, industrial, and agricultural wastewaters since they were first proposed before 1960 by Oswald and colleagues at University of California-Berkeley. The term "high rate pond" was first used by Oswald to describe agitated shallow raceway ponds which, by maintaining microalgae in suspension, promote algal blooming and a profuse production of photosynthetic oxygen which is used by aerobic heterotrophic bacteria to reduce the biological oxygen demand (BOD) of the wastewater (Oswald *et al.*, 1957; Young *et al.*, 2017). The first designs were mostly targeted to transform solar energy into electrical power by first using wastewater to produce algal biomass, and further processing to produce biogas by anaerobic digestion and then into electrical power (Oswald and Golueke, 1960). However, in the following decades, HRAPs were continuously optimized to produce algal biomass from wastewater for a diversity of purposes, mostly for the production of biofuels (biogas, biodiesel, bioethanol, and bio-crude oil) (Park *et al.*, 2011; Young *et al.*, 2017).

Wastewater treatment HRAPs are normally part of an advanced pond system (APS) which typically comprises a sequence of advanced facultative ponds, anaerobic digestion pits, HRAPs, algal settling ponds, and maturation ponds. APSs require approximately 50 times more land area than activated sludge systems for an equivalent BOD removal but about half of the capital costs for construction (Park *et al.*, 2011) which could be advantageous in rural regions of low-income countries.

A recent comprehensive review by Young and colleagues on the efficiency of HRAPs run over the years in a broad variety of geographical regions and seasons indicated BOD removal rates of 22–93.4% (median of 59%), total N removal of 26.6–75.7% (a median of 61.23%), and P removal of 10.48–97.2% (median of 42.73%). The removal of N is considered to be mainly through incorporation into algal biomass, and via pH-dependent ammonia volatilization and nitrification to a minor extent. P is removed through incorporation into the algal biomass and pH-dependent precipitation. (Young *et al.*, 2017 and references therein).

There is comparatively less information on the removal of heavy metals by HRAPs, but the few existing studies suggested effective removal mainly through adsorption in algal and microbial biomass, including bound and/or detached EPS (Cuellar-Bermudez *et al.*, 2017).

On the other hand, the removal of infectious agents in HRAPs is believed to be mainly dependent on solar irradiance, which is a function of weather, pond depth, and pH. Reported $\log_{10}$ reduction values for *Escherichia coli* to range between 1 and 3.01 $\log_{10}$ *E. coli* most probable number (MPN)/100 mL (median of 1.4 $\log_{10}$ *E. coli* MPN/100 mL) (Young *et al.*, 2017 and references therein).

Emerging pollutants such as pharmaceuticals and person-care products in wastewater can also be detoxified by algae. Pharmaceuticals such as endocrine disruptors and antibiotics are of a rising concern (Arnold *et al.*, 2013). Microalgae and cyanobacteria can remove from negligible amounts to up to 99%, depending on the physicochemical characteristics of the compound, and the cultivation operation parameters such as the hydraulic retention time and temperature (Matamoros *et al.*, 2016).

A comparative study showed that PBR outperformed HRAP in the removal of all standard wastewater parameters, and also produced a higher concentration of algal biomass (Arbib *et al.*, 2013). Nevertheless, the overall performance comparison must take into account the techno-economic variables already discussed in the previous section.

Mixed communities of periphytic microalgae and cyanobacteria, including filamentous cyanobacteria such as *Oscillatoria*, *Phormidium*, *Lyngbya*, green algae, diatoms, bacteria, etc., can also grow at the expense of wastewater, and efficiently remove nutrients therein, even when immobilized onto surfaces. These and similar observations led to the development of the algal turf scrubber, which proved to produce up to 35 g dry weight/m^2/d, which is similar to that most commonly observed in HRAPs. However, ATSs make biomass harvesting largely simpler and leaves fewer cells in the treated water (Adey *et al.*, 1993; Hoffmann, 1998). Optimization of algal biofilms is an active field of research, especially for recycling N and P for animal manure in rural regions. For example, Rajendran and colleagues, recently showed the combination of anaerobic digestion of manure together with wastewater generated in corn ethanol process to culture a surface-attached composite biofilm, mostly composed of polyphosphate accumulating fungi and N-accumulating microalgae. They showed an efficient recovery of nutrients in easily harvestable biomass ready to return nutrients to the soil as a green manure fertilizer. It was proposed that this technology would offer a sustainable alternative to treat the digested manure with energy production. Additionally, it would upgrade the fertilizer properties and value of manure (and wastewater) by facilitating storage and transportation from feedlot to crop areas and by attaining a better relative ratio of nutrients (Rajendran *et al.*, 2018).

Controlling microalgal and cyanobacterial-specific populations, and especially maintaining monoalgal cultures at the expense of wastewater is extremely hard (if possible) to achieve. Most attempts aimed at stabilizing populations of filamentous and/or faster sinking cells to improve biomass harvesting, with limited success (Hoffmann, 1998). The initial composition of the wastewater, especially N, (Beuckels *et al.*, 2015), the hydraulic retention time in HRAPs, or PBRs run in a continuous mode, influence the algal population (Arias *et al.*, 2020). Heterocyst-forming N_2-fixing filamentous cyanobacteria were often observed in wastewater cultures; however, they seldom become a dominant alga, especially in HRAPs operated at low hydraulic retention time. A recent laboratory-scale study, in which the filamentous strains (*Nostoc* sp., *Tolypothrix* sp., and *Calothrix* sp.) were inoculated into wastewater and run continuously in PBRs at different hydraulic retention times, confirmed a higher dominance of the inoculated cyanobacteria at higher retention times. Nevertheless, fragmentation of trichomes and destruction of heterocysts in the *Nostoc* sp. strain, suggested nonfunctional N_2-fixation under the used growing conditions (Arias *et al.*, 2020). This was not surprising as NH_4^+ as the main source of inorganic N in wastewater exerts a strong control to suppress heterocysts differentiation and *nif* genes expression in cyanobacteria (Herrero *et al.*, 2001).

Recent studies showed successful cultivation of desert cyanobacteria at the expense of wastewater, which resulted in the efficient removal of nutrients and suggested a great potential for transferring excessive nutrients from wastewaters to desert areas to increase soil fertility (Wu *et al.*, 2018; Wu *et al.*, 2020). Interestingly this strategy would allow marginal land reclamation for environmental restoration and agricultural extensification.

5.7 Downstream process: harvesting and drying processes

Another bottleneck in the development of large-scale cyanobacterial biomass production systems is the harvesting and drying process. Given the cyanobacterial small size (0.5–60 μm) and the low densities in current cultivation systems (0.3–0.5 dry biomass/L in raceway ponds), the process of harvesting and drying of biomass requires a high amount of energy to eliminate the large volume of culture medium and to obtain algae paste. These processes are complex and together represent more than 40% of the total production cost. A combination of choices for the cyanobacterial strain (large and filamentous strains such as *Anabaena*, *Nostoc*, *Tolypothrix*, etc. are preferred because they sediment faster and/or self-flocculate), culture system for high density of biomass (as discussed before), and methods for biomass harvesting and drying largely determine the economic viability of the whole process (Velu *et al.*, 2015; Kadir *et al.*, 2018; Do Nascimento *et al.*, 2019).

Usually, the collection of cyanobacteria and solid-liquid separation is a two-phase process. The first one, or bulk harvesting, aims at separating the biomass from the suspension, in which it is possible to concentrate the biomass up to 2–7% of total solid matter from the original culture. It includes one or a combination of processes for gravity sedimentation, flocculation, or flotation. The second phase is intended to further concentrate the mixture (thickening), which is achieved by centrifugation, filtration, and ultrasound (or a combination) for a final concentration of solid matter of more than 90% (de Souza *et al.*, 2019).

Once the biomass has been harvested, it is recommended that it be dried for better conservation and durability as a feedstock. Currently, there are several methods among which freeze-drying, sun drying, oven drying, and spray drying are the most common.

Freeze-drying allows better maintenance of cell integrity, but costs for scaling up are very high. Spray drying is one of the most used techniques at an industrial scale. It is characterized by high efficiency, and higher productivity than freeze-drying; however, this type of drying consumes large amounts of energy. On the other hand, direct sun drying is a simple and less expensive method, although it generates some deterioration of the biomass. It is the most recommended and profitable method for small-scale or rural enterprises (Muylaert *et al.*, 2017; Fasaei *et al.*, 2018).

Dry cyanobacterial biomass, without further processing, would be the base of stable green manure, which can be stored and transported from fertilizer production to application locations. Such a product could be applied together with live plant growth–promoting microorganisms, including the same or another N_2-fixing cyanobacterium. This combination would take advantage of the most desired aspects of the mineralization pathway (dry biomass) and continuous release of plant stimulants by live cyanobacteria and/or other microbes.

5.8 Large-scale project for cyanobacterial- or microalgal biomass–based fertilizers

The use of algae has a long history, with the first records of natural populations that used algae for food and medicinal purposes (e.g. *Spirulina* in Mexico and *Nostoc* in Asia). From a commercial perspective, algae have been used since the 1960s, mainly for the development of high value-added products (Borowitzka, 2006; Milledge, 2011).

In recent years, the energy crisis has led to a resurgence of interest in microalgae metabolites, mainly as a feedstock for biofuels. However, the current scenario shows that it is not economically profitable to produce microalgae biomass on a large scale for this sole purpose (Olaizola and Grewe, (2019). Currently, there are some companies still working on the development of biofuels from algal biomass, but the reality indicates that, in order for the process to be profitable, it will have to be accompanied by the production of additional value-added products. The main companies on algae biotechnology have initiated conversion processes by diversifying their research objectives, mainly with the aim of producing biomass for food and nutritional supplements (Su *et al.*, 2017).

Today, a few strains of microalgae are cultivated on an industrial scale for product development, already available in commercial channels, and only high-value applications in these specialized markets are viable due to the high cost of microalgae biomass production, which lies on average between 5 US$/kg and 120 US$/kg of dry matter (Acién Fernández *et al.*, 2019).

Worldwide production of algae biomass exceeds 30,000 t/year and is dominated by the species *Spirulina* (*Arthrospira*), which is used to obtain pigments and food applications, as wells as *Chlorella*. *Haematococcus* and *Dunaliella* are used for their pigments with antioxidant properties (astaxanthin and β-carotene, respectively), while *Nannochloropsis* biomass is used in aquaculture, both for its proteins and its high levels of eicosapentaenoic acid (EPA) and *Crypthecodinium* for docosahexaenoic acid (DHA) production. Worldwide turnover, for the market of products derived from these species, amounts to 700 million US$ per year (Walker *et al.*,2005; Sharma and Sharma, 2017; Camacho *et al.*, 2019).

Although, as mentioned above, cyanobacteria and microalgae biomass have great potential as a feedstock material in emerging markets (bulk chemicals, aquatic food, biofertilizers and wastewater treatment, among others), commercially viable exploitation of these products is currently limited. In addition to cost, this is also due to current limitations to produce the large amount of biomass required by these markets, more than 100 kt/year for a single product with a value below 1.0 €/kg (Acién Fernández *et al.*, 2019).

Table 5.1 shows some representative cyanobacteria and microalgae-based biofertilizer companies with products currently on the market. Only companies with products composed of cyanobacteria and microalgae were considered, leaving out seaweed-based fertilizers.

There is a broad consensus that the development of an algal biorefinery that includes wastewater treatment for the production of cyanobacteria and microalgae is one of the best alternatives to produce large quantities of biomass at low prices, as required for emerging markets (Mehta *et al.*, 2018; Acién Fernández *et al.*, 2019; Chisti, 2020).

TABLE 5.1 Companies currently selling microalgae and/or cyanobacteria-based fertilizers.

Company	web	Country	Since	Product	Composition	Recommended dose
Agroplasma	www.agroplasma.es	Spain	1983	Ferticell	Plant stimulant based on 16 strains of freshwater algae extracts. Provide protein, natural amino acids, microelements, and vitamins, cytokinin, auxin, algae/soil cells extract 25% algae/plant extract solution, 1% potassium	Irrigation: 1 L / 1000 L Foliar spray: 1 L / 300 L Drip irrigation: 0.5 L/ 1000 L
AlgaEnergy	www.algaenergy.com	Spain	2009	AgriAlgae Original and some other variants	Free L-amino acids 4.1 (% w/w); Total nitrogen 1.7 (% w/w); Organic nitrogen 1,4 (% w/w); P_2O_5 0.5 (% w/w) K_2O 8.0 (% w/w); Cu 0.8 (% w/w); phytohormones, vitamins, minerals, pigments, peptides, polyunsaturated fatty acids and polysaccharides	5–7 L · ha. 3–5 applications throughout the cycle
AlgEnerg Pvt. Ltd.	www.algenerg.co.in	India	2019	Blue-green algae biofertilizer	100% powder, presumably from *Anabaena, Nostoc, Oscillatoria*	Not indicated
Algeternal	https://algeternal.com https://elixearth.com	USA	2018	ElixEarth	*Chlorella vulgaris* and *Scenedesmus acuminatus* in approximately equal proportions. 10–15 million cells per milliliter (mL)	1 gallon/10 acres. Apply to the soil every 3 months on row crops or just before planting
Allgrow AB	www.allgrow.se	Sweden	2013	AllGrow	ALLGROW is a high-tech microalgae-based biofertilizer (*Chlorella*). Nitrogen 0.15%, Phosphorus 0.05%, Potassium 0.16%, Amino acid 12%	For cereals foliage fertilization 10–15 L/ha. Concentration max 3%

BioFlora	www.bioflora.com	USA/Mexico	2016	GOgreen IsoGreen	Blend of green microalgae cells (*Chlorella*, *Nannochloris*, and *Scenedesmus*). Enhance the buffering capacity of soils, reducing soil compaction and increasing nutrient uptake.	10 gallons/acre. Applied three times after fumigation at 2–4 week intervals
Biohorizon	www.biorizon.es	Spain		Algafert	*Spirulina* Total N 0.75 % w/w, P_2O_5 7.42% w/w, K_2O 3.2 % w/w), Amino acid 5.6 % w/w	5–10 L/ ha soil application
Heliae	www.phyco-terra.com	USA	2008	Phycoterra	Proprietary whole-cell pasteurized liquid product containing algal-based solid produced from microalgae (*Chlorella*). (9% Organic matter derived from algae/1.5% Phosphoric acid, 0.3% total nitrogen)	1 pint to 1 gallon per acre in a minimum of 5 gallons of finished solution
Junespirulina	www.junespir-ulina.com	Myanmar	1989	Shwe Awzar Spirulina Biofertilizer	*Spirulina*	Not indicated
MDT Denge Tarim	www.dengeta-rim.com.tr	Turkey	2003	Natural Bioplasma	This product contains the suspension of viable green algae of *Chlorella* group, and the following macro and microelements, vitamins, and amino acids	See: http://www.dengetarim.com.tr/en/product/natural-bioplasma,1/4/
Natur agro	https://naturagro.eu/	Hungary	2002	Natur Plasma / Natur Plasma "T"	A rich concentrate of algae enriched by macro, mezo and microelements	5 –10 L/ha

(continued)

TABLE 5.1 (Cont'd)

Company	web	Country	Since	Product	Composition	Recommended dose
Neoalgae Micro Seaweeds Products SL	https://neoalgae.es	Spain	2012	Spiragro Other variants	Enzymatic hydrolysate of *Spirulina.* It is a 100% natural product, with a high content of free amino acids, vitamins, polysaccharides, phytohormones, trace elements and antioxidants	30 mL/10 L of water. Weekly application
Qingdao Seawin Biotech Group Co., Ltd	www.seawin-bio.com	China	2000	Seawinner Extra 30	Microalgae extract powder (brown and green algae), containing 30%, polysaccharides, 35%, Organic matter, 0.5 % Nitrogen, 15%, K_2O)	1–2 kg/ha
Snap Natural and Alginate Products PVT Limited	www.snapalgi-nate.com	India	1979	Kdal Pasi Gel	Seaweed extract, enhanced with *Chroococcus turgidus* and *Chlamydomonas* spp.	3–10 kg per acre depending on the crop
Soil Technologies corporation	www.soiltech-corp.com	USA	1982	Microp 4XL	*Chlamydomonas* and Cyanobacteria	The standard application is 70 grams per hectare, and can be repeated one to four times per year depending on crop conditions
TrueAlgae	www.truealgae.com	USA	2018	Chlopia-P Liquid Fertilizer	100% from microalgae (*Chlorella vulgaris)*	Not indicated

Many national research programs on large-scale cultivation and processing of algal technology have been initiated with the support of local governments and private investment.

United States created programs for the development of the third-generation biofuels (Aquatic Species Program (ASP)) during 1978–1996. US Department of Energy (DOE) have been a major driver of algae research since 2007 and funded four algal biomass biorefineries: Solazyme, Inc, Sapphire Energy, Algenol Biotech ,and BioProcess Algae (Su *et al.*,2017). China and Israel have also built plants for the cultivation of microalgae and the production of blue-green algae also began in India (Spolaore *et al.*, 2006; Milledge, 2011).

The European Union (EU) also has a policy of incorporating and developing green-blue economy strategies at different levels. EU built large capital investments, initially in R+D+i (Research, Development and Innovation) projects related to the production of biofuels based on microalgae. (Lauritano and Ianora, 2018)

In 2008, the Carbon Trust promoted a public investment project, with an approximate investment of £30 million, focused on the study of microalgae biofuels with the aim of reducing production costs. In 2009, England and Ireland started the BioMara project, with an initial budget of €6 million, provided by the EU, with the aim of producing the third-generation biofuels from marine biomass (Su *et al.*,2017). In the last 10 years, the EU has placed as a priority issue, the development of bioproducts capable of increasing crop yields, aquaculture, and also nutrient recovery in the framework of a circular economy, promoting the development of algal biotechnology in these areas (Enzing *et al.*, 2014).

As a result of these policies, several facilities for large-scale algae biomass development have been established in different European countries. Many excellent reviews are already available in the literature (Su *et al.*, 2017; Lauritano and Ianora, 2018; Schiano di Visconte *et al.*, 2019).

In Spain, the company AlgaEnergy was created in 2007 and has recently been selected by the European Commission as one of the Small and medium-sized enterprises with the greatest potential. It has been granted 1.7 M € to finance its business plan between 2017 and 2018. The company is dedicated to the manufacture and marketing of products for industries such as agriculture, nutrition (both human and animal), cosmetics, or aquaculture (Table 5.1).

Other relevant projects on this field include the SABANA project. This project is funded by the European Union under the "Blue Growth" program. The objective is focused on the sustainable use of marine resources for developing a large-scale integrated microalgae-based biorefinery to produce biofertilizers, biopesticides, feed additives, and aquafeed from wastewater. The project has a budget of 10.6 M € and it is led by the University of Almería. The consortium comprises relevant companies as FCC Aqualia and Biorizon Biotech in Spain, as well as prestigious research groups from Italy, Spain, Germany, Hungary, and the Czech Republic. It also has the support of Fundación Cajamar and several other Spanish institutions.

The commercialization of algae-based products still has many challenges ahead to achieve sustainable and profitable industrial development, for which government support is crucial.

Nevertheless, there is a general trend for algae biomass companies to become increasingly available and to offer well-diversified products, taking advantage of advances in biorefineries and circular economy concepts.

Acknowledgments

This work was supported by Grants from the Agencia Nacional de Promoción Científica y Tecnológica, Argentina PICT-3559 and PICT-3382 to LC and Midwestern University Intramural Funds to JH.

References

Abinandan, S., Subashchandrabose, S.R., Venkateswarlu, K., Megharaj, M., 2019. Soil microalgae and cyanobacteria: the biotechnological potential in the maintenance of soil fertility and health. Crit. Rev. Biotechnol. 39 (8), 981–998.

Abboud, Al, A., M., Ghany, T.A., Alawlaqi, M.M, 2014. Role of biofertilizers in agriculture: a brief review. Mycopath 11 (2).

Acién, F.G., Molina, E., Reis, A., Torzillo, G., Zittelli, G.C., Sepúlveda, C., Masojídek, J., 2017. Photobioreactors (PBRs) for the production of microalgae. In Microalgae-based biofuels and bioproducts. Woodhead Publishing, pp. 1–44.

Acién Fernández, F.G., Fernández Sevilla, J.M., Molina Grima, E., 2019. Costs analysis of microalgae production. In: Biofuels from Algae, Second ed. Elsevier B.V. pp. 551–566.

Acién Fernández, F.G., Fernández Sevilla, J.M., Llamas Moya, B., Molina Grima, E., 2020. Microalgae production systems. In: Jacob-Lopes, E., Maroneze, M.M., Queiroz, M.I., Zepka, L.Q. (Eds.), Handbook of Microalgae-Based Processes and Products. Academic Press, Cambridge, MA, pp. 127–163.

Adam, B., Klawonn, I., Svedén, J.B., Bergkvist, J., Nahar, N., Walve, J., Littmann, S., Whitehouse, M.J., Lavik, G., Kuypers, M.M.M., Ploug, H, 2016. N_2-fixation, ammonium release and N-transfer to the microbial and classical food web within a plankton community. ISME J. 10 (2), 450–459.

Adessi, A., de Carvalho, R.C., De Philippis, R., Branquinho, C., da Silva, J.M., 2018. Microbial extracellular polymeric substances improve water retention in dryland biological soil crusts. Soil Biol. Biochem. 116, 67–69.

Adey, W., Luckett, C., Jensen, K., 1993. Phosphorus removal from natural waters using controlled algal production. Restor. Ecol. 1 (1), 29–39.

Aliverti, A., Faber, R., Finnerty, C.M., Ferioli, C., Pandini, V., Negri, A., Karplus, P.A., Zanetti, G., 2001. Biochemical and crystallographic characterization of ferredoxin– NADP+ reductase from nonphotosynthetic tissues. Biochemistry 40 (48), 14501–14508.

Allen, J.F., Thake, B., Martin, W.F., 2019. Nitrogenase inhibition limited oxygenation of earth's Proterozoic atmosphere. Trends Plant Sci. 24 (11), 1022–1031.

Almon, H., Böger, P., 1988. Nitrogen and hydrogen metabolism: induction and measurement. Methods Enzymol. *167*, 459–467.

Almon, H., Böhme, H., 1982. Photophosphorylation in isolated heterocysts from the blue-green alga *Nostoc muscorum*. Biochim. Biophysic. Acta-Bioenergetics *679* (2), 279–286.

Alobwede, E., Leake, J.R., Pandhal, J., 2019. Circular economy fertilization: testing micro and macro algal species as soil improvers and nutrient sources for crop production in greenhouse and field conditions. Geoderma 334, 113–123.

Angert, E.R., 2005. Alternatives to binary fission in bacteria. Nat. Rev. Microbiol. *3* (3), 214–224.

Apel, A.C., Pfaffinger, C.E., Basedahl, N., Mittwollen, N., Göbel, J., Sauter, J., Brück, T., Weuster-Botz, D., 2017. Open thin-layer cascade reactors for saline microalgae production evaluated in a physically simulated Mediterranean summer climate. Algal Res. 25, 381–390.

Apte, S.K., Rowell, P., Stewart, W.D.P, 1978. Electron donation to ferredoxin in heterocysts of the N_2-fixing alga *Anabaena cylindrica*, Proc. R. Soc. Lond. B. Biol. Sci., 200, 1–25.

Arbib, Z., Ruiz, J., Álvarez-Díaz, P., Garrido-Pérez, C., Barragan, J., Perales, J.A., 2013. Long term outdoor operation of a tubular airlift pilot photobioreactor and a high rate algal pond as tertiary treatment of urban wastewater. Ecol. Eng. 52, 143–153.

Arias, D.M., Uggetti, E., García, J., 2020. Assessing the potential of soil cyanobacteria for simultaneous wastewater treatment and carbohydrate-enriched biomass production. Algal Res. 51, 102042.

Arnold, K.E., Boxall, A.B., Brown, A.R., Cuthbert, R.J., Gaw, S., Hutchinson, T.H., Jobling, S., Madden, J.C., Metcalfe, C.D., Naidoo, V., Shore, R.F., Smits, J.E., Taggart, M.A., Thompson, H.M., 2013. Assessing the exposure risk and impacts of pharmaceuticals in the environment on individuals and ecosystems. Biol. Lett. 9, 20130492.

Awasthi, M.K., Chen, H., Liu, T., Awasthi, S.K., Wang, Q., Ren, X., Duan, Y., Zhang, Z., 2019. Respond of clay amendment in chicken manure composts to understand the antibiotic resistant bacterial diversity and its correlation with physicochemical parameters. J. Clean. Prod. 236, 117715.

Bailey-Serres, J., Parker, J.E., Ainsworth, E.A., Oldroyd, G.E.D, 2019. Genetic strategies for improving crop yields. Nature 575, 109–118.

Balkhair, K.S., Ashraf, M.A., 2016. Field accumulation risks of heavy metals in soil and vegetable crop irrigated with sewage water in western region of Saudi Arabia. Saudi J. Biol. Sci. 23 (1), S32–S44.

Banerjee, S., Ramaswamy, S., 2019. Comparison of productivity and economic analysis of microalgae cultivation in open raceways and flat panel photobioreactor. Bioresour. Technol. Rep. 8, 100328.

Barney, B.M., 2020. Aerobic nitrogen-fixing bacteria for hydrogen and ammonium production: current state and perspectives. Appl. Microbiol. Biotechnol. 104 (4), 1383–1399.

Bauer, C.C., Buikema, W.J., Black, K., Haselkorn, R., 1995. A short-filament mutant of *Anabaena* sp. strain PCC 7120 that fragments in nitrogen-deficient medium. J. Bacteriol. 177 (6), 1520–1526.

Bauer, C.C., Scappino, L., Haselkorn, R., 1993. Growth of the cyanobacterium *Anabaena* on molecular nitrogen: NifJ is required when iron is limited, Proc. Natl. Acad. Sci., 90, 8812–8816.

Bebout, B.M., Paerl, H.W., Crocker, K.M., Prufert, L.E., 1987. Diel interactions of oxygenic photosynthesis and N_2 fixation (acetylene reduction) in a marine microbial mat community. Appl. Environ. Microbiol. 53 (10), 2353–2362.

Belnap, J., 2001. Factors influencing nitrogen fixation and nitrogen release in biological soil crusts. In: Belnap, J., Lange, O.L. (Eds.), Biological Soil Crusts: Structure, Function, and Management. Springer, Berlin, Heidelberg, pp. 241–261.

Belnap, J., Büdel, B., Lange, O.L., 2001. Biological soil crusts: characteristics and distribution. In: Belnap, J., Lange, O.L. (Eds.), Biological Soil Crusts: Structure, Function, and Management. Springer, Berlin, Heidelberg, pp. 3–30.

Benavides, A.M.S., Ranglová, K., Malapascua, J.R., Masojídek, J., Torzillo, G, 2017. Diurnal changes of photosynthesis and growth of *Arthrospira platensis* cultured in a thin-layer cascade and an open pond. Algal Res. 28, 48–56.

Benemann, J.R., 1979. Production of nitrogen fertilizer with nitrogen-fixing blue-green algae. Enzyme Microb. Technol. 1, 83–90.

Benemann, J.C.Van Olst, Massingill, M.J., Carlberg, J.A., Weissman, J.C., Brune, D.E., 2003. The controlled eutrophication process: using microalgae for CO_2 utilization and agricultural fertilizer recycling. In: Gale, J., Kaya, Y. (Eds.), Greenhouse Gas Control Technologies, In: Proc. of the 6th International Conference. Pergamon, Oxford, II, 1433–1438.

Berg, I.A., 2011. Ecological aspects of the distribution of different autotrophic CO_2 fixation pathways. Appl. Environ. Microbiol. 77 (6), 1925–1936.

Bergman, B., Gallon, J.R., Rai, A.N., Stal, L.J., 1997. N_2 fixation by non-heterocystous cyanobacteria. FEMS Microbiol. Rev. 19 (3), 139–185.

Bergman, B., Sandh, G., Lin, S., Larsson, J., Carpenter, E.J., 2013. Trichodesmium–a widespread marine cyanobacterium with unusual nitrogen fixation properties. FEMS Microbiol. Rev. 37 (3), 286–302.

Beuckels, A., Smolders, E., Muylaert, K., 2015. Nitrogen availability influences phosphorus removal in microalgae-based wastewater treatment. Water Res. 77, 98–106.

Bloch, S.E., Ryu, M.H., Ozaydin, B., Broglie, R., 2020. Harnessing atmospheric nitrogen for cereal crop production. Curr. Opin. Biotechnol. 62, 181–188.

Böhm, I., Halbherr, A., Smaglinski, S., Ernst, A., Böger, P., 1992. In vitro activation of dinitrogenase reductase from the cyanobacterium *Anabaena variabilis* (ATCC 29413). J. Bacteriol. 174 (19), 6179–6183.

Böhme, H., 1998. Regulation of nitrogen fixation in heterocyst-forming cyanobacteria. Trends Plant Sci. 3 (9), 346–351.

Böhme, H., Schrautemeier, B., 1987. Electron donation to nitrogenase in a cell-free system from heterocysts of *Anabaena variabilis*. Biochim. Biophys. Acta 891, 115–120.

Bongiovanni, R., Lowenberg-DeBoer, J., 2004. Precision agriculture and sustainability. Precis. Agric. 5 (4), 359–387.

Borowitzka, M.A., Vonshak, A., 2017. Scaling up microalgal cultures to commercial scale. Eur. J. Phycol. 52 (4), 407–418.

Borowitzka, M., Borowitzka, L., 1988. Micro-algal biotechnology, Cambridge University Press. ISBN 0-521-323495.

Borowitzka, M.A., 2006. In: Biotechnological & Environmental Applications of Microalgae. Biotechnological & Environmental Applications of Microalgae [Online] 2006. Murdoch University.

Bothe, H., Neuer, G., Kalbe, I., Eisbrenner, G., 1980. Electron donors and hydrogenase in nitrogen-fixing microorganisms. In Stewart, W.D.P., Gallon, J.R. (Eds.), Nitrogen fixation. Academic Press Ltd. London, pp. 83–112.

Bothe, H., Schmitz, O., Yates, M.G., Newton, W.E., 2010. Nitrogen fixation and hydrogen metabolism in cyanobacteria. Microbiol. Mol. Biol. Rev. 74 (4), 529–551.

Brennan, L., Owende, P., 2010. Biofuels from microalgae—a review of technologies for production, processing, and extractions of biofuels and co-products. Renew. Sustain. Energy Rev. 14 (2), 557–577.

Brusca, J.S., Hale, M.A., Carrasco, C.D., Golden, J.W., 1989. Excision of an 11-kilobase-pair DNA element from within the nifD gene in *Anabaena variabilis* heterocysts. J. Bacteriol. 171 (8), 4138–4145.

Buikema, W.J., Haselkorn, R.O.B.E.R.T, 1991. Isolation and complementation of nitrogen fixation mutants of the cyanobacterium *Anabaena* sp. strain PCC 7120. J. Bacteriol. 173 (6), 1879–1885.

Burén, S., Rubio, L.M., 2018. State of the art in eukaryotic nitrogenase engineering. FEMS Microbiol. Lett. 365 (2), fnx274.

Burén, S., Jiménez-Vicente, E., Echavarri-Erasun, C., Rubio, L.M., 2020. Biosynthesis of nitrogenase cofactors. Chem. Rev 120 (12), 4921–4968.

Burén, S., López-Torrejón, G., Rubio, L.M., 2018. Extreme bioengineering to meet the nitrogen challenge. Proc. Natl. Acad. Sci. 115 (36), 8849–8851.

Burnat, M., Herrero, A., Flores, E., 2014. Compartmentalized cyanophycin metabolism in the diazotrophic filaments of a heterocyst-forming cyanobacterium, 111, 3823–3828.

Camacho, F., Macedo, A., Malcata, F., 2019. Potential industrial applications and commercialization of microalgae in the functional food and feed industries: a short review. Marine Drugs 17 (6), 312.

Capone, D.G., Carpenter, E.J., 1982. Nitrogen fixation in the marine environment. Science 217 (4565), 1140–1142.

Capone, D.G., O'Neil, J.M., Zehr, J., Carpenter, E.J., 1990. Basis for diel variation in nitrogenase activity in the marine planktonic cyanobacterium *Trichodesmium thiebautii*. Appl. Environ.Microbiol. 56 (11), 3532–3536.

Capra, A., Scicolone, B., 2007. Recycling of poor quality urban wastewater by drip irrigation systems. J. Clean. Prod. 15 (16), 1529–1534.

Carbonell, V., Vuorio, E., Aro, E.M., Kallio, P., 2019. Enhanced stable production of ethylene in photosynthetic cyanobacterium *Synechococcus elongatus* PCC 7942. World J. Microbiol. Biotechnol. 35 (5), 77.

Carpenter, E.J., Capone, D.G., 1992. Nitrogen fixation in Trichodesmium blooms. In: Carpenter, E.J., Capone, D.G., Rueter, J.G. (Eds.), Marine Pelagic Cyanobacteria: Trichodesmium and Other Diazotrophs. Springer, Dordrecht, pp. 211–217.

Carr, N.G., 1983. Biochemical aspects of heterocyst differentiation and function. In: Papageorgiou, G.C., Paker, L. (Eds.), Biochemical aspects of heterocyst differentiation and function. Photosynthetic Prokaryotes: Cell Differentiation and Function, 265–280.

Carrasco, C.D., Holliday, S.D., Hansel, A., Lindblad, P., Golden, J.W., 2005. Heterocyst-specific excision of the Anabaena sp. strain PCC 7120 hupL element requires xis C. J. Bacteriol. 187 (17), 6031–6038.

Chamizo, S., Adessi, A., Certini, G., De Philippis, R., 2020. Cyanobacteria inoculation as a potential tool for stabilization of burned soilsRestor. Ecol28, S106–S114.

Chen, T., Zhang, S., Yuan, Z., 2020. Adoption of solid organic waste composting products: a critical review. J. Clean. Prod. 272, 122712.

Chen, Y.B., Dominic, B., Mellon, M.T., Zehr, J.P., 1998. Circadian rhythm of nitrogenase gene expression in the diazotrophic filamentous nonheterocystous *Cyanobacterium Trichodesmium* sp. Strain IMS 101. J. Bacteriol. 180 (14), 3598–3605.

Chisti, Y., 2016. Large-scale production of algal biomass: raceway ponds. In: Bux, F., Chisti, Y. (Eds.), Algae Biotechnology. Springer, Cham, pp. 21–40.

Chisti, Y., 2020. Microalgae biotechnology: a brief introduction. In: Jacob-Lopes, E., Maroneze, M.M., Queiroz, M.I., Zepka, L.Q. (Eds.), Handbook of Microalgae-Based Processes and Products. Academic Press, pp. 3–23.

Chittora, D., Meena, M., Barupal, T., Swapnil, P., 2020. Cyanobacteria as a source of biofertilizers for sustainable agriculture. Biochem. Biophys. Rep. 22 January, 100737.

Chojnacka, K, Witek-Krowiak, A, Moustakas, K, Skrzypczak, D, Mikula, K, Loizidou, M, 2020. A transition from conventional irrigation to fertigation with reclaimed wastewater: prospects and challenges. Renew. Sustain. Energy Rev 130, 109959.

Choudhury, A.T.M.A., Kennedy, I.R, 2004. Prospects and potentials for systems of biological nitrogen fixation in sustainable rice production. Biol. Fertil. Soils 39 (4), 219–227.

Ciais, P., Sabine, C., Bala, G., Bopp, L., Brovkin, V., Canadell, J., Chhabra, A., DeFries, R., Galloway, J., Heimann, M., Jones, C., Quéré, C.Le, Myneni, R.B., Piao, S., Thornton, P., 2013. The Physical Science Basis. Contribution of Working Group I to the Fifth Assessment Report of the Intergovernmental Panel on Climate Change. Carbon and

Other Biogeochemical Cycles. In: Stocker, T.F., Qin, D., Plattner, G.-K., Tignor, M., Allen, S.K., Boschung, J., Nauels, A., Xia, Y., Bex, V., Midgley, P.M. (Eds.), Climate Change 2013. Cambridge University Press, Cambridge, United Kingdom and New York, NY, USA, pp. 465–570.

Ciceri, D., Allanore, A., 2019. Local fertilizers to achieve food self-sufficiency in Africa. Sci. Total Environ. 648, 669–680.

Clippinger, J.N., Davis, R.E., 2019. Techno-economic analysis for the production of algal biomass via closed photobioreactors: future cost potential evaluated across a range of cultivation system designs (No. NREL/TP-5100-72716). National Renewable Energy Lab. (NREL).

Cohen, Z., 1986. Products from Microalgae. In: Richmond A (Ed.), Handbook of Microalgal Mass Culture. CRC Press, Boca-Raton, Fl, pp. 421–454.

Colica, G., De Philippis, R., 2013. Exopolysaccharides from cyanobacteria and their possible industrial applications. In: Sharma, N.K., Rai, A.K., Stal, L.J. (Eds.), Cyanobacteria: An Economic Perspective. John Wiley & Sons, NJ, pp. 197–208.

Colón-López, M.S., Sherman, D.M., Sherman, L.A., 1997. Transcriptional and translational regulation of nitrogenase in light-dark-and continuous-light-grown cultures of the unicellular cyanobacterium *Cyanothece* sp. strain ATCC 51142. J. Bacteriol. 179 (13), 4319–4327.

Coppens, J., Grunert, O., Van Den Hende, S., Vanhoutte, I., Boon, N., Haesaert, G., De Gelder, L., 2016. The use of microalgae as a high-value organic slow-release fertilizer results in tomatoes with increased carotenoid and sugar levels. J. Appl. Phycol. 28 (4), 2367–2377.

Costa, J.A.V., Freitas, B.C.B., Santos, T.D., Mitchell, B.G., Morais, M.G, 2019. Open pond systems for microalgal culture. In: Pandey, A., Lee, D-J., Chisti, Y., Soccol, C.R. (Eds.), Biofuels from Algae. Elsevier, Amsterdam, pp. 199–223.

Crowther, T.W., Van den Hoogen, J., Wan, J., Mayes, M.A., Keiser, A.D., Mo, L., Averill, C., Maynard, D.S., 2019. The global soil community and its influence on biogeochemistry. Science 365 (6455), eaav0550.

Cuellar-Bermudez, S.P., Aleman-Nava, G.S., Chandra, R., Garcia-Perez, J.S., Contreras-Angulo, J.R., Markou, G., Muylaert, K., Rittmann, B.E., Parra-Saldivar, R., 2017. Nutrients utilization and contaminants removal. A review of two approaches of algae and cyanobacteria in wastewater. Algal Res. 24, 438–449.

Curatti, L., Rubio, L.M., 2014. Challenges to develop nitrogen-fixing cereals by direct nif-gene transfer. Plant Sci. 225, 130–137.

Curatti, L., Flores, E., Salerno, G., 2002. Sucrose is involved in the diazotrophic metabolism of the heterocyst-forming cyanobacterium *Anabaena* sp. FEBS Lett. 513 (2-3), 175–178.

Davis, R., Markham, J., Kinchin, C., Grundl, N., Tan, E.C., Humbird, D., 2016. Process Design and Economics for the Production of Algal biomass: Algal Biomass Production in Open Pond Systems and Processing Through Dewatering for Downstream Conversion (No. NREL/TP-5100-64772). National Renewable Energy Lab. (NREL), Golden, CO, United States.

De Pauw, N., Persoone, G., 1988. Micro-algae for aquaculture. In Borowitzka, M.A., Borowitzka, L.J. (Eds.), Micro-algal Biotechnology. Cambridge University Press, Cambridge, pp 197–221.

de Souza, M.P., Hoeltz, M., Brittes Benitez, L., Machado, Ê.L., de Souza Schneider, R.D.C, 2019. Microalgae and clean technologies: a review. CLEAN–Soil, Air, Water, 47 (11), 1800380.

Di Pippo, F., Ellwood, N.T., Gismondi, A., Bruno, L., Rossi, F., Magni, P., De Philippis, R., 2013. Characterization of exopolysaccharides produced by seven biofilm-forming cyanobacterial strains for biotechnological applications. J. Appl. Phycol. 25 (6), 1697–1708.

Do Nascimento, M., Battaglia, M.E., Rizza, L.S., Ambrosio, R., Di Palma, A.A., Curatti, L., 2019. Prospects of using biomass of N_2-fixing cyanobacteria as an organic fertilizer and soil conditioner. Algal Res. 43, 101652.

Drechsel, P., Heffer, P., Magen, H., Mikkelsen, R., Singh, H., Wichelns, D., 2015. Managing water and nutrients to ensure global food security, while sustaining ecosystem services. In Drechsel, P., Heffer, P., Magen, H., Mikkelsen, R., Wichelns, D. (Eds.), Managing Water and Fertilizer for Sustainable Agricultural Intensification. International Fertilizer Industry Association (IFA), International Water Management Institute (IWMI), International Plant Nutrition Institute (IPNI), and International Potash Institute (IPI). First edition, Paris, France, pp. 1–7. ISBN 979-10-92366-02-0.

Drechsel, P., Keraita, B., Amoah, P., Karg, H., 2014. Health risk management for safe vegetable irrigation. In Drechsel, P., Keraita, B. (Eds.) Irrigated urban vegetable production in Ghana: characteristics, benefits and risk mitigation. Colombo, Sri Lanka: International Water Management Institute (IWMI), pp. 180–198.

Drews, G., Weckesser, J., 1982. Function, structure and composition of cell wall and external layers. In: Carr, N.G., Whitton, B.A. (Eds.), The Biology of Cyanobacteria (Vol. 19). University of California Press. Berkeley and Los Angeles, United States, p. 333–359.

Durner, J., Böhm, I., Hilz, H., Böger, P., 1994. Posttranslational modification of nitrogenase: differences between the purple bacterium *Rhodospirillum rubrum* and the cyanobacterium *Anabaena variabilis*. Eur. J. Biochem. 220 (1), 125–130.

Dutton, P.L., 1986. Energy transduction in anoxygenic photosynthesis. In: Staehelin, L.A., Arntzen, C.J. (Eds.), Photosynthesis III. Springer, Berlin, Heidelberg, pp. 197–237.

Egener, T., Martin, D.E., Sarkar, A., Reinhold-Hurek, B., 2001. Role of a ferredoxin gene cotranscribed with thenifHDK operon in N_2 fixation and nitrogenase "Switch-Off" of *Azoarcus* sp Strain BH72. J. Bacteriol. 183 (12), 3752–3760.

Elhai, J., Wolk, C.P., 1990. Developmental regulation and spatial pattern of expression of the structural genes for nitrogenase in the cyanobacterium *Anabaena*. EMBO J. 9 (10), 3379–3388.

Emerson, S.R., Huested, S.S., 1991. Ocean anoxia and the concentrations of molybdenum and vanadium in seawater. Marine Chem. 34 (3–4), 177–196.

Enzing, C., Ploeg, M., Barbosa, M., Sijtsma, L., 2014. Microalgae-based products for the food and feed sector: an outlook for Europe. JRC Sci. Policy Rep., 19–37.

Ernst, A., Black, T., Cai, Y., Panoff, J.M., Tiwari, D.N., Wolk, C.P., 1992. Synthesis of nitrogenase in mutants of the cyanobacterium *Anabaena* sp. strain PCC 7120 affected in heterocyst development or metabolism. J. Bacteriol. 174 (19), 6025–6032.

Ernst, A., Liu, Y.D., Reich, S., Böger, P., 1990. Diurnal nitrogenase modification in the cyanobacterium *Anabaena variabilis*. Bot. Acta 103 (2), 183–189.

Espenberg, M., Truu, M., Mander, Ü., Kasak, K., Nõlvak, H., Ligi, T., Oopkaup, K., Maddison, M., Truu, J., 2018. Differences in microbial community structure and nitrogen cycling in natural and drained tropical peatland soils. Sci. Rep. 8 (1), 4742.

Esteves-Ferreira, A.A., Cavalcanti, J.H.F., Vaz, M.G.M.V., Alvarenga, L.V., Nunes-Nesi, A., Araújo, W.L, 2017. Cyanobacterial nitrogenases: phylogenetic diversity, regulation and functional predictions. Genet. Mol. Biol. 40 (1), 261–275.

Falkowski, P.G., 1997. Evolution of the nitrogen cycle and its influence on the biological sequestration of CO_2 in the ocean. Nature 387 (6630), 272–275.

FAO, IFAD, UNICEF, WFP and WHO, 2019. The State of Food Security and Nutrition in the World 2019. Safeguarding against economic slowdowns and downturns. Rome, FAO. Licence: CC BY-NC-SA 3.0 IGO.

FAO, IFAD, UNICEF, WFP and WHO, 2020. The State of Food Security and Nutrition in the World 2020. Transforming Food Systems for Affordable Healthy Diets, Rome, FAO.

FAO, 2019. The State of Food and Agriculture 2019. Moving Forward on Food Loss and Waste Reduction, Rome.

Fasaei, F., Bitter, J.H., Slegers, P.M., Van Boxtel, A.J.B, 2018. Techno-economic evaluation of microalgae harvesting and dewatering systems. Algal Res. 31, 347–362.

Fasahati, P., Wu, W., Maravelias, C.T., 2019. Process synthesis and economic analysis of cyanobacteria biorefineries: a superstructure-based approach. Appl. Energy 253, 113625.

Fay, P., 1992. Oxygen relations of nitrogen fixation in cyanobacteria. Microbiol. Molecul. Biol. Rev. 56 (2), 340–373.

Finkelstein, R.R., Gampala, S.S., Rock, C.D., 2002. Abscisic acid signaling in seeds and seedlings. Plant Cell 14 (suppl 1), S15–S45.

Flores, E., Herrero, A., 2010. Compartmentalized function through cell differentiation in filamentous cyanobacteria. Nat. Rev. Microbiol. 8 (1), 39–50.

Flores, E., Herrero, A., Forchhammer, K., Maldener, I., 2016. Septal junctions in filamentous heterocyst-forming cyanobacteria. Trends Microbiol. 24 (2), 79–82.

Flores, E., Herrero, A., Wolk, C.P., Maldener, I., 2006. Is the periplasm continuous in filamentous multicellular cyanobacteria? Trends Microbiol. 14 (10), 439–443.

Flores, E., Muro-Pastor, A.M., Herrero, A., 1999. Cyanobacterial nitrogen assimilation genes and NtcA-dependent control of gene expression. In: Peschek, G.A., Löffelhardt, W., Schmetterer, G. (Eds.), The Phototrophic Prokaryotes. Springer, Boston, MA, pp. 463–477.

Flores, E., Pernil, R., Muro-Pastor, A.M., Mariscal, V., Maldener, I., Lechno-Yossef, S., Fan, Q., Wolk, C.P., Herrero, A., 2007. Septum-localized protein required for filament integrity and diazotrophy in the heterocyst-forming cyanobacterium *Anabaena* sp. strain PCC 7120. J. Bacteriol. 189 (10), 3884–3890.

Foster, R.A., Zehr, J.P., 2019. Diversity, genomics, and distribution of phytoplankton-cyanobacterium single-cell symbiotic associations. Annu. Rev. Microbiol. 73, 435–456.

Fredriksson, C., Bergman, B., 1995. Nitrogenase quantity varies diurnally in a subset of cells within colonies of the non-heterocystous cyanobacteria *Trichodesmium* spp. Microbiology 141 (10), 2471–2478.

Fuchs, G., 2011. Alternative pathways of carbon dioxide fixation: insights into the early evolution of life? Annu. Rev. Microbiol. 65, 631–658.

Fuller, W.H., Rogers, R.N., 1952. Utilization of the phosphorus of algal cells as measured by the Neubauer technique. Soil Sci. 74 (6), 417–430.

Gallon, J.R., 1980. Nitrogen fixation by photoautotrophs, Proc. Phytochemical Society Of Europe Symposium. Sussex.

Gallon, J.R., Hamadi, A.F., 1984. Studies on the effects of oxygen on acetylene reduction (nitrogen fixation) in *Gloeothece* sp. ATCC 27152. Microbiology 130 (3), 495–503.

Garcia-Gonzalez, J., Sommerfeld, M., 2016. Biofertilizer and biostimulant properties of the microalga *Acutodesmus dimorphus*. J. Appl. Phycol. 28 (2), 1051–1061.

Garcia-Pichel, F., Belnap, J., Neuer, S., Schanz, F., 2003. Estimates of global cyanobacterial biomass and its distribution. Algol. Stud. 109 (1), 213–227.

Garcia-Pichel, F., Zehr, J.P., Bhattacharya, D., Pakrasi, H.B., 2020. What's in a name? The case of cyanobacteria. J. Phycol. 56 (1), 1–5.

Garlapati, D., Chandrasekaran, M., Devanesan, A.A., Mathimani, T., Pugazhendhi, A., 2019. Role of cyanobacteria in agricultural and industrial sectors: an outlook on economically important byproducts. Appl. Microbiol. Biotechnol. 103 (12), 4709–4721.

Gennaro, G., Hübner, P., Sandmeier, U., Yakunin, A.F., Hallenbeck, P.C., 1996. Cloning, characterization, and regulation of nifF from *Rhodobacter capsulatus*. J. Bacteriol. 178 (13), 3949–3952.

Giani, D., Krumbein, W.E., 1986. Growth characteristics of non-heterocystous cyanobacterium *Plectonema boryanum* with N_2 as nitrogen source. Arch. Microbiol. 145 (3), 259–265.

Giddings, T.H., Staehelin, L.A., 1978. Plasma membrane architecture of *Anabaena cylindrica*: occurrence of microplasmodesmata and changes associated with heterocyst development and the cell cycle. Cytobiologie 16, 235–249.

Gonzalez López, C.V., Fernández, F.A., Sevilla, J.F., Fernández, J.S., García, M.C., Grima, E.M., 2009. Utilization of the cyanobacteria *Anabaena* sp. ATCC 33047 in CO_2 removal processes. Bioresour. Technol. 100 (23), 5904–5910.
Griggs, D., Stafford-Smith, M., Gaffney, O., Rockström, J., Öhman, M.C., Shyamsundar, P., Steffen, W., Glaser, G., Kanie, N., Noble, I., 2013. Sustainable development goals for people and planet. Nature 495, 305–307.
Grivalský, T., Ranglová, K., da Câmara Manoel, J.A., Lakatos, G.E., Lhotský, R., Masojídek, J., 2019. Development of thin-layer cascades for microalgae cultivation: milestones. Folia Microbiol. 64, 603–6142.
Gutekunst, K., Chen, X., Schreiber, K., Kaspar, U., Makam, S., Appel, J., 2014. The bidirectional NiFe-hydrogenase in *Synechocystis* sp. PCC 6803 is reduced by flavodoxin and ferredoxin and is essential under mixotrophic, nitrate-limiting conditions. J. Biol. Chem. 289 (4), 1930–1937.
Haroun, S.A., Hussein, M.H., 2003. The promotive effect of algal biofertilizers on growth, protein pattern and some metabolic activities of *Lupinus termis* plants grown in siliceous soil. Asian J. Plant Sci 13 (2), 944–951.
Hartung, W., 2010. The evolution of abscisic acid (ABA) and ABA function in lower plants, fungi and lichen. Funct. Plant Biol. 37 (9), 806–812.
Haselkorn, R., 1978. Heterocysts. Annu. Rev. Plant Physiol. 29 (1), 319–344.
Haselkorn, R., 1992. Developmentally regulated gene rearrangements in prokaryotes. Annu. Rev. Genet. 26 (1), 113–130.
Haselkorn, R., Buikema, W.J., 1992. Nitrogen Fixation in Cyanobacteria. In: Stacey, G., Burris, R.H., Evans, H.J. (Eds.), Biological Nitrogen Fixation. Chapman & Hall, New York, pp. 166–190.
Hashtroudi, M.S., Ghassempour, A., Riahi, H., Shariatmadari, Z., Khanjir, M., 2013. Endogenous auxins in plant growth-promoting Cyanobacteria—*Anabaena vaginicola* and *Nostoc calcicola*. J. Appl. Phycol. 25 (2), 379–386.
Hernandez, J.A., George, S.J., Rubio, L.M., 2009. Molybdenum trafficking for nitrogen fixation. Biochemistry 48 (41), 9711–9721.
Herrero, A., Muro-Pastor, A.M., Flores, E., 2001. Nitrogen control in cyanobacteria. J. Bacteriol. 183 (2), 411–425.
Herrero, A., Picossi, S., Flores, E., 2013. Gene expression during heterocyst differentiation-Adv. Bot. Res65, 281–329.
Herrero, A., Stavans, J., Flores, E., 2016. The multicellular nature of filamentous heterocyst-forming cyanobacteria. FEMS Microbiol. Rev. 40 (6), 831–854.
Herridge, D.F., Peoples, M.B., Boddey, R.M., 2008. Global inputs of biological nitrogen fixation in agricultural systems. Plant Soil 311 (1–2), 1–18.
Heyer, H., Stal, L., Krumbein, W.E., 1989. Simultaneous heterolactic and acetate fermentation in the marine cyanobacterium *Oscillatoria limosa* incubated anaerobically in the dark. Arch. Microbiol. 151 (6), 558–564.
Hill, DR, Belbin, TJ, Thorsteinsson, M V, Bassam, D, Brass, S, Ernst, A, Böger, P, Paerl, H, Mulligan, ME, Potts, M, 1996. GlbN (cyanoglobin) is a peripheral membrane protein that is restricted to certain *Nostoc* spp. J. Bacteriol. 178, 6587–6598.
Hoffmann, J.P., 1998. Wastewater treatment with suspended and nonsuspended algae. J. Phycol. 34 (5), 757–763.
Hoiczyk, E., Hansel, A., 2000. Cyanobacterial cell walls: news from an unusual prokaryotic envelope. J. Bacteriol. 182 (5), 1191–1199.
Howarth, R.W., Marino, R., Cole, J.J., 1988. Nitrogen fixation in freshwater, estuarine, and marine ecosystems. 2. Biogeochemical controls 1. Limnol. Oceanogr. 33 (4part2), 688–701.
Huang, T.C., Chow, T.J., 1986. New type of N_2-fixing unicellular cyanobacterium (blue-green alga). FEMS Microbiol. Lett. 36 (1), 109–110.

Huang, T.C., Chow, T.J., 1988. Comparative studies of some nitrogen-fixing unicellular cyanobacteria isolated from rice fields. Microbiology 134 (12), 3089–3097.

Huang, T.C., Lin, R.F., Chu, M.K., Chen, H.M., 1999. Organization and expression of nitrogen-fixation genes in the aerobic nitrogen-fixing unicellular cyanobacterium *Synechococcus* sp. strain RF-1. Microbiology 145 (3), 743 753.

Husain, T., Fatima, A., Suhel, M., Singh, S., Sharma, A., Prasad, S.M., Singh, V.P., 2020. A brief appraisal of ethylene signaling under abiotic stress in plants. Plant Signal Behav. 15 (9), 1782051.

Hussain, A., Hasnain, S., 2011. Interactions of bacterial cytokinins and IAA in the rhizosphere may alter phytostimulatory efficiency of rhizobacteria. World J. Microbiol. Biotechnol. 27 (11), 2645.

Hussain, A., Hamayun, M., Shah, S.T., 2013. Root colonization and phytostimulation by phytohormones producing entophytic *Nostoc* sp. AH-12. Curr. Microbiol. 67 (5), 624–630.

Inomura, K., Bragg, J., Follows, M.J., 2017. A quantitative analysis of the direct and indirect costs of nitrogen fixation: a model based on *Azotobacter vinelandii*. ISME J. 11 (1), 166–175.

Issa, A.A., Abd-Alla, M.H., Ohyama, T., 2014. Nitrogen fixing cyanobacteria: future prospect. In: Ohyama, T. (Ed.), Nitrogen fixing cyanobacteria: future prospect. Advances in Biology and Ecology of Nitrogen Fixation, 24–48.

Jagannathan, B., Golbeck, J.H., 2009. Photosynthesis: microbial. In: Schaechter, M. (Ed.), Encyclopedia of Microbiology. Elsevier Inc, Amsterdam, pp. 325–341.

Janson, S., Edward, J., Carpenter, E.J., Birgitta Bergman, B., 1994. Compartmentalisation of nitrogenase in a non-heterocystous cyanobacterium: *Trichodesmium contortum*. FEMS Microbiol. Lett. 118, 9–14.

Jones, K., 1992. Diurnal nitrogen fixation in tropical marine cyanobacteria: a comparison between adjacent communities of non-heterocystous *Lyngbya* sp. and heterocystous *Calothrix* sp. Br. Phycol. J. 27 (2), 107–118.

Joshi, H., Shourie, A., Singh, A., 2020. Cyanobacteria as a source of biofertilizers for sustainable agriculture. In: Singh, P., Kumar, A., Singh, V.K., Shrivistava, A. (Eds.), Advances in Cyanobacterial Biology. Elsevier, Amsterdam, pp. 385–396.

Kadir, W.N.A., Lam, M.K., Uemura, Y., Lim, J.W., Lee, K.T, 2018. Harvesting and pre-treatment of microalgae cultivated in wastewater for biodiesel production: a review. Energy Convers. Manag. 171, 1416–1429.

Kannaiyan, S., Aruna, S.J., Kumari, S.M.P., Hall, D.O, 1997. Immobilized cyanobacteria as a biofertilizer for rice crops, Intl. Conference on Applied Algology, Knysna, South Africa, April 1996. J. Appl. Phycol. 9 (2), 167–174.

Karl, D.M., Letelier, R., Hebel, D.V., Bird, D.F., Winn, C.D., 1992. Trichodesmium blooms and new nitrogen in the North Pacific gyre. In: Carpenter, E.J., Capone, D.G., Rueter, J.G. (Eds.), Marine Pelagic Cyanobacteria: Trichodesmium and Other Diazotrophs. Springer, Dordrecht, pp. 219–237.

Kesaano, M., Sims, R.C., 2014. Algal biofilm based technology for wastewater treatment. Algal Res. 5, 231–240.

Khanna, N., Lindblad, P., 2015. Cyanobacterial hydrogenases and hydrogen metabolism revisited: recent progress and future prospects. Int. J. Mol. Sci. 16 (5), 10537–10561.

Kleijn, D., Bommarco, R., Fijen, T.P., Garibaldi, L.A., Potts, S.G., van der Putten, W.H., 2019. Ecological intensification: bridging the gap between science and practice. Trends Ecol. Evol. 34 (2), 154–166.

Kuraganti, G., Edla, S., Pallaval, V.B., 2020. Cyanobacteria as biofertilizers: current research, commercial aspects, and future challenges. In: Yadav, A., Rastegari, A., Yadav, N., Kour, D. (Eds.). Advances in Plant Microbiome and Sustainable Agriculture. Microorganisms for Sustainability, 20. Springer, Singapore.

Lakaniemi, A.M., Intihar, V.M., Tuovinen, O.H., Puhakka, J.A., 2012. Growth of *Chlorella vulgaris* and associated bacteria in photobioreactors. Microb. Biotechnol. 5 (1), 69–78.

Langhans, T.M., Storm, C., Schwabe, A., 2009. Community assembly of biological soil crusts of different successional stages in a temperate sand ecosystem, as assessed by direct determination and enrichment techniques. Microb. Ecol. 58 (2), 394–407.

Latorre, C., Lee, J.H., Spiller, H., Shanmugam, K.T., 1986. Ammonium ion-excreting cyanobacterial mutant as a source of nitrogen for growth of rice: a feasibility study. Biotechnol. Lett. 8 (7), 507–512.

Laurens, L.M., Chen-Glasser, M., McMillan, J.D., 2017. A perspective on renewable bioenergy from photosynthetic algae as feedstock for biofuels and bioproducts. Algal Res. 24, 261–264.

Lauritano, C., Ianora, A., 2018. Grand challenges in marine biotechnology: Overview of recent EU-funded projects. In: Rampelotto, P.H., Trincone, A. (Eds.), Grand Challenges in Marine Biotechnology. Springer Nature, Switzerland AG, pp. 425–449.

Li, S., Peng, C., Cheng, T., Wang, C., Guo, L., Li, D., 2019. Nitrogen-cycling microbial community functional potential and enzyme activities in cultured biofilms with response to inorganic nitrogen availability. J. Environ. Sci. 76, 89–99.

Li, S., Wang, C., Qin, H., Li, Y., Zheng, J., Peng, C., Li, D., 2016. Influence of phosphorus availability on the community structure and physiology of cultured biofilms. J. Environ. Sci. 42, 19–31.

Li, W-W., Yu, H-Q., Rittmann, B.E., 2015. Chemistry: Reuse water pollutants. Nature 528, 29–31.

Liu, D., Liberton, M., Yu, J., Pakrasi, H.B., Bhattacharyya-Pakrasi, M., 2018. Engineering nitrogen fixation activity in an oxygenic phototroph. MBio 9 (3), e01029 –18.

Liu, J., Pemberton, B., Lewis, J., Scales, P.J., Martin, G.J., 2020. Wastewater treatment using filamentous algae–a review. Bioresour. Technol. 298, 122556.

Lopez-Gomollon, S., Hernandez, J.A., Pellicer, S., Angarica, V.E., Peleato, M.L., Fillat, M.F., 2007. Cross-talk between iron and nitrogen regulatory networks in Anabaena (Nostoc) sp. PCC 7120: identification of overlapping genes in FurA and NtcA regulons. J. Mol. Biol. 374 (1), 267–281.

MacFarlane, D.R., Choi, J., Suryanto, B.H., Jalili, R., Chatti, M., Azofra, L.M., Simonov, A.N., 2020. Liquefied sunshine: transforming renewables into fertilizers and energy carriers with electromaterials. Adv. Mater. 32 (18), 1904804.

Mager, D.M., Thomas, A.D., 2011. Extracellular polysaccharides from cyanobacterial soil crusts: a review of their role in dryland soil processes. J. Arid Environ. 75 (2), 91–97.

Magnuson, A., 2019. Heterocyst thylakoid bioenergetics. Life 9 (1), 13.

Mahanty, T., Bhattacharjee, S., Goswami, M., Bhattacharyya, P., Das, B., Ghosh, A., Tribedi, P., 2017. Biofertilizers: a potential approach for sustainable agriculture development. Environ. Sci. Pollut. Res. 24 (4), 3315–3335.

Mahapatra, D.M., Chanakya, H.N., Joshi, N.V., Ramachandra, T.V., Murthy, G.S., 2018. Algae-based biofertilizers: a biorefinery approach. In: Panpatte, D., Jhala, Y., Shelat, H., Vyas, R. (Eds.). Microorganisms for Green Revolution. Microorganisms for Sustainability, 7. Springer, Singapore.

Manning, D.A.C, 2010. Mineral sources of potassium for plant nutrition. A review.Agron. Sustain. Dev.30, 281–294.

Margulis, L., 1981. Symbiosis in Cell Evolution: Life and Its Environment on the Early Earth. W.H. Freeman & Co. Boston University Boston, MA, United States. ISBN 9780716712558.

Mariscal, V., 2014. Cell-cell joining proteins in heterocyst-forming cyanobacteria. In: Flores, E., Herrero, A. (Eds.), The Cell Biology of Cyanobacteria. Caister Academic Press, U.K., pp. 293–304.

Mariscal, V., Herrero, A., Flores, E., 2007. Continuous periplasm in a filamentous, heterocyst-forming cyanobacterium. Mol. Microbiol. 65 (4), 1139–1145.

Maršálek, B., Šimek, M., 1992. Abscisic acid and its synthetic analog in relation to growth and nitrogenase activity of *Azotobacter chroococcum* and *Nostoc muscorum*. Folia Microbiol. 37 (2), 159–160.

Martin, A.E., Burgess, B.K., Iismaa, S.E., Smartt, C.T., Jacobson, M.R., Dean, D.R., 1989. Construction and characterization of an *Azotobacter vinelandii* strain with mutations in the genes encoding flavodoxin and ferredoxin I. J. Bacteriol. *171* (6), 3162–3167.
Martín-Figueroa, E., Navarro, F., Florencio, F.J., 2000. The GS-GOGAT pathway is not operative in the heterocysts. Cloning and expression of glsF gene from the cyanobacterium *Anabaena* sp. PCC 7120. FEBS Lett. 476 (3), 282–286.
Maryan, P.S., Eady, R.R., Chaplin, A.E., Gallon, J.R., 1986. Nitrogen fixation by the unicellular cyanobacterium Gloeothece. Nitrogenase synthesis is only transiently repressed by oxygen. FEMS Microbiol. Lett. 34 (3), 251–255.
Mata, T.M., Martins, A.A., Caetano, N.S., 2010. Microalgae for biodiesel production and other applications: a review. Renew. Sustain. Energy Rev. 14 (1), 217–232.
Matamoros, V., Rodríguez, Y., Albaigés, J., 2016. A comparative assessment of intensive and extensive wastewater treatment technologies for removing emerging contaminants in small communities. Water Res. 88, 777–785.
Mazhar, S., Cohen, J.D., Hasnain, S., 2013. Auxin producing non-heterocystous cyanobacteria and their impact on the growth and endogenous auxin homeostasis of wheat. J. Basic Microbiol. 53 (12), 996–1003.
Mazid, M., Khan, T.A., 2015. Future of bio-fertilizers in Indian agriculture: an overview. Int. J. Agric. Food Res. 3 (3).
Medlicott, K.O., Bos, R., 2014. Public Health Measures: Safe Use of Wastewater for Agricultural Production. Encyclopedia of Food Safety, 4, 107–114.
Mehta, P., Singh, D., Saxena, R., Rani, R., Gupta, R.P., Puri, S.K., Mathur, A.S., 2018. High-value coproducts from algae—an innovational way to deal with advance algal industry. In: Singhania, R.R., Agarwal, R.A., Kumar, R.P., Sukumaran, R.K. (Eds.), Waste to Wealth. Springer, Singapore, pp. 343–363.
Milledge, J.J., 2011. Commercial application of microalgae other than as biofuels: a brief review. Rev. Environ. Sci. Bio/Technol. 10 (1), 31–41.
Mishra, P., Dash, D., 2014. Rejuvenation of biofertilizer for sustainable agriculture and economic development. Consilience (11), 41–61.
Molina-Grima, E., Fernández Sevilla, J.M., Acién Fernández, F.G., 2009. Microalgae, mass culture methods. In: Flickinger, M.C. (Ed.), Encyclopedia of Industrial Biotechnology: Bioprocess, Bioseparation, and Cell Technology. John Wiley & Sons, pp. 1–24.
Moreno, J., Vargas, M.Á., Rodrıguez, H., Rivas, J., Guerrero, M.G., 2003. Outdoor cultivation of a nitrogen-fixing marine cyanobacterium, *Anabaena* sp. ATCC 33047. Biomol. Eng. 20 (4–6), 191–197.
Mukherjee, C., Chowdhury, R., Ray, K., 2015. Phosphorus recycling from an unexplored source by polyphosphate accumulating microalgae and cyanobacteria—A step to phosphorus security in agriculture. Front. Microbiol. 6, 1421.
Mulbry, W., Westhead, E.K., Pizarro, C., Sikora, L., 2005. Recycling of manure nutrients: use of algal biomass from dairy manure treatment as a slow release fertilizer. Bioresour. Technol. 96 (4), 451–458.
Mullineaux, C.W., Nürnberg, D.J., 2014. Tracing the path of a prokaryotic paracrine signal. Mol. Microbiol. 94 (6), 1208–1212.
Mus, F., Colman, D.R., Peters, J.W., Boyd, E.S., 2019. Geobiological feedbacks, oxygen, and the evolution of nitrogenase. Free Radic. Biol. Med. 140, 250–259.
Mus, F., Crook, M.B., Garcia, K., Costas, A.G., Geddes, B.A., Kouri, E.D., Paramasivan, P., Ryu, M.H., Oldroyd, G.E.D., Poole, P.S., Udvardi, M.K., Voigt, C.A., Ané, J.M., Peters, J.W, 2016. Symbiotic nitrogen fixation and the challenges to its extension to nonlegumes. Appl. Environ. Microbiol. 82, 3698–3710.
Musatenko, L.I., 2001. Algal hormones. Algologiya 11 (1), 37–51.
Muylaert, K., Bastiaens, L., Vandamme, D., Gouveia, L., 2017. Harvesting of microalgae: overview of process options and their strengths and drawbacks. In:

GonzalezFernandez, C., Muñoz, R. (Eds.), Microalgae-based Biofuels and Bioproducts. Woodhead Publishing, Cambridge, pp. 113–132.

Narala, R.R., Garg, S., Sharma, K.K., Thomas-Hall, S.R., Deme, M., Li, Y., Schenk, P.M., 2016. Comparison of microalgae cultivation in photobioreactor, open raceway pond, and a two-stage hybrid system. Front. Energy Res. 4, 29.

Nelson, J.M., Hauser, D.A., Gudiño, J.A., Guadalupe, Y.A., Meeks, J.C., Salazar Allen, N., Villarreal, J.C., Li, F.W., 2019. Complete genomes of symbiotic cyanobacteria clarify the evolution of vanadium-nitrogenase. Genome Biol. Evol. 11 (7), 1959–1964.

Nierzwicki-Bauer, S.A., Balkwill, D.L., Stevens, S.E., 1984. Heterocyst differentiation in the cyanobacterium *Mastigocladus laminosus*. J. Bacteriol. 157 (2), 514–525.

Nieva-Gomez, D., Roberts, G.P., Klevickis, S., Brill, W.J., 1980. Electron transport to nitrogenase in *Klebsiella pneumoniae*, 77, 2555–2558.

Nikaido, H., 2003. Molecular basis of bacterial outer membrane permeability revisited. Microbiol. Mol. Biol. Rev. 67 (4), 593–656.

Nisha, R., Kaushik, A., Kaushik, C.P., 2007. Effect of indigenous cyanobacterial application on structural stability and productivity of an organically poor semi-arid soil. Geoderma 138 (1-2), 49–56.

Njoroge, R., Otinga, A.N., Okalebo, J.R., Pepela, M., Merckx, R., 2018. Maize (*Zea mays* L.) response to secondary and micronutrients for profitable N, P and K fertilizer use in poorly responsive soils. Agronomy 8 (4), 49.

Novoveská, L., Zapata, A.K., Zabolotney, J.B., Atwood, M.C., Sundstrom, E.R., 2016. Optimizing microalgae cultivation and wastewater treatment in large-scale offshore photobioreactors. Algal Res. 18, 86–94.

Nwoba, E.G., Vadiveloo, A., Ogbonna, C.N., Ubi, B.E., Ogbonna, J.C., Moheimani, N.R., 2020. Algal cultivation for treating wastewater in African developing countries: a review. CLEAN–Soil, Air, Water, 48 (3), 2000052.

Oelze, J., 2000. Respiratory protection of nitrogenase in Azotobacter species: is a widely held hypothesis unequivocally supported by experimental evidence? FEMS Microbiol. Rev. 24 (4), 321–333.

Ohki, K., Zehr, J.P., Fujita, Y., 1992. Regulation of nitrogenase activity in relation to the light-dark regime in the filamentous non-heterocystous cyanobacterium *Trichodesmium* sp. NIBB 1067. Microbiology 138 (12), 2679–2685.

Olaizola, M., Grewe, C., 2019. Commercial microalgal cultivation systems. In: Hallmann, A., Rampelotto, P.H. (Eds.), Grand Challenges in Algae Biotechnology. Springer, Cham, pp. 3–34.

Omairi-Nasser, A., Haselkorn, R., Austin, J., 2014. Visualization of channels connecting cells in filamentous nitrogen-fixing cyanobacteria. FASEB J. 28 (7), 3016–3022.

Osman, M.E.H., El-Sheekh, M.M., El-Naggar, A.H., Gheda, S.F, 2010. Effect of two species of cyanobacteria as biofertilizers on some metabolic activities, growth, and yield of pea plant. Biol. Fertil Soils 46 (8), 861–875.

Oswald, W.J., Golueke, C.G., 1960. Biological transformation of solar energyAdv. Appl. Microbiol.2, 223–262.

Oswald, W.J., Gotaas, H.B., Golueke, C.G., Kellen, W.R., Gloyna, E.F., Hermann, E.R., 1957. Algae in waste treatment [with discussion]. Sewage Ind. Wastes 29 (4), 437–457.

Paerl, H.W., Prufert, L.E., 1987. Oxygen-poor microzones as potential sites of microbial N_2 fixation in nitrogen-depleted aerobic marine waters. Appl. Environ. Microbiol. 53 (5), 1078–1087.

Paerl, H.W., Bebout, B.M., Prufert, L.E., 1989. Naturally occurring patterns of oxygenic photosynthesis and N_2 fixation in a marine microbial mat: physiological and ecological ramifications. In: Cohen, Y, Rosenberg, E (Eds.), Microbial Mats: Physiological Ecology of Benthic Microbial Communities. American Society for Microbiology, Washington, DC, pp. 326–341.

Paerl, H.W., Pinckney, J.L., Kucera, S.A., 1995. Clarification of the structural and functional roles of heterocysts and anoxic microzones in the control of pelagic nitrogen fixation. Limnol. Oceanogr. 40 (3), 634–638.

Paerl, H.W., Prufert, L.E., Ambrose, W.W., 1991. Contemporaneous N_2 fixation and oxygenic photosynthesis in the nonheterocystous mat-forming cyanobacterium *Lyngbya aestuarii*. Appl. Environ. Microbiol. 57 (11), 3086–3092.

Park, J.B.K., Craggs, R.J., Shilton, A.N, 2011. Wastewater treatment high rate algal ponds for biofuel production. Bioresour. Technol. 102 (1), 35–42.

Park, J.B., Craggs, R.J., Tanner, C.C., 2018. Eco-friendly and low-cost enhanced pond and wetland (EPW) system for the treatment of secondary wastewater effluent. Ecol. Eng. 120, 170–179.

Pathak, J., Maurya, P.K., Singh, S.P., Häder, D.P., Sinha, R.P., 2018. Cyanobacterial farming for environment friendly sustainable agriculture practices: innovations and perspectives. Front. Environ. Sci. 6, 7.

Paul, E.A., 1978. Contribution of nitrogen fixation to ecosystem functioning and nitrogen fluxes on a global basis. Ecol. Bull. (Stockholm). 26, 282–293.

Pearson, H.W., Howsley, R., Kjeldsen, C.K., Walsby, A.E., 1979. Aerobic nitrogenase activity associated with a non-heterocystous filamentous cyanobacterium. FEMS Microbiol. Lett. 5 (3), 163–167.

Picossi, S., Valladares, A., Flores, E., Herrero, A., 2004. Nitrogen-regulated genes for the metabolism of cyanophycin, a bacterial nitrogen reserve polymer expression and mutational analysis of two cyanophycin synthetase and cyanophycinase gene clusters in the heterocyst-forming *Cyanobacterium Anabaena* SP. PCC 7120. J. Biol. Chem. 279 (12), 11582–11592.

Potts, M., Angeloni, S.V., Ebel, R.E., Bassam, D., 1992. Myoglobin in a cyanobacterium. Science 256 (5064), 1690–1692.

Potts, M., Krumbein, W.E., Metzger, J., 1978. Nitrogen fixation rates in anaerobic sediments determined by acetylene reduction, a new N^{15} field assay, and simultaneous total N^{15} determination. In: Krumbein, W.E. (Ed.), Environmental biogeochemistry and geomicrobiology (Vol 3) Methods, Metals and Assessment. Ann. Arbor. Science, Ann Arbor, pp. 753–769.

Poudel, S., Colman, D.R., Fixen, K.R., Ledbetter, R.N., Zheng, Y., Pence, N., Seefeldt, L.C., Peters, J.W., Harwood, C.S., Boyd, E.S., 2018. Electron transfer to nitrogenase in different genomic and metabolic backgrounds. J. Bacteriol. 200, e00757 -17.

Prado, R., de M., Caione, G., Campos, C.N.S., 2013. Filter cake and vinasse as fertilizers contributing to conservation agriculture. Appl. Environ. Soil Sci., 2013, Article ID 581984. https://doi.org/10.1155/2013/581984.

Prasanna, R., Sharma, E., Sharma, P., Kumar, A., Kumar, R., Gupta, V., Pal, R.K., Shivay, Y.S., Nain, L., 2013. Soil fertility and establishment potential of inoculated cyanobacteria in rice crop grown under non-flooded conditions. Paddy Water Environ. 11 (1–4), 175–183.

Prufert-Bebout, L., Paerl, H.W., Lassen, C., 1993. Growth, nitrogen fixation, and spectral attenuation in cultivated *Trichodesmium* species. Appl. Environ. Microbiol. 59 (5), 1367–1375.

Pufahl, P.K., Groat, L.A., 2017. Sedimentary and igneous phosphate deposits: formation and exploration: an invited paper. Econ. Geol. 112 (3), 483–516.

Qadir, M., Wichelns, D., Raschid-Sally, L., McCornick, P.G., Drechsel, P., Bahri, A., Minhas, P.S., 2010. The challenges of wastewater irrigation in developing countries. Agric. Water Manag. 97 (4), 561–568.

Qin, L., Alam, M.A., Wang, Z., 2019. Open pond culture systems and photobioreactors for microalgal biofuel production. In: Alam, M.A., Wang, Z. (Eds.), Microalgae Biotechnology for Development of Biofuel and Wastewater Treatment. Springer, Singapore, pp. 45–74.

Rai, A.N., Borthakur, M., Bergman, B., 1992. Nitrogenase derepresssion, its regulation and metabolic changes associated with diazotrophy in the non-heterocystous cyanobacterium *Plectonema boryanum* PCC 73110. Microbiology 138 (3), 481–491.

Rai, A.N., Singh, A.K., Syiem, M.B., 2019. Plant growth-promoting abilities in cyanobacteria. In: Mishra, A.K., Tiwari, D.N., Rai, A.N. (Eds.), Cyanobacteria. Academic Press, CA, pp. 459–476.

Rajendran, A., Fox, T., Reis, C.R., Wilson, B., Hu, B., 2018. Deposition of manure nutrients in a novel mycoalgae biofilm for nutrient management. Biocatal. Agric. Biotechnol. 14, 120–128.

Reich, S., Böger, P., 1989. Regulation of nitrogenase activity in Anabaena variabilis by modification of the Fe protein. FEMS Microbiol. Lett. 58 (1), 81–86.

Renuka, N., Guldhe, A., Prasanna, R., Singh, P., Bux, F., 2018. Microalgae as multi-functional options in modern agriculture: current trends, prospects and challenges. Biotechnol. Adv. 36 (4), 1255–1273.

Rey, F.E., Oda, Y., Harwood, C.S., 2006. Regulation of uptake hydrogenase and effects of hydrogen utilization on gene expression in *Rhodopseudomonas palustris*. J. Bacteriol. 188 (17), 6143–6152.

Rippka, R., Stanier, R.Y., 1978. The effects of anaerobiosis on nitrogenase synthesis and heterocyst development by nostocacean cyanobacteria. Microbiology 105 (1), 83–94.

Rippka, R., Waterbury, J.B., 1977. The synthesis of nitrogenase by nonheterocystous cyanobacteria. FEMS Microbiol. Lett. 2 (2), 83–86.

Rippka R., Herdman, M., 1985. Division patterns and cellular differentiation in cyanobacteria. Ann. Inst. Pasteur Microbiol. 136A (1), 33–39.

Ris, H., Singh, R.N., 1961. Electron microscope studies on blue-green algae. J. Cell Biol. 9 (1), 63–80.

Risch, A.C., Zimmermann, S., Ochoa-Hueso, R., Schütz, M., Frey, B., Firn, J.L., Fay, P.A., Hagedorn, F., Borer, E.T., Seabloom, E.W., Harpole, W.S., Knops, J.M.H., McCulley, R.L., Broadbent, A.A.D., Stevens, C.J., Silveira, M.L., Adler, P.B., Báez, S., Biederman, L.A., Blair, J.M., Brown, C.S., Caldeira, M.C., Collins, S.L., Daleo, P., di Virgilio, A., Ebeling, A., Eisenhauer, N., Esch, E., Eskelinen, A., Hagenah, N., Hautier, Y., Kirkman, K.P., MacDougall, A.S., Moore, J.L., Power, S.A., Prober, S.M., Roscher, C., Sankaran, M., Siebert, J., Speziale, K.L., Tognetti, P.M., Virtanen, R., Yahdjian, L., Moser, B, 2019. Soil net nitrogen mineralisation across global grasslands. Nat. Commun. 10 (1), 1–10.

Rockström, J., Williams, J., Daily, G., Noble, A., Matthews, N., Gordon, L., Wetterstrand, H., DeClerck, F., Shah, M., Steduto, P., de Fraiture, C., Hatibu, N., Unver, O., Bird, J., Sibanda, L., Smith, J., 2017. Sustainable intensification of agriculture for human prosperity and global sustainability. Ambio Feb 46 (1), 4–17. doi:10.1007/s13280-016-0793-6. Epub 2016 Jul 12. PMID: 27405653; PMCID: PMC5226894.

Rockström, J., Edenhofer, O., Gaertner, J., et al., 2020. Planet-proofing the global food system. Nat. Food 1, 3–5.

Roger, P.A., Santiago-Ardales, S., Watanabe, I., 1986. Nitrogen fixing blue-green algae in rice soils of Northern Luzon (Philippines)Philipp. Agri.69, 589–598.

Roger, P.A., Voggesberger, M., Margraf, J., 1986. Nitrogen-fixing phototrophs in the Ifugao rice terraces (Philippines). Philipp. Agri. 69 (4B), 599–609.

Rubio, L.M., Ludden, P.W., 2008. Biosynthesis of the iron-molybdenum cofactor of nitrogenase. Annu. Rev. Microbiol. 62, 93–111.

Saeki, K., 2004. Electron transport to nitrogenase: diverse routes for a common destination. In: Klipp, W., Masepohl, B., Gallon, J.R., Newton, W.E. (Eds.), Genetics and Regulation of Nitrogen Fixation in Free-Living Bacteria. Springer, Dordrecht, pp. 257–290.

Sahu, A., Bhattacharjya, S., Mandal, A., Thakur, J.K., Atoliya, N., Sahu, N., Manna, M.C., Patra, A.K., 2018. Microbes: a sustainable approach for enhancing nutrient availability in agricultural soils. In: Meena, VS (Ed.), Role of Rhizospheric Microbes in Soil:

Volume 2: Nutrient Management and Crop Improvement. Springer Singapore, Singapore, pp. 47–75.

Sandh, G., Xu, L., Bergman, B., 2012. Diazocyte development in the marine diazotrophic cyanobacterium *Trichodesmium*. Microbiology 158 (2), 345–352.

Sandmann, G., Peleato, M.L., Fillat, M.F., Lázaro, M.C., Gómez-Moreno, C., 1990. Consequences of the iron-dependent formation of ferredoxin and flavodoxin on photosynthesis and nitrogen fixation on Anabaena strains. Photosynthesis Res. 26 (2), 119–125.

Schädler, T., Neumann-Cip, A.C., Wieland, K., Glöckler, D., Haisch, C., Brück, T., Weuster-Botz, D., 2020. High-density microalgae cultivation in open thin-layer cascade photobioreactors with water recycling. Appl. Sci. 10 (11), 3883.

Schiano di Visconte, G., Spicer, A., Chuck, C.J., Allen, M.J., 2019. The microalgae biorefinery: A perspective on the current status and future opportunities using genetic modification. Appl. Sci. 9, 4793. https://doi.org/10.3390/app9224793.

Schmitz, O., Boison, G., Hilscher, R., Hundeshagen, B., Zimmer, W., Lottspeich, F., Bothe, H., 1995. Molecular biological analysis of a bidirectional hydrogenase from cyanobacteria. Eur. J. Biochem. *233* (1), 266–276.

Schrautemeier, B., Böhme, H., 1985. A distinct ferredoxin for nitrogen fixation isolated from heterocysts of the cyanobacterium *Anabaena variabilis*. FEBS Lett. 184 (2), 304–308.

Schrautemeier, B., Neveling, U., Schmitz, S., 1995. Distinct and differently regulated Mo-dependent nitrogen-fixing systems evolved for heterocysts and vegetative cells of *Anabaena variabilis* ATCC 29413: characterization of the fdxH1/2 gene regions as part of the nif1/2 gene clusters. Mol. Microbiol. 18 (2), 357–369.

Schroeder, P., Anggraeni, K., Weber, U., 2019. The relevance of circular economy practices to the sustainable development goals. J. Ind. Ecol. 23, 77–95. doi:10.1111/jiec.12732.

Shanmugam, K.T., Valentine, R.C., 1975. Microbial production of ammonium ion from nitrogen. Proc. Natl. Acad. Sci. 72 (1), 136–139.

Sharma, P., Sharma, N., 2017. Industrial and biotechnological applications of algae: a review. J. Adv. Plant Biol. 1 (1), 01.

Sherman, D.M., Tucker, D., Sherman, L.A., 2000. Heterocyst development and localization of cyanophycin in N_2-fixing cultures of *Anabaena* sp. PCC 7120 (Cyanobacteria). J. Phycol. 36 (5), 932–941.

Siddiqui, P.J., Carpenter, E.J., Bergman, B., 1992. Ultrastructure and immunolocalization of phycobiliproteins and ribulose 1, 5-bisphosphate carboxylase/oxygenase in the marine cyanobacterium *Trichodesmium thiebautii*. J. Phycol. 28 (3), 320–327.

Sikora, L.J., Yakovchenko, V., 1996. Soil organic matter mineralization after compost amendment. Soil Sci. Soc. Am. J. 60 (5), 1401–1404.

Singh, J.S., Kumar, A., Rai, A.N., Singh, D.P., 2016. Cyanobacteria: a precious bio-resource in agriculture, ecosystem, and environmental sustainability. Front. Microbiol. 7, 1–19.

Singh, N.K., Dhar, D.W., Tabassum, R., 2016. Role of cyanobacteria in crop protection, Proc. Natl. Acad. Sci. India Section B: Biol. Sci. 86, 1–8.

Singh, R.N., Sharma, S., 2012. Development of suitable photobioreactor for algae production–a review. Renew. Sustain. Energy Rev. 16 (4), 2347–2353.

Singh, R., Parihar, P., Singh, M., Bajguz, A., Kumar, J., Singh, S., Singh, V.P., Prasad, S.M., 2017. Uncovering potential applications of cyanobacteria and algal metabolites in biology, agriculture and medicine: current status and future prospects. Front. Microbiol. 8, 515.

Singh, S., 2014. A review on possible elicitor molecules of cyanobacteria: their role in improving plant growth and providing tolerance against biotic or abiotic stress. J. Appl. Microbiol. 117 (5), 1221–1244.

Siville, B., Boeing, W.J., 2020. Optimization of algal turf scrubber (ATS) technology through targeted harvest rate. Bioresour. Technol. Rep. 9, 100360.

Smith, V.H., Joye, S.B., Howarth, R.W., 2006. Eutrophication of freshwater and marine ecosystems. Limnol. Oceanogr. 51 (1part2), 351–355.

Soni, R.A., Sudhakar, K., Rana, R.S., 2017. Spirulina–from growth to nutritional product: a review. Trends Food Sci. Technol. 69, 157–171.

Soo, R.M., Skennerton, C.T., Sekiguchi, Y., Imelfort, M., Paech, S.J., Dennis, P.G., Steen, J.A., Parks, D.H., Tyson, G.W., Hugenholtz, P., 2014. An expanded genomic representation of the phylum cyanobacteria. Genome Biol. Evol. 6, 1031–1045.

Soto-Urzúa, L., Baca, B.E., 2001. Protection mechanisms of nitrogenase to inactivation by oxygen. Rev. Latinoam. Microbiol. 43 (1), 37–49.

Spence, D.W., Stewart, W.D., 1987. Heterocystless mutants of *Anabaena* PCC7120 with nitrogenase activity. FEMS Microbiol. Lett. 40 (1), 119–122.

Spencer, D.S., 1996. Infrastructure and technology constraints to agricultural development in the humid and subhumid tropics of Africa. Afr. Dev. Rev. 8 (2), 68–93.

Spiller, H., Latorre, C., Hassan, M.E., Shanmugam, K.T., 1986. Isolation and characterization of nitrogenase-derepressed mutant strains of cyanobacterium *Anabaena variabilis*. J. Bacteriol. 165 (2), 412–419.

Spolaore, P., Joannis-Cassan, C., Duran, E., Isambert, A., 2006. Commercial applications of microalgae. J. Biosci. Bioeng. 101 (2), 87–96.

Stal, L.J., Grossberger, S., Krumbein, W.E., 1984. Nitrogen fixation associated with the cyanobacterial mat of a marine laminated microbial ecosystem. Marine Biol. 82 (3), 217–224.

Steffen, W., Richardson, K., Rockström, J., Cornell, S.E., Fetzer, I., Bennett, E.M., Biggs, R., Carpenter, S.R., de Vries, W., de Wit, C.A., Folke, C., Gerten, D., Heinke, J., Mace, G.M., Persson, L.M., Ramanathan, V., Reyers, B., Sörlin, S., 2015. Planetary boundaries: guiding human development on a changing planet. Science 347 (80), 1259855.

Steunou, A.-S., Bhaya, D., Bateson, M.M., Melendrez, M.C., Ward, D.M., Brecht, E., Peters, J.W., Kühl, M., Grossman, A.R., 2006. In situ analysis of nitrogen fixation and metabolic switching in unicellular thermophilic cyanobacteria inhabiting hot spring microbial mats. Proc. Natl. Acad. Sci. 03 (7), 2398–2403.

Stewart, W.D., Lex, M., 1970. Nitrogenase activity in the blue-green alga *Plectonema boryanum* strain 594. Arch. Mikrobiol. 73 (3), 250–260.

Stöckel, J., Jacobs, J.M., Elvitigala, T.R., Liberton, M., Welsh, E.A., Polpitiya, A.D., Gritsenko, M.A., Nicora, C.D., Koppenaal, D.W., Smith, R.D., Pakrasi, H.B., 2011. Diurnal rhythms result in significant changes in the cellular protein complement in the cyanobacterium *Cyanothece* 51142. PloS One 6 (2), e16680.

Su, Y., Song, K., Zhang, P., Su, Y., Cheng, J., Chen, X., 2017. Progress of microalgae biofuel's commercialization. Renew. Sust. Energy Rev. 74, 402–411.

Summers, M.L., Wallis, J.G., Campbell, E.L., Meeks, J.C., 1995. Genetic evidence of a major role for glucose-6-phosphate dehydrogenase in nitrogen fixation and dark growth of the cyanobacterium *Nostoc* sp. strain ATCC 29133. J. Bacteriol. 177 (21), 6184–6194.

Summit, A.F., 2006. Africa fertilizer summit proceedings. IFDC, Muscle Shoals, 182.

Sutton, M., Oenema, O., Erisman, J., et al., 2011. Too much of a good thing. Nature 472, 159–161.

Tan, J.S., Lee, S.Y., Chew, K.W., Lam, M.K., Lim, J.W., Ho, S.H., Show, P.L., 2020. A review on microalgae cultivation and harvesting, and their biomass extraction processing using ionic liquids. Bioengineered 11 (1), 116–129.

Therien, J.B., Artz, J.H., Poudel, S., Hamilton, T.L., Liu, Z., Noone, S.M., Adams, M.W.W., King, P.W., Bryant, D.A., Boyd, E.S., Peters, J.W, 2017. The physiological functions and structural determinants of catalytic bias in the [FeFe]-hydrogenases CpI and CpII of *Clostridium pasteurianum* strain W5. Front. Microbiol. 8, 1305.

Thiel, T., 2019. Organization and regulation of cyanobacterial nif gene clusters: implications for nitrogenase expression in plant cells. FEMS Microbiol. Lett. 366 (7), 1–8.

Thiel, T., Pratte, B.S., 2014. Regulation of three nitrogenase gene clusters in the cyanobacterium *Anabaena variabilis* ATCC 29413. Life 4 (4), 944–967.

Thiel, T., Lyons, E.M., Erker, J.C., Ernst, A., 1995. A second nitrogenase in vegetative cells of a heterocyst-forming cyanobacterium, Proc. Natl Acad. Sci., 92, 9358–9362.

Thomas, J., Meeks, J.C., Wolk, C.P., Shaffer, P.W., Austin, S.M., 1977. formation of glutamine from [13n] ammonia,[13n] dinitrogen, and [14C] glutamate by heterocysts isolated from *Anabaena cylindrica*. J. Bacteriol. 129 (3), 1545–1555.

Thompson, A.W., Zehr, J.P., 2013. Cellular interactions: lessons from the nitrogen-fixing cyanobacteria. J. Phycol. 49 (6), 1024–1035.

Tiwari, P., Singh, R.P., Singh, R., Suman, S., 2016. Effect of biofertilizer, liquid bio-manure, organic nutrients, polymorphic growth hormones and multi-micronutrients on yield attributes of mung bean under custard apple-based agri-horti system. Int. J. Farm Sci. 6 (1), 61–65.

Tomitani, A., Knoll, A.H., Cavanaugh, C.M., Ohno, T., 2006. The evolutionary diversification of cyanobacteria: molecular–phylogenetic and paleontological perspectives, 103, 5442–5447.

Trivedi, P., Delgado-Baquerizo, M., Anderson, I.C., Singh, B.K., 2016. Response of soil properties and microbial communities to agriculture: implications for primary productivity and soil health indicators. Front. Plant Sci. 7, 990.

Tsujimoto, R., Kotani, H., Yokomizo, K., Yamakawa, H., Nonaka, A., Fujita, Y., 2018. Functional expression of an oxygen-labile nitrogenase in an oxygenic photosynthetic organism. Sci. Rep. 8 (1), 1–10.

UN, 2019. The Sustainable Development Goals Report 2019. UN, New York. https://doi.org/10.18356/55eb9109-en.

Valladares, A., Herrero, A., Pils, D., Schmetterer, G., Flores, E., 2003. Cytochrome c oxidase genes required for nitrogenase activity and diazotrophic growth in *Anabaena* sp. PCC 7120. Mol. Microbiol. 47 (5), 1239–1249.

Vargas, M.A., Rodriguez, H., Moreno, J., Olivares, H., Campo, J.D., Rivas, J., Guerrero, M.G., 1998. Biochemical composition and fatty acid content of filamentous nitrogen-fixing cyanobacteria. J. Phycol. 34 (5), 812–817.

Varshney, G.K., Pei, W., LaFave, M.C., Idol, J., Xu, L., Gallardo, V., Carrington, B., Bishop, K., Jones, M., Li, M., Harper, U., Huang, S.C., Prakash, A., Chen, W., Sood, R., Ledin, J., Burgess, S.M., 2015. High-throughput gene targeting and phenotyping in zebrafish using CRISPR/Cas9. Genome Res. 25 (7), 1030–1042.

Velu, C., Cirés, S., Alvarez-Roa, C., Heimann, K., 2015. First outdoor cultivation of the N_2-fixing cyanobacterium *Tolypothrix* sp. in low-cost suspension and biofilm systems in tropical Australia. J. Appl. Phycol. 27 (5), 1743–1753.

Venkataraman, G.S., 1975. The role of blue-green algae in tropical rice cultivation. In: Stewart, W.D.P. (Ed.), Nitrogen Fixation by Free-living Microorganisms. Cambridge University Press, New York, pp. 207–218.

Villbrandt, M., Stal, L.J., Krumbein, W.E., 1990. Interactions between nitrogen fixation and oxegenic photosynthesis in a marine cyanobacterial mat. FEMS Microbiol. Lett. 74 (1), 59–71.

Wade, J., Culman, S.W., Logan, J.A., Poffenbarger, H., Demyan, M.S., Grove, J.H., Mallarino, A.P., McGrath, J.M., Ruark, M., West, J.R., 2020. Improved soil biological health increases corn grain yield in N fertilized systems across the corn belt. Sci. Rep. 10 (1), 1–9.

Wahl, R.C., Orme-Johnson, W.H., 1987. Clostridial pyruvate oxidoreductase and the pyruvate-oxidizing enzyme specific to nitrogen fixation in *Klebsiella pneumoniae* are similar enzymes. J. Biol. Chem. 262 (22), 10489–10496.

Walker, T.L., Purton, S., Becker, D.K., Collet, C., 2005. Microalgae as bioreactors. Plant Cell Rep. 24 (11), 629–641.

Walsby, A.E., 2007. Cyanobacterial heterocysts: terminal pores proposed as sites of gas exchange. Trends Microbiol. 15 (8), 340–349.

Wang, Y, Ying, H, Yin, Y, Zheng, H, Cui, Z, 2019. Estimating soil nitrate leaching of nitrogen fertilizer from global meta-analysis. Sci. Total Environ. 657, 96–102.

Wang, L., Xia, M., Wang, H., Huang, K., Qian, C., Maravelias, C.T., Ozin, G.A., 2018. Greening ammonia toward the solar ammonia refinery. Joule 2 (6), 1055–1074.

Wang, Y., Li, C., Tu, C., Hoyt, G.D., DeForest, J.L., Hu, S., 2017. Long-term no-tillage and organic input management enhanced the diversity and stability of soil microbial community. Sci. Total Environ. 609, 341–347.

Weare, N.M., Benemann, J.R., 1974. Nitrogenase activity and photosynthesis in *Plectonema boryanum*. J. Bacteriol. 119 (1), 258–265.

Weisshaar, H., Böger, P., 1983. Nitrogenase activity of the non-heterocystous cyanobacterium *Phormidium foveolarum*. Arch. Microbiol. 136 (4), 270–274.

Wenz, J., Davis, J.G., Storteboom, H., 2019. Influence of light on endogenous phytohormone concentrations of a nitrogen-fixing *Anabaena* sp. cyanobacterium culture in open raceways for use as fertilizer for horticultural crops. J. Appl. Phycol. 31 (6), 3371–3384.

Whitton, B.A., Grainger, S.L.J., Hawley, G.R.W., Simon, J.W, 1991. Cell-bound and extracellular phosphatase activities of cyanobacterial isolates. Microb. Ecol. 21 (1), 85–98.

Wildon, D.C., Mercer, F.V., 1963. The ultrastructure of the heterocyst and akinete of the blue-green algae. Arch. Mikrobiol. 47 (1), 19–31.

Williams, P.J.L.B., Laurens, L.M, 2010. Microalgae as biodiesel & biomass feedstocks: review & analysis of the biochemistry, energetics & economics. Energy Environ. Sci. 3 (5), 554–590.

Win, T.T., Barone, G.D., Secundo, F., Fu, P., 2018. Algal biofertilizers and plant growth stimulants for sustainable agriculture. Ind. Biotechnol. 14 (4), 203–211.

WMO, 2018. WMO Statement on the State of the Global Climate in 2019. World Meteorological Organisation (WMO), Switzerland ISBN 978-92-63-11248-4.

WMO, 2020. WMO Statement on the State of the Global Climate in 2019. World Meteorological Organization (WMO), Switzerland. WMO-No. 1248. ISBN 978-92-63-11248-4.

Wolk, C.P., 1968. Movement of carbon from vegetative cells to heterocysts in *Anabaena cylindrica*. J. Bacteriol. 96 (6), 2138–2143.

Wolk, C.P., Ernst, A., Elhai, J., 1994. Heterocyst metabolism and development. In: Bryant, D.A. (Ed.), The Molecular Biology of Cyanobacteria. Springer, Dordrecht, pp. 769–823.

Wolk, C.P., Thomas, J., Shaffer, P.W., Austin, S.M., Galonsky, A., 1976. Pathway of nitrogen metabolism after fixation of 13N-labeled nitrogen gas by the cyanobacterium, *Anabaena cylindrica*. J. Biol. Chem. 251 (16), 5027–5034.

Woodward, A.W., Bartel, B., 2005. Auxin: regulation, action, and interaction. Ann. Bot. 95 (5), 707–735.

World Health Organization, 2010. World Health Statistics 2010. World Health Organization.

Wu, L., Qian, L., Deng, Z., Zhou, X., Li, B., Lan, S., Yang, L., Zhang, Z., 2020. Temperature modulating sand-consolidating cyanobacterial biomass, nutrients removal and bacterial community dynamics in municipal wastewater. Bioresour. Technol. 301, 122758.

Wu, L., Zhu, Q., Yang, L., Li, B., Hu, C., Lan, S., 2018. Nutrient transferring from wastewater to desert through artificial cultivation of desert cyanobacteria. Bioresour. Technol. 247, 947–953.

WWAP (United Nations World Water Assessment Programme), 2016. The United Nations World Water Development Report. Paris, France, pp. 1–148.

Xu, X., Elhai, J., Wolk, C.P., 2008. Transcriptional and developmental responses by Anabaena to deprivation of fixed nitrogen. In: Herrero, A., Flores, E. (Eds.), The Cyanobacteria. Molecular Biology, Genomics and Evolution. *Caister Academic Press, Norfolk, United Kingdom*, pp. 383–422.

Yakunin, A.F., Hallenbeck, P.C., 1998. Purification and characterization of pyruvate oxidoreductase from the photosynthetic bacterium *Rhodobacter capsulatus*. Biochim. Biophys. Acta -Bioenergetics 1409 (1), 39–49.

Yandigeri, M.S., Meena, K.K., Srinivasan, R., Pabbi, S., 2011. Effect of mineral phosphate solubilization on biological nitrogen fixation by diazotrophic cyanobacteria. Indian J. Microbiol. 51 (1), 48–53.

Yang, Z.Y., Ledbetter, R., Shaw, S., Pence, N., Tokmina-Lukaszewska, M., Eilers, B., Guo, Q., Pokhrel, N., Cash, V.L., Dean, D.R., Antony, E., Bothner, B., Peters, J.W., Seefeldt, L.C., 2016. Evidence that the Pi release event is the rate-limiting step in the nitrogenase catalytic cycle. Biochemistry 55 (26), 3625–3635.

Yen, H.W., Hu, I.C., Chen, C.Y., Nagarajan, D., Chang, J.S., 2019. Design of photobioreactors for algal cultivation. In: Pandey, A., Lee, D-J, Chisti, Y., Soccol, C.R. (Eds.), Biofuels from Algae. Elsevier, Amsterdam, pp. 23–45.

Yoon, K.S., Nguyen, N.T., Tran, K.T., Tsuji, K., Ogo, S., 2017. Nitrogen fixation genes and nitrogenase activity of the non-heterocystous cyanobacterium Thermoleptolyngbya sp. O-77. Microbes Environ. 32 (4), 324–329.

Young, P., Taylor, M., Fallowfield, H.J., 2017. Mini-review: high rate algal ponds, flexible systems for sustainable wastewater treatment. World J. Microbiol. Biotechnol. 33 (6), 1C7.

Zehr, J.P., Capone, D.G., 2020. Changing perspectives in marine nitrogen fixation. Science 368 (6492), 1–9. https://doi.org/10.1126/science.aay9514.

Zhang, X., Davidson, E.A., Zou, T., Lassaletta, L., Quan, Z., Li, T., Zhang, W., 2020. Quantifying nutrient budgets for sustainable nutrient management. Global Biogeochem. Cycles 34 (3), e2018GB006060.

Zhang, Y.P., Grunwald, S.K., Lies, D., Halbleib, C., Ma, Y., Roberts, G.P., Burris, R.H., Ludden, P.W., 1995. Posttranslational regulation of nitrogenase activity by reversible ADP-ribosylation: how are the regulatory enzymes DRAT and DRAG regulated? In: Tikhonovich, I.A., Provorov, N.A., Romanov, V.I., Newton, W.E. (Eds.), Nitrogen Fixation: Fundamentals and Applications. Springer, Dordrecht, pp. 177–182.

Ziegler, K., Stephan, D.P., Pistorius, E.K., Ruppel, H.G., Lockau, W., 2001. A mutant of the cyanobacterium *Anabaena variabilis* ATCC 29413 lacking cyanophycin synthetase: growth properties and ultrastructural aspects. FEMS Microbiol. Lett. 196 (1), 13–18.

CHAPTER

6

Thermophilic and thermotolerant cyanobacteria: Environmental and biotechnological perspectives

Gajanan T. Mehetre[a], *Zothanpuia*[b], *Purbajyoti Deka*[a], *William Carrie*[a], *Lalrokimi*[a], *Bhim Pratap Singh*[c]

[a]Department of Biotechnology, Mizoram University, Mizoram, India
[b]Department of Biotechnology/Life sciences, Mizoram University, Pachhunga University College, Aizawl, Mizoram, India
[c]Department of Agriculture and Environmental Sciences, National Institute of Food Technology Entrepreneurship & Management (NIFTEM), Sonepat, Haryana, India

6.1 Introduction

Microbial communities inhabiting extreme environments perform a very crucial role in energy metabolism and matter cycling. Cyanobacteria are one of the ecological diversifications reflecting their capacity to acclimate to various extreme environments. They carry out an extremely significant role in sustaining the productivity of such extreme ecosystems through photosynthesis, nitrogen fixation, and accumulation of phosphorus (Seckbach et al., 2007). Therefore, they remain ideal microbial systems to investigate processes such as photosynthesis, respiration, circadian rhythm, photoregulation of gene expression, developmental gene rearrangements. In addition, most cyanobacteria possess various mechanisms to acclimatize to extreme conditions of diverse environments (Wilde and Dienst, 2011). Various cyanobacterial groups demonstrate adaptive abilities to

Cyanobacterial Lifestyle and its Applications in Biotechnology.
DOI: https://doi.org/10.1016/B978-0-323-90634-0.00014-7

sustain in adverse environmental conditions such as elevated temperatures, extreme pH, cold, osmotic stress, high salinity, desiccation, and light. Remarkably, their ancient origin and survival under stressful conditions during evolution shows their ability to prevail under various environmental and human-made stresses (Bahl et al., 2011, Curren and Leong, 2020). They are distributed in most of the inhabitable environments and contribute for around 50% of the global net primary productivity. They found dominance in various extreme habitats such as deserts (Garcia-Pichel et al., 2001), salt lakes, solar salterns, hypersaline lagoons, and hypersaline sulfur springs (Oren, 2012), hot springs (Brenes-Guillén et al., 2021; Singh et al., 2018), and many other hostile environments.

Tolerance to high-temperature conditions is one of the most significant properties among cyanobacterial groups that inhabit a wide range of high-temperature ecosystems. However, remarkably few species were reported from such environments with water warmer than 60 °C, as the maximum limit for cyanobacteria is to be 73 °C (Strunecký et al., 2019). Thermophilic cyanobacteria were widely investigated in thermal springs such as Yellowstone National Park (YNP), USA, Indian subcontinent, Kamchatka from distinct parts around the world (Brock, 1967; Mehetre et al., 2016; Merkel et al., 2017; Singh et al., 2018). In such ecological niches, they stand as the dominant primary producers capable of oxygenic photosynthesis at temperatures below 73°C. Several studies invariably showed that thermophilic members exhibiting remarkable diversity of thermotolerant lineages are evolved from a moderately thermophilic ancestor (Miller et al., 2019). Among various thermophilic cyanobacterial groups, *Synechococcus* sp. represents an ecologically diverse group of cyanobacteria found in hot-spring microbial mats. Group members are engaged in photosynthesis and aerobic respiration during the daytime and carry out fermentative metabolism and nitrogen fixation during the night in diverse thermal ecologies (Pedersen and Miller, 2017). Therefore, the *Synechococcus* radiation establishes a model system for investigating microbial severance while adapting to novel environments. Several studies demonstrated that thermophilic members of cyanobacteria adopt various mechanisms for survival under extreme temperature conditions. For example, *Synechocystis* induces the expression of heat shock proteins (HSPs) in response to heat stress. In *Synechocystis* genome several such genes such as *htpG, dnaK, groEL, groES, hspA* were observed for the biosynthesis of HSPs (Červený et al., 2015).

The well-adapted cyanobacterial members evolve with a genetic makeup to produce a range of indispensable biomolecules and secondary metabolites that support their growth and survival under hostile conditions (Babele et al., 2019). Therefore, most cyanobacterial strains are exploited to generate several value-added agricultural, pharmaceutical, and industrial products. Research studies also focus on increasing efficiencies and production levels of such compounds in cyanobacteria using several approaches. This chapter provides an overview of the biodiversity, high-temperature tolerance, and status of biotechnological applications of thermophilic and thermotolerant cyanobacteria communities with an outlook on future prospects. Furthermore, the chapter equally describes the significant advances in metabolic engineering used in harnessing the metabolic potential of thermophilic and thermotolerant cyanobacteria.

6.2 Thermophilic cyanobacteria diversity

Thermophilic cyanobacteria are of interest due to their ability to grow in high temperatures and in other extreme environments. The existing distributions of cyanobacteria in high-temperature habitats are shaped largely by various factors that mark the biogeographic patterns and environmental heterogeneity. The hot spring ecosystem supports diverse microbial communities, including cyanobacteria that remain dominant as the primary producers in these habitats. The eco-physiological feature of cyanobacteria could be better examined in hot springs due to natural gradients of geochemistry, pH, and temperature (Bennett et al., 2020). YNP hot spring was extensively studied for thermophilic cyanobacteria, and decades of research have been dedicated on phototrophs in YNP (Brock 1967; Ward et al., 1998). In earlier studies, identification and characterization were majorly performed based on culture-based methods that basically rely on classical morphology-based identification. This has led to identifying some genera's prominent presence, such as *Synechococcus*, *Phormidium*, *Calothrix*, or *Mastigocladus* from hot springs (Ward et al., 2012). Dominant cyanobacteria genera in microbial mat have distinct temperature ranges, however, optima for growth were found consistent with their mat distributions. It has been observed that temperature effects on photosynthesis in cyanobacteria may not significantly affect the growth. Therefore, cyanobacteria have ecological diversification that is invasive to several hot springs across the world. In general, hot spring outflows exhibit striking temperature gradients at temperatures ranging from 45 to 73 °C. In this context, a substantial cyanobacterial molecular diversity has been reported from different hot springs worldwide (Table 6.1).

Based on the 16S rRNA sequences, various genera were identified in such mats. Peary and Castenholz (1964) have demonstrated the presence of four thermophilic strains of *Synechococcus* in Hunter's Hot Springs in Oregon. Cyanobacteria represent several species as the dominant primary producers in hot springs. To date, various research studies have been dedicated in analyzing cyanobacteria taxonomy and the evolution of cyanobacterial genomic adaptations to high temperature conditions.

TABLE 6.1 Taxonomic diversity of cyanobacteria present in hot spring environments.

Hot spring name	Location/ country	Temperature range (°C)	Cyanobacterial species	Reference
Las Lilas thermal spring	Fortuna hill, Costa Rica	47–74	*Leptolyngbya* sp., *Synechococcus* sp.	Brenes Guillén et al. (2021)
Yellowstone National Park (YNP)	Rabbit Creek, YNP, USA	45–72	*Leptococcus* sp., *Chloroflexus* sp., *Roseiflexus* sp.	Bennett et al. (2020)
Chin-Lun hot spring	Taitung, Taiwan	62	*Thermosynechoc-occus* sp.	Cheng et al. (2020)
El Tatio geyser field	Atacama, Chile	32–75	*Fischerella* sp., *Thermosynecho-coccus* sp., *Synechococcus* sp., *Chroogloeo-cystis* sp., *Chlorogloeopsis* sp., *Calothrix* sp.	Alcorta et al. (2020)
Cahuelmó hot spring	Northern Patagonia, Chile			
Kroner Lake	Deception Island, Antarctica			
Alkaline hot springs	Eastern Asia	50–65	*Thermosynechoc-occus* sp.	Cheng et al. (2020)
Chin-Lun hot spring	Taiwan	62	*Thermosynechoc-occus sp*	
Lianhua Lake	Sichuan, China	67.2	*Thermosynechoc-occus elongatus*	Liang et al. (2019)
Tattapani, Manikaran, Kheer Ganga, Vashisht, Kasol and Kalath, Surya Kund, Janki Chatti and Gauri Kund	Himalayan Geothermal Province, India	40–90	*Synechococcus, Gloeocapsa, Gloeocapsopsis, Pseudanabaena, Geitlerinema, Leptolyngbya, Phormidium, Microcoleus, Chlorogloeopsis, Fischerella and Mastigocladus*	Singh et al. (2018)
Taptapani hot spring	Eastern Ghats, India	48	*Arthronema* sp.	Subudhi et al. (2018)
Atri hot spring		58	*Leptolyngbya* sp.	
Sirch hot springs	Kerman, Iran	44–45	*Oscillatoria Microcoleus, Aphanocapsa, Chroococcus,*	Ansary et al. (2017)

Hot spring name	Location/ country	Temperature range (°C)	Cyanobacterial species	Reference
Erdaoqiao	Western China	40.8	*Cyanobacteria* sp.	(Tang et al., (2018))
Lotus lake	Western China	42.7	*Cyanobacteria* sp.	Tang et al. (2018)
Danba hot spring		64.6		
Akkuş-Gazlıgöl hot spring	Afyonkarahisar, Western Turkey	64	*Geitlerinema* sp.	Cankilic and Berg (2016)
Ömerli spring		98	*Geitlerinema* sp., *Cyanobacterium* sp.	
Hüdai-Sandıklı spring		68	*Geitlerinema* sp., *Phormidium* sp.	
Algerian hot springs	Eastern Algeria	45–70	*Leptolyngbya foveolarum, L. laminosa, L. amplivaginata*	Amarouche-Yala et al. (2014)
Nakabusa hot spring	Nagano, Japan	52–60	*Thermosynechococcus* sp.	Stolyar et al. (2014)
Euganean spring	Padua, Italy	26–59	*L. boryana*	Moro et al. (2010)
Algerian hot springs	Eastern Algeria	45–70	*Leptolyngbya foveolarum, L. laminosa, L. amplivaginata*	Amarouche-Yala et al. (2014)
Beppu hot spring	Beppu, Japan	55	*Thermosynechococcus elongatus*	Nakamura et al. (2002)
Wakayama hot spring	Wakayama, Japan	50–65	*Thermosynechococcus vulcanus*	
Hunter's Hot Springs	North Lakeview, Oregon	45–73	*Synechococcus sp.*	Miller and Castenholz (2000)
Yellowstone National Park (YNP)	United States	52	*Synechococcus lividus*	Dyer and Gafford (1961)

Consequentially, substantial cyanobacterial diversity has been identified from such environments based on the 16S rRNA gene sequences from various hot springs. In Octopus hot spring, *Synechococcus* population was represented by several genetically distinct 16S rRNA sequences that are distantly related to the isolated unicellular cyanobacterium *Synechococcus lividus* strains (Allewalt et al., 2006).

The taxonomic diversity studies are performed basically by using two different approaches, that is, culture-dependent and culture-independent methods. Nine Algerian hot springs with a 39–93 °C temperature range were investigated for the cyanobacterial diversity based on morphological and the 16S rRNA gene sequencing analysis. The authors reported 19 cyanobacterial morphotypes with highest dominance of *Leptolyngbya* followed by *Gloeocapsa, Gloeocapsopsis, Stigonema, Fischerella, Synechocystis, Microcoleus, Cyanobacterium, Chroococcus,* and *Geitlerinema* (Amarouche-Yala et al. 2014). Isolating and characterizing cyanobacteria from hot springs has practical importance, as the isolated cultures may have biotechnological potential in terms of the production of bioactive molecules and thermostable enzymes. However, the conventional culture-based methods need isogenic populations of model organism that is not suitable for gene-by-gene analysis. Moreover, due to limitation in culturing some rare taxa and a limited range of morphological features limits the taxonomic diversity studies using these methods. In recent years, culture-independent methods that use advances in high-throughput sequencing techniques were found more promising to identify the uncultivable cyanobacteria and their metabolites.

Recent advances in high-throughput sequencing technologies have driven more intensive cyanobacteria diversity surveys over different thermal habitats. This has led to the development of a deeper understanding of cyanobacteria communities of high-temperature environments. Microbial mats are highly organized communities with different organisms splitting up the work in the hot spring environment, with cyanobacteria acting as the leading photosynthetic primary producers. Studies on cyanobacteria in hot spring microbial mats were targeted to analyze taxonomy and understating the evolution of genomic adaptations to high temperature. Currently, genomic information of various genera reported being dominant such as *Fischerella* and *Synechococcus* is available. Recently, Alcorta et al. (2020) performed genomic landscape studies on hot spring cyanobacteria to trace the phylum-wide genomic consequences of evolution in hot spring habitats. This study recovered 57 cyanobacterial metagenome-assembled genomes from 21 hot springs with temperatures range from 32 °C to 75 °C that are globally distributed. Thermophilic unicellular cyanobacterium, *Thermosynechococcus,* has a wide distribution in thermal habitats and was reported in abundance in microbial mats in alkaline hot springs of eastern Asia, having a temperature range from 50 °C to 65 °C (Tang et al., 2018a). *Thermosynechococcus elongatus* TA-1 strain recovered from Taian hot springs grown in the presence of 10–20% CO_2 at 50 °C and using 6000 lx light intensity demonstrated a high growth rate. The strain also showed phycocyanin production (C-PC) under these growth conditions (Leu et al., 2013).

Thermophilic cyanobacterium *Thermosynechococcus* has been found dominant in hot springs and represented by various species. For example, *Thermosynechococcus elongates* has been reported to grow optimally at 55 °C and different strains such as *T. elongates* (BP-1) isolated from various hot springs from Japan (Stolyar et al., 2014), *T. elongatus* PKUAC-SCTE542 from the Lianhua Lake microbial mat of western Sichuan, China (Tang et al., 2018b). In addition, other species such as *Thermosynechococcus vulcanus* (NIES-2134) and *Thermosynechococcus* sp. (NK55) were also cultivated from hot spring environments (Stolyar et al., 2014). Thermophilic member of *S. lividus*, strain PCC6715 was cultured from YNP hot spring (Liang et al., 2019). *Thermosynechococcus* strains isolated from hot springs are used widely for photosynthesis studies as well as exploration in other scientific applications. Some other common filamentous cyanobacteria, such as *Oscillatoria* and *Phormidium*, typically were isolated in abundant hot springs (Tang et al., 2018a). Another cyanobacterial member, *Nostoc commune*, a cosmopolitan cyanobacterium, majorly inhabits nutrient-poor or bare rock surfaces and soil that receive very less water from precipitation periodically. It has been observed that the temperature tolerance of *N. commune* is high, and a few species has a broad temperature tolerance range (-60 °C to 25 °C) under Arctic conditions and under temperate conditions able to grow in -30 °C to 40 °C and above (Sand-Jensen et al., 2010). A metagenomic analysis has also revealed the functional gene repertoire of the predominant cyanobacteria and associated communities from various hot springs.

6.3 Temperature stress responses in thermophiles cyanobacteria

Thermophilic cyanobacteria that thrive under extreme environmental conditions have several mechanisms to respond to environmental stresses (Fig. 6.1).

The composition of the cell membrane is one of most important factors that helps survival at high-temperature conditions. Effect of temperature on cell membrane molecules such as lipids and fatty acid vary among different species. The magnitude of temperature effect on different types of microbial species has not been thoroughly characterized. It has been observed that variation in membrane lipids occurs to maintain the chemical stability and membrane permeability even at high temperatures (Koga, 2012). Thermophilic microorganism responds to high temperature condition through homeoviscous adaptation mechanism. In such types of adaptive responses, microbial cells change the membrane lipid composition in response to changing temperature condition to maintain the membrane fluidity. The modification in the fatty acid composition response to

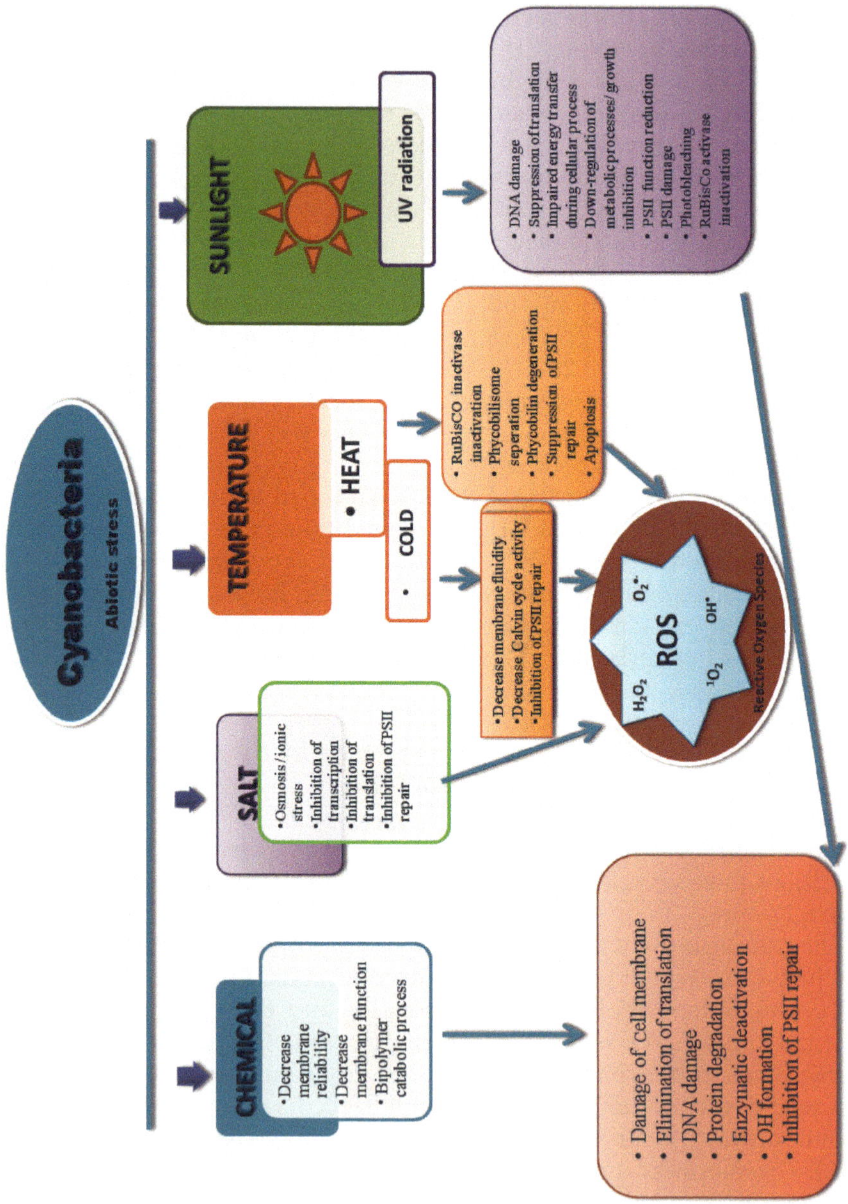

FIG. 6.1 Various mechanisms used in response to abiotic stresses in cyanobacteria.

changing temperature is one of the mechanisms used by thermophilic organisms to maintain normal cellular function across temperature (Ranawat and Rawat, 2017). For example, in a thermophilic strain of *Synechococcus* species, a decrease in the ratio of saturated and unsaturated fatty acids fatty acid was observed at the elevated temperature (Maslova et al., 2004). Therefore, membrane lipids are the main players in such type of adaptive responses and also perform other function such as maintaining membrane fluidity, help in transportation, and even protein molecules' stability (Daniel and Cowan, 2000). In maintaining the ratio of saturated and unsaturated fatty acids, other molecules such as hexadecenoic acids also play an essential role in cyanobacteria. Several cyanobacteria species serve as the model organisms and test the hypotheses about their evolution of tolerance. Nishiyama et al. (1997) reported that in the cyanobacterial strain of *Synechococcus* sp Photosystem II stability is the most crucial in responding to high temperature condition. In another study, it was revealed that Rubisco is the main limiting factor in a thermophilic cyanobacteria strain of *Chroococcidiopsis* subjected to elevated CO_2 concentration and temperatures (Hayashi et al., 1995). *Nostoc commune* is a cosmopolitan cyanobacterium that has a wide range of temperature tolerance growing on rock surfaces with annual temperature ranges of about -60 °C to 25°C under Arctic conditions and -30 °C to more than 40 °C under temperate conditions (Sand-Jensen et al., 2010). *Phormidium* species, a hot spring isolates of San Miquel from Azore, showed the production of extracellular substances and grows faster at 41 °C than 30 °C with reduced production of growth-inhibitory production at the higher temperature. Hongsthong et al. (2009) stated that *Spirulina platensis* when grown at higher temperature showed upregulation of several proteins such as soluble-8, plasma membrane-52(PM), and thylakoid membrane-39. A comparative analysis of the *S. platensis* subcellular proteome in response to low- and high-temperature stresses, performed by Kurdrid et al. (2011), revealed that high-temperature stress is linked with nitrogen ammonia assimilation. The study also indicated that the signaling proteins such as Hik14, Hik26, and Hik28 have potential interactions with other deregulated proteins in cold and high-temperature stress conditions.

6.4 Biotechnological application of thermophilic cyanobacteria

Cyanobacteria are efficient in the assimilation of organic matter and have the potential for the production of various bioactive compounds and biodegradation and transformation capability. Recent studies have shown that cyanobacterial species are potential sources for biofuels, biofertilizers, polysaccharide production, etc., making them a viable and sustainable approach in various industrial applications (Mona et al., 2020). The overwhelming diversity of cyanobacteria is the most attractive source for isolating valuable metabolites with antimicrobial and anticancer activities. Recent studies on cyanobacteria have also effectively demonstrated their use in bioremediation and wastewaters' treatment (Abed and Köster, 2005). In addition, cyanobacteria species are also explored in food and colorants, vitamins, enzymes, and pharmaceuticals industries (Abed et al., 2009).

Among the several bioactive compounds produced by cyanobacteria, lipopeptides are important surface-active agents with potential applications. Lipopeptides obtained from cyanobacteria have showed promising biological activities such as antibacterial, antifungal, antialgal, and antiprotozoal. Apart from lipopeptides, cyanobacteria produce several other metabolites with broad-spectrum biological activities. For example, compounds produced by *Phormidium* sp. have a broad-spectrum antibacterial activity that can inhabit the growth of both the Gram-positive and Gram-negative bacteria. In addition, these compounds also exhibited antifungal potential active against yeasts and fungi (Nainangu et al., 2020). Numerous other chemicals such as polyketides, cyclic peptides, and other active small molecules include protein kinase C activators, and many other anticancerous, antimicrobial compounds are produced by cyanobacteria (Abed et al., 2009; Swain et al., 2017). Peptides produced by nonribosomal peptide synthetase or polyketide synthase enzyme systems and both the enzymes have frequent and complex integration in gene clusters of cyanobacteria (Dittmann et al., 2015).

Thermophilic cyanobacteria that inhabit hot springs and other thermal habitats are also known to produce various pigments such as carotenoids, phycobiliproteins (PBPs), and several nutrition supplements such as omega-3 and omega-6 polyunsaturated fatty acids (Patel et al., 2019). The uses of colorful pigments in food and cosmetic industries are essential, as synthetic dyes are harmful and have carcinogenic effects on humans. Pigments, such as lutein, phycocyanin, are medicinally used in the treatments of eye, cardiovascular, and renal diseases. Cyanobacteria are the natural source of various compounds with therapeutic and industrial importance and used in diverse biotechnological applications (Fig. 6.2 and Table 6.2).

Various approaches are used to enhance the production of these important molecules, as these could be useful in various food preparations, in cosmetics, and also showed promising medicinal values (Mandal et al., 2020). The photosynthetic apparatus of cyanobacteria contains several water-soluble PBPs that are organized in phycobilisomes. These molecules act as free-radical scavengers that enhances their use in pharmaceutical industries, and also as the fluorochromes in biomedical research. Also, PBPs are reported as the effective anticancer, neuroprotective, antiinflammatory, antiallergic, and hepatoprotective agents (Galetović et al., 2020).

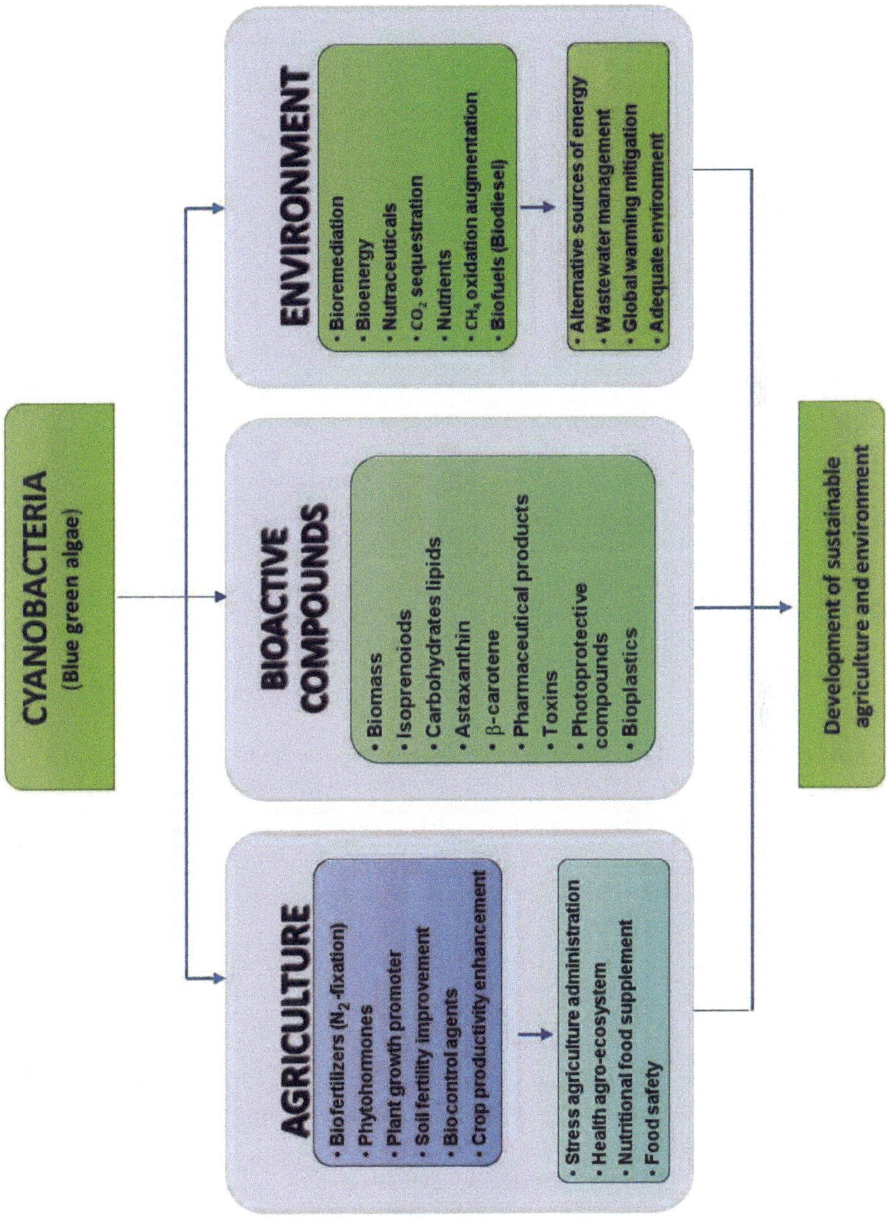

FIG. 6.2 Biotechnological application of cyanobacteria and their metabolites.

TABLE 6.2 Thermophilic and thermotolerant cyanobacteria and their biotechnological applications.

Sampling site	Cyanobacteria	Application	References
Hot spring, Azores, Portugal	*Phormidium* sp.	Antimicrobial agents	(Fish and Codd, 1994)
Coiba National Park, Panama	*Lyngnya* sp.	Active against HeLaS3 epithelial carcinoma and SNB-78 central nervous system	Vijayakumar and Menakha (2015)
Water reservoir, Savannah River Site	*Fischerella* sp.	Nutrient removal and waste heat utilization	Radway et al. (1994)
Shallow water, Bocos del Toro, Panama	*Lyngbya majuscula*	Antiparasitic agents	Balunas et al. (2010)
Volcanic rock, Japan	*Synechococcus* sp.	Poly-β-hydroxybutyrate (PHB) production	Miyake et al. (1996)
Savannah river, South Carolina, USA	*Mastigocladus laminosus*	Effluent's treatment	Weissman et al. (1998)
Basque Country, Bilbao, Spain	*Phormidium laminosum*	Removal of nitrate and phosphate ions from water	Sawayama et al. (1998)
Culture Collection of Algae, Austin	*Fischerella muscicola*	Antialgal, antifungal	Dahms et al. (2006)
Taian hot springs, Taiwan	*Thermosynechococcus elongatus*	CO_2 mitigation from industrial flue gas	Leu et al. (2013)
Yellowstone National Park (YNP), USA	*Chlorogleopsis* sp.	CO_2 biofixation	Ono and Cuello (2007)
Kazusa DNA Research Institute	*Thermosynechococcus elongatus*	Production of dipeptides	Sato et al. (2007)
Hot springs, the Western Ghats, India	*Limnothrixredekei sp. Planktolyngbya crassa, Geitlerinema sulphureum*	CO_2 sequestration	Manjre and Deodhar (2013)
Chin-Lun hot spring, Taiwan	*Thermosynechococcus* sp.	Carbon biofixation and carbohydrate production	Hsueh et al. (2009)
Chin-Lun hot spring, Taiwan	*Thermosynechococcus* sp.	Biomass and bioenergy production	Su et al. (2013)
Chin-Lun hot spring, Taiwan	*Thermosynechococcus* sp.	Bioenergy production	Hsueh et al., (2007)

Sampling site	Cyanobacteria	Application	References
Volcanic rock, Japan	*Synechococcus sp.*	PHB production	Nishioka et al. (2001), Nishioka et al. (2002)
Octopus spring, YNP, USA	*Synechococcus ecotypes*	*In-situ* analysis of nitrogen fixation	Steunou et al. (2006)
Hot springs, Ayas, Turkey	*Synechococcus* sp. and *Phormidium* sp.	Dye bioaccumulation	Sadettin and Dönmez (2006)
Hot springs, Ayas, Turkey	*Phormidium* sp.	Dye-rich wastewater treatment	Ertuğrul et al. (2008)
Thermal springs, Zerka, Jordan	*Thermosynechococcus, Chroogloeocystis, Fischerella, Scytonem*	Production of nitrogenase reductase	Oren et al. (2009)
Hot springs, Ayas, Turkey	*Phormidium* sp.	Biosorption of chromium	Aksu et al. (2009)
Jakrem hot spring, India	*Mastigocladus* sp. *Microcoleus* sp.	Medicinal application	Siangbood and Ramanujam (2011)
Geno hot spring, Bandar Abbas, Iran	*Oscillatoria subbrevis, Oscillatoria tenuis, O.limentica, O.angusta, O. articulate, Synechocystis aquatilis, S. cerdorum*	Antimicrobial agents	Heidari et al. (2012)
Hazen, Monitor hot spring, Nevada	*Aphanocapsa* sp, *Leptolyngbya* sp.	Lipid production	Bywaters and Fritsen (2014)

Phycocyanins are an important light-harvesting accessory pigment of cyanobacteria and have many commercial applications in the food and pharmaceutical industries. Ferraro et al. (2020) carried out the C-phycocyanin (C-PC) purification from the extremophilic species of *Galdieri*. The compound is characterized using several advanced techniques such as SDS-PAGE (Sodium Dodecyl Sulphate–Polyacrylamide Gel Electrophoresis), mass spectrometry, etc. The identified proteins exhibited interesting antioxidant properties even at high-temperature treatments and are found to be cytotoxic for A431 and SVT2 cancer cells but nontoxic to nonmalignant cells. Such studies, therefore, show that C-phycocyanin can act as the multitasking protein that can be used in the food industry as an antioxidant and anticancer agent. *Thermosynechococcaceae* members were known to produce thermostable C-PC. Liang et al., (2018) studied three different strains when grown in two different types of media and at two different temperatures (45 °C and 50 °C). The study showed that C-PC synthesized by *Synechococcus* sp. was found to be more stable than the C-PC synthesized by *Spirulina* species. In another study, strains of *Galdieria phlegrea* isolated from different geographic areas were also tested for their thermostable C-PC during autotrophic and heterotrophic cultivation (Carfagna et al., 2018).

Cyanobacteria also produce polyhydroxyalkanoates, which is the best raw material for the production of bioplastics. One such raw biopolymer produced by cyanobacteria is the polyhydroxy butyrate (PHB). It is synthesized by many cyanobacterial strains and gets accumulates inside the cells, which limit their efficient production as it needs more expensive cultivation and harvesting equipment. Several research groups are inovled in improving the viability of cyanobacterial PHB production (Price et al., 2020). In the past few years, significant advances have arisen in cyanobacterial biotechnology. Worldwide, research focuses on cyanobacteria for their possible use in various industries, including energy sectors and biofuels. However, till now, only a few cyanobacterial species were exploited commercially. With advancements in new sequencing techniques and bioinformatics techniques, novel strains of cyanobacteria are being identified. A deep understanding of their genomics and metabolites could help produce the synthesis of high-value products.

6.5 Metabolic engineering in cyanobacteria

In recent years, cyanobacteria have been utilized for critical metabolic engineering applications. Many strains have been engineered successfully to produce a variety of industrially important compounds. One of the important factors in balancing flux via a pathway is the kinetic efficiency of the pathway that can be done by optimizing enzyme expression levels, availability of precursor metabolites, and cofactor equilibrium within the cell system (Zahra et al., 2020). Presently, efforts are being taken to underpin the biosynthesis mechanisms of a diverse range of natural molecules produced by cyanobacteria. Advances in genomics could help to link the biosynthetic gene clusters to natural products synthesis by cyanobacteria using genome mining approaches. Genomics studies of the few of the cyanobacterial genera have demonstrated a vast majority of the cryptic gene clusters of natural products. Various novel molecular-based approaches are being employed in developing genetically modified strains. Approaches such as homologous recombination and small interfering RNAs, CRISPR/Cas adaptive immune system and random insertion mutagenesis have been adapted to genetically modify the phototrophic bacteria (Gomaa et al., 2016). The productivity of cyanobacteria depends on their carbon fixation rate and light utilization efficiency, which differ in cyanobacteria strains (Rosgaard et al., 2012). Therefore, attempts on utilization of genetic engineering tools to modify genes responsible for CO_2 fixation and light utilization have been reported. One of the important factors that limit the higher accumulation of the desired products is the inherent metabolic capacity of organisms. Therefore, studies are also being initiated using genetic engineering tools to overcome these limitations. Cyanobacteria have wide genetic diversity that could be useful for gathering the genomic information of metabolic pathway maps.

Nevertheless, very few studies have focused on gene annotation and functional analysis in cyanobacteria. In the future, further development in bioinformatics tools and techniques will help in making precision in genome assembly and annotation in cyanobacteria (Kumar et al., 2020). With more such precious information, metabolic engineering in cyanobacteria will be become easier with higher productivity. With the technological advances in modifying biochemical pathways for the efficient production of cyanobacterial metabolites in biotechnological exploitations, metabolic engineering is recently gaining more attention. However, such studies are not well explored for thermophile members that produce valuable metabolites.

6.6 Conclusion

Significant work on cyanobacterial diversity and potential biotechnological application has attracted much of the research attention. Studying cyanobacterial diversity and understanding cyanobacteria's environmental role is essential for their exploration in the sustainable production of bioactive compounds, value-added molecules, in food industry, and for green energy sources such as biodiesel and bioremediation. Promising advancements in the cultivation of cyanobacteria and high-throughput analysis using molecular tools will undoubtedly help better understand cyanobacteria taxonomy and their metabolic potentials. Studies on thermophilic cyanobacteria demonstrated considerable interest due to their metabolic activities at higher temperatures. Various tools and techniques have contributed to clarifying cyanobacteria's metabolic potential, such as genetic and metabolic engineering, to develop the inducing autolysis and secretory product systems in cyanobacteria. With further improvement in cultivation methods, purification and harvesting of product techniques will achieve the production process's economic feasibility.

Acknowledgments

Authors are thankful to the Department of Science and Technology (DST), New Delhi, Government of India for financial support. The work was carried out under financial support from a joint international grant from the Department of Science and Technology (DST) of the Government of India; project number C/1756/IFD/2019-20, and the Russian Science Foundation, project number 19-46-02004.The authors would like to extend their sincere appreciation to the Vice Chancellor of Mizoram University, Mizoram, India and Principal, Pachhunga University College, Aizawl, Mizoram, India, for their continuous support.

References

Abed, R.M., Dobretsov, S., Sudesh, K., 2009. Applications of cyanobacteria in biotechnology. J. Appl. Microbiol 106, 1–2.

Abed, R., Köster, J., 2005. The direct role of aerobic heterotrophic bacteria associated with cyanobacteria in the degradation of oil compounds. Int. Biodeterior. Biodegrad 55 (1), 29–37.

Aksu, Z., Ertuğrul, S., Dönmez, G., 2009. Single and binary chromium (VI) and Remazol Black B biosorption properties of *Phormidium* sp. J. Hazard. Mater. 168, 310–318.

Alcorta, J., Alarcón-Schumacher, T., Salgado, O, Díez, B., 2020. Taxonomic novelty and distinctive genomic features of hot spring cyanobacteria. Front. Genet. 11, 568223.

Allewalt, J.P., Bateson, M.M., Revsbech, N.P., Slack, K., Ward, D.M., 2006. Effect of temperature and light on growth of and photosynthesis by *Synechococcus* isolates typical of those predominating in the octopus spring microbial mat community of Yellowstone National Park. Appl. Environ. Microbiol 72, 544–550.

Amarouche-Yala, S., Benouadah, A., El Ouahab Bentabet, A., Lopez-Garcıa, P, 2014. Morphological and phylogenetic diversity of thermophilic cyanobacteria in Algerian hot springs. Extremophiles 18, 1035–1047.

Ansary, M.M., Ahmadimoghadam, A., Mirtadzadini, S.M., 2017. Distribution of cyanobacteria in two sirch hot springs with regards to the physicochemical traits of water. Banat's J. Biotechnol. 8, 83.

Babele, P.K., Kumar, J., Chaturvedi, V., 2019. Proteomic de-regulation in cyanobacteria in response to abiotic stresses. Front. Microbiol. 10, 1315.

Bahl, J., Lau, M.C., Smith, G.J., Vijaykrishna, D., Cary, S.C., Lacap, D.C., Lee, C.K., Papke, R.T., Warren-Rhodes, K.A., Wong, F.K., McKay, C.P., 2011. Ancient origins determine global biogeography of hot and cold desert cyanobacteria. Nat. Commun 2, 1–6.

Balunas, M.J., Linington, R.G., Tidgewell, K., Fenner, A.M., Urena, L.D., Togna, G, D., et al., 2010. Dragonamide E, a modified linear lipopeptide from *Lynngbya* majuscule with antileishmanial activity. J. Nat. Prod 73, 60–66.

Bennett, A.C., Murugapiran, S.K., Hamilton, T.L., 2020. Temperature impacts community structure and function of phototrophic *Chloroflexi* and cyanobacteria in two alkaline hot springs in Yellowstone National Park. Environ. Microbiol. Rep 12, 503–513.

Brenes-Guillén, L., Vidaurre-Barahona, D., Morales, S., Mora-López, M., Sittenfeld, A., Uribe-Lorío, L., 2021. Novel cyanobacterial diversity found in Costa Rican thermal springs associated with Rincon de la Vieja and Miravalles Volcanoes: a polyphasic approach. J. Phycol 57, 183–198.

Brock, T.D., 1967. Microorganisms adapted to high temperatures. Nature 214, 882–885.

Bywaters, K.F., Fritsen, C., 2014. Biomass and neutral lipid production in geothermal microalgal consortia. Front. Bioeng. Biotechnol. 2, 1–11.

Cankilic, Y., Berk., A, 2016. Analysis of cyanobacterial diversity of some hot springs in Afyonkarahisar,Turkey. Appl. Ecol. Environ. Res. 14, 463–484.

Carfagna, S., Landi, V., Coraggio, F., Salbitani, G., Vona, V., Pinto, G., Pollio, A., Ciniglia, C., 2018. Different characteristics of C-phycocyanin (C-PC) in two strains of the extremophilic Galdieria phlegrea. Algal Res. 31, 406–412.

Červený, J., Sinetova, M.A., Zavřel, T., Los, D.A., 2015. Mechanisms of high temperature resistance of *Synechocystis* sp. PCC 6803: an impact of histidine kinase 34. Life 5, 676–699.

Cheng, Y.I., Chou, L., Chiu, Y.F., Hsueh, H.T., Kuo, C.H, Chu, H.A., 2020. Comparative genomic analysis of a novel strain of Taiwan hot-spring cyanobacterium *Thermosynechococcus* sp. CL-1. Front. Microbiol. 11, 82.

Curren, E., Leong, S.C., 2020. Natural and anthropogenic dispersal of cyanobacteria: a review. Hydrobiologia 847, 2801–2822.

Dahms, H.U., Xu, Y., Pfeiffer, C., 2006. Antifouling potential of cyanobacteria: a mini-review. Biofouling 22, 317–327.

Daniel, R.M., Cowan, D.A., 2000. Biomolecular stability and life at high temperatures. Cell Mol. Life Sci. 57 (2), 250–264.

Dittmann, E., Gugger, M., Sivonen, K., Fewer, D.P., 2015. Natural product biosynthetic diversity and comparative genomics of the cyanobacteria. Trends Microbiol. 23, 642–652.

Dyer, D.L., Gafford, R.D., 1961. Some characteristics of a thermophilic blue-green alga. Science 134, 616–617.

Ertuğrul, S., Bakir, M., Dönmez, G., 2008. Treatment of dye-rich wastewater by an immobilized thermophilic cyanobacterial strain: *Phormidium* sp. Ecol. Eng. 32, 244–248.

Ferraro, G., Imbimbo, P., Marseglia, A., Illiano, A., Fontanarosa, C., Amoresano, A, Olivieri, G., Pollio, A., Monti, D.M., Merlino, A., 2020. A thermophilic C-phycocyanin with unprecedented biophysical and biochemical properties. Int. J. Biol. Macromol 150, 38–51.

Fish, S.A., Codd, G.A., 1994. Bioactive compound production by thermophilic and thermotolerant cyanobacteria (blue-green algae). World J. Microbiol. Biotechnol. 10, 338–341.

Galetović, A, Seura, F, Gallardo, V, Graves, R, Cortés, J, Valdivia, C, Núñez, J, Tapia, C, Neira, I, Sanzana, S, Gómez-Silva, B., 2020. Use of phycobiliproteins from Atacama cyanobacteria as food colorants in a dairy beverage prototype. Foods 9, 244.

Garcia-Pichel, F., López-Cortés, A., Nübel, U, 2001. Phylogenetic and morphological diversity of cyanobacteria in soil desert crusts from the Colorado Plateau. Appl. Environ. Microbiol 67, 1902–1910.

Gomaa, M.A., Al-Haj, L., Abed, R.M., 2016. Metabolic engineering of cyanobacteria and microalgae for enhanced production of biofuels and high-value products. J. Appl. Microbiol 121, 919–931.

Hayashi, N.R., Ishida, T., Peerapornpisal, Y., et al., 1995. Effect of carbon dioxide concentration on the growth and Rubisco activity of a thermophilic cyanobacterium, *Chroococcidiopsis* sp. strain TS-821. J. Ferment. Bioeng 80, 507–509.

Heidari, F., Riahi, H., Yousefzadi, M., Asadi, M., 2012. Antimicrobial activity of cyanobacteria isolated from hot spring of geno. Middle-East J. Sci. Res. 12, 336–339.

Hongsthong, A., Sirijuntarut, M., Yutthanasirikul, R., Senachak, J., Kurdrid, P., Cheevadhanarak, S., Tanticharoen, M., 2009. Subcellular proteomic characterization of the high-temperature stress response of the cyanobacterium *Spirulina platensis*. Proteom. Sci 7, 33.

Hsueh, H.T.T., Chu, H., Yu, S.T.T, 2007. A batch study on the bio-fixation of carbon dioxide in the absorbed solution from a chemical wet scrubber by hot spring and marine algae. Chemosphere 66, 878–886.

Hsueh, H.T.T., Li, W.J., Chen, H.H., Chu, H, 2009. Carbon bio-fixation by photosynthesis of *Thermosynechococcus* sp. CL-1 and *Nannochloropsis oculta*. J. Photochem. Photobiol. B. Biol. 95, 33–39.

Koga, Y., 2012. Thermal adaptation of the archaeal and bacterial lipid membranes. Archaea 789652.

Kumar, G., Shekh, A., Jakhu, S., Sharma, Y., Kapoor, R., Sharma, T.R., 2020. Bioengineering of microalgae: recent advances, perspectives, and regulatory challenges for industrial application. Front. Bioeng. Biotechnol. 8, 914.

Kurdrid, P., Senachak, J., Sirijuntarut, M., Yutthanasirikul, R., Phuengcharoen, P., Jeamton, W., et al., 2011. Comparative analysis of the *Spirulina* platensis subcellular proteome in response to low-and high-temperature stresses: uncovering cross-talk of signaling components. Proteom. Sci 9, 39.

Leu, J.-Y., Lin, T.-H., Selvamani, M.J.P., Chen, H.-C., Liang, J.-Z., Pan, K.-M, 2013. Characterization of a novel thermophilic cyanobacterial strain from Taian hot springs in Taiwan for high CO_2 mitigation and C-phycocyanin extraction. Process. Biochem. 48, 41–48.

Liang, Y., Tang, J., Luo, Y., Kaczmarek, M.B., Li, X., Daroch, M., 2019. *Thermosynechococcus* as a thermophilic photosynthetic microbial cell factory for CO_2 utilization. Bioresour. Technol 278, 255–265.

Liang., Y., Kaczmarek, M.B., Kasprzak, A.K., Tang, J., Shah, M.M., Jin, P., Klepacz-Smółka, A., Cheng, J.J, Ledakowicz, S., Daroch, M., 2018. *Thermosynechococcaceae* as a source of thermostable C-phycocyanins: properties and molecular insights. Algal Res 35, 223–235.

Mandal, M.K., Chanu, N.K., Chaurasia, N., 2020. Cyanobacterial pigments and their fluorescence characteristics: applications in research and industry. Adv. Cyanobacterial Biol., 55–72.

Manjre, S.D., Deodhar, M.A., 2013. Screening of thermotolerant microalgal species isolated from Western Ghats of Maharashtra, India for CO_2 sequestration. J. Sustain. Energy Environ. 4, 61–67.

Maslova, I.P., Mouradyan, E.A., Lapina, S.S., Klyachko-Gurvich, G.L., Los, D.A., 2004. .Lipid fatty acid composition and thermophilicity of cyanobacteria. Russ. J. Plant Physiol., 51, 353–360.

Mehetre, G.T., Paranjpe, A., Dastager, S.G., Dharne, M.S., 2016. Investigation of microbial diversity in geothermal hot springs in Unkeshwar, India, based on 16S rRNA Amplicon Metagenome Sequencing. Genome Announc 4, 4–5.

Merkel, A.Y., Pimenov, N.V., Rusanov, II, Slobodkin, A.I., Slobodkina, G.B., Tarnovetckii, I.Y., Frolov, E.N., Dubin, A.V., Perevalova, A.A., Bonch-Osmolovskaya, E.A., 2017. Microbial diversity and autotrophic activity in Kamchatka hot springs. Extremophiles 21, 307–317.

Miller, S.R., Carvey, D., 2019. Ecological divergence with gene flow in a thermophilic cyanobacterium. Microbol. Ecol 78, 33–41.

Miller, S.R., Castenholz, R.W., 2000. Evolution of thermotolerance in hot spring cyanobacteria of the genus *Synechococcus*. Appl. Environ. Microbiol. 66, 4222–4229.

Miller, S.R., Castenholz, R.W., 2000. The evolution of thermotolerance in hot spring cyanobacteria of the genus *Synechococcus*. J. Phycol 36, 48.

Miyake, M., Erata, M., Asada, Y., 1996. A thermophilic cyanobacterium, *Synechococcus* sp. MA19, capable of accumulating poly-B-hydroxybutyrate. J. Ferment. Bioeng. 82, 512–514.

Mona, S., Kumar, V., Deepak, B., Kaushik, A, 2020.) Cyanobacteria: the eco-friendly tool for the treatment of industrial wastewaters. In: Saxena, G., Bharagava, R.N. (Eds.), Bioremediation of industrial waste for environmental safety 2020. Springer, Singapore, pp. 389–413.

Moro, I., Rascio, N., La Rocca, N., Sciuto, K., Albertano, P., Bruno, L., Andreoli, C., 2010. Polyphasic characterization of a thermo-tolerant filamentous cyanobacterium isolated from the Euganean thermal muds (Padua, Italy). Eur. J. Phycol. 45, 143–154.

Nainangu, P., Antonyraj, A.P., Subramanian, K., Kaliyaperumal, S., Gopal, S., Renuka, PS., 2020. In vitro screening of antimicrobial, antioxidant, cytotoxic activities, and characterization of bioactive substances from freshwater cyanobacteria *Oscillatoria* sp. SSCM01 and *Phormidium* sp. SSCM02. Biocatal. Agri. Biotechnol 29, 101772.

Nakamura, Y., Kaneko, T., Sato, S., Ikeuchi, M., Katoh, H., Sasamoto, S., et al., 2002. Complete genome structure of the thermophilic cyanobacterium *Thermosynechococcus elongatus* BP-1. DNA Res 9, 123–130.

Nishioka, M., Nakai, K., Miyake, M., Asada, Y., Taya, M., 2001. Production of poly-β-hydroxybutyrate by thermophilic cyanobacterium, *Synechococcus* sp. MA19, under phosphate under phosphate-limited conditions. Biotechnol. Lett. 23, 1095–1099.

Nishioka, M., Nishiuma, H., Miyake, M., Asada, Y., Shimizu, K., Taya, M., 2002. Metabolic flux analysis of a poly-β-hydroxybutyrate producing cyanobacterium, *Synechococcus* sp. MA19, grown under photoautotrophic conditions. Biotechnol. Bioprocess. Eng. 7, 295–302.

Nishiyama, Y., Los, D.A., Hayashi, H., Murata, N., 1997. Thermal protection of the oxygen-evolving machinery by PsbU, an extrinsic protein of photosystem II, in *Synechococcus* species PCC 7002. Plant. Physiol 115, 1473–1480.

Ono, E., Cuello, J.L., 2007. Carbon dioxide mitigation using thermophilic cyanobacteria. Biosyst. Eng. 96, 129–134.

Oren, A., Ionescu, D., Hindiyeh, M., Malkawi, H., 2009. Morphological, phylogenetic and physiological diversity of cyanobacteria in the hot springs of Zerka Ma. BioRisk 3, 69–82.

Oren, A., 2012. Salts and brines. In: Whitton, B.A. (Ed.), Ecology of Cyanobacteria II: Their Diversity in Space and Time. Springer, Dordrecht, pp. 401–426.

Patel, A., Matsakas, L., Rova, U., Christakopoulos, P., 2019. A perspective on biotechnological applications of thermophilic microalgae and cyanobacteria. Bioresour. Technol. 278, 424–434.

Peary, J.A., Castenholz, R.W., 1964. Temperature strains of a thermophilic blue-green alga. Nature 202, 720–721.

Pedersen, D., Miller, S.R., 2017. Photosynthetic temperature adaptation during niche diversification of the thermophilic cyanobacterium *Synechococcus* A/B clade. ISME J 2017 (11), 1053–1057.

Price, S., Kuzhiumparambil, U., Pernice, M., Ralph, P.J., 2020. Cyanobacterial polyhydroxybutyrate for sustainable bioplastic production: critical review and perspectives. J. Environ. Chem. Eng., 104007.

Radway, J.C., Weissman, J.C., Wilde, E.W., Benemann, J.R., 1994. Nutrient removal by thermophilic Fischerella (*Mastigocladus laminosus)* in a simulated algaculture process. Bioresour. Technol. 50, 227–233.

Ranawat, P., Rawat, S., 2017. Stress response physiology of thermophiles. Arch. Microbiol 199, 391–414.

Rosgaard, L., Porcellinis, A.J., Jacobsen, J.H., Frigaard, N., Sakuragi, Y., 2012. Bioengineering of carbon fixation, biofuels, and biochemical in cyanobacteria and plants. Journal of Biotechnology 162, 134–147.

Sadettin, S., Dönmez, G., 2006. Bioaccumulation of reactive dyes by thermophilic cyanobacteria. Process. Biochem 41, 836–841.

Sand-Jensen, K., Baastrup-Spohr, L., Winkel, A., M0ller, C.L., Borum, J., Brodersen, K., Lindell, T., Staehr, P.A., 2010. Plant distribution patterns and adaptation in a limestone quarry on 01and. Sv. Bot. Tidsskr. 104, 23–31.

Zahra, Z., Choo, D.H., Lee, H., Parveen, A., 2020. Cyanobacteria: review of current potentials and applications. Environments 7, 13.

Sato, M., Masuda, Y., Kirimura, K., Kino, K., 2007. Thermostable ATP regeneration system using polyphosphate kinase from *Thermosynechococcus elongatus* BP-1for d-amino acid dipeptide synthesis. J. Biosci. Bioeng. 103, 179–184.

Sawayama, S., Rao, K.K., Hall, D.O., 1998. Nitrate and phosphate ion removal from water by *Phormidium laminosum* immobilized on hollow fibres in a photobioreactor. Appl. Microbiol. Biotechnol. 49, 463–468.

Seckbach, J., Chapman, D.J., Garbary, D., Oren, A., Reisser, W., 2007. Algae and cyanobacteria under environmental extremes. In: Seckbach, J. (Ed.), Algae and Cyanobacteria in Extreme Environments. Springer, Dordrecht, pp. 781–786.

Siangbood, H., Ramanujam, P., 2011. A report on thermophilic cyanophyta (cyanobacteria) from Jakrem Hotspring. Meghalaya. Int. J. Algae 13, 178–185.

Singh, Y., Gulati, A., Singh, D.P., Khattar, J.I., 2018. Cyanobacterial community structure in hot water springs of Indian North-Western Himalayas: a morphological, molecular and ecological approach. Algal Res 29, 179–192.

Steunou, A.S., Bhaya, D., Bateson, M.M., Melendrez, M.C., Ward, D.M., Brecht, E., Peters, J.W., Kühl, M., Grossman, A.R., 2006. In situ analysis of nitrogen fixation and metabolic switching in unicellular thermophilic cyanobacteria inhabiting hot spring microbial mats. Proc. Natl. Acad. Sci 103, 2398–2403.

Stolyar, S., Liu, Z., Thiel, V., Tomsho, L.P., Pinel, N., Nelson, W.C., Fredrickson, J.K., 2014. Genome Sequence of the Thermophilic Cyanobacterium Thermosynechococcus sp. Strain NK55a.Genome Announc.2, e01060–e01113.

Strunecký, O., Kopejtka, K., Goecke, F., Tomasch, J., Lukavský, J., Neori, A., Kahl, S., Pieper, D.H., Pilarski, P., Kaftan, D., Koblížek, M., 2019. High diversity of thermophilic cyanobacteria in Rupite hot spring identified by microscopy, cultivation, single-cell PCR and amplicon sequencing. Extremophiles 23, 35–48.

Su, C.M., Hsueh, H.T., Chen, H.H., Chu, H., 2013. Effects of nitrogen availability on the bioenergy production potential and CO_2 fixation of *thermosynechococcus* CL-1 under continuous cultivation. Aerosol Air Qual. Res. 13, 1321–1330.

Subudhi, E., Sahoo, R.K., Gaur, M., Singh, M., Das, A., 2018. Shift cyanobacteria community diversity in hot springs of India. Geomicrobiol. J 35, 141–147.

Swain, S.S., Paidesetty, S.K., Padhy, R.N., 2017. Antibacterial, antifungal, and antimycobacterial compounds from cyanobacteria. Biomed. Pharmacother. 90, 760–776.

Tang, J., Jiang, D., Luo, Y., Liang, Y., Li, L., Shah, M.M., Daroch, M., 2018. Potential new genera of cyanobacterial strains isolated from thermal springs of western Sichuan. China. Algal Res. 31, 14–20.

Tang, J., Liang, Y., Jiang, D., Li, L., Luo, Y., Shah, M.M.R., Daroch, M, 2018. Temperature-controlled thermophilic bacterial communities in hot springs of western Sichuan. China. BMC Microbiol. 18, 134.

Vijayakumar, S., Menakha, M., 2015. Pharmaceutical applications of cyanobacteria-a review. J. Acute. Med 5, 15–23.

Ward, D.M., Ferris, M.J., Nold, S.C., Bateson, M.M., 1998. A natural view of microbial biodiversity within hot spring cyanobacterial mat communities. Microbiol. Mol. Biol. Rev 62, 1353–1370.

Ward, D.M., Castenholz, R.W., Miller, S.R., 2012. Cyanobacteria in geothermal habitats. In: Whitton, BE (Ed.), Ecology of Cyanobacteria II. Springer, Dordrecht, TheNetherlands.

Weissman, J.C., Radway, J.C., Wilde, E.W., Benemann, J.R., 1998. Growth and production of thermophilic cyanobacteria in a simulated thermal mitigation process. Bioresour. Technol. 65, 87–95.

Wilde, A., Dienst, D., 2011. Tools for genetic manipulation of cyanobacteria. In: Peschek, G.A., Obinger, C., Renger, G. (Eds.), Bioenergetic Processes of Cyanobacteria. Springer, Dordrecht, pp. 685–703.

CHAPTER

7

Exploring the ability of cyanobacterial ferric uptake regulator (FUR) proteins to increase yeast tolerance to abiotic stresses

Violeta C. Sein-Echaluce[a], *José Miguel Mulet*[b], *María V. Barja*[c], *M. Luisa Peleato*[a], *María F. Fillat*[a]

[a]Departamento de Bioquímica y Biología Molecular y Celular, Universidad de Zaragoza, Zaragoza, Spain

[b]Instituto de Biología Molecular y Celular de Plantas, Universidad Politécnica de Valencia-CSIC, Valencia, Spain

[c]Centre for Research in Agricultural Genomics, Barcelona, Spain

7.1 Introduction

Heterologous expression of proteins leading to improvement of cell performance at harsh environments has been used in the last decades for the development of novel strains useful for agriculture and industry (Cletus, Balasubramanian, Vashisht, & Sakthivel, 2013; Tanaka, Ishii, Ogawa, & Shima, 2012; Tognetti et al., 2006). Oxidative damage is the ultimate factor affecting fermentation processes, crop development, and productivity, among other operations related to food processing or health (Auesukaree, 2017; Feki et al., 2015). Therefore, the expression of exogenous proteins involved in the defense against oxidative stress has been widely used to increase tolerance of industrial fungi, bacteria, and plants (Martinez et al., 2015; Zhang et al., 2014). The expression of antioxidant proteins from

Cyanobacterial Lifestyle and its Applications in Biotechnology.
DOI: https://doi.org/10.1016/B978-0-323-90634-0.00012-3

cyanobacteria, such as metallothionein from *Synechococcus*, improves cadmium tolerance and bioaccumulation in *Deinococcus radiodurans* R1 (Chaturvedi & Archana, 2014). Furthermore, *Anabaena* sp. PCC 7120 alkyl hydroperoxidases (Shrivastava, Singh, Singh, Pandey, & Rai, 2015), phytochelatin synthase (Chaurasia, Mishra, & Rai, 2008), arsenate reductase Alr1105 (Pandey, Shrivastava, Rai, & Rai, 2013), as well as the hypothetical protein Alr0882 (Shrivastava, Pandey, Singh, Rai, & Rai, 2012) and the Dps-like All3940 (Narayan, Kumari, & Rai, 2010) confer transgenic *Escherichia coli* tolerance to oxidative stresses. The expression of flavodoxin from *Anabaena* sp. PCC 7120 also increases tolerance to oxidants in *E. coli* and is beneficial to soil bacteria (Coba de la Pena, Redondo, Fillat, Lucas, & Pueyo, 2013). This flavoprotein has been successfully used to produce multiresistant tobacco plants (Tognetti, et al., 2006) and to obtain nodules more tolerant to abiotic stresses and exhibiting delayed senescence (Redondo, de la Pena, Morcillo, Lucas, & Pueyo, 2009). Notably, the characterization of those transgenic organisms has also provided new clues to decipher novel pathways contributing to relieve oxidative pressure in biotechnologically relevant organisms (Tognetti, et al., 2006; Zurbriggen et al., 2008).

Ferric uptake regulator (FUR) proteins are prokaryotic transcriptional regulators that control a plethora of genes, mainly involved in the control of metal homeostasis and the defense against oxidative stress (Fillat, 2014). The cyanobacterium *Anabaena* sp. PCC 7120 expresses three paralogs, namely FurA (Fur), FurB (Zur, Zinc uptake regulator), and FurC (Hernandez, Lopez-Gomollon, Bes, Fillat, & Peleato, 2004). Under certain conditions, the overexpression of FurA in *E. coli* increases its tolerance to oxidative damage (Lopez-Gomollon, Sevilla, Bes, Peleato, & Fillat, 2009), while enhanced expression in *Anabaena* slightly decreases intracellular reactive oxygen species in spite of lowering catalase and superoxide dismutase activities (Gonzalez, Bes, Barja, Peleato, & Fillat, 2010). Unlike most Fur proteins described so far, FurA does not contain structural zinc and instead exhibits disulfide reductase activity (Botello-Morte et al., 2014; Hernández, Bes, Fillat, Neira, & Peleato, 2002), working as a cysteine-based redox sensor (Botello-Morte et al., 2015). Similar to FurA, FurB plays a dual role in *Anabaena* and, in addition to regulating zinc homeostasis, increased levels of FurB expression elicit tolerance of the cyanobacterium to oxidants (Lopez-Gomollon, et al., 2009; Sein-Echaluce et al., 2014). Besides, the ability of FurB to protect DNA against oxidative cleavage and DNaseI digestion through unspecific binding led to propose that, as well as controlling zinc homeostasis, FurB might work as a DpsA protein protecting the chromosome from oxidative damage *in vivo*.

The budding yeast *Saccharomyces cerevisiae* is a powerful model organism that has been traditionally used in baking, brewing, wine production, and, more recently, in ethanol production (Da Silva & Srikrishnan, 2012).

The cost-effective production depends on many factors related to the improvement of yeast performance and productivity under stress conditions, being oxidative stress one of the most common drawbacks (Dogan, Demirci, Aytekin, & Sahin, 2014; Snowdon, Schierholtz, Poliszczuk, Hughes, & van der Merwe, 2009). Moreover, *S. cerevisiae* is often used for studying fundamental aspects of eukaryotic cell biology, such as the mechanisms of the stress response that are very similar to those in plants (Mulet, Alemany, Ros, Calvete, & Serrano, 2004). Accordingly, yeasts can be regarded as a previous step to infer the effect of a transgene in plant cells. As the heterologous expression of cyanobacterial proteins has been successfully used to manage improved biotechnological organisms (Coba de la Pena, Redondo, Manrique, Lucas, & Pueyo, 2010; He et al., 2011; Lieman-Hurwitz, Rachmilevitch, Mittler, Marcus, & Kaplan, 2003; Miyagawa, Tamoi, & Shigeoka, 2001; Tognetti, et al., 2006), we sought to investigate the effect of the heterologous expression of FurA and FurB in the tolerance of *S. cerevisiae* to several abiotic stresses.

7.2 Materials and methods

7.2.1 Strains and growth conditions

Bacterial strains used in this study are described in Table 7.1. *Anabaena* sp. PCC 7120 was grown photoautotrophically in BG-11 medium (Rippka, Deruelles, Waterbury, Herdman, & Stanier, 1979) at 28 °C and constant

TABLE 7.1 Bacterial and yeast strains used in this study.

Strain	Characteristics and use	Source
Anabaena sp. PCC 7120	Wild type	Pasteur Institute
E. coli DH5α	F- ϕ80*lac*ZΔM15 Δ(*lac*ZYA-*arg*F)U1697 *rec*Al *end*A1 *hsd*R17(rK-, mK+) *pho*A *sup*E44 *thi*-1 *gvr*A96*re*/Al λ- Plasmid conservation	*In-vitro* gen
S. cerevisiaeyhbl	BY4741 (MATa his3Δl leu2Δ0 metl 5Δ0 ura3Δ0 yhbl::kanMX4	Euroscarf collection. Frankfurt
S. cerevisiae [pYPGE15]	*yhbl* [pYPGE15] control strain	This work
S. cerevisiae pYPGE15: *furA*	*yhbl* [pYPGE15: *furA*] FurA-expressing strain	This work
S. cerevisiae pYPGE15:*furB*	*yhbl* [pYPGE15:*furB*] FurB-expressing strain	This work
S. cerevisiae FurB-GFP	[pYPGE: *furA*-GTY'] FurA-GFP-expressing strain	This work

shaking of 120 rpm. *Escherichia coli* strains used for cloning procedures were grown in Luria Bertani medium at 37 °C and constant shaking of 180 rpm. When necessary, the medium was supplemented with 50 µg/mL ampicillin. *Saccharomyces cerevisiae* strains were grown in yeast extract–peptone–dextrose (YPD) medium (2% glucose, 2% peptone, 1% yeast extract) at 28 °C. For growth tests under different stress conditions, serial dilutions of saturated cultures were spotted on YPD plates containing the reagents and incubated for 48 h at 28 °C.

7.2.2 Cloning and transformation procedures

For FurA and FurB expression in *S. cerevisiae*, genomic DNA was extracted from *Anabaena* (Cai & Wolk, 1990) and the *furA* and *furB* genes were amplified with oligonucleotides pYPGE-furA-up/dw and pYPGE-furB-up/dw as shown in Table 7.2. PCR products were digested and cloned into the EcoRI and XhoI sites of pYPGE15 plasmid (2 µm origin, URA3 marker, a constitutive promoter from the *PGK* gene) (Brunelli & Pall, 1993). Constructions were verified by sequencing and transferred to *S. cerevisiae* by electroporation. Empty pYPGE15 plasmid was also introduced in *S. cerevisiae* and the resulting strain was used as a control in the following experiments. The mutant strain *yhb1* (MATa his3Δ1 leu2Δ0 met15Δ0 ura3Δ0 yhb1::kanMX4) was used because the antioxidant activity of the yeast flavohemoglobin YHB1 could mask the effect of FurA and FurB proteins. A similar strategy has been previously used to characterize antioxidant proteins in yeast (Sainz et al., 2013).

TABLE 7.2 Oligonucleotides used in this study.

Name	Sequence (5′→3′)	Purpose
pYPGE-furA-up	GTATGACTGTC TACAC AAATACTTOGCTC	*fur A* cloning in EcoRI-XhoI sites from pYPGE15
pYPGE-furA-dw	TGCTAAAGTGGCATGAGCGCAC GTTG	
pYPGE-furB-up	CCGGAATTC TCTATGAGAGCCATAC GCACC	*furB* cloning in EcoRI-XhoI sites from pYPGE15
pYPGE-furB-dw	CCGCTCGAGTGAC-TAAATTTCACTCGC GTG	
pYPGEFURBGFPD	TTGTGTGGTAAGTGCCAAATGAAT-CACGCGAGT GAAATTATGGTGAGC AAGGGC GAG	GFP tagging in the carb oxyl terminal of *furB* in the pYPGE plasmid.
pYPGEFURBGFPR.	CCACCAAAGGCCATCTTGGAAC-CGGGCCCCCCCT CGAGC TTGTACAGCTCGTCCAT	

7.2.3 Western blot

For Western blotting of *S. cerevisiae* protein extracts, cells were lysed as follows. Cells from a 50-mL exponentially growing culture were collected by centrifugation and washed three times with phosphate buffered saline (PBS) buffer. Glass beads of 0.5 mm were added (3 g beads/g humid pellet) and the mixture was frozen with liquid nitrogen prior to treatment in a bead beater at 30 Hz for 30 s\. One volume of cold lysis buffer (50 mM Tris-HCl pH 8, 1% dimethyl sulfoxide, 50 mM ethylenediaminetetraacetic acid, 100 mM phenylmethylsulfonyl fluoride) was added and the sample was frozen again with liquid nitrogen. Cells were lysed in the bead beater with three cycles of 3 min, with cooling intervals of 1 min between cycles. Samples were centrifuged for 10 min at 20,000*g* and 4 °C so that cell debris and beads could be removed. Protein concentration in the resulting extracts was determined by using the bicinchoninic acid method (BCA Protein Assay Kit, Thermo Fischer Scientific). Samples containing 8 μg of total proteins were loaded into 17% SDS (sodium dodecyl sulfate) - polyacrylamide gel electrophoresis (PAGE) gels and subjected to reducing electrophoresis. Proteins were transferred to a polyvinyl difluoride membrane (0.45 μm pore size Immobilon transfer membrane from Millipore) and immunodetection was carried out with rabbit polyclonal antibodies raised against FurA or FurB.

7.2.4 Construction of the Green fluorescent protein (GFP)-tagged *S. cerevisiae*

Plasmid pYPGE15::furB-GFP was constructed using a homologous recombination strategy. For this purpose, the sGFP sequence was amplified with oligonucleotides pYPGEFURBGFPD and pYPGEFURBGFPR (Table 7.2). These primers amplify the sGFP sequence and include in both borders a sequence homologous to the Eco RI site of the pYPGE15::FurB plasmid. The yeast strain *yhb1* was then co-transformed with the described PCR product and with plasmid pYPGE::*furB* digested with Eco RI. When natural yeast recombination reconstituted the plasmid, the ability to synthesize uracil was recovered and colonies appeared in a plate with synthetic defined (SD) medium without uracil. Three positive colonies were selected and sequenced to confirm the absence of mutations.

7.2.5 Fluorescence microscopy

Images were obtained for live cells grown to exponential phase in minimal SD medium using a Leica TCS SL inverted confocal microscope with a 40 × /1.25-numerical-aperture planapochromat oil objective. Samples were excited at 488 nm and GFP emission was detected between 500 and 530 nm.

FIG. 7.1 Verification of FurB and FurA expression in yeast by Western blot. All lanes contain 8 μg of protein extracts from *S. cerevisiae* YHB1 (harbors empty plasmid pYPGE15), FurB (*S. cerevisiae* with plasmid pYPGE15::*furB*), and FurA (*S. cerevisiae* with plasmid pYPGE15::*furA*).

7.3 Results

7.3.1 Generation of *S. cerevisiae* strains expressing FurA and FurB from *Anabaena* sp. PCC 7120

Cloning of *furA* and *furB* at the polylinker region of vector pYPGE15 under control of the strong and constitutive promoter PGK1 led to the heterologous expression of FurA and FurB in *S. cerevisiae*. The correct expression of both proteins was verified by Western blot of protein extracts from the derivative strains (Fig. 7.1). Although the amount of protein detected was slightly higher for FurA than for FurB, in both cases the expression levels in *S. cerevisiae* were in the same order of magnitude than the expression in *Anabaena* under standard culture conditions (Hernández, et al., 2002).

7.3.2 Expression of FurB in *S. cerevisiae* increases its sensitivity to copper and manganese

As the best-known roles for FurA and FurB are related to their function as metalloregulators, we sought to investigate the effect of an excess of divalent metals (Co^{2+}, Mn^{2+}, Cu^{2+}, and Zn^{2+}) in *S. cerevisiae* pYPGE15::*furA* and pYPGE15::*furB*. In both strains, the growth under cobalt and zinc treatments was similar to the growth observed in the control strain *yhb1* strain transformed with the empty plasmid. Although pYPGE15::*furA* also exhibited a similar behavior than the control strain under manganese and copper treatments, *S. cerevisiae* expressing FurB displayed higher sensitivity to the challenge with these two metals (Fig. 7.2).

7.3.3 The presence of FurB in *S. cerevisiae* enhances the effects of membrane-damaging compounds and saline stress

To find out whether the expression of FurA and FurB affected yeast tolerance to other harmful conditions such as treatment with membrane-damaging compounds or saline stress, derivative strains were grown in

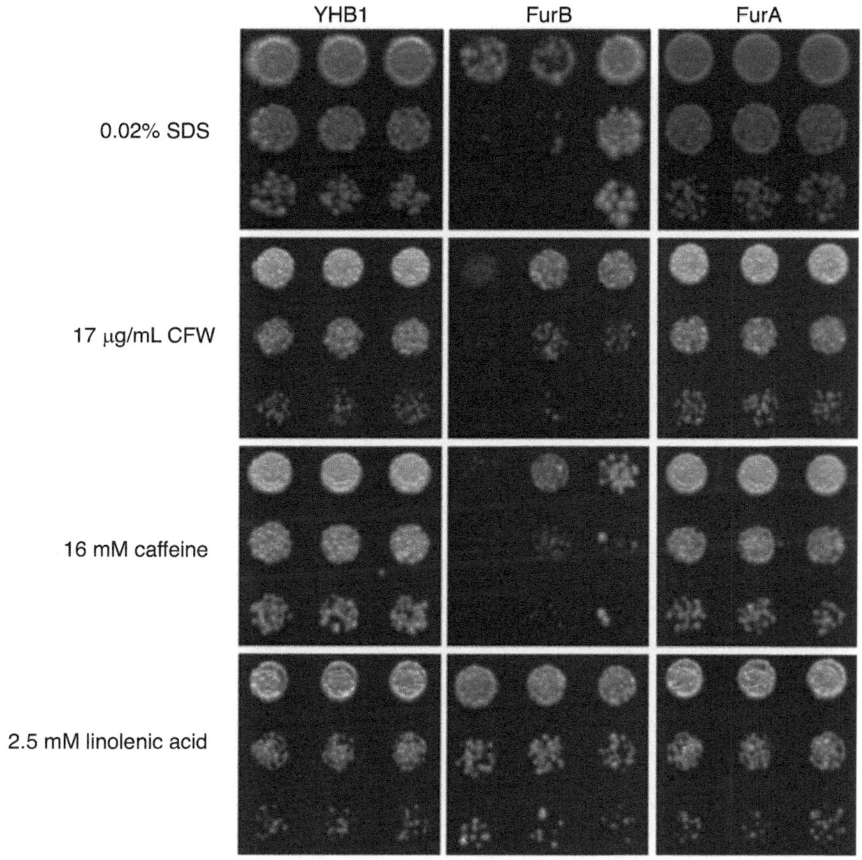

FIG. 7.2 Tolerance against metals in *S. cerevisiae* derivative strains. The strains were grown on YPD plates for 48 h at 28 °C with different metal cations: 10 mM $CuSO_4$, 8 mM $CoCl_2$, 10 mM $MnSO_4$, and 10 mM $ZnSO_4$. Serial dilutions (10^{-1}, 10^{-2}, and 10^{-3}) of saturated cultures (top to bottom) and three replicates (left to right) are shown on the plates.

the presence of SDS, CFW (CalcoFluor White), caffeine, and linolenic acid (Fig. 7.3), as well as in media containing LiCl and NaCl (Fig. 7.4). The behavior of the FurA-expressing strain was similar to that of the control under exposure to both membrane-damaging reagents and saline stress. However, SDS, CFW, and caffeine had a clear deleterious effect on the pYPGE15::*furB* strain, whose growth was severely impaired. Noteworthy, pYPGE15::*furB* growth was not affected by the treatment with linolenic acid (Fig. 7.3). *Saccharomyces cerevisiae* pYPGE15::*furB* viability was also impaired under saline stress conditions (Fig. 7.4).

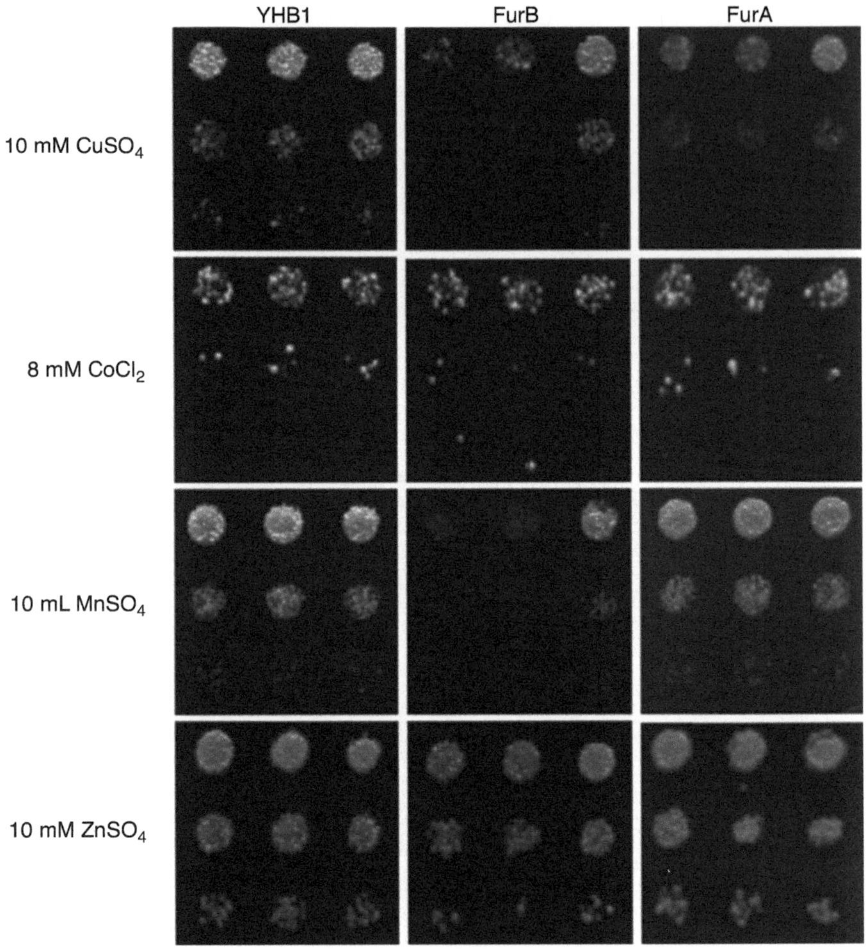

FIG. 7.3 Tolerance against membrane damage inducers in *S. cerevisiae* derivative strains. The strains were grown on YPD plates for 48 h at 28 °C with membrane damage inducers: 0.02% SDS, 17 μg/mL CFW, 16 mM caffeine, and 2.5 mM linolenic acid. Serial dilutions (10^{-1}, 10^{-2}, and 10^{-3}) of saturated cultures (top to bottom) and three replicates (left to right) are shown on the plates.

7.3.4 Fur proteins confer increased tolerance to oxidative stress in *S. cerevisiae*

Growth of *S. cerevisiae yhb1* control strain transformed with the empty plasmid, pYPGE15::*furA* and pYPGE15::*furB* strains in the presence of different oxidants unveiled that expression of both FurA and FurB increased yeast tolerance to oxidative stress under some of the conditions tested (Fig. 7.5). Specifically, the FurA-expressing strain presented a higher growth than the control when treated with the lowest concentrations of

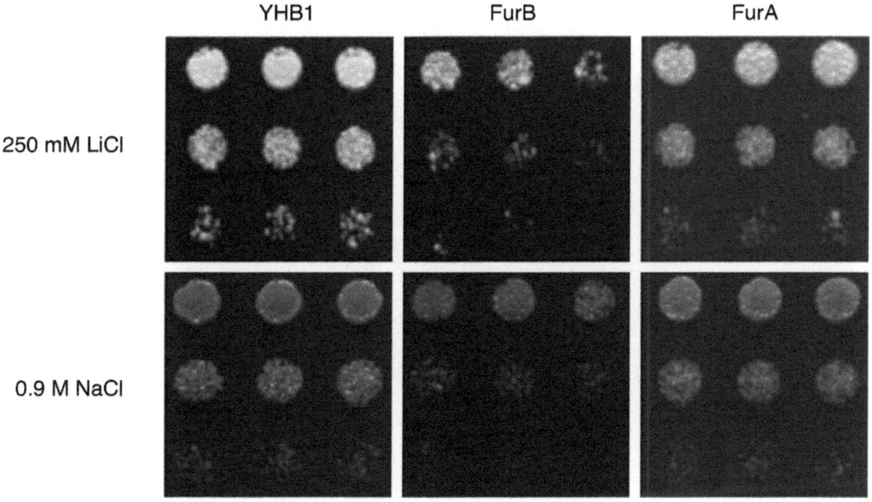

FIG. 7.4 Tolerance against salt stress in *S. cerevisiae* derivative strains. The strains were grown on YPD plates for 48 h at 28 °C with different salts: 250 mM LiCl and 0.9 M NaCl. Serial dilutions (10^{-1}, 10^{-2}, and 10^{-3}) of saturated cultures (top to bottom) and three replicates (left to right) are shown on the plates.

hydrogen peroxide and t-butyl hydroperoxide (0.25 and 0.5 mM, respectively). The FurB-expressing strain exhibited an increased tolerance even at the highest concentrations of the above-mentioned reagents (1.5 and 1 mM, respectively). Finally, neither the FurA-expressing strain nor the FurB-expressing one seemed to be affected by the treatment with methyl viologen, a superoxide anion generating reagent, exhibiting the same growth than the control.

7.3.5 Recombinant FurB is located in the cytosol of *S. cerevisiae*

It has been proposed that the mechanisms underlying the tolerance to oxidative stress of cyanobacteria and *E. coli* with enhanced expression of FurB could be related to the ability of this protein to coat and protect DNA (Lopez-Gomollon, et al., 2009; Sein-Echaluce et al., 2018). While large molecules require transport receptors to facilitate their movement through the nuclear pore complex, metabolites, ions, and small proteins can pass through the nuclear pore complex freely (Gorlich & Kutay, 1999). Hence, to find out whether the mechanism of oxidative stress tolerance provided by FurB in *S. cerevisiae* was similar to the one proposed in *E. coli* and *Anabaena*, we investigated the location of this regulator in the yeast cell. Fluorescence analyses of several colonies of recombinant yeasts carrying a FurB-GFP fusion protein showed that FurB was mainly localized in the

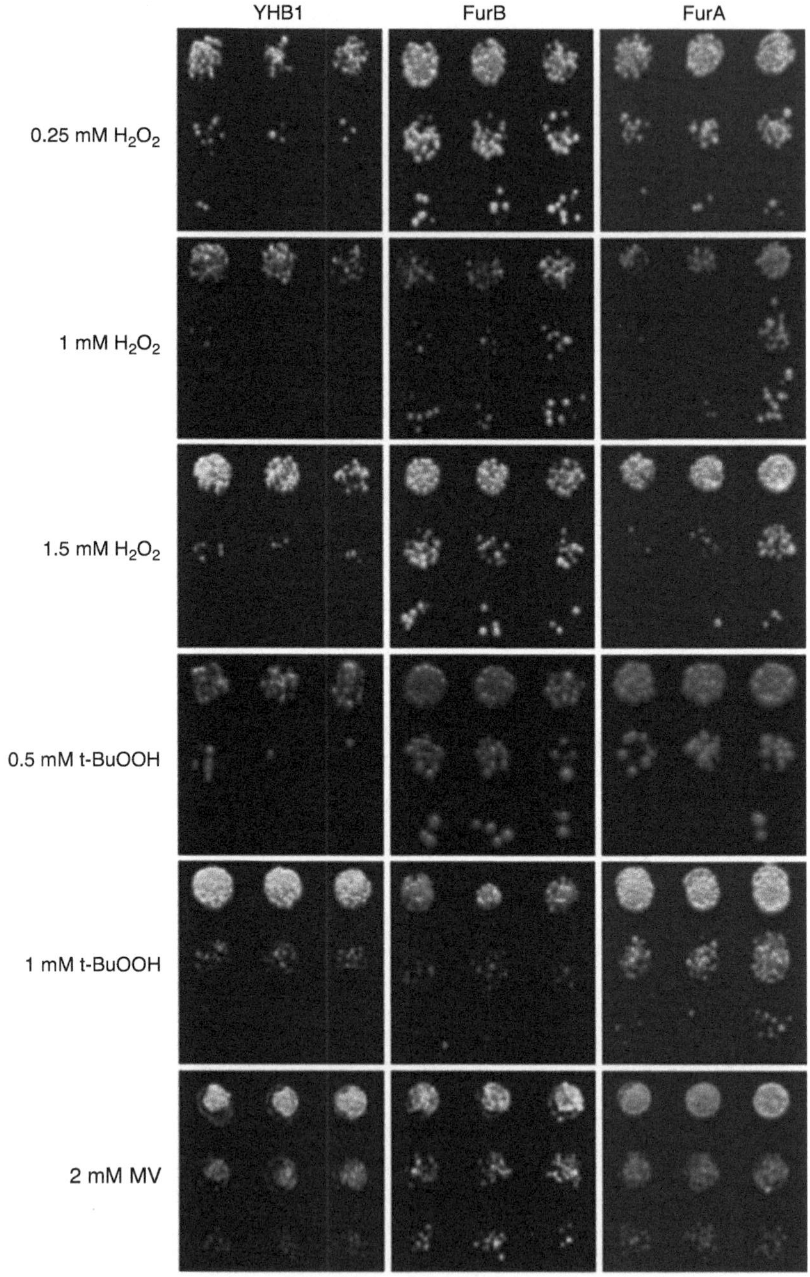

FIG. 7.5 Tolerance against oxidative stress in *S. cerevisiae* derivative strains. The strains were grown on YPD plates for 48 h at 28 °C with different oxidative stress inducers: 0.25, 1, and 1.5 mM H_2O_2, 0.25- and 1-mM t-butyl hydroperoxide (t-BuOOH) and 2 mM methyl viologen. Serial dilutions (10^{-1}, 10^{-2}, and 10^{-3}) of saturated cultures (top to bottom) and three replicates (left to right) are shown on the plates.

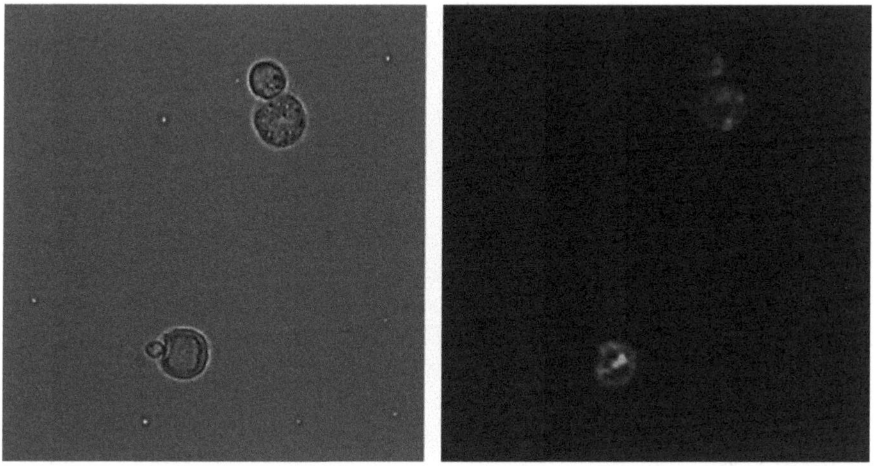

FIG. 7.6 Localization of FurB-GFP in *S. cerevisiae*. A yeast *yhb1* mutant strain was transformed with pYPGE-FURB-GFPm grown in SD and observed using laser scanning microscopy. Signal was only detected in the cytoplasm.

cytosol, indicating that FurB is unable to enter the nuclear pore (Fig. 7.6). These results indicate that the resistance acquired by the FurB-expressing recombinant yeasts is not due to its role as a DNA coating agent, and points to a different mechanism, probably related to a direct or an indirect function as an antioxidant.

7.4 Discussion

Saccharomyces cerevisiae is broadly used in ethanol production because of its high efficiency and resistance to the multiple stresses associated with the fermentation process (decreasing pH, osmotic pressure, and ethanol concentration, among others) (Dogan, et al., 2014; Gibson, Lawrence, Leclaire, Powell, & Smart, 2007). Although each stress involves specific molecular mechanisms from the yeast adaptative response (Auesukaree, 2017), most of them lead to an enhancement of reactive oxygen species (ROS) levels inside the cell (Du & Takagi, 2007; Kitichantaropas et al., 2016). Consequently, all strategies focused on the decrease of oxidative stress to get improved *S. cerevisiae* strains are desirable.

In the present work, we have expressed FurA and FurB from the cyanobacterium *Anabaena* sp. PCC 7120 in *S. cerevisiae* in an attempt to increase yeast tolerance to different abiotic stresses and, especially, to oxidative stress. Both proteins have been demonstrated to improve cell growth under oxidative challenge when expressed in *E. coli* (Lopez-Gomollon, et al., 2009) and, in the case of FurB, also in *Anabaena* itself (Sein-Echaluce, et al., 2014). Hence, we considered interest to evaluate the consequences of the expression of these FUR proteins in *S. cerevisiae*. We achieved the correct heterologous expression by cloning the genes into a pYPGE15 vector, which is commonly used for the expression of proteins in yeast (Mulet, et al., 2004; Perez-Valle et al., 2010; Yenush, Mulet, Arino, & Serrano, 2002) and tested the tolerance of the recombinant strains against different abiotic stresses.

As in other organisms, although divalent metals are essential for the correct performance of a wide variety of biological processes, an excess of these species is detrimental to yeast viability (Bird, 2015; Ehrensberger & Bird, 2011). Although the pYPGE15::*furA* strain presented a similar tolerance than the control under all metal treatments, pYPGE::*furB* exhibited impaired growth under copper and manganese treatment but not when treated with Zn^{2+} and Co^{2+}. It is noticeable that FurB is able to bind up to three zinc ions per molecule (Sein-Echaluce, et al., 2018), so that the protein could be binding some of the Zn^{2+} ions in excess when *S. cerevisiae* was subjected to that challenge, relieving cellular stress in that manner. Regarding Co^{2+}, it has been described to present comparable coordination chemistry than Zn^{2+} and bind to protein ligands in a similar fashion (Eide, 2003). Therefore, FurB could be also binding the excess of Co^{2+} ions. Another possibility is that FurA and FurB are affecting yeast proteins. For instance, a yeast mutant for the multidrug and copper transporter *QDR2* is sensitive to high copper but tolerant to high cobalt (Rios et al., 2013). It is also known that glucose or hydrogen peroxide treatments can induce programmed cell death in yeast (Hoeberichts et al., 2010). Hence, we cannot discard that the observed phenotypes are due to the effect of the Fur proteins on these yeast processes.

Interestingly, only the FurB derivative strain was visibly affected when treated with cell wall damaging reagents and salt stress inducers. One could speculate that the binding of endogenous Zn^{2+} by FurB could lead to a mild Zn^{2+} limitation. A typical *Saccharomyces* response is the activation of low-affinity Zrt2 transporters, whose activity requires an intact plasma membrane (Eide, 2003). Therefore, its disruption in pYPGE::*furB* by the compounds presented in Fig. 7.3, could impair Zrt2 activity and in turn influence yeast viability. Similarly, Na^{+}/K^{+} ATPases exporting excess Na^{+} and Li^{+} ions are located in the plasma membrane (Cyert & Philpott, 2013) so that an alteration in this structure due to a higher content of transporters due to FurB overexpression could affect ATPases's functionality and compromise tolerance to salt stress. Recently it has been shown that Na/K ATPase Ena1 is transported through the exomer complex in a polarized way. If Fur proteins are interacting with this process, this could affect the function of ENA1 and therefore explain the observed sensitivity phenotype (Anton et al., 2017), as well as the effects of CFW, caffeine, and SDS, which affect cell wall and plasma membrane integrities through different mechanisms (Berthelet et al., 2010; Kuranda, Leberre, Sokol, Palamarczyk, & Francois, 2006; Roncero & Duran, 1985), and would also be more pronounced if those structures were already affected.

As derived from the results obtained under linolenic acid treatment, the higher susceptibility of the pYPGE::*furB* derivative to the cell wall and plasma membrane–damaging reagents is not caused by a defect in the protection against lipid peroxidation. This polyunsaturated fatty acid cannot be synthesized by yeasts but can be incorporated into the membranes, making them more sensitive to peroxidation. In cells with effective protection systems against lipid peroxidation, the effects of linolenic acid may not be visible unless cells are also treated with an oxidative stress–inducing reagent (Avery & Avery, 2001).

Both strains, but especially the FurB-expressing derivative, exhibited increased tolerance against most of the oxidative stress situations tested. In prokaryotic organisms such as *Anabaena* and *E. coli*, FurB was proposed to protect cells by unspecifically binding DNA, preventing damage of the nucleic acid by ROS. However, the functioning of FurB as a DNA protecting protein in a eukaryote such as *S. cerevisiae* would require translocation of the protein from the cytosol to the cell nucleus. Transport of large molecules through the nuclear pore complexes is an active process coupled to energy consumption and, in the case of proteins, requires the presence of nuclear localization sequences (NLSs) (Kabachinski & Schwartz, 2015). As FurB was cloned into the pYPGE15 vector without any NLS, the only manner for the protein to enter the nucleus would be passive diffusion.

Although for many years only small molecules, ions, and proteins of up to 60 kDa were thought to enter the nucleus by passive diffusion (Gorlich & Kutay, 1999), in the past years, even proteins with sizes from 90 to 100 kDa have been reported to passively enter into the nucleus (Wang & Brattain, 2007). In our case, experiments with the derivative strain expressing FurB fused with GFP showed that the protein was localized in the cytosol, suggesting that the oxidative stress resistance acquired by the FurB-expressing yeast was due to a different mechanism, not related to the unspecific binding of the protein to the DNA. Regarding the increased tolerance of the FurA-expressing strain to oxidative stress, it is remarkable that a disulfide reductase activity has been described for this protein and its involvement in dithiol/disulfide exchange processes, such as redox signalling or maintaining redox homeostasis, has been suggested (Botello-Morte, et al., 2014). Thus, it is possible that FurA could be contributing to relieve oxidative stress in the derivative strain by mediating in redox homeostasis, which could explain many of the observed phenotypes, including tolerance to several transition metals (Rios, et al., 2013).

Taken altogether, the results here are the first evidence of cyanobacterial FUR proteins conferring tolerance against oxidative stress when heterologously expressed in a eukaryotic organism. Although further studies need to be done to establish the precise mechanisms whereby FurA and FurB confer this tolerance; these results are promising and, hence, the possibility of expressing FurA and FurB in more complex organisms, such as plants, can be considered.

References

Anton, C., Zanolari, B., Arcones, I., Wang, C., Mulet, J.M., Spang, A., Roncero, C., 2017. Involvement of the exomer complex in the polarized transport of Ena1 required for *Saccharomyces cerevisiae* survival against toxic cations. Mol. Biol. Cell 28 (25), 3672–3685. doi:10.1091/mbc.E17-09-0549.

Auesukaree, C., 2017. Molecular mechanisms of the yeast adaptive response and tolerance to stresses encountered during ethanol fermentation. J. Biosci. Bioeng. 124 (2), 133–142. doi:10.1016/j.jbiosc.2017.03.009.

Avery, A.M., Avery, S.V., 2001. *Saccharomyces cerevisiae* expresses three phospholipid hydroperoxide glutathione peroxidases. J. Biol. Chem. 276 (36), 33730–33735. doi:10.1074/jbc.M105672200.

Berthelet, S., Usher, J., Shulist, K., Hamza, A., Maltez, N., Johnston, A., Baetz, K., 2010. Functional genomics analysis of the *Saccharomyces cerevisiae* iron-responsive transcription factor Aft1 reveals iron-independent functions. Genetics 185 (3), 1111–1128. doi:10.1534/genetics.110.117531.

Bird, A.J., 2015. Cellular sensing and transport of metal ions: implications in micronutrient homeostasis. J. Nutr. Biochem. 26 (11), 1103–1115. doi:10.1016/j.jnutbio.2015.08.002.

Botello-Morte, L., Bes, M.T., Heras, B., Fernandez-Otal, A., Peleato, M.L., Fillat, M.F., 2014. Unravelling the redox properties of the global regulator FurA from *Anabaena* sp. PCC 7120: disulfide reductase activity based on its CXXC motifs. Antioxid. Redox Signal. 20 (9), 1396–1406.

Botello-Morte, L., Pellicer, S., Sein-Echaluce, V.C., Contreras, L.M., Neira, J.L., Abian, O., Bes, M.T., 2015. Cysteine mutational studies provide insight into a thiol-based redox switch mechanism of metal and DNA binding in FurA from *Anabaena* sp. PCC 7120. Antioxid. Redox. Signal. doi:10.1089/ars.2014.6175.

Brunelli, J.P., Pall, M.L., 1993. A series of yeast/*Escherichia coli* lambda expression vectors designed for directional cloning of cDNAs and cre/lox-mediated plasmid excision. Yeast 9 (12), 1309–1318. doi:10.1002/yea.320091204.

Cai, Y.P., Wolk, C.P., 1990. Use of a conditionally lethal gene in *Anabaena* sp. strain PCC 7120 to select for double recombinants and to entrap insertion sequences. J. Bacteriol. 172 (6), 3138–3145.

Cletus, J., Balasubramanian, V., Vashisht, D., Sakthivel, N., 2013. Transgenic expression of plant chitinases to enhance disease resistance. Biotechnol. Lett. 35 (11), 1719–1732. doi:10.1007/s10529-013-1269-4.

Coba de la Pena, T., Redondo, F.J., Fillat, M.F., Lucas, M.M., Pueyo, J.J, 2013. Flavodoxin overexpression confers tolerance to oxidative stress in beneficial soil bacteria and improves survival in the presence of the herbicides paraquat and atrazine. J. Appl. Microbiol. 115 (1), 236–246. doi:10.1111/jam.12224.

Coba de la Pena, T., Redondo, F.J., Manrique, E., Lucas, M.M., Pueyo, J.J, 2010. Nitrogen fixation persists under conditions of salt stress in transgenic *Medicago truncatula* plants expressing a cyanobacterial flavodoxin. Plant Biotechnol. J. 8 (9), 954–965. doi:10.1111/j.1467-7652.2010.00519.x.

Cyert, M.S., Philpott, C.C., 2013. Regulation of cation balance in *Saccharomyces cerevisiae*. Genetics 193 (3), 677–713. doi:10.1534/genetics.112.147207.

Chaturvedi, R., Archana, G., 2014. Cytosolic expression of synthetic phytochelatin and bacterial metallothionein genes in *Deinococcus radiodurans* R1 for enhanced tolerance and bioaccumulation of cadmium. Biometals 27 (3), 471–482. doi:10.1007/s10534-014-9721-z.

Chaurasia, N., Mishra, Y., Rai, L.C., 2008. Cloning expression and analysis of phytochelatin synthase (PCS) gene from *Anabaena* sp. PCC 7120 offering multiple stress tolerance in *Escherichia coli*. Biochem. Biophys. Res. Commun. 376 (1), 225–230. doi:10.1016/j.bbrc.2008.08.129.

Da Silva, N.A., Srikrishnan, S, 2012. Introduction and expression of genes for metabolic engineering applications in *Saccharomyces cerevisiae*. FEMS Yeast Res. 12 (2), 197–214. doi:10.1111/j.1567-1364.2011.00769.x.

Dogan, A., Demirci, S., Aytekin, A.O., Sahin, F., 2014. Improvements of tolerance to stress conditions by genetic engineering in *Saccharomyces cerevisiae* during ethanol production. Appl. Biochem. Biotechnol. 174 (1), 28–42. doi:10.1007/s12010-014-1006-z.

Du, X., Takagi, H., 2007. N-Acetyltransferase Mpr1 confers ethanol tolerance on *Saccharomyces cerevisiae* by reducing reactive oxygen species. Appl. Microbiol. Biotechnol. 75 (6), 1343–1351. doi:10.1007/s00253-007-0940-x.

Ehrensberger, K.M., Bird, A.J., 2011. Hammering out details: regulating metal levels in eukaryotes. Trends Biochem. Sci. 36 (10), 524–531. doi:10.1016/j.tibs.2011.07.002.

Eide, D.J., 2003. Multiple regulatory mechanisms maintain zinc homeostasis in *Saccharomyces cerevisiae*. J. Nutr. 133 (5 Suppl 1), 1532S–1535S.

Feki, K., Kamoun, Y., Ben Mahmoud, R., Farhat-Khemakhem, A., Gargouri, A., Brini, F., 2015. Multiple abiotic stress tolerance of the transformants yeast cells and the transgenic *Arabidopsis* plants expressing a novel durum wheat catalase. Plant Physiol. Biochem. 97, 420–431. doi:10.1016/j.plaphy.2015.10.034.

Fillat, M.F., 2014. The FUR (ferric uptake regulator) superfamily: diversity and versatility of key transcriptional regulators. Arch. Biochem. Biophys. 546, 41–52.

Gibson, B.R., Lawrence, S.J., Leclaire, J.P., Powell, C.D., Smart, K.A., 2007. Yeast responses to stresses associated with industrial brewery handling. FEMS Microbiol. Rev. 31 (5), 535–569. doi:10.1111/j.1574-6976.2007.00076.x.

Gonzalez, A., Bes, M.T., Barja, F., Peleato, M.L., Fillat, M.F., 2010. Overexpression of FurA in *Anabaena* sp. PCC 7120 reveals new targets for this regulator involved in photosynthesis, iron uptake and cellular morphology. Plant Cell Physiol. 51 (11), 1900–1914.

Gorlich, D., Kutay, U., 1999. Transport between the cell nucleus and the cytoplasm. Annu. Rev. Cell. Dev. Biol. 15, 607–660. doi:10.1146/annurev.cellbio.15.1.607.

He, Y., He, C., Li, L., Liu, Z., Yang, A., Zhang, J., 2011. Heterologous expression of *ApGSMT2* and *ApDMT2* genes from *Aphanothece halophytica* enhanced drought tolerance in transgenic tobacco. Mol. Biol. Rep. 38 (1), 657–666. doi:10.1007/s11033-010-0152-9.

Hernández, J.A., Bes, M.T., Fillat, M.F., Neira, J.L., Peleato, M.L., 2002. Biochemical analysis of the recombinant Fur (ferric uptake regulator) protein from *Anabaena* PCC 7119: factors affecting its oligomerization state. Biochem. J. 366 (Pt 1), 315–322.

Hernandez, J.A., Lopez-Gomollon, S., Bes, M.T., Fillat, M.F., Peleato, M.L., 2004. Three fur homologues from *Anabaena* sp. PCC7120: exploring reciprocal protein-promoter recognition. FEMS Microbiol. Lett. 236 (2), 275–282.

Hoeberichts, F.A., Perez-Valle, J., Montesinos, C., Mulet, J.M., Planes, M.D., Hueso, G., Serrano, R., 2010. The role of K(+) and H(+) transport systems during glucose- and H(2)O(2)-induced cell death in *Saccharomyces cerevisiae*. Yeast 27 (9), 713–725. doi:10.1002/yea.1767.

Kabachinski, G., Schwartz, T.U., 2015. The nuclear pore complex–structure and function at a glance. J. Cell Sci. 128 (3), 423–429.

Kitichantaropas, Y., Boonchird, C., Sugiyama, M., Kaneko, Y., Harashima, S., Auesukaree, C., 2016. Cellular mechanisms contributing to multiple stress tolerance in *Saccharomyces cerevisiae* strains with potential use in high-temperature ethanol fermentation. AMB Express 6 (1), 107. doi:10.1186/s13568-016-0285-x.

Kuranda, K., Leberre, V., Sokol, S., Palamarczyk, G., Francois, J., 2006. Investigating the caffeine effects in the yeast *Saccharomyces cerevisiae* brings new insights into the connection between TOR, PKC and Ras/cAMP signalling pathways. Mol. Microbiol. 61 (5), 1147–1166. doi:10.1111/j.1365-2958.2006.05300.x.

Lieman-Hurwitz, J., Rachmilevitch, S., Mittler, R., Marcus, Y., Kaplan, A., 2003. Enhanced photosynthesis and growth of transgenic plants that express *ictB*, a gene involved in HCO_3- accumulation in cyanobacteria. Plant Biotechnol. J. 1 (1), 43–50. doi:10.1046/j.1467-7652.2003.00003.x.

Lopez-Gomollon, S., Sevilla, E., Bes, M.T., Peleato, M.L., Fillat, M.F., 2009. New insights into the role of Fur proteins: FurB (All2473) from *Anabaena* protects DNA and increases cell survival under oxidative stress. Biochem. J. 418 (1), 201–207.

Martinez, F., Arif, A., Nebauer, S.G., Bueso, E., Ali, R., Montesinos, C., Serrano, R., 2015. A fungal transcription factor gene is expressed in plants from its own promoter and improves drought tolerance. Planta 242 (1), 39–52. doi:10.1007/s00425-015-2285-5.

Miyagawa, Y., Tamoi, M., Shigeoka, S., 2001. Overexpression of a cyanobacterial fructose-1,6-/sedoheptulose-1,7-bisphosphatase in tobacco enhances photosynthesis and growth. Nat. Biotechnol. 19 (10), 965–969. doi:10.1038/nbt1001-965.

Mulet, J.M., Alemany, B., Ros, R., Calvete, J.J., Serrano, R., 2004. Expression of a plant serine O-acetyltransferase in *Saccharomyces cerevisiae* confers osmotic tolerance and creates an alternative pathway for cysteine biosynthesis. Yeast 21 (4), 303–312. doi:10.1002/yea.1076.

Narayan, O.P., Kumari, N., Rai, L.C., 2010. Heterologous expression of Anabaena PCC 7120 all3940 (a Dps family gene) protects *Escherichia coli* from nutrient limitation and abiotic stresses. Biochem. Biophys. Res. Commun. 394 (1), 163–169. doi:10.1016/j.bbrc.2010.02.135.

Pandey, S., Shrivastava, A.K., Rai, R., Rai, L.C., 2013. Molecular characterization of Alr1105 a novel arsenate reductase of the diazotrophic cyanobacterium *Anabaena* sp. PCC7120 and decoding its role in abiotic stress management in *Escherichia coli*. Plant Mol. Biol. 83 (4-5), 417–432. doi:10.1007/s11103-013-0100-x.

Perez-Valle, J., Rothe, J., Primo, C., Martinez Pastor, M., Arino, J., Pascual-Ahuir, A., Yenush, L., 2010. Hal4 and Hal5 protein kinases are required for general control of carbon and nitrogen uptake and metabolism. Eukaryot. Cell 9 (12), 1881–1890. doi:10.1128/EC.00184-10.

Redondo, F.J., de la Pena, T.C., Morcillo, C.N., Lucas, M.M., Pueyo, J.J., 2009. Overexpression of flavodoxin in bacteroids induces changes in antioxidant metabolism leading to delayed senescence and starch accumulation in alfalfa root nodules. Plant Physiol. 149 (2), 1166–1178. doi:10.1104/pp.108.129601.

Rios, G., Cabedo, M., Rull, B., Yenush, L., Serrano, R., Mulet, J.M., 2013. Role of the yeast multidrug transporter Qdr2 in cation homeostasis and the oxidative stress response. FEMS Yeast Res. 13 (1), 97–106. doi:10.1111/1567-1364.12013.

Rippka, R., Deruelles, J., Waterbury, J.B., Herdman, M., Stanier, R.Y., 1979. Generic assignments, strain histories and properties of pure cultures of *Cyanobacteria*. J. Gen. Microbiol. 111, 1–61.

Roncero, C., Duran, A., 1985. Effect of Calcofluor white and Congo red on fungal cell wall morphogenesis: in vivo activation of chitin polymerization. J. Bacteriol. 163 (3), 1180–1185.

Sainz, M., Perez-Rontome, C., Ramos, J., Mulet, J.M., James, E.K., Bhattacharjee, U., Becana, M., 2013. Plant haemoglobins may be maintained in functional form by reduced flavins in the nuclei, and confer differential tolerance to nitro-oxidative stress. Plant J. 76 (5), 875–887. doi:10.1111/tpj.12340.

Sein-Echaluce, V.C., Gonzalez, A., Napolitano, M., Luque, I., Barja, F., Peleato, M.L., Fillat, M.F., 2014. Zur (FurB) is a key factor in the control of the oxidative stress response in *Anabaena* sp. PCC 7120. Environ. Microbiol. 17 (6), 2006–2017. doi:10.1111/1462-2920.12628.

Sein-Echaluce, V.C., Pallares, M.C., Lostao, A., Yruela, I., Velazquez-Campoy, A., Luisa Peleato, M., Fillat, M.F, 2018. Molecular basis for the integration of environmental signals by FurB from *Anabaena* sp. PCC 7120. Biochem. J. 475 (1), 151–168. doi:10.1042/BCJ20170692.

Shrivastava, A.K., Pandey, S., Singh, P.K., Rai, S., Rai, L.C., 2012. alr0882 encoding a hypothetical protein of *Anabaena* PCC7120 protects *Escherichia coli* from nutrient starvation and abiotic stresses. Gene 511 (2), 248–255. doi:10.1016/j.gene.2012.09.033.

Shrivastava, A.K., Singh, S., Singh, P.K., Pandey, S., Rai, L.C., 2015. A novel alkyl hydroperoxidase (AhpD) of *Anabaena* PCC7120 confers abiotic stress tolerance in *Escherichia coli*. Funct. Integr. Genom. 15 (1), 77–92. doi:10.1007/s10142-014-0407-y.

Snowdon, C., Schierholtz, R., Poliszczuk, P., Hughes, S., van der Merwe, G., 2009. ETP1/YHL010c is a novel gene needed for the adaptation of *Saccharomyces cerevisiae* to ethanol. FEMS Yeast Res. 9 (3), 372–380. doi:10.1111/j.1567-1364.2009.00497.x.

Tanaka, K., Ishii, Y., Ogawa, J., Shima, J., 2012. Enhancement of acetic acid tolerance in *Saccharomyces cerevisiae* by overexpression of the HAA1 gene, encoding a transcriptional activator. Appl. Environ. Microbiol. 78 (22), 8161–8163. doi:10.1128/AEM.02356-12.

Tognetti, V.B., Palatnik, J.F., Fillat, M.F., Melzer, M., Hajirezaei, M.R., Valle, E.M., Carrillo, N., 2006. Functional replacement of ferredoxin by a cyanobacterial flavodoxin in tobacco confers broad-range stress tolerance. Plant Cell 18 (8), 2035–2050. doi:10.1105/tpc.106.042424.

Wang, R., Brattain, M.G., 2007. The maximal size of protein to diffuse through the nuclear pore is larger than 60kDa. FEBS Lett. 581 (17), 3164–3170. doi:10.1016/j.febslet.2007.05.082.

Yenush, L., Mulet, J.M., Arino, J., Serrano, R., 2002. The Ppz protein phosphatases are key regulators of K+ and pH homeostasis: implications for salt tolerance, cell wall integrity and cell cycle progression. EMBO J 21 (5), 920–929. doi:10.1093/emboj/21.5.920.

Zhang, J., Zhang, M., Tian, S., Lu, L., Shohag, M.J., Yang, X., 2014. Metallothionein 2 (SaMT2) from *Sedum alfredii* Hance confers increased Cd tolerance and accumulation in yeast and tobacco. PLoS One 9 (7), e102750. doi:10.1371/journal.pone.0102750.

Zurbriggen, M.D., Tognetti, V.B., Fillat, M.F., Hajirezaei, M.R., Valle, E.M., Carrillo, N., 2008. Combating stress with flavodoxin: a promising route for crop improvement. Trends. Biotechnol. 26 (10), 531–537. doi:10.1016/j.tibtech.2008.07.001.

CHAPTER

8

Exploring ecological diversity and biosynthetic potential of cyanobacteria for biofuel production

Galyna Kufryk

Grand Canyon University, Phoenix, AZ, United States

Cyanobacteria are Gram-negative, prokaryotic organisms having the ability to perform oxygenic plant-like photosynthesis and some – to fix atmospheric N_2. It has been estimated that cyanobacteria are one of the oldest organisms on the Earth dating back to about 3.5 billion years ago (Schopf, Packer, 1987; Schopf, 1993). The oxygen-rich atmosphere can be attributed to cyanobacteria's photosynthetic activity (Bekker et al., 2004). Cyanobacteria are the oldest oxygenic photosynthesizers currently known. They are the precursors to essentially all oxygenic photosynthetic organisms. A large number of cyanobacteria can use sunlight to extract electrons from water to form strong reducing agents that play a part in molecular hydrogen production via hydrogenases and nitrogenases. For a long time, the oxygen produced by these organisms was thought to be held in iron deposits rather than being released into the atmosphere. Eventually, oxygen started being released into the atmosphere and about 2.4 billion years ago sufficient amount of oxygen was accumulated in the Earth atmosphere so that the development of living species that require oxygen for respiration was made possible (Falkowski, 2006; Kump, 2008). Photosynthesis involves electrons being transferred from water through Photosystem II (PSII) down the photosynthetic electron transport chain and Photosystem I (PSI) to NADP and other molecules (Fig. 8.1). These systems also play a part in driving ATP production within the cell. Early prokaryotes were known to only have one of these two photosystems; however, cyanobacteria contain both photosystems that are physiologically

Cyanobacterial Lifestyle and its Applications in Biotechnology.
DOI: https://doi.org/10.1016/B978-0-323-90634-0.00010-X

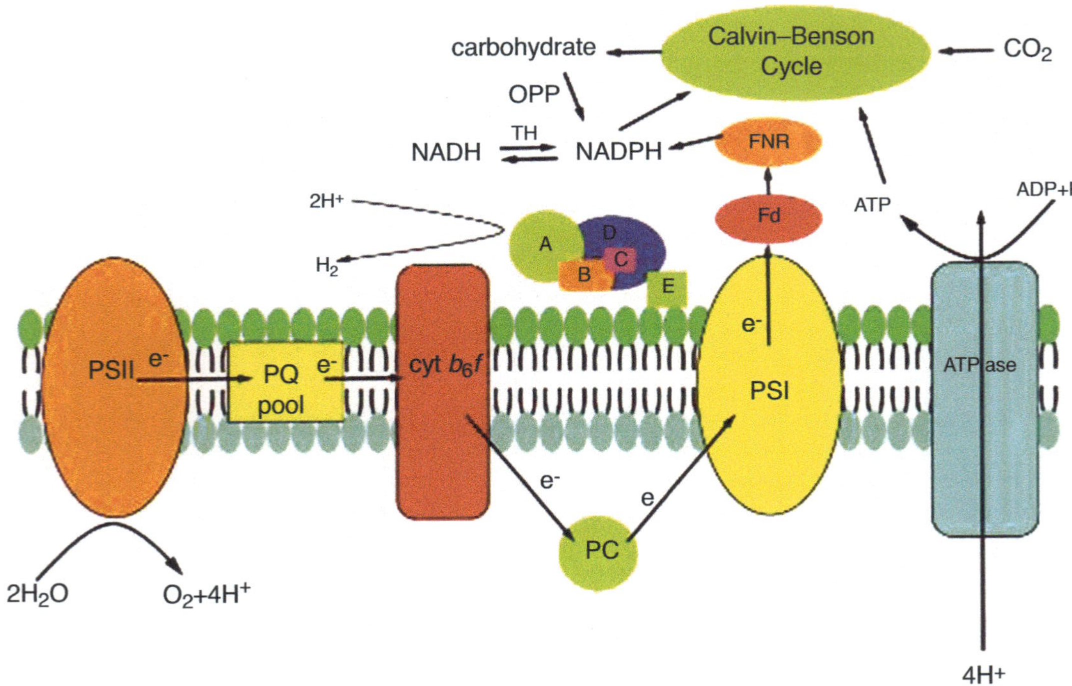

FIG. 8.1 Electron transport in thylakoid membranes of cyanobacteria
Bidirectional hydrogenase subunits:
(A) Hox H- [NiFe] large subunit, catalytic
(B) Hox Y- [FeS] small subunit
(C) Hox U- small subunit for diaphorase, used for electron transfer
(D) Hox F- large subunit for diaphorase, used for electron transfer
(E) Hox E- subunit used for attaching to membrane and electron transfer
Credit: Galyna Kufryk.

connected. Cyanobacteria have a comparatively high photosynthetic efficiency, but the yield is still less than 10%.

Approximately 2.3 billion years ago cyanobacteria started to become the dominant photosynthetic organisms in the environment (Blankenship et al., 2007). Coincidentally, this was also the time at which a large amount of oxygen started accumulating in the atmosphere. This accumulation of oxygen forced anaerobic organisms to develop ways to live in an oxygen-rich environment, otherwise, they faced extinction. Many organisms survived this environmental change by taking on some form of endosymbiosis.

Cyanobacteria play an essential role in homeostasis, even in small natural populations (Kozak et al., 2018). Being the only known prokaryotes capable of oxygenic photosynthesis, cyanobacteria are unique in their contribution to the aerobic food chain (Ortega-Ramos et al., 2014). Cyanobacterial strains with similar taxonomic affiliations are related in their niche occupancy (Walter et al., 2017). The taxonomic scheme can be utilized to predict their niche and function.

Cyanobacteria have great taxonomic diversity, with some species able to survive nutrient-poor/oxygen-rich (oligotrophic) waters while others reside in nutrient-rich/oxygen-poor (eutrophic) environments. Some cyanobacteria are also extremophilic and thermophilic and can survive at different pH ranges (Singh, 2014). Due to their genetic diversity and range of habitats, cyanobacteria have been able to make ecological interactions with other species that have enabled the production of secondary metabolites (Foster et al., 2011; Freeman, Thacker, 2011).

Cyanobacteria can be used to create biofertilizers rich in nitrogen, and can increase the phosphorus content in the soil which increases the amount of nutrients available for the plants. Natural fertilizers are sought after to improve agricultural yield. Cyanobacteria can increase the availability of phosphorus. By solubilizing insoluble compounds such as iron phosphate and calcium phosphate, through the aid of phosphatase enzymes, cyanobacteria increase the amount of nutrients available for the surrounding plants in the environment (Nozzi et al., 2013). The energy and resources needed to produce sustainable and successful agricultural products are provided with the natural fertilizers offered by cyanobacteria. Traditional phosphate-based fertilizers rely on limited natural resources. With the use of genetic engineering and intentional intervention, enhanced biological phosphate removal has been proven to increase phosphate yield in polyphosphate accumulating organisms (Nozzi et al., 2013). Cyanobacteria offer a sustainable option for fertilizer through aiding the active processes in crop growth.

Cyanobacteria are unique in that they can survive in many harsh conditions. They are found in a multitude of places, in a wide range of temperatures, and at various pH levels. This resiliency allows them to be utilized for the reclamation of salt-ridden and arid soils. Soils high in the salt content struggle to support crops due to a dense upper layer that prevents water permeation. Furthermore, the salt content increases osmotic resistance within the soil that roots must pull against to acquire water. Finally, the density of these soils prevents necessary aeration from occurring. Cyanobacteria, however, can form a thick stratum on the soil surface, increasing the retention of organic carbon, nitrogen, phosphorous, and moisture. They also convert sodium clay into calcium clay, decreasing the soil salt content. Cyanobacteria could also be used to recultivate salt-affected soils. The addition of organic molecules and nitrogen promotes binding of the soil, increasing aeration and permeability as well (Singh et al., 2016).

Cyanobacteria excrete substances that promote plant growth. These can be classified as hormones, vitamins, amino acids, antibiotics, or toxins. Cyanobacteria have been shown to positively affect the plant growth of both rice and wheat (Singh et al., 2016).

Many cyanobacteria produce biocontrol agents. These biocontrol compounds can have antibacterial, antifungal, and antiglial (Teuscher et al., 1992; Dahms et al., 2006) properties; they diffuse through the soil to prevent contamination and infection of surrounding organisms. These bioactive compounds inhibit pathogenic growth and potentially control disease carried in soil, water, and plants.

Cyanobacteria contribute to the aquatic bioremediation with their ability to oxidize oil and complex organic compounds as well as accumulate metal ions. Cyanobacteria contain functional groups that bind metals, giving them the ability to reduce metal pollution of water. Cyanobacteria can also degrade organophosphorus insecticides in water that harm soil ecology. However, the inability to separate biomass from the effluent before discharge presents a difficulty for researchers. To remedy this, immobilization of cyanobacteria in matrices can help with harvesting as well as increase the efficiency of metal removal. Cyanobacteria can help to degrade hydrocarbons in oil by providing oxygen and nutrients to oil-degrading bacteria.

Their high rate of growth and high metal sorption capacity gives the cyanobacteria the ability to detoxify and prevent contamination (Singh et al., 2016). In an industrial context, this can be used to purify the oils and pharmaceuticals. While living in polluted and toxic environments, they can provide the means to balance pH, treat water, degrade toxic compounds, and support healthy microbial growth (Singh et al., 2016). Bioremediation and intentional control of the environment are a potential use of cyanobacteria. Their mechanisms of toxin degradation and bioremediation can be a greener solution to the maintenance of current environmental dynamics.

Oil spills bioremediation is another large problem that cyanobacteria could potentially help with. Consortia of cyanobacteria and aerobic organotrophic bacteria can degrade oils and other complex organic materials

(Radwan, Al-Hasan, 2000; Raghukumar et al., 2001; Mansy, El-Bestway, 2002; Abed, Ko¨ster, 2005). Cyanobacterial exopolysaccharides (EPS) may function as surfactants and emulsifiers that aid in oil spills bioremediation (Paniagua-Michel et al., 2013; Marti et al., 2014). Genetically engineered cyanobacteria can provide valuable resources for bioremediation.

Cyanobacteria are useful in oil recovery through the production of high levels of EPS. These decrease oil surface tension and increase its solubility and mobility. EPS may also be used in the removal of dyes from textile effluent. EPS yield can be greatly increased through the immobilization of light-diffusing optical fibers (Abed et al., 2009).

Cyanobacteria can also potentially diminish the atmospheric CO_2 concentration which could aid in the halt of global warming. They can play a role in aquatic bioremediation due to their ability to remove nitrogen and phosphate ions from wastewater, and their potential to absorb some metals (Hashimoto, Furukawa, 1989; Pouliot et al., 1989; Balaji et al., 2014). *Phormidium, Aphanocapsa, Oscillatoria,* and *Westiellopsis* have been seen to remove nitrogenous and phosphate ions from wastewater, whereas several *Arthrospira* (formerly *Spirulina*) species have been reported to remove zinc and nickel using functional groups useful for metal binding (Mohapatra, Gupta, 2005). Cyanobacteria have also been reported to absorb and detoxify pesticides in soil, as well as to break down harmful organophosphorus or organochlorine insecticides in aquatic environments (Jha et al., 2005; Singh et al., 2011). Cyanobacteria can be immobilized in polysaccharides such as agarose, carrageenan, chitosan, or alginate to increase their efficiency in waste removal (El-Bestawy, 2008).

Cyanobacteria can help mitigate global warming by decreasing the amount of CO_2 in the air. It is shown that cyanobacteria fix CO_2 approximately 10–50 times faster than terrestrial plants, making their use one of the most effective approaches to decrease environmental CO_2 and thus slow global warming. Cyanobacteria store CO_2 as organic molecules that can be used for a variety of things. Flue gas is a conglomeration of N_2, CO_2, O_2, and water vapors, and it is produced with the combustion of many fossil fuels. Cyanobacterial strains resistant to high temperatures may serve as sequestrants of flue gas, as the flue gas has a very high temperature when released from power plants. The bacteria can combine the CO_2 from flue gas with calcium and thereby be used in the process of calcification. Methane is another greenhouse gas that must be mitigated. Its main source is the human use of fossil fuels, landfilling, livestock farming, and biomass burning. Natural sources can also emit methane such as flooded rice paddies. If no alternative to carbon based fuel is employed, methane levels will continue to rise with population numbers. Cyanobacteria can minimize the production of methane by providing an alternative option to fossil fuels as well as enhancing the methane uptake of methanotrophs.

In the case of flooded rice paddies, the oxygen released by cyanobacterial metabolism creates an aerobic environment that antagonizes methane production (Singh et al., 2016).

Cyanobacteria are contributing to the sustainable and environmentally friendly solutions for multiple ecological, agricultural, and industrial processes.

8.1 The biosynthetic potential of cyanobacteria

Cyanobacteria can carry out fermentation to acquire energy when oxygen and light are not available for use. While many industries today are based on fermentation of heterotrophic bacteria for products such as enzymes, amino acids, and vitamins, the economic cost of this methodology proves lofty due to the carbon substrates needed for the fermentation process. This makes cyanobacteria an ideal vessel for the fixation of CO_2 as it eliminates the need to supply carbon sources and complex growth media by employing photosynthesis to reduce carbon dioxide—only sunlight, water, CO_2 and minimal nutrients are required. In addition to their cost-effective use of atmospheric carbon dioxide, cyanobacteria can convert 3–9% of solar energy into biomass in contrast to their plant counterparts which only utilize about 0.25–3% of solar energy. Cyanobacteria present many more benefits such as smaller space requirements than crops, easy manipulation due to simple genetic background, and recyclable residual biomass. Due to these factors, cyanobacteria provide a promising tool for the synthesis of industrial products in the future (Abed et al., 2009).

Secondary metabolites, once thought to be a product of higher organisms, have been associated with marine cyanobacteria (Dittmann et al., 2015). Many predators of the marine ecosystem are deterred by secondary metabolites released by cyanobacteria as defense mechanisms. One of these defense mechanisms is the release of quorum-sensing inhibitors, which keep harmful organisms from proliferating in cyanobacterial mats. Another method of self-defense used by benthic cyanobacteria is the secretion of chemicals that decrease predators' palatability, preventing them from feeding on cyanobacteria. Lastly, cyanobacteria use symbiotic relationships with higher organisms, which provide safety in return for beneficial cyanobacterial secondary metabolites. To protect themselves from harmful ultraviolet radiation, cyanobacteria produce photoprotective metabolites. One of the most prevalent of these photoprotective substances is carotenoids. Additionally, cyanobacteria produce protective mycosporine and mycosporine-like amino acids that can be transferred to higher levels in the food chain upon cyanobacterial consumption (Mazard et al., 2016).

Many bioactive compounds have been found in cyanobacteria (Abarzua et al. 1999; Shimizu 2003; Dahms et al., 2006). Cyanobacterial strains have been shown to inhibit the growth of some organisms via their natural presence or by the compounds they produce. Antibacterials, antifungals, antivirals, and antialgals are all examples of bioactive compounds produced by cyanobacterial strains (Østensvik et al., 1998; Dahms et al., 2006; Volk, Furkert, 2006). It is believed that cyanobacteria may excrete these compounds to outcompete other surrounding organisms due to their allelochemicals' ability to inhibit many functions of the targeted cell (i.e., photosynthesis, respiration, growth, carbon uptake, and enzymatic activity). Cyanobacteria have significant biomedical application potential. Several genes encode secondary metabolites with bacteriocin antimicrobial properties, presenting a treatment option for multidrug-resistant microorganisms. In addition, cyanobacteria produce compounds that are antitrypanosomal, antileishmanial, antiinflammatory, and neurotoxic. Certain antiproliferative activities seem to promise an alternative cancer treatment as well. The biosynthetic potential of cyanobacteria is very diverse and can be dramatically improved by genomic and metabolomics engineering (Knoot et al., 2018). Genes for novel bacteriocins and antimicrobials, such as lantipeptides, have been identified in many cyanobacterial genomes (Wang et al., 2011; Li et al., 2010). Many of these compounds can be potentially developed as new drugs. The nonribosomal peptide synthase and polyketide synthase of cyanobacteria are responsible for the biosynthesis of the majority of their bioactive secondary metabolites (Nunnery et al., 2010; Calteau et al., 2014).

Commercial usage of the fine chemicals produced by cyanobacteria is an emerging possibility. For instance, phycobiliproteins and carotenoids have high functionality in bioindustry and high value commercially. Carotenoids can be used as food additives, food colorants, and supplements for human and animal consumption. They have antioxidant properties and potential to benefit health conditions such as heart disease, cataracts, and muscular dystrophy. Several marine cyanobacteria, as well as *Spirulina*, produce vitamins B and E. Additionally, cyanobacteria produce a wide variety of enzymes that can be utilized in the food processing industry and the starch industry. Cyanobacteria can also produce sugars, lipids, and amino acids (Shimizu, 2003). Genomic analysis showed that the two-thirds of cyanobacterial peptides, polyketides, alkaloids, lipids, and terpenes are produced by four cyanobacterial genera: *Lyngbya*, *Microcystis*, *Hapalosiphon*, and *Nostoc* (Dittman et al., 2015). Cyanobacteria also contribute to the commercial market of nutraceuticals, pigments, vitamins, and enzymes (Guedes et al., 2011; Watanabe et al., 1999). They are useful as nutritional supplements and are a rich source of pigments and chromophores. Tetrapyrrole chromophores are known to have beneficial health effects, providing macronutrients and micronutrients and aiding in digestion. *Arthrospira* (formerly *Spirulina*) is a popular cyanobacterial genus that reportedly has high nutritional value, digestibility, and protein content (Shanmugapriya, Ramanathan, 2012). It is the most widely commercially cultivated cyanobacterial species with the annual production estimated to be 5000–15,000 tons of dry weight (Lu et al., 2011). Being a rich source of carotenoids and other pigments, cyanobacteria have been used as coloring agents and color enhancers over chemical coloring agents that have harmful effects. Cyanobacteria can also secrete monosaccharides such as fructose and glucose (Niederholtmeyer et al., 2010).

With over 300 genome sequences publicly available in CyanoBase alone and over 400 in total across all cyanobacterial databases (Fujisawa et al., 2017; Alvarenga et al., 2017), as well as the relative ease of genetic manipulations with cyanobacteria, they have been extensively used in genetic engineering. Modifications by genetic engineering in the *Synechococcus elongatus* PCC 7942 fatty acid synthesis pathway can lead to the production and secretion of omega-3 fatty acid in this strain (Santos-Merino et al., 2018).

Cyanobacteria also produce isoprene—an energy-rich hydrocarbon that can be utilized for biofuel, feedstock in the synthetic chemistry industry, and the production of synthetic rubber. While isoprene is produced naturally by many plants, the harvest of isoprene from these plants proves impractical. Photosynthetic cyanobacteria, however, provide a much more effective and efficient way to accumulate isoprene. The expression of isoprene synthase from the plant *Pueraria montana* in *Synechocystis* sp. PCC 6803 enables it to produce isoprene, a hydrocarbon especially relevant in synthetic chemistry (Lindberg et al., 2010). Further introduction of the heterologous bacterial mevalonic acid pathway in this strain showed the 2.5-fold improvement in the production yield of isoprene (Bentley et al., 2014).

High yield of limonene was achieved by the genetic engineering of *S. elongatus* PCC 7942 coupled with the computational modeling of changes in limonene flux as a response to photosynthetic output (Wang et al., 2016).

Cyanobacterial strains can produce and accumulate polyhydroxyalkanoates (PHAs) bioplastics. In the presence of excess carbon, cyanobacteria store PHA granules or water-insoluble inclusions. PHA has properties resembling polyethylene and polypropylene, as well as the ability to be mineralized into water and CO_2 by microorganisms, making it a candidate for biodegradable plastic. In addition, the use of cyanobacterial PHA would provide a carbon-neutral method of plastic production that is fueled by readily available environmental CO_2 and sunlight. While the amounts of PHA produced by each cyanobacterial cell is very low, the production can be increased with mixotrophic conditions and high concentrations of free coenzyme A (CoA), which inhibits the regulatory effects of 3-ketothiolase. In cyanobacteria, PHA functions to remove excess reducing equivalents resulting from an unbalanced production of ATP and NADPH by photosynthesis (Abed et al., 2009).

Cyanobacteria can fix CO_2 and turn it into PHAs under nitrogen-limiting conditions. PHAs are the bioplastics that can be utilized as a biodegradable renewable alternative to the traditional oil-based nonbiodegradable plastics, as well as the biocompatible material for pharmaceutical and biomedical applications (Sudesh, 2004; Williams et al., 1999). The most common type of PHAs that cyanobacteria produce and accumulate is poly(3-hydroxybutyrate) (PHB) (Vincenzini et al., 1990; Sudesh et al., 2002). Genetically modified *Synechocystis* sp. PCC6803 was capable of accumulating PHB up to 26% of cell dry weight when grown photoautotrophically with nitrogen limitation; and the PHB accumulation would reach 35% of cell dry weight upon addition of 0.4% (w/v) acetate (Khetkorn et al., 2016). Thermophilic cyanobacterium *Synechococcus* sp. MA19 reached the 55% level of PHB accumulation per cell dry weight under phosphate-limited conditions (Nishioka et al., 2001). Photobioreactors for commercial production of PHB by cyanobacteria are being explored in commercial settings (Troschl et al., 2017). The commercial applications of cyanobacterial PHAs are increasingly diverse (Gomes Gradíssimo et al., 2020).

A multitude of metabolic pathways and genetic manipulations in cyanobacteria and their abilities to photosynthesize and ferment under anoxic conditions makes them a better substitute for both heterotrophic bacteria and autotrophic eukaryotes, having proven to produce ~3–36 times more biomass than certain crops (Dismukes et al., 2008).

8.2 Genomic diversity and genetic tools for cyanobacteria

Cyanobacteria are photoautotrophic bacteria that have greatly contributed to the existence of atmospheric oxygen the Earth has today. Cyanobacteria also produce organic matter due to photosynthesis, fix nitrogen, produce bioactive compounds, and serve as a potential source of renewable energy. These bacteria are highly diverse morphologically and physiologically: some have heterocysts while others do not; some are facultative heterotrophs (Stanier, Bazine, 1977). These qualities make cyanobacteria an interesting candidate for genetic engineering, which has been performed extensively in the past few decades.

DNA sequencing can reveal the genome size and plasmid count of the cyanobacterial strain, thereby allowing investigators to infer how cyanobacteria adapt and behave in different environments (Alvarenga et al., 2017; Prabha et al., 2016). Deeper study in cyanobacterial genomics (i.e., metagenomics and ecogenomics) also divulges possible problems that may arise before or while culturing and/or sequencing, such as gene loss and duplication, horizontal gene transfer, phage susceptibility, collection of non-axenic cultures, and other ecological variables that affect the final outcome (Prabha et al., 2016; Walter et al., 2017). The advent of -omics analytical tools for the creation of cyanobacterial databases have engendered developments in cyanobacterial genomics, which has flourished prolifically in the past decade.

With over 300 genome sequences publicly available in CyanoBase alone and over 400 in total across all cyanobacterial databases (Fujisawa et al., 2017; Alvarenga et al., 2017) compared to the tens of thousands of genomes sequenced for other prokaryotic phyla, cyanobacteria have great potential for further research. For instance, *Synechocystis* sp. PCC 6803, the first cyanobacterial genome sequenced in 1996, has since been the subject of multifarious experiments apropos of genetic manipulation (Alvarenga et al., 2017; Eckert et al., 2012). Had it not been for the genomic sequencing of *Synechocystis*, genetic and comparative analyses of significant enzymes such as Hox or bidirectional hydrogenase would have proven more difficult, considering that the primers for these genes were derived from genomic DNA (Eckert et al., 2012).

Generally, genomic and genetic analyses are conducted using 16s rRNA gene sequences, average amino acid identity, and genome-to-genome distance (Walter et al., 2017). On the other hand, cyanobacteria may taxonomically be divided based on their ecological niche rather than using the traditional method of classifying based on their phylogenetic tree (Walter et al., 2017; Komárek, 2016). Research by Walter et al. (2017) established a novel taxonomic framework for cyanobacteria distinguishing them by their temperature and environmental preferences, that is, (1) low temperature, (2) low-temperature copiotroph, and (3) high-temperature oligotroph. In doing so, they found four areas that required taxonomic revision, notably among those within the *Oscillatoria*, *Arthrospira*, and *Synechococcus* groups. More relevant to feedstock production, however, was the initial goal to divide cyanobacteria by temperature and environmental preference: while members of the low-temperature group had positive correlations with nitrogen and phosphorus source concentration as well as minimum generation time, they had negative correlations with autotrophic microbial cell abundance, oxygen availability, and salinity; low-temperature copiotrophs had the same positive and negative correlations as those in the low-temperature group; high-temperature oligotrophs had positive correlations with temperature and cell abundance and negative correlations with nitrogen and phosphorus concentrations (Walter et al., 2017). Genomic information obtained from niche preference may shed some light on ultrastructure, nutrient uptake, nutrient utilization, and efficiency, most of which comprise the biosynthetic capacity of cyanobacteria (Walter et al., 2017; Gonzalez-Esquer et al., 2016). The polyphasic approach of categorizing cyanobacteria factors the following when classifying cyanobacteria into cryptotaxa, morphotaxa, and ecotaxa: 16s rRNA gene sequencing, the threshold for phylogenetic similarity (~95%), unique morphological markers, ecological separation, and basic nomenclature (Komarek, 2016). Comparatively, some strains have sequenced genomes driven to completion while others are more challenging to sequence due to the aforementioned reasons, such as *Fischerella* sp. PCC 9605 has completeness score of 100 as opposed to *Geitlerinema* sp. PCC 7105's score of 93.76 (Walter et al., 2017). Additionally, ecogenomic grouping can impart important comparisons, such as how strains collected from the open ocean do not contain any *hox* genes, which are vital for the hydrogenase enzymes that facilitate hydrogen production in these photosynthetic prokaryotes (Puggioni et al., 2016). This knowledge can save time and resources in the examination of strains. In any case, cyanobacterial groups are diverse and numerous, and much remains to be explored across genomes to optimize cyanobacterial biosynthetic capacity.

Apropos of biosynthetic capacities, a salient instance in which genomic data were utilized to ascertain biosynthetic capacity include the regulatory nature of AbrB2 in *Synechocystis* sp. PCC 6803, a repressor that binds to the promoter of the *hox* operon, thereby obviating the encoding of the hydrogenase enzyme (Leplat et al., 2013); finding that two-thirds of peptides, polyketides, alkaloids, lipids, and terpenes are produced by four cyanobacterial genera (namely *Lyngbya*, *Microcystis*, *Hapalosiphon*, and *Nostoc*) (Dittmann et al., 2015) and probing into orphan gene clusters, or known pathways with unknown natural products (Dittmann et al., 2015; Micallef et al., 2015) are discoveries owed to genomics. In terms of searching across genomes for enzymes that expedite hydrogen production, ecogenomics contributes tremendously to the identification of cyanobacteria that are able to produce the bidirectional (Hox) and uptake (Hup) enzymes: although the distribution of hydrogen metabolism seems to be wide among cyanobacteria, many open ocean strains seems to lack the ability to synthesize said enzymes (Puggioni et al., 2016). More crucially, sifting through genomes for these enzymes uncover both enzyme-encoding genes of interest as well as certain promoters that help express other genes, such as those of the O_2-tolerant-hydrogenase-encoding gene found in *Lyngbya confervoides* BDU14195 (although potentially inactive) (Wang et al., 2012; Puggioni et al., 2016).

Cyanobacterial physiology makes way for biotechnological applications (Leister, 2019) as their photosynthetic capabilities can be manipulated to produce certain compounds (Zhu et al., 2010). Additionally, cyanobacteria may also be targeted for its direct coupling of metabolic pathways (Mellor et al., 2017) and other methods of redirecting electron flow (Saar et al., 2018; Longatte et al., 2015; Fu et al., 2017).

In the past few years, there has been significant progress in the development of synthetic biology technologies. Cyanobacterial cell machinery has been lauded as especially amenable to being engineered. These prokaryotes also grow very quickly. For example, *S. elongatus* UTEX 2973 has been identified as the fastest-growing cyanobacteria known to date while also being one of the fastest-growing photosynthetic microorganisms (Yu et al., 2015; Ungerer et al., 2018). Furthermore, certain strains are able to withstand extremes in salinity, pH, and temperature, such as *Arthrospira* sp. due to their metabolic versatility (Xiong et al., 2015; Jiménez et al., 2003). Other opportunities for synthetic biology applications in cyanobacteria include carbon fixation using carboxysomes, which have the ability to host heterologous enzymes (Gonzalez-Esquer et al., 2016). Moreover, cyanobacteria do not rely on organelles to conduct photosynthesis, thereby allowing them to integrate heterologous metabolic reactions into their photosystems (Berepiki et al., 2016; Wlodarczyk et al., 2016; Mellor et al., 2017). A stumbling block in exploiting this feature is the gaps of knowledge in how cyanobacteria use their electron transport chain for respiration and photosynthesis. Other interesting instances of using synthetic biology techniques in cyanobacteria include the introduction of *Zymomonas mobilis* pyruvate decarboxylase and alcohol dehydrogenase II genes in *S. elongatus* PCC 7942 for ethanol production (Deng and Coleman, 1999), as well as energy-storing and energy-generating 3D-printed cyanobacteria (Nielsen et al., 2016; Matson, Atsumi, 2018; Sengupta et al., 2018).

Bioengineering of cyanobacteria provides the venue for biosynthetic optimization necessary to produce the maximum product yield. Typically, pathways are transferred to cyanobacteria after being tested in heterotrophic hosts; however, the search for naturally transformable bacteria has made genetic manipulation easier (Carroll et al., 2018). Some strains such as *S. elongatus* PCC 7942, *Synechocystis* sp. PCC 6803, and *Synechococcus* sp. PCC 7002 are naturally transformable. However, the introduction of exogenous DNA in other cyanobacteria may be complicated by some restriction-modification enzymes (Elhai, 1994). DNA can be introduced by electroporation and conjugation; regardless, selection markers are utilized while assessing success of DNA integration: antibiotic resistance genes (with the exceptions of rifampicin and tetracycline due to their light sensitivity) are commonly used. In addition, markerless systems may be helpful when there are multiple gene mutations to be constructed. This is due to the limited number of convenient markers. Thus, a markerless system can be made by the insertion of the sacB-antibiotic resistance cassette fragment into the chromosome, allowing for positive selection; when this fragment is replaced in a mutant, the addition of sucrose allows for negative selection as the sacB-containing cells convert sucrose into levans and thus cannot survive in sucrose, whereas the mutant cells survive in the presence of sucrose (Dexter, Fu, 2009; Vermaas, 1996).

There are considerations to be made for gene expression when genetically engineering cyanobacteria: gene copy number, promoter strength, induced versus constitutive expression and codon usage. Plasmid and chromosomal gene expression direct gene copy number; integration sites in the chromosome may be complicated as each cyanobacterial cell can contain multiple copies of a chromosome (Schneider et al., 2007), leading to longer growth times and genome segregation for the isolation of homozygous recombinant strains. The rewards of this, however, are multiple copies of the target gene for enhanced gene expression. When the efficiency of this method is low, plasmid shuttle vectors may be exploited.

Advances in metabolic engineering have furthered potential use of cyanobacteria as a host organism using technology such as CRISPR-based and sRNA-based tools to manipulate "BioBricks," that is, endogenous promoters, heterologous promoters, transcriptional regulators, induction and repression systems, and ribosome binding sites in cyanobacteria (Case, Atsumi, 2016; Carroll et al., 2018; Wang et al., 2012; Behler et al., 2018; Gordon, Pfleger, 2018).

Many cyanobacteria are unique in that they can naturally transform by integrating transgenes into their chromosomes with homologous recombination. A large selection of inducible, and repressible promoters and riboswitches are available to cyanobacteria. Also, ribosome-binding sites and riboregulators have been characterized. Antibiotic resistance has been successfully used for marking successful genetic alterations, as well as a two-step technique that provides for markerless mutant strains. Reporter genes can be utilized to show the behavior of and characterize other genetic components, especially promoters such as *Lac I*-repressible and metal-inducible promoters (Ma et al., 2014; Ohbayashi et al., 2016; Videau et al., 2016; Immethun, Moon, 2018; Qiao et al., 2018; Wegelius et al., 2018). Common reporter genes include bacterial luciferase and fluorescent proteins. LacI behaves differently in cyanobacteria than eukaryotes in that it has reduced efficiency of repression, deeming tightly controlled on/off transcriptional circuits in cyanobacteria impossible. Inducible promoters activated by materials such as copper, iron, and zinc, are often used in synthetic biology for intermittent controlled expression. However, these promoters are not suitable for large-scale production due to toxicity and cost. Other expression systems include the rhamnose-inducible expression system, riboswitches, and the combination of phycocyanin C promoter with 14 transcription factor recognition sequences (Prabha et al., 2016).

The use of ectopic self-replicating vectors has several benefits. First, they provide more efficient transformation through conjugation and electroporation. Second, they do not require segregation to create the transgenic sequence in all copies of the genome. Third, they allow increased orthogonality of the heterologous sequences. Also, hybrid strategies can be employed that both use a shuttle vector and have the stability of genomic insertion by using native cyanobacterial plasmids as insertion sites or shuttle vectors after modification. Shuttle vectors that enable assembly of constructs in cyanobacteria are advantageous for several reasons: there would be an increase of efficiency in transformation, faster incorporation of the transgene due to avoidance of segregation, and more accurate insertion into neutral sites. Native plasmid cyanobacterial vectors serve double duty as they can be both insertion sites and true shuttle vectors (Ng et al., 2015; Jin et al., 2018). Golden Gate–based techniques relying on type II restriction enzymes (Engler et al., 2008) offer possibility of genetic assembly either through streamlined insertion (Taton et al., 2014) or through the modular cloning assembly standard (Weber et al., 2011; Vasudevan et al., 2019). Golden Gate–based techniques provide a way to generate large numbers of plasmids, streamline strain generation, and characterize them. This tool can be used across cyanobacterial phylum.

As a genome editing tool, the clustered regularly interspaced short palindromic repeats (CRISPR) technology may be utilized for transformation and interference (Li et al., 2016; Ungerer et al., 2018; Gordon et al., 2016; Wendt et al., 2016; Yao et al., 2016). Using CRISPR for the gene deletion of *glgC* to drive ethanol titers up, for instance, lends to the possibility that such technology can be used to drive hydrogen production up as well (Hagemann, Hess, 2018).

This technology may offer a new way to knock out multiple genes, giving researchers much to look forward to. A markerless system using CRISPR has yet to be found, however.

Despite these advances, several factors challenge the effective genetic manipulation of cyanobacteria. One of these is the fact the cyanobacteria harbor multiple genome copies, making it difficult and sometimes impossible to ensure multiple genes are edited on the same copy. While CRISPR systems are effective, they require markers and cannot perform multiplex transformations. Another obstacle lies in the insufficiently known properties of cyanobacteria that make it difficult to predict the capability of models (Vavitsas et al., 2019).

8.3 Cyanobacterial biofuels

Sustainability and consistent energy sourcing are a concern to meet the demands of the exponential growth of the population. Natural sources for energy and commercial product are sought after as a realistic solution to this problem. Photosynthetic organisms provide a possible outlet; by offering clean energy, photosynthesis is an ideal mechanism to aid in sustainability. As photosynthetic prokaryotes, cyanobacteria offer clean biofuel production and simple mechanisms that can be manipulated for efficiency. CO_2 fixation naturally occurs through ribulose-1,5-bisphosphate carboxylase/oxygenase (RuBisCO) enzyme. Because it lacks specificity, the mechanism for carbon fixation can be slowed down by other binding agents. Carbon fixation can be optimized by genetically engineering RuBisCO and making RuBisCO more efficient (Tabita et al., 2007). Similarly, because of the simplicity of cyanobacteria, their production of biofuels can be enhanced by adjusting and improving the metabolic pathways.

Using carbon fixation as the primary carbon source, photosynthetic organisms eliminate the need for fermentable sugar for energy production. Additionally, cyanobacteria are abundant in water, allowing for more productive carbon fixation due to the excess CO_2 present. Minimal nutrients are needed, and the microorganisms are abundant. Thus, cyanobacteria may be the optimal species for the production of biofuels.

Cyanobacteria offer a significant range of options in biofuel and clean energy production (Nozzi et al., 2013). They can be genetically engineered to produce ethanol, 1-butanol, isobutanol, hydrogen, and alkanes or alkenes.

For bioethanol production, the addition of pyruvate decarboxylase and alcohol dehydrogenase II genes was required in *S. elongatus* sp. PCC 7942 (Deng and Coleman, 1999). Gao et al. (2012) achieved a further increase in ethanol productivity of cyanobacteria by genetic engineering. Pyruvate decarboxylase acts as a catalyst for nonoxidative decarboxylation of pyruvate, which yields CO_2 and acetaldehyde. Alcohol dehydrogenase converts the acetaldehyde to ethanol (Deng and Coleman, 1999).

As for isobutanol, which is less corrosive, less volatile, and higher in energy density, production of up to 450 mg/L can be achieved with the introduction of a heterologous isobutanol synthesis pathway (Ducat et al., 2011; Atsumi et al., 2008).

Isobutyraldehyde was successfully produced by diverting carbon flux from the valine biosynthesis pathway in *S. elongatus* PCC7942 through the addition of a ketoacid decarboxylase (Atsumi et al., 2009). Isobutyraldehyde has a higher energy density than ethanol making it a more efficient biofuel than ethanol. Isobutanol can be converted from isobutyraldehyde and is a substitute for gasoline (Lau et al., 2015).

Heterologous expression of acetolactate synthase, 2-acetolactate decarboxylase, and secondary alcohol dehydrogenase, in *S. elongatus* PCC7942, were necessary for it to produce 2,3-butanediol (Oliver et al., 2013).

Synechocystis sp. PCC6803 was genetically modified to produce and secrete fatty acids (Liu et al., 2011). Subsequently, similar genetic engineering of *S. elongatus* PCC 7942 produced a more modest yield of free fatty acids (Ruffing, Jones, 2012). *Synechococcus* sp. PCC 7002 proved to be the most promising producer of these compounds, both in terms of productivity, and tolerance to the damaging effects of free fatty acids on host physiology (Ruffing, 2014).

The modified *efe* gene from *Pseudomonas syringae pv. Phaseolicola* was expressed in *Synechocystis* sp. PCC 6803 with a successful continuous ethylene production so that up to 5.5% of the fixed carbon was directed to ethylene synthesis (Ungerer et al., 2012). Synthesis of hydrocarbons such as alkanes and alkenes by cyanobacteria is facilitated by acyl-acyl carrier protein reductase and aldehyde decarbonylase (Mendez-Perez et al., 2011; Steen et al, 2010).

8.4 Hydrogen biofuel

Within the next 30 years, the global population is expected to rise by nearly 2 billion. This will result in greater use of natural resources, fossil fuels for energy, and greater output of CO_2 into the environment. An alternative source of clean and renewable energy is needed as soon as possible to reduce the average carbon footprint of an individual, and to meet the resource needs in the environmentally sensible way.

Current fossil fuel alternatives such as ethanol, 1-butanol, isobutanol, hydrogen gas, and alkanes have been introduced over the years. Through genetic manipulation, cyanobacteria have the potential to produce all of these alternative energy sources. Some strains of cyanobacteria can produce small amounts of ethanol as a byproduct of fermentation. Many experiments have been successful in increasing the amount of ethanol, isobutanol, 1-butanol, and alkanes in various genera of cyanobacteria. Hydrogen biofuel, however, could have some of the greatest potentials as, in essence, it is the cleanest form of biofuel. Due to the increasing amounts of CO_2 emissions from fossil fuels, their replacement with H_2 is becoming more and more favorable because H_2 production has the potential to be carbon negative (Karthic, Joseph, 2012). It has a high energy density, is still just as clean as solar, but it does require a bit more work for harvesting, in comparison to solar energy.

Cyanobacteria demonstrate oxygenic photosynthesis as well as expression of functional hydrogenase (Rexroth et al., 2015; Lindblad et al., 2012; Lindblad, Khanna, 2015). More than 40 cyanobacterial strains have been found to produce hydrogen gas (Dutta et al., 2005). Hydrogen metabolism in cyanobacteria appears to be dependent on physiology and structure. Nitrogen fixation and photosynthesis occur in the same cell in most unicellular, filamentous, and nonheterocystous cyanobacteria through temporal separation, whereas heterocystous cyanobacteria may rely on spatial separation (Bothe et al., 2010; Tamagnini et al., 2002, 2007). There are two types of photolysis in cyanobacteria that result in H_2 production: indirect water photolysis and direct water photolysis.

Indirect photolysis starts with solar energy being harvested by the cell and transformed into chemical energy via carbohydrates. These carbohydrates are then used as substrates for an H_2 production cycle. Indirect biophotolysis does not require a light source for the hydrogen evolution making the need for reduction-associated enzymes less significant (Khanna, Lindblad, 2015).

Cyanobacteria rely on nitrogenase and hydrogenase for hydrogen production. There are many limiting factors to H_2 production in cyanobacterial cells, such as light/dark periods for the cells, the presence of heterocysts, as well as the potential for O_2 to inactivate nitrogenase, causing it to not be able to function correctly.

Direct photolysis uses electrons from light-driven water-splitting reactions of PSII so that the reductant produced from photosynthesis is used as electron donor by hydrogenase (Khanna, Lindblad, 2015).

Biophotolysis occurs in a range of environments and in the different strains of cyanobacteria. Heterocystous cyanobacteria allow oxygen-sensitive nitrogenase to function even in environments abundant with oxygen (Khanna, Lindblad, 2015). Though the process can be slow, the expression of nitrogenase despite the aerobic environment renders heterocystous cyanobacteria as valuable hydrogen-producing organisms. In environments with no light, fermentable sugars are required to produce hydrogen. Nonheterocystous cyanobacteria contain a reversible NAD-linked hydrogenase that performs biophotolysis when light is available, as well as in the dark fermentative conditions (Khanna, Lindblad, 2015).

There are more than 14 genera of cyanobacteria that are capable of producing hydrogen under various conditions, most notably *Anabaena, Calothrix, Oscillatoria, Cyanothece, Nostoc, Synechococcus, Microcystis, Gloeobacter, Aphanocapsa, Chroococcidiopsis*, and *Microcoleus* (Masukawa et al., 2001). Among cyanobacteria, nitrogen-fixing heterocystous species are the strongest contestant for hydrogen production (Schütz et al., 2004). Molecular hydrogen is generated by the nitrogenase enzyme as a byproduct of nitrogen fixation (Tsygankov, 2007). This is usually a recycled product by uptake of hydrogenase and the uptake may serve as another source of electrons for the nitrogenase enzyme (Tamagnini et al., 2002). As a result, the majority of heterocystous strains isolated from the environment show a small net gain of H_2 under regular air conditions. Hydrogen production is affected by low heterocyst-to-vegetative-cell ratios, high energy requirements for the nitrogenase reaction, and oxygenic inactivation of hydrogenase (Bandyopadhyay et al., 2010; Kruse et al., 2005; Pilon, Halil, 2014).

The nitrogenase enzyme performs this process by following the equation below:

$$N_2 + 12H^+ + 12e^- + 40ATP \rightarrow 2NH_3 + 40ADP + 40Pi + 3H_2$$

(Tsygankov, 2007).

Nitrogen fixation is the active reduction of N_2 into ammonia. Because this process has many energetically unstable intermediates, nitrogenase forms a complex that assists in the reduction of nitrogen (Berg et al., 2002). Nitrogenase produces a hydrogen side reaction at the same time as nitrogen fixation. Because of the specific active site formed by the complex, hydrogen side reactions are not ideally efficient (Hallenbeck, 2012). Genetic engineering can be utilized to create nitrogenase which is less specific and thus producing more hydrogen side reactions.

Specific protein components of nitrogenase, Mo-Fe protein and the Fe protein, are necessary for hydrogen production by nitrogenase (Khanna, Lindblad, 2015).

Site-directed mutagenesis of the active site of nitrogenase as well as in the vicinity of FeMo-cofactor were shown to have a dramatic effect on hydrogen production in *Anabaena* sp. PCC 7120. For instance, mutagenesis of the R284 residue in the NifD subunit of *Anabaena* MoFe protein exhibited rates of hydrogen production in an N_2 atmosphere similar to hydrogen production in an Ar atmosphere (Masukawa et al., 2010).

Deficiency of nitrogen creates a suitable environment for the reduction of protons to molecular hydrogen by the hydrogenase enzyme (Sveshnikov et al., 1997; Kosourov et al., 2014). *Gloeocapsa alpicola* increases its hydrogen production when starved of sulfur (Antal, Lindblad, 2005), and *Spirulina platensis* can produce hydrogen in the complete darkness and anoxia (Aoyama et al., 1997).

In the process of enhancing the yield of H_2, two approaches have been discovered specifically for the heterocystous nitrogen-fixing cyanobacteria within thin alginate films immobilized on a hard surface. One approach allowed CO_2 supplements to alginate-entrapped, nitrogen-deprived cells. The deprivation of nitrogen led to a decrease in overall hydrogen production from the inhibition of photosynthetic activity in the vegetative cells. The alginate-entrapped Δ*hupL* cells were more sensitive to light, nitrogen-deficiency, and an imbalance of carbon and nitrogen ratios. In the second approach, a Δ*hupL* mutant was supplemented with nitrogen in air periodically or by carbon dioxide. These treatments revived the photosynthetic activity of the cells and increased hydrogen production (Kosourov et al., 2014).

Bidirectional hydrogenase is a membrane-associated enzyme many cyanobacteria have which is capable of hydrogen evolution. Referred to as Hox, bidirectional hydrogenase is able to reduce protons or oxidize molecular hydrogen.

The [NiFe]-bidirectional hydrogenase consists of a large catalytic subunit and a small subunit with a [FeS] cluster that functions to transfer electrons to and from the large subunit (Fig. 8.2). The maturation of the

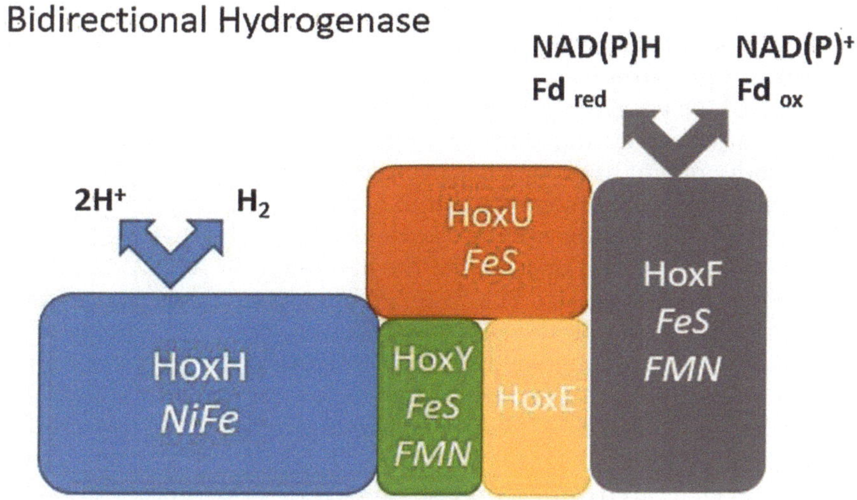

FIG. 8.2 Structure of cyanobacterial bidirectional hydrogenase
Bidirectional hydrogenase subunits:
(A) Hox H- [NiFe] large subunit, catalytic
(B) Hox Y- [FeS] small subunit
(C) Hox U- small subunit for diaphorase, used for electron transfer
(D) Hox F- large subunit for diaphorase, used for electron transfer
(E) Hox E- subunit used for attaching to membrane and electron transfer
Credit: Galyna Kufryk.

[NiFe]-hydrogenase involves six proteins used for the assembly of the [NiFe] active site. The maturation of Hox hydrogenase was studied in cyanobacterium *Synechocystis* sp. PCC 6803. Using this strain, the function of each subunit in hydrogenase activity was established, and the evidence shows the complexity of the maturation and assembly process (Eckert et al., 2012).

The role of Hox hydrogenase is dependent on the organism; however, the baseline function of this enzyme is to catalyze H_2 production in dark, fermentative conditions. This enzyme also functions in light in the presence of thiosulfate (Rákhely et al., 2007).

Two similar enzymes, [FeFe]-hydrogenases and molybdenum (Mo)-nitrogenase, have unique complex iron-sulfur cofactors within their active sites. Specific maturation for synthesis and insertion into the structural enzymes is required for the H-cluster of [FeFe]-hydrogenases and the FeMo cofactor of Mo-nitrogenase. These cofactors are similar biosynthetically, as both have iron-sulfur cluster precursors modified by the radical S-adenosylmethionine enzymes (Peters, Broderick, 2012).

Hox acts to remove excess electrons in anoxic conditions and generate electrons in hydrogen oxidation (Khanna, Lindblad, 2015).

Additionally, Hox hydrogenase can produce H_2 both in the dark and when light sources are present (Eckert et al., 2012). Because of this bidirectional and environmental nonspecificity, producing molecular hydrogen by Hox has a great potential. In the unicellular *Synechocystis* sp. PCC6803 strain of cyanobacteria, the controlled growth and biological photoproduction of hydrogen has been observed (Ortega-Ramos, 2014). To increase the hydrogenase activity in *Synechocystis*, the light-inducible promoter of the photosynthetic gene was used to overexpress the endogenous *hoxEFUYH* operon and the heterologous *NoshypABCPLOS-DEF* operon from *Nostoc* sp. PCC7120 (Germer et al, 2009). The observed increase in hydrogen production was minimal, potentially because the expression of the *hoxEFUYH* and the *NoshypABCDEF* genes was more active under high light, leading to an increase in the photosynthetic production of O_2, which inhibits hydrogenase activity. An overproduction of proteins HoxEFUYH and HypABCDEF coupled with an increase in nickel availability led to an approximately 20-fold increase of hydrogenase level. This has greater significance for the engineering of cyanobacteria for the bioproduction of hydrogen using solar energy (Ortega-Ramos, 2014).

The uptake of H_2 and reduction of NAD+ by endogenous bidirectional [NiFe] hydrogenase of *S. elongatus* has been examined in cell extracts and determined to be low (Schmitz, Bothe, 1996). More molecular hydrogen would be produced if the electrons were redirected to the bidirectional hydrogenase, rather than flowing to the uptake hydrogenase (Ducat et al., 2011).

Oxygen sensitivity is a factor that greatly limits the production of H_2. Not only it inhibits the hydrogenase function, but it also limits transcription and protein maturation (Cohen et al., 2005). Genetic engineering methods have been used to reduce the O_2 sensitivity of the hydrogenase enzyme. The methods include random mutagenesis, as well as targeted mutagenesis of the catalytic site, which restricts O_2 access (Stiebritz, Reiher, 2012). Reduced O_2 sensitivity would allow for a direct path to H_2 production from oxygenic water splitting. However, this leads to the coproduction of O_2 which requires gas separation.

The expression of very active nonnative [FeFe] hydrogenases is an effective way to improve hydrogen production by cyanobacteria. The [FeFe] hydrogenase from *Clostridium acetobutylicum* was overexpressed in *S. elongatus* PCC 7942 along with the accessory operon genes including *hydE*, *hydF*, and *hydG* which are required for maturation, and assembly and insertion of the di-iron subcluster into the active site of the [FeFe] hydrogenases (Ducat et al., 2011). This forms the complete six-iron prosthetic group and the catalytically active protein. The accessory genes were placed in the neutral site I under the regulation of the constitutive *psbA2* promoter and the structural *hydA* gene was placed in the neutral site II under the regulation of the isopropyl β-D-1-thiogalactopyranoside (IPTG)-induced *lac* promoter. In this strain, hydrogen production was studied under both anoxic light and dark conditions. It was observed that higher hydrogen production was under the light conditions, while the production was greatly reduced in the dark conditions. There was a 500-fold increase in hydrogen production when compared to the wild type (Ducat et al., 2011).

Heterologous expression of the [NiFe] hydrogenase from the marine bacterium *Alteromonas macleodii* "Deep ecotype" that is partially tolerant to oxygen in cyanobacteria was successful (Yonemoto et al., 2013). The authors hypothesized that this enzyme could be mutated to be oxygen tolerant in cyanobacteria with the right cluster ligations, residue replacements, and substitutions. The individual effects and the combined effects of these substitutions still need verification.

Other strategies to improve hydrogen production include optimization of light capture efficiency using genetically- or proteomically engineered mutants, improving electron flow by engineering reduced ferredoxin (PETF), ferredoxin-NADP+ reductase, and hydrogenase (Lubner et al., 2011) and competitors for electron flow; alteration of the thylakoid proton gradient under carbonyl cyanide *m*-chlorophenyl hydrazone (CCCP) (Lee, 2013; Lee, Greenbaum, 2003) or the development of a leaky ATPase to reduce ATP production (Das et al., 2014); and curtailing oxygen sensitivity of hydrogenase (Melnicki et al., 2012).

Uptake hydrogenase, a heterodimeric enzyme, consists of a large subunit (HupL) and a small subunit (HupS). The large subunit contains the active site, while the small subunit corresponds to an active electron transport chain. Cyanobacteria contain both photosynthesis and the electron transport chain on the same membrane (Mullineaux, 2014). Both the large HupL and small HupS subunits have no transmembrane domains (Tamagnini et al., 2007). However, uptake hydrogenase is hypothesized to have a membrane-bound component that is further supported by the analysis of available genomes. There is no definitive proof to rule out the existence of both soluble and membrane-bound forms of uptake hydrogenase in cyanobacteria (Tamagnini et al., 2007).

Uptake hydrogenases have three functions: to remove oxygen from nitrogenase using Knall gas reaction, make ATP consumed during nitrogen fixation, and prevent feedback inhibition of nitrogenase by excess hydrogen (Singh et al., 2016). There are three different types of uptake hydrogenases: [NiFe], [FeFe], and Hdm which have different properties. The [NiFe] hydrogenase is more oxygen tolerant and thermotolerant than [FeFe], while [FeFe] is more catalytically efficient. The Hmd is most recently discovered and is FeS cluster free (Vignais, 2007). These comparisons are general due to species having varying enzyme biases. Catalytic bias affects the efficiency of an enzyme. The catalytic bias for the hydrogenase produced by A. *macleodii* and *Thiocapsa persicina* is highly similar, while the hydrogen oxidation by hydrogenase in *Desulfovibrio fructosovorans* is 10 times more favored than in *Allochromatium vinosum* (Stal, 2017). The factor that correlates with the enzyme bias for the uptake hydrogenase is the reactivity of the active site. The higher activity correlates with higher rates of hydrogen oxidation. The hydrogenase found in cyanobacteria is the [NiFe] version, which is composed of two subunits: large and small. The large subunit contains the active site, while the small subunit contains the electron transport chain which is composed of three FeS clusters: proximal, medial, and distal. The proximity is in comparison to the active site. The distal cluster has a midpoint potential of -340 mV, the medial -70 mV, and the proximal -290 mV. The electron enters the distal site and is quickly transferred to the medial site due to a less negative reduction potential. The proximal cluster has a higher reduction potential to keep the electron from reentering the electron transport chain after entering the active site. Variations in reduction potential cause different species to have different catalytic bias. The reduction potential of each cluster can be changed by mutation; which has been shown to alter the activity of the enzyme. In *D. fructosovorans*, the reduction potential of the medial FeS cluster changed from +65 mV to -250 mV (Tamagnini et al., 2007). This increased the hydrogen evolution by twofold only by changing the amino acid from proline to cystine. Two genes code for the subunits of the uptake hydrogenase enzyme, HupS and HupL. The *hupS* functional gene is upstream of the *hupL* and they are under the same promoter. There is a long section of short tandem repeat sequences between the two genes that can form a secondary hairpin structure. The formation of this structure may have positive or negative effects of the transcript. The type and length of these repeats vary for each species. The small subunit contains 320 amino acids and the large subunit contains 531 amino acids (Tamagnini et al., 2007). The uptake hydrogenase is thought to be anchored to the cell membrane to a third subunit, HupC. Other genes involved in the posttranslational modifications and expression are *hypABCEDF* and *hupW*. The *hupW* gene encodes for the endopeptidase that assists in the maturation of the uptake hydrogenase. The *hypABCDEF* genes are usually clustered close to a hydrogenase gene, either uptake or bidirectional. These modification and expression genes vary in different organisms. Each gene is under a different promoter. There is a -35 and -10 consensus region located 17 to 250 bp upstream of the *hupS* start codon (Tamagnini et al., 2007). The NctA binding is located farther upstream with an integrated host factor (IHF) located between the NctA and the consensus regions. The IHF is thought to fold the DNA to allow binding of the NctA regulator to the RNA polymerase complex. This suggests a similar regulator to the mechanism of σ factor binding in *Escherichia coli*. The *hupSL* promoter is also controlled by the NtcA regulators, which is important in the regulation of nitrogenase (Holmqvist et al., 2009). This, as well as the presence of hydrogenase in all nitrogen-fixing cyanobacteria, suggests the cooperativity of the two enzymes. The uptake hydrogenase minimizes the energy loss from hydrogen reduction by nitrogenase through recycling of hydrogen. Therefore, as shown in experiments, the accumulation of molecular hydrogen in the cell increases the amount of uptake hydrogenase. Increased levels of *hupSL* transcripts occur during the dark hours when nitrogenase is active. During maturation of the uptake hydrogenase, three main events occur: the NiFe metallo-center is added to the large subunit, a C-terminus extension is cleaved, and the FeS clusters form. The FeS clusters are formed with help of the eight cysteine residues in the small subunit. The metallo-center is coordinated by four cysteine residues. The C-terminus extension is cleaved by an endopeptidase only when the NiFe metallo-center is added to the large subunit (Senger et al., 2017). This suggests that the C-terminal extension is necessary for the recognition of the maturation machinery. This cleavage allows only mature enzymes containing the [FeNi] cluster continue through maturation. The cleavage of the C-terminus allows the two subunits to dimerize. The promoter for the *hupL* gene was studied through deletion mutations. The shortest fragment had the greatest expression, which leads to the belief that the regulatory fragment is located within a small region within the promoter or downstream of the gene. The active site of the [NiFe] uptake hydrogenase has structural and catalytic capabilities that can be altered through amino acid substitutions. The conserved Arg (509) is thought to act as the acid during the oxidation of H_2. The metals act as the base and allow the hydrogen to bond. Changing the Arg (509) to lysine maintains the structural component of the active site, but reduces the catalytic capacity by more than 100-fold. Cyanobacteria exhibit differences in hydrogenase gene activity in heterocysts and vegetative cells (Carrieri et al., 2011). The open reading frame (ORF) in the heterocyst allows transcription of the *hupSL* gene cluster. In vegetative cells, the ORF transcription is blocked by a large unknown element (Tamagnini et al., 2007).

Hup hydrogenase is found to be coregulated with nitrogenase (Khanna, Lindblad, 2015). Hup hydrogenase does work complimentary to nitrogenase by recovering some of the energy lost in the nitrogenase reduction process. For the uptake of hydrogen to occur, the hydrogen-derived electrons must be able to be redirected to alternate electron-dependent metabolic processes.

Molecular hydrogen can be used to produce biogas in fermentation and anaerobic digestion. During these processes, CO_2 is removed. As an energy source, molecular hydrogen is optimal in both the amount of energy produced and the sustainability of the process. Fermentable sugars are needed to supply the carbon in carbon-consuming processes like this. However, cyanobacteria offer the solution to eliminate the need for fermentable sugars (Nozzi et al., 2013).

Cyanobacteria grow to high densities and produce significant levels of desired metabolites essential in large-scale industrial applications. Different types of bioreactors are built to optimize cyanobacterial hydrogen production and reduce costs. Genetic and metabolic engineering and synthetic biology are also in place to improve hydrogen yields. By combining these technologies, biohydrogen production can be maximized. Further research in this area can lead to optimizing biofuel production by using nonarable land and abundant saline water resources.

Being one of the most substantial greenhouse gases, CO_2 accumulation is very dangerous and increases the rate of climate change significantly. Cyanobacteria have the ability to sequester CO_2 through photosynthetic mechanisms. Because of their low nutrient and environmental needs, these microorganisms are efficient for optimized reversal of CO_2 production. Similarly, cyanobacteria can reduce methane emissions. By release of O_2 during photosynthesis, the environment created by the cyanobacteria is not suitable for CH_4 accumulation and leads to minimizing the effect of CH_4 as a greenhouse gas (Berg et al., 2002). Cyanobacteria are significant in providing clean energy and fuel sources; while naturally producing clean energy through photosynthesis and formation of molecular hydrogen, cyanobacteria also sequester CO_2 and benefit the environment, thus playing a promising role in the reversal of climate change.

As the cyanobacterial production of biomass is linked to the CO_2 sequestration, industrial use of cyanobacteria may reduce the carbon footprint.

Large-scale use of algae and cyanobacteria for renewable energy production has been a recent focus with proposals to form algae farms incorporated into landscape inside and outside of cities. Also, self-fueling "green" buildings have already been erected that receive energy and heat from surrounding microalgae cultivation panels. The largest obstacle in the utilization of cyanobacteria for the large-scale industry is the difficulty in growing the cyanobacteria in densities high enough for optimal production and at reduced costs (Mazard et al., 2016). To achieve this, -omic technologies are being used to identify the relevant genes, mRNA, proteins, and metabolites, while genetic engineering is being used to alter them. Summarily, the genetic engineering of cyanobacteria is the key to increasing cyanobacterial hydrogen yield and creating a more sustainable solution to our energy needs.

References

Abarzua, S., Jakubowski, S., Eckert, S., Fuchs, P., 1999. Biotechnological investigation for the prevention of marinebiofouling II. Blue-green algae as potential producers of biogenic agents for the growth inhibition of microfoulingorganisms. Bot. Mar. 42, 459–465.

Abed, R., Dobretsov, S., Sudesh, K., 2009. Applications of cyanobacteria in biotechnology. J. Appl. Microbiol. 106 (1), 1–12. doi:10.1111/j.1365-2672.2008.03918.x.

Abed, R.M.M., Ko¨ster, J., 2005. The direct role of aerobic heterotrophic bacteria associated with cyanobacteria in the degradation of oil compounds. Int. Biodeterior. Biodegrad. 55, 29–37.

Alvarenga, D.O., Fiore, M.F., Varani, A.M., 2017. A metagenomic approach to cyanobacterial genomics. Front. Microbiol. 8, 809. doi:10.3389/fmicb.2017.00809.

Antal, T.K., Lindblad, P., 2005. Production of H_2 by sulphur-deprived cells of the unicellular cyanobacteria *Gloeocapsa alpicola* and *Synechocystis* sp. PCC 6803 during dark incubation with methane or at various extracellular pH. J. Ferment. Bioeng. 98, 114–120.

Aoyama, K., Uemura, I., Miyake, J., Asada, Y., 1997. Fermentation metabolism to produce hydrogen gas and organic compounds in a cyanobacterium *Spirulina platensis*. J. Ferment. Bioeng. 83, 17–20.

Atsumi, S., Hanai, T., Liao, J.C., 2008. Non-fermentative pathways for synthesis of branched-chain higher alcohols as biofuels. Nature 451 (7174), 86–89.

Atsumi, S., Higashide, W., Liao, J.C., 2009. Direct photosynthetic recycling of carbon dioxide to isobutyraldehyde. Nat. Biotechnol. 27, 1177–1180. doi:10.1038/nbt.1586.

Balaji, S., Kalaivani, T., Rajasekaran, C., 2014. Biosorption of zinc and nickel and its effect on growth of different *Spirulina* strains. CLEAN—Soil Air Water 42 (4), 507–512.

Bandyopadhyay, A., Stockel, J., Min, H., Sherman, L.A., Pakrasi, H.B., 2010. High rates of photobiological H_2 production by a cyanobacterium under aerobic conditions. Nat. Commun. 1, 139.

Behler, J., Vijay, D., Hess, W.R., Akhtar, M.K., 2018. CRISPR-based technologies for metabolic engineering in cyanobacteria. Trends Biotechnol. 36, 996–1010.

Bekker, A., Holland, H.D., Wang, P.L., Rumble, D., Stein, H.J., Hannah, J.L., Coetzee, L.L., Beukes, N.J., 2004. Dating the rise of atmospheric oxygen. Nature 427, 117–120.

Bentley, F.K., Zurbriggen, A., Melis, A., 2014. Heterologous expression of the mevalonic acid pathway in cyanobacteria enhances endogenous carbon partitioning to isoprene. Mol. Plant 7 (1), 71–86.

Berepiki, A., Hitchcock, A., Moore, C.M., Bibby, T.S., 2016. Tapping the unused potential of photosynthesis with a heterologous electron sink. ACS Synth. Biol. 5, 1369–1375.

Berg, J.M., Tymoczko, J.L., Stryer, L., 2002. BiochemistrySection 24.1, Nitrogen Fixation: microorganisms use atp and a powerful reductant to reduce atmospheric nitrogen to ammonia, 5th edition. W H Freeman, New York. Available from https://www.ncbi.nlm.nih.gov/books/NBK22522/.

Blankenship, R.E., Sadekar, S., Raymond, J., 2007. The evolutionary transition from anoxygenic to oxygenic photosynthesis. In: Falkowski, P.G., Knoll, A.N. (Eds.), Evolution of Primary Producers in the Sea. Academic Press, New York, pp. 21–35. https://doi.org/10.1016/B978-012370518-1/50004-7.

Bothe, H., Schmitz, O., Yates, M.G., Newton, W.E., 2010. Nitrogen fixation and hydrogen metabolism in cyanobacteria. Microbiol. Mol. Biol. Rev. 74, 529–551.

Calteau, A., Fewer, D.P., Latifi, A., Coursin, T., Laurent, T., Jokela, J., Kerfeld, C.A., Sivonen, K., Piel, J., Gugger, M., 2014. Phylum-wide comparative genomics unravel the diversity of secondary metabolism in cyanobacteria. BMC Genom 15, 977.

Carrieri, D., Wawrousek, K., Eckert, C., Yu, J., Maness, P., 2011. The role of the bidirectional hydrogenase in cyanobacteria. Bioresour. Technol. 102 (18), 8368–8377. doi:10.1016/j.biortech.2011.03.103.

Carroll, A.L., Case, A.E., Zhang, A., Atsumi, S., 2018. Metabolic engineering tools in model cyanobacteria. Metab. Eng. 50, 47–56.

Case, A.E., Atsumi, S., 2016. Cyanobacterial chemical production. J. Biotechnol. 231, 106–114.

Cohen, I., Knopf, J.A., Irihimovitch, V., Shapira, M., 2005. A proposed mechanism for the inhibitory effects of oxidative stress on rubisco assembly and its subunit expression. Plant Physiol 137, 738–746.

Dahms, H.U., Xu, Y., Pfeiffer, C., 2006. Antifouling potential of cyanobacteria: a mini-review. Biofouling 22, 317–327. doi:10.1080/08927010600967261.

Das, D., Khanna, N., Dasgupta, C.N., 2014. Biohydrogen Production: Fundamentals and Technology Advances. CRC Press, Boca Raton, USA.

Deng, M-D., Coleman, J.R., 1999. Ethanol synthesis by genetic engineering in cyanobacteria. Appl. Environ. Microbiol. 65 (2), 523–528.

Dexter, J., Fu, P., 2009. Metabolic engineering of cyanobacteria for ethanol production. Energy Environ. Sci. 2, 857–864.

Dismukes, G.C., Carrieri, D., Bennette, N., Ananyev, G.M., Posewitz, M.C., 2008. Aquatic phototrophs: efficient alternatives to land-based crops for biofuels. Curr. Opin. Biotechnol. 19 (3), 235–240.

Dittmann, E., Gugger, M., Sivonen, K., Fewer, D.P., 2015. Natural product biosynthetic diversity and comparative genomics of the cyanobacteria. Trends Microbiol 23 (10), 642–652. doi:10.1016/j.tim.2015.07.008.

Ducat, D.C., Sachdeva, G., Silver, P.A., 2011. Rewiring hydrogenase-dependent redox circuits in cyanobacteria. Proc. Natl. Acad. Sci. 108 (10), 3941–3946. doi:10.1073/pnas.1016026108.

Ducat, D.C., Way, J.C., Silver, P.A., 2011. Engineering cyanobacteria to generate high-value products. Trends Biotechnol. 29 (2), 95–103.

Dutta, D., De, D., Chaudhuri, S., Bhattacharya, S.K., 2005. Hydrogen production by cyanobacteria. Microb. Cell Fact. 4. doi:10.1186/1475-2859-4-36.

Eckert, C., Boehm, M., Carrieri, D., Yu, J., Dubini, A., Nixon, P.J., Maness, P.C., 2012. Genetic analysis of the Hox hydrogenase in the cyanobacterium *Synechocystis* sp. PCC 6803 reveals subunit roles in association, assembly, maturation, and function. J. Biol. Chem 287 (52), 43502–43515. doi:10.1074/jbc.M112.392407.

El-Bestawy, E., 2008. Treatment of mixed domestic-industrial wastewater using cyanobacteria. J. Ind. Microbiol. Biotechnol. 35 (11), 1503–1516.

Elhai, J., 1994. Genetic techniques appropriate for the biotechnological exploitation of cyanobacteria. J. Appl. Phycol. 6, 177–186. doi:10.1007/bf02186071.

Engler, C., Kandzia, R., Marillonnet, S., 2008. A one pot, one step, precision cloning method with high throughput capability. PLoS One 3, e3647.

Falkowski, P.G., 2006. Tracing oxygen's imprint on Earth's metabolic evolution. Science 311, 1724–1725.

Foster, R.A., Kuypers, M.M., Vagner, T., Paerl, R.W., Musat, N., Zehr, J.P., 2011. Nitrogen fixation and transfer in open ocean diatom-cyanobacterial symbioses. ISME J 5, 1484–1493.

Freeman, C.J., Thacker, R.W., 2011. Complex interactions between marine sponges and their symbiotic microbial communities. Limnol. Oceanogr. 56, 1577–1586.

Fu, H.-Y., Picot, D., Choquet, Y., Longatte, G., Sayegh, A., Delacotte, J., Guille-Collignon, M., Lemaître, F., Rappaport, F., Wollman, F-A., 2017. Redesigning the QA binding site of photosystem II allows reduction of exogenous quinones. Nat. Commun. 8, 15274.

Fujisawa, T., Narikawa, R., Maeda, S.I., Watanabe, S., Kanesaki, Y., Kobayashi, K., Nomata, J., Hanaoka, M., Watanabe, M., Ehira, S., Suzuki, E., Awai, K., Nakamura, Y., 2017. CyanoBase: a large-scale update on its 20th anniversary. Nucleic Acids Res. 45 (D1), D551–D554. doi:10.1093/nar/gkw1131.

Gao, Z.X., Zhao, H., Li, Z.M., Tan, X.M., Lu, X.F., 2012. Photosynthetic production of ethanol from carbon dioxide in genetically engineered cyanobacteria. Energy. Environ. Sci. 5, 9857–9865. doi:10.1038/nbt.1586.

Germer, F., Zebger, I., Saggu, M., Lendzian, F., Schulz, R., Appel, J., 2009. Overexpression, isolation, and spectroscopic characterization of the bidirectional [NiFe] hydrogenase from *Synechocystis* sp. PCC 6803. J. Biol. Chem. 284 (52), 36462–36472. https://doi.org/10.1074/jbc.M109.028795.

Gomes Gradíssimo, D., Pereira Xavier, L., Valadares Santos, A., 2020. Cyanobacterial polyhydroxyalkanoates: a sustainable alternative in circular economy. Molecules 25 (18), 4331. https://doi.org/10.3390/molecules25184331.

Gonzalez-Esquer, C.R., Smarda, J., Rippka, R., Axen, S.D., Guglielmi, G., Gugger, M., Kerfeld, C.A., 2016. Cyanobacterial ultrastructure in light of genomic sequence data. Photosynth. Res. 129 (2), 147–157. doi:10.1007/s11120-016-0286-2.

Gordon, G.C., Korosh, T.C., Cameron, J.C., Markley, A.L., Begemann, M.B., Pfleger, B.F., 2016. CRISPR interference as a titratable, trans-acting regulatory tool for metabolic engineering in the cyanobacterium *Synechococcus* sp. strain PCC 7002. Metab. Eng. 38, 170–179.

Gordon, G.C., Pfleger, B.F., 2018. Regulatory tools for controlling gene expression in cyanobacteria. Adv. Exp. Med. Biol. 1080, 281–315. doi:10.1007/978-981-13-0854-3_12.

Guedes, A.C., Amaro, H.M., Malcata, F.X., 2011. Microalgae as sources of carotenoids. Marine Drugs 9 (4), 625–644.

Hagemann, M., Hess, W.R., 2018. Systems and synthetic biology for the biotechnological application of cyanobacteria. Curr. Opin. Biotechnol. 49, 94–99. https://doi.org/10.1016/j.copbio.2017.07.008.

Hallenbeck, P.C., 2012. Hydrogen production by cyanobacteria. In: Hallenbeck, P. (Ed.), Microbial Technologies in Advanced Biofuels Production. Springer, Boston, MA, pp. 15–28.

Hashimoto, S., Furukawa, K., 1989. Nutrient removal from secondary effluent by filamentous algae. J. Ferment. Bioeng 67 (1), 62–69.

Holmqvist, M., Stensjö, K., Oliveira, P., Lindberg, P., Lindbald, P., 2009. Characterization of the *hupSL* promoter activity in *Nostoc punctiforme* ATCC 29133. BMC Microbiol 9 (54). https://doi.org/10.1186/1471-2180-9-54.

Immethun, C.M., Moon, T.S., 2018. Synthetic gene regulation in cyanobacteria. Adv. Exp. Med. Biol. 1080, 317–355.

Jha, M.N., Mishra, S.K., 2005. Biological Responses of cyanobacteria to insecticides and their insecticide degrading potential. Bull. Environ. Contam. Toxicol. 75, 374–381.

Jiménez, C., Cossıo, B.R., Niell, F.X., 2003. Relationship between physicochemical variables and productivity in open ponds for the production of *Spirulina*: a predictive model of algal yield. Aquaculture 221, 331–345.

Jin, H., Wang, Y., Idoine, A., Bhaya, D., 2018. Construction of a shuttle vector using an endogenous plasmid from the cyanobacterium *Synechocystis* sp. PCC6803. Front. Microbiol. 9, 1662.

Karthic, P., Joseph, S., 2012. Comparison and limitations of biohydrogen production processes. Res. J. Biotechnol. 7, 59–71.

Khanna, N., Lindblad, P., 2015. Cyanobacterial hydrogenases and hydrogen metabolism revisited: recent progress and future prospects. Int. J. Mol. Sci. 16 (5), 10537–10561. doi:10.3390/ijms160510537.

Khetkorn, W., Incharoensakdi, A., Lindblad, P., Jantaro, S., 2016. Enhancement of poly-3-hydroxybutyrate production in *Synechocystis* sp. PCC 6803 by overexpression of its native biosynthetic genes. Bioresour. Technol 214, 761–768. doi:10.1016/j.biortech.2016.05.014.

Knoot, C.J., Ungerer, J., Wangikar, P.P., Pakrasi, H.B., 2018. Cyanobacteria: promising biocatalysts for sustainable chemical production. J. Biol. Chem. 293 (14), 5044–5052. doi:10.1074/jbc.R117.815886.

Komárek, J., 2016. A polyphasic approach for the taxonomy of cyanobacteria: principles and applications. Eur. J. Phycol. 51, 346–353.

Kosourov, S., Leino, H., Murukesan, G., Lynch, F., Sivonen, K., Tsygankov, A.A., Aro, E.M., Allahverdiyeva, Y., 2014. Hydrogen photoproduction by immobilized n2-fixing cyanobacteria: understanding the role of the uptake hydrogenase in the long-term process. Appl. Environ. Microbiol. 80 (18), 5807–5817. https://doi.org/10.1128/AEM.01776-14.

Kozak, A., Celewicz, S., Kuczyńska-Kippen, N., 2018. Cyanobacteria in small water bodies: the effect of habitat and catchment area conditions. Sci. Total Environ 646. doi:10.1016/j.scitotenv.2018.07.330.

Kruse, O., Rupprecht, J., Bader, K.P., Thomas-Hall, S., Schenk, P.M., Finazzi, G., Hankamer, B., 2005. Improved photobiological H_2 production in engineered green algal cells. J. Biol. Chem. 280, 34170–34177.

Kump, L.P., 2008. The rise of atmospheric oxygen. Nature 451, 277–278.

Lau, N.-S., Matsui, M., Abdullah, A.A., 2015. Cyanobacteria: photoautotrophic microbial factories for the sustainable synthesis of industrial products. BioMed. Res. Int., 1–9. doi:10.1155/2015/754934.

Lee, J., 2013. Designer transgenic algae for photobiological production of hydrogen from water. In: Lee, J.W. (Ed.), Advanced Biofuels and Bioproducts. Springer US, New York, pp. 371–404.

Lee, J.W., Greenbaum, E., 2003. A new oxygen sensitivity and its potential application in photosynthetic H_2 production. Appl. Biochem. Biotechnol. 105, 303–313.

Leister, D., 2019. Genetic engineering, synthetic biology and the light reactions of photosynthesis. Plant Physiol. 179, 778–793.

Leplat, C., Champeimont, R., Saenkham, P., Cassier-Chauvat, C., Jean-Christophe, A., Chauvat, F., 2013. Genome-wide transcriptome analysis of hydrogen production in the cyanobacterium *Synechocystis*: towards the identification of new players. Int. J. Hydrogen Energy 38 (4), 1866–1872.

Li, B., Sher, D., Kelly, L., Shi, Y., Huang, K., Knerr, P.J., Joewono, I., Rusch, D., Chisholm, S.W., van der Donk, W.A., 2010. Catalytic promiscuity in thebiosynthesis of cyclic peptide secondary metabolites in planktonic marine cyanobacteria. Proc. Natl. Acad. Sci. USA 107, 10430–10435.

Li, H., Shen, C.R., Huang, C-H., Sung, L-Y., Wu, M-Y, Hu, Y-C., 2016. CRISPRCas9 for the genome engineering of cyanobacteria and succinate production. Metab. Eng. 38, 293–302.

Lindberg, P., Park, S., Melis, A., 2010. Engineering a platform for photosynthetic isoprene production in cyanobacteria, using *Synechocystis* as the model organism. Metab. Eng. 12 (1), 70–79.

Lindblad, P., Khanna, N., 2015. Engineering of cyanobacteria for increased hydrogen production. In: Rögner, M (Ed.), Biohydrogen. Walter de Gruyter GmbH, Berlin, Germany, pp. 175–190 2015.

Lindblad, P., Lindberg, P., Oliveira, P., Stensjö, K., Heidorn, T., 2012. Design, engineering, and construction of photosynthetic microbial cell factories for renewable solar fuel production. Ambio 41, 163–168.

Liu, X., Sheng, J., Curtiss III, R., 2011. Fatty acid production in genetically modified cyanobacteria. Proc. Natl. Acad. Sci. USA 108, 6899–6904. doi:10.1073/pnas.1103014108.

Longatte, G., Fu, H-Y., Buriez, O., Labbé, E., Wollman, F-A., Amatore, C., Rappaport, F., Guille-Collignon, M., Lemaître, F., 2015. Evaluation of photosynthetic electrons derivation by exogenous redox mediators. Biophys. Chem. 205, 1–8.

Lu, Y.M., Xiang, W.Z., Wen, Y.H., 2011. *Spirulina* (*Arthrospira*) industry in inner Mongolia of China: current status and prospects. J. Appl. Phycol 23 (2), 265–269.

Lubner, C.E., Applegate, A.M., Knorzer, P., Ganago, A., Bryant, D.A., Happe, T., Golbeck, J.H., 2011. Solar hydrogen-producing bionanodevice outperforms natural photosynthesis. Proc. Natl. Acad. Sci. USA 108, 20988–20991.

Ma, A.T., Schmidt, C.M., Golden, J.W., 2014. Regulation of gene expression in diverse cyanobacterial species by using theophylline-responsive riboswitches. Appl. Environ. Microbiol. 80, 6704–6713.

Mansy, A.E., El-Bestway, E., 2002. Toxicity and biodegradation of fluometuron by selected cyanobacterial species. World J. Microbiol. Biotechnol. 18, 125–131. doi:10.1023/A:1014490811121.

Marti, M.E., Colonna, W.J., Patra, P., Zhang, H., Green, C., Reznik, G., Pynn, M., Jarrell, K., Nyman, J.A., Somasundaran, P., et al., 2014. Production and characterization of microbial biosurfactants for potential use in oil-spill remediation. Enzyme Microb. Technol. 55, 31–39.

Masukawa, H., Inoue, K., Sakurai, H., Wolk, C.P., Hausinger, R.P., 2010. Site-directed mutagenesis of the *Anabaena* sp. strain PCC 7120 nitrogenase active site to increase photobiological hydrogen production. Appl. Environ. Microbiol. 76 (20), 6741–6750.

Masukawa, H., Nakamura, K., Mochimaru, M., Sakurai, H., 2001. Photobiological hydrogen production and nitrogenase activity in some heterocystous cyanobacteria. In: Miyake, J., Matsunaga, T., Pietro, A.S. (Eds.), BioHydrogenII: An Approach to Environmentally Acceptable Technology. Elsevier, Oxford, UK, pp. 63–66.

Matson, M.M., Atsumi, S., 2018. Photomixotrophic chemical production in cyanobacteria. Curr. Opin. Biotechnol. 50, 65–71.

Mazard, S., Penesyan, A., Ostrowski, M., Paulsen, I., Egan, S., 2016. Tiny microbes with a big impact: the role of cyanobacteria and their metabolites in shaping our future. Marine Drugs 14 (5), 97. doi:10.3390/md14050097.

Mellor, S.B., Vavitsas, K., Nielsen, A.Z., Jensen, P.E., 2017. Photosynthetic fuel for heterologous enzymes: the role of electron carrier proteins. Photosynth. Res. 134, 329–342.

Melnicki, M.R., Pinchuk, G.E., Hill, E.A., Kucek, L.A., Fredrickson, J.K., Konopka, A., Beliaev, A.S., 2012. Sustained H_2 production driven by photosynthetic water splitting in a unicellular cyanobacterium. mBio 3, e00197–e00212.

Mendez-Perez, D., Begemann, M.B., Pfleger, B.F., 2011. Modular synthase-encoding gene involved in α-olefin biosynthesis in *Synechococcus* sp. strain PCC 7002. Appl. Environ. Microbiol. 77 (12), 4264–4267.

Micallef, M.L., D'Agostino, P.M., Al-Sinawi, B., Neilan, B.A., Moffitt, M.C., 2015. Exploring cyanobacterial genomes for natural product biosynthesis pathways. Marine Genom 21, 1–12. doi:10.1016/j.margen.2014.11.009.

Mohapatra, H., Gupta, R., 2005. Concurrent sorption of Zn(II), Cu(II) and Co(II) by *Oscillatoria angustissima* as a function of pH in binary and ternary metal solutions. Bioresour. Technol. 96 (12), 1387–1398.

Mullineaux, C.W., 2014. Electron transport and light-harvesting switches in cyanobacteria. Front. Plant Sci 5, 7. doi:10.3389/fpls.2014.00007.

Ng, A.H., Berla, B.M., Pakrasi, H.B., 2015. Fine-tuning of photoautotrophic protein production by combining promoters and neutral sites in the cyanobacterium *Synechocystis* sp. strain PCC 6803. Appl. Environ. Microbiol. 81, 6857–6863.

Niederholtmeyer, H., Wolfstadter, B.T., Savage, D.F., Silver, P.A., Way, J.C., 2010. Engineering cyanobacteria to synthesize and export hydrophilic products. Appl. Environ. Microbiol. 76, 3462–3466. doi:10.1128/AEM.00202-10.

Nielsen, A.Z., Mellor, S.B., Vavitsas, K., Wlodarczyk, A.J., Gnanasekaran, T., Perestrello Ramos H de Jesus, M., King, B.C., Bakowski, K., Jensen, P.E., 2016. Extending the biosynthetic repertoires of cyanobacteria and chloroplasts. Plant J. 87, 87–102.

Nishioka, M., Nakai, K., Miyake, M., Asada, Y., Taya, M., 2001. Production of poly-β-hydroxybutyrate by thermophilic cyanobacterium, *Synechococcus* sp. MA19, under phosphate-limited conditions. Biotechnol. Lett. 23, 1095–1099. doi:10.1023/A:1010551614648.

Nozzi, N.E., Oliver, J.W., Atsumi, S., 2013. Cyanobacteria as a platform for biofuel production. Front. Bioeng. Biotechnol 1, 7. doi:10.3389/fbioe.2013.00007.

Nunnery, J.K., Mevers, E., Gerwick, W.H., 2010. Biologically active secondary metabolites from marine cyanobacteria. Curr. Opin. Biotechnol. 21, 787–793.

Ohbayashi, R., Akai, H., Yoshikawa, H., Hess, W.R., Watanabe, S., 2016. A tightly inducible riboswitch system in *Synechocystis* sp. PCC 6803. J. Gen. Appl. Microbiol. 62, 154–159.

Oliver, J.W.K., Machado, I.M.P., Yoneda, H., Atsumi, S., 2013. Cyanobacterial conversion of carbon dioxide to 2,3-butanediol. Proc. Natl. Acad. Sci. USA 110, 1249–1254.

Ortega-Ramos, M., Jittawuttipoka, T., Saenkham, P., Czarnecka-Kwasiborski, A., Bottin, H., Cassier-Chauvat, C., Chauvat, F., 2014. Engineering *Synechocystis* PCC6803 for hydrogen production: influence on the tolerance to oxidative and sugar stresses. PLoS. One 9 (2). doi:10.1371/journal.pone.0089372.

Østensvik, Ø., Skuberg, O.M., Underdak, B., Hormazabal, V., 1998. Antibacterial properties of extracts from selectedplanktonic cyanobacteria – a comparative study of bacterial bioassays. J. Appl. Microbiol. 84, 1117–1124.

Paniagua-Michel, J.J., Olmos-Soto, J., Morales-Guerrero, E.R., 2013. Algal and microbial exopolysaccharides: new insights as biosurfactants and bioemulsifiers. Adv. Food Nutr. Res. 73, 221–257.

Peters, J.W., Broderick, J.B., 2012. Emerging paradigms for complex iron-sulfur cofactor assembly and insertion. Annu. Rev. Biochem. 81 (1), 429–450. doi:10.1146/annurev-biochem-052610-094911.

Pilon, L., Halil, B.l., 2014. Photobiological hydrogen production. In: Sherif, S.A., Goswami, Yogi, D., Stefanakos, (Lee), E.K., Steinfeld, A (Eds.), Handbook of Hydrogen Energy. CRC Press, Boca Raton, pp. 369–418.

Pouliot, Y., Buelna, G., Racine, C., de la Noue, J., 1989. Culture of cyanobacteria for tertiary wastewater treatment and biomass production. Biol. Wastes 29 (2), 81–91.

Prabha, R., Singh, D.P., Somvanshi, P., Rai, A., 2016. Functional profiling of cyanobacterial genomes and its role in ecological adaptations. Genom. data 9, 89–94. doi:10.1016/j.gdata.2016.06.005.

Puggioni, V., Tempel, S., Latifi, A., 2016. Distribution of hydrogenases in cyanobacteria: a phylum-wide genomic survey. Front. Genet. 7. doi:10.3389/fgene.2016.00223.

Qiao, C., Duan, Y., Zhang, M., Hagemann, M., Luo, Q., Lu, X., 2018. Effects of reduced and enhanced glycogen pools on salt-induced sucrose production in a sucrose-secreting strain of *Synechococcus elongatus* PCC 7942. Appl. Environ. Microbiol. 84, e02023–e02117.

Radwan, S.S., Al-Hasan, R.H., 2000. Oil pollution and cyanobacteria. In: Whitton, B.A., Potts, M. (Eds.), The Ecology of Cyanobacteria. Kluwer Academic Publishers, Berlin, pp. 307–319.

Raghukumar, C., Vipparty, V., David, J.J., Chandramohan, D., 2001. Degradation of crude oil by marine cyanobacteria. Appl. Microbiol. Biotechnol. 57, 433–436. doi:10.1007/s002530100784.

Rákhely, G., Laurinavichene, T.V., Tsygankov, A.A., Kovács, K.L., 2007. The role of Hox hydrogenase in the H_2 metabolism of *Thiocapsa roseopersicina*. Biochim. Biophys. Acta 1767, 671–676.

Rexroth, S., Weigand, K., Rögner, M., 2015. Cyanobacterial design cell for the production of hydrogen from water. In: Rögner, M. (Ed.), Biohydrogen. Walter de Gruyter GmbH, Berlin, Germany, pp. 1–14.

Ruffing, A.M., 2014. Improved free fatty acid production in cyanobacteria with *Synechococcus* sp. PCC 7002 as Host. Front. Bioeng. Biotechnol. 2 (17). https://doi.org/10.3389/fbioe.2014.00017.

Ruffing, A.M., Jones, H.D.T., 2012. Physiological effects of free fatty acid production in genetically engineered *Synechococcus elongatus* PCC 7942. Biotechnol. Bioeng. 109, 2190–2199. doi:10.1002/bit.24509.

Saar, K.L., Bombelli, P., Lea-Smith, D.J., Call, T., Aro, E-M., Müller, T., Howe, C.J., Knowles, T.P.J., 2018. Enhancing power density of biophotovoltaics by decoupling storage and power delivery. Nat. Energy 3, 75–81.

Santos-Merino, M., Garcillán-Barcia, M.P., de la Cruz, F., 2018. Engineering the fatty acid synthesis pathway in *Synechococcus elongatus* PCC 7942 improves omega-3 fatty acid production. Biotechnol. Biofuels 11 (239). https://doi.org/10.1186/s13068-018-1243-4.

Schmitz, O., Bothe, H., 1996. The diaphorase subunit HoxU of the bidirectional hydrogenase as electron transferring protein in cyanobacterial respiration? Naturwissenschaften 83, 525–527.

Schneider, D., Fuhrmann, E., Scholz, I., Hess, W.R., Graumann, P.L., 2007. Fluorescence staining of live cyanobacterial cells suggest non-stringent chromosome segregation and absence of a connection between cytoplasmic and thylakoid membranes. BMC Cell Biol. 8 (39). https://doi.org/10.1186/1471-2121-8-39.

Schopf, J.W., 1993. Microfossils of the early Archean apex chert: new evidence of the antiquity of life. Science 260, 640–646.

Schopf, J.W., Packer, B.M, 1987. Early Archean (3.3-billion to 3.5-billion-year-old) microfossils from Warrawoona Group. Aust. Sci. 237, 70–73.

Schütz, K., Happe, T., Troshina, O., Lindblad, P., Leitao, E., Oliveira, P., Tamagnini, P., 2004. Cyanobacterial H_2 production—a comparative analysis. Planta 218, 350–359.

Senger, M., Stripp, S.T., Soboh, B., 2017. Proteolytic cleavage orchestrates cofactor insertion and protein assembly in [NiFe]-hydrogenase biosynthesis. J. Biol. Chem. 292, 11670–11681. doi:10.1074/jbc.M117.788125.

Sengupta, A., Pakrasi, H.B., Wangikar, P.P., 2018. Recent advances in synthetic biology of cyanobacteria. Appl. Microbiol. Biotechnol. 102, 5457–5547.

Shanmugapriya, R., Ramanathan, T., 2012. Antioxidant and radical scavenging effect of blue-green alga *Spirulina platensis*. Int. J. Pharma. Biol. Arch. 3 (5), 1086–1090.

Shimizu, Y., 2003. Microalgal metabolites. Curr. Opin. Microbiol. 6, 236–243.

Singh, D.P., Khattar, J.I., Nadda, J., Singh, Y., Garg, A., Kaur, N., Gulati, A., 2011. Chlorpyrifos degradation by the cyanobacterium *Synechocystis* sp. strain PUPCCC 64. Environ. Sci. Pollut. Res. 18 (8), 1351–1359. doi:10.1007/s11356-011-0472-x.

Singh, J.S., 2014. Cyanobacteria: a vital bio-agent in eco-restoration of degraded lands and sustainable agriculture. Clim. Change Environ. Sustain. 2, 133–137.

Singh, J.S., Kumar, A., Rai, A.N., Singh, D.P., 2016. Cyanobacteria: a precious bio-resource in agriculture, ecosystem, and environmental sustainability. Front. Microbiol. 7, 529. doi:10.3389/fmicb.2016.00529.

Stal, L.J., 2017. The effect of oxygen concentration and temperature on nitrogenase activity in the heterocystous cyanobacterium *Fischerella* sp. Sci. Rep. 7 (1). doi:10.1038/s41598-017-05715-0.

Stanier, R.Y., Bazine, G.C., 1977. Phototrophic prokaryotes: the cyanobacteria. Annu. Rev. Microbiol 31, 225–274. doi:10.1146/annurev.Mi.31.100177.001301.

Steen, E.J., Kang, Y., Bokinsky, G., Hu, Z., Schirmer, A., McClure, A., del Cardayre, S., Keasling, J.D., 2010. Microbial production of fatty-acid-derived fuels and chemicals from plant biomass. Nature 463 (7280), 559–562.

Stiebritz, M.T., Reiher, M., 2012. Hydrogenases and oxygen. Chem. Sci. 3, 6.

Sudesh, K., 2004. Microbial polyhydroxyalkanoates (PHAs): an emerging biomaterial for tissue engineering and thera-peutic applications. Med. J. Malaysia 59 (Suppl B), 55–56.

Sudesh, K., Taguchi, K., Doi, Y., 2002. Effect of increased PHA synthase activity on polyhydroxyalkanoates biosynthesis in *Synechocystis* sp. PCC6803. Int. J. Biol. Macromol. 30, 97–104.

Sveshnikov, D.A., Sveshnikova, N.V., Rao, K.K., Hall, D.O., 1997. Hydrogen metabolism of mutant forms of *Anabaena variabilis* in continuous cultures and under nutritional stress. FEBS Microbiol. Lett 147, 297–301.

Tabita, F.R., Hanson, T.E., Li, H., Satagopan, S., Singh, J., Chan, S., 2007. Function, structure, and evolution of the RubisCO-like proteins and their RubisCO homologs. Microbiol. Mol. Biol. Rev. 71 (4), 576–599. doi:10.1128/MMBR.00015-07.

Tamagnini, P., Axelsson, R., Lindberg, P., Oxelfelt, F., Wunschiers, R., Lindblad, P., 2002. Hydrogenases and hydrogen metabolism of cyanobacteria. Microbiol. Mol. Biol. Rev. 66 (25), 1–20.

Tamagnini, P., Leita, E., Oliveira, P., Ferreira, D., Pinto, F.A.L., Harris, D.J., Heidorn, T., Lindblad, P., 2007. Cyanobacterial hydrogenases diversity, regulation and application. FEMS Microbiol. Rev. 31, 692–720.

Taton, A., Unglaub, F., Wright, N.E., Zeng, W.Y., Paz-Yepes, J., Brahamsha, B., Palenik, B., Peterson, T.C., Haerizadeh, F., Golden, S.S., 2014. Broad host-range vector system for synthetic biology and biotechnology in cyanobacteria. Nucleic Acids Res. 42, e136.

Teuscher, E., Lindequist, U., Mundt, S., 1992. Cyanobakterien, Quellen biogener Wirkstoffe. Pharm. Ztg. Wiss. 137, 57–69.

Troschl, C., Meixner, K., Drosg, B., 2017. Cyanobacterial PHA production-review of recent advances and a summary of three years' working experience running a pilot plant. Bioengineering. (Basel) 4 (2), 26. https://doi.org/10.3390/bioengineering4020026.

Tsygankov, A.A., 2007. Nitrogen-fixing cyanobacteria: a review. Appl. Biochem. Microbiol. 43, 279–288. doi:10.1134/S0003683807030040.

Ungerer, J., Lin, P-C., Chen, H-Y., Pakrasi, H.B., 2018. Adjustments to photosystem stoichiometry and electron transfer proteins are key to the remarkably fast growth of the cyanobacterium *Synechococcus elongatus* UTEX 2973. MBio 9, e02327–e02417.

Ungerer, J., Tao, L., Davis, M., Ghirardi, M., Maness, P.-C., Yu, J., 2012. Sustained photosynthetic conversion of CO_2 to ethylene in recombinant cyanobacterium *Synechocystis* 6803. Energy Environ. Sci. 5, 8998. doi:10.1039/c2ee22555g.

Vasudevan, R., Gale, G.A.R., Schiavon, A.A., Puzorjov, A., Malin, J., Gillespie, M.D., Vavitsas, K., Zulkower, V., Wang, B., Howe, C.J., et al., 2019. Cyano- gate: a modular cloning suite for engineering cyanobacteria based on the plant MoClo syntax. Plant Physiol. 180, 39–55.

Vavitsas, K., Crozet, P., Vinde, M., Davies, F., Lemaire, S., Vickers, C., 2019. The synthetic biology toolkit for photosynthetic microorganisms. Plant Physiol. 181 (1), 14–27. doi:10.1104/pp.19.00345.

Vermaas, W., 1996. Molecular genetics of the cyanobacterium *Synechocystis* sp. PCC 6803: principles and possible biotechnology applications. J. Appl. Phycol. 8, 263–273. doi:10.1007/bf02178569.

Videau, P., Wells, K.N., Singh, A.J., Gerwick, W.H., Philmus, B., 2016. Assessment of *Anabaena* sp. strain PCC 7120 as a heterologous expression host for cyanobacterial natural products: production of lyngbyatoxin A. ACS. Synth. Biol. 5, 978–988.

Vignais, P.M., 2007. Hydrogenases and H+-reduction in primary energy conservation. In: Schäfer, G., Penefsky, H.S. (Eds.). Bioenergetics. Results and Problems in Cell Differentiation, 45. Springer, Berlin, Heidelberg. https://doi.org/10.1007/400_2006_027.

Vincenzini, M., Sili, C., Philippis, R., Ena, A., Materassi, R., 1990. Occurrence of poly-b-hydroxybutyrate in *Spirulina* species. J. Bacteriol. 172, 2791–2792.

Volk, R-B., Furkert, F.H., 2006. Antialgal, antibacterialand antifungal activity of two metabolites produced andexcreted by cyanobacteria during growth. Microbiol. Res. 161, 180–186.

Walter, J.M., Coutinho, F.H., Dutilh, B.E., Swings, J., Thompson, F.L., Thompson, C.C., 2017. Ecogenomics and taxonomy of cyanobacteria phylum. Front. Microbiol. 8, 2132. doi:10.3389/fmicb.2017.02132.

Wang, B., Wang, J., Zhang, W., Meldrum, D.R., 2012. Application of synthetic biology in cyanobacteria and algae. Front. Microbiol. 3. doi:10.3389/fmicb.2012.00344.

Wang, H., Fewer, D.P., Sivonen, K., 2011. Genome mining demonstrates the widespread occurrence of gene clusters encoding bacteriocins in cyanobacteria. PLoS ONE 6, e22384.

Wang, X., Liu, W., Xin, C., Zheng, Y., Cheng, Y., Sun, S., Li, R., Zhu, X-G., Dai, S.Y., Rentzepis, P.M., Yuan, J.S., 2016. Efficient cyanobacterial limonene biosynthesis. Proc. Natl. Acad. Sci. 113 (50), 14225–14230. doi:10.1073/pnas.1613340113.

Watanabe, F., Katsura, H., Takenaka, S., Fujita, T., Abe, K., Tamura, Y., Nakatsuka, T., Nakano, Y., 1999. Pseudovitamin B12 is the predominant cobamide of an algal health food, spirulina tablets. J. Agric. Food Chem 47 (11), 4736–4741.

Weber, E., Engler, C., Gruetzner, R., Werner, S., Marillonnet, S., 2011. A modular cloning system for standardized assembly of multigene constructs. PLoS One 6, e16765.

Wegelius, A., Li, X., Turco, F., Stensjö, K., 2018. Design and characterization of a synthetic minimal promoter for heterocyst-specific expression in filamentous cyanobacteria. PLoS One 13, e0203898.

Wendt, K.E., Ungerer, J., Cobb, R.E., Zhao, H., Pakrasi, H.B., 2016. CRISPR/Cas9 mediated targeted mutagenesis of the fast-growing cyanobacterium *Synechococcus elongatus* UTEX 2973. Microb. Cell. Fact. 15, 115.

Williams, S.F., Martin, D.P., Horowitz, D.M., Peoples, O.P., 1999. PHA applications: addressing the price performance issue. I. Tissue engineering. Int. J. Biol. Macromol. 25, 111–121.

Wlodarczyk, A., Gnanasekaran, T., Nielsen, A.Z., Zulu, N.N., Mellor, S.B., Luckner, M., Thøfner, J.F.B., Olsen, C.E., Mottawie, M.S., Burow, M., et al., 2016. Metabolic engineering of light-driven cytochrome P450 dependent pathways into *Synechocystis* sp. PCC 6803. Metab. Eng. 33, 1–11.

Xiong, W., Lee, T-C., Rommelfanger, S., Gjersing, E., Cano, M., Maness, P-C., Ghirardi, M., Yu, J., 2015. Phosphoketolase pathway contributes to carbon metabolism in cyanobacteria. Nat. Plants 2, 15187.

Yao, L., Cengic, I., Anfelt, J., Hudson, E.P., 2016. Multiple gene repression in cyanobacteria using CRISPRi. ACS. Synth. Biol. 5, 207–212.

Yonemoto, I.T., Matteri, C.W., Nguyen, T.M., Smith, H.M., Weyman, P.D., 2013. Dual organism design cycle reveals small subunit substitutions that improve [NiFe] hydrogenase hydrogen evolution. J. Biol. Eng. 7, 17. doi:10.1186/1754-1611-7-17.

Yu, J., Liberton, M., Cliften, P.F., Head, R.D., Jacobs, J.M., Smith, R.D., Koppenaal, D.W., Brand, J.J., Pakrasi, H.B., 2015. *Synechococcus elongatus* UTEX 2973, a fast-growing cyanobacterial chassis for biosynthesis using light and CO_2. Sci. Rep. 5, 8132.

Zhu, X-G., Long, S.P., Ort, D.R., 2010. Improving photosynthetic efficiency for greater yield. Annu. Rev. Plant. Biol. 61, 235–261.

CHAPTER

9

Cyanobacterial availability for CRISPR-based genome editing: Current and future challenges

Sandeep Kumar Singh[i], Ajay Kumar[b], Mukesh Kumar Yadav[a], Garima Singh[d], Punuri Jayasekhar Babu[a], Zothanpuia[a], Liansangmawii Chhakchhuak[a], Naveen Chandra Joshi[e], Avinash Chandra Rai[b], Avinash Singh[c], Kaushalendra[f], Rosie Lalmuanpuii[g], Esther Lalnunmawii[h], Bendangchuchang Longchar[j], Prashant Kumar Singh[a]

[a]Department of Biotechnology, Mizoram University (A Central University), Pachhunga University College Campus, Aizawl, Mizoram, India
[b]Agriculture Research Organization (ARO), The Volcani Center, Rishon-Lezion, Israel
[c]Department of Biochemistry and Molecular Pharmacology, New York University, New York, USA
[d]Department of Botany, Mizoram University (A Central University), Pachhunga University College Campus, Aizawl, Mizoram, India
[e]Amity Institute of Microbial Technology (AIMT), Amity University, Noida, Uttar Pradesh, India
[f]Department of Zoology, Mizoram University (A Central University), Pachhunga University College Campus, Aizawl, Mizoram, India
[g]Department of Botany, Mizoram University, (A Central University), Aizawl, Mizoram, India
[h]Department of Botany, Mizoram University, (A Central University), Aizawl, Mizoram, India
[i]Department of Botany, Banaras Hindu University, Varanasi, Uttar Pradesh, India
[j]Department of Life Science, Mizoram University (A Central University), Pachhunga University College Campus, Aizawl, Mizoram, India

Cyanobacterial Lifestyle and its Applications in Biotechnology.
DOI: https://doi.org/10.1016/B978-0-323-90634-0.00009-3

9.1 Introduction

Cyanobacteria evolved around 2.4 billion years ago and prepared a background for life evolution by converting the anoxygenic earth environment to an oxygenic one (Shih, 2015). Cyanobacteria epitomize a diverse group of photosynthetic prokaryotes found in a wide range of environmental habitats (Hu et al. 2008) and have been regarded as attractive laboratory models for genetic studies (Khanna et al. 2015). Cyanobacteria are a morphologically diverse group of organisms characterized by blue-green color (Singh et al., 2018). Cyanobacteria lack membrane-bound organelles and possess the ability to fix atmospheric nitrogen as well as carbon. The process of energy conversion and CO_2 fixation occurs in the specialized internal membranous system known as photosynthetic lamellae (equivalent to plants thylakoids) and carboxysomes, respectively (Ripka et al., 1974). The genetic material (chromosomal DNA) is usually in multiple copies per cell, and hence polyploidy is common in these organisms (Watanabe et al., 2015; Zerulla et al., 2016; Watanabe, 2020). Some vegetative cells in filamentous cyanobacterium are converted into specialized cells known as heterocysts, which convert atmospheric nitrogen into organic compounds (Kumar et al., 2010). The photosynthetic capabilities of these tiny creatures empowered genetic engineers to exploit cyanobacteria to produce valuable chemicals.

These photosynthetic prokaryotes are investigated by different scientists worldwide because of their capabilities to sustain in extreme environmental vagaries and plant photosynthesis research as a host for bioproduction of various biotechnologically crucial products (Singh et al., 2020). Cyanobacteria are blessed with photosynthesis and biological nitrogen fixation abilities to efficiently harvest CO_2 from the atmosphere and fix atmospheric dinitrogen (Singh et al., 2018). The interesting fact is that they are amenable to genetic manipulation through the tools available for other prokaryotes and could be exploited for biotechnological applications (Porter, 1986). Cyanobacterial molecular biology perceived colossal progress in the past two decades with cyanobacterial transformation system development (Golden and Sherman, 1984; Golden et al., 1987). With the availability of a complete genome sequence of model cyanobacterium *Synechocystis* PCC 6803, the field of cyanobacteria molecular biology bloomed as the most promising research field (Vermaas, 1996). To date >300 cyanobacterial genomes have been sequenced, making cyanobacterial genetic manipulations easier (Fujisawa et al., 2017).

Scientists worldwide now try to unlock the cyanobacterial potential by developing various molecular toolboxes for genetic manipulations (Singh et al., 2020). It is warranted to mention that most of these manipulations primarily set to engineer metabolic pathways (Behler et al., 2018). Usually, metabolic engineering emphasizes the production of industrially important chemicals, bioremediation, and biosensing (Haro et al., 2001; Keasling, 2010; Libis et al., 2016) and in earlier decades, by employing classical genetic approaches, a large number of chemicals were synthesized from cyanobacteria (Liu et al., 2011; Kiyota et al., 2014; Hirokawa et al., 2016; van der Woude et al., 2016; Veetil et al., 2017; Yoshikawa et al., 2017; Weiss et al., 2017). Again, it is pertinent to mention that polyploidy has been advocated by cyanobacteria and requires a tedious need to segregate the homogenous lines (Watanabe et al., 2020). Nevertheless, the advent of clustered regularly interspaced short palindromic repeats (CRISPR) based systems added novel revolutions in the cyanobacterial metabolic engineering.

9.2 CRISPR/Cas9-based genome editing in cyanobacteria

The genetic manipulations in cyanobacteria are labor and time-intensive process (Behler et al., 2018). Furthermore, the ploidy (oligo and polyploidy) does not allow easy gene segregation. Hence, the generation of homozygous mutants requires culture streaking for several rounds of antibiotic selection, a time-consuming process (Watanabe et al., 2015; Zerulla et al., 2016; Watanabe, 2020). With the arrival of CRISPR-based technology, genome editing and/or engineering in cyanobacteria revolutionized. Clustered regularly interspaced short palindromic repeats (CRISPR) are the prokaryotic DNA segments with short and repetitive sequences disseminated by foreign DNA elements originated spacer sequences. These repeat spacer sequences collectively generate CRISPR arrays (Rath et al., 2015). These sequences are reported from bacteria and archaea as a defense mechanism against viruses and other mobile genetic elements. CRISPR-linked genes are located proximally from the CRISPR array and encode for the Cas proteins, which serve as nuclease to abolish invading nucleic acids.

Notwithstanding the above, the Cas proteins also integrate new protospacers into the CRISPR array and contribute to the short crRNA's maturation. For cyanobacterial genome engineering, CRISPR/CRISPR-associated (Cas) systems have been employed in combination with homologous recombination-dependent tools (Hou et al., 2019). CRISPR/Cas introduce a double-strand break in the desired gene along the genome, and homology-directed repair stimulated the DNA repair for improved genome editing. Fig. 9.1 shows the different CRISPR systems employed for cyanobacterial genome editing, and Table 9.1 shows the details of distinct gears of three major CRISPR systems.

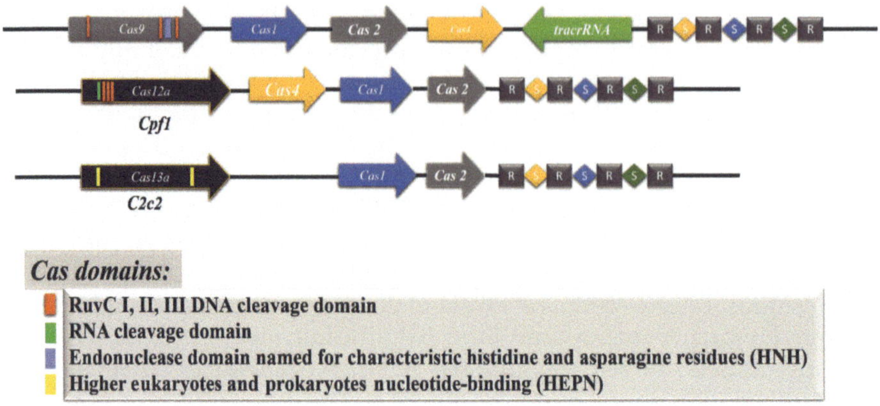

FIG. 9.1 Loci arrangement and domain structures of the effective CRISPR/Cas systems have been exploited for cyanobacterial genome editing. The Cas proteins, in combination with the CRISPR arrays and any additional RNA molecules, constitute the CRISPR/Cas bacterial immune system's core functional unit. The CRISPR/Cas9 for CRISPR processing relies on an additional trans-activating crRNA (tracrRNA). However, the CRISPR/Cas12a and CRISPR/Cas13a lack tracrRNA.

9.3 CRISPR/Cas9-mediated cyanobacterial genome editing

Cas9 is a type II (Class 2) RNA-guided DNA endonuclease exploited to introduce blunt-ended double-stranded breaks (DSBs) in target DNA in a 50-NGG-30 PAM–dependent manner. Wendt et al. (2016), for the first time, demonstrated 70% of *nonbleaching A* (*nblA*) gene mutants of cyanobacteria does not require any antibiotic selection (nblA protein serves as a reporter gene for segregation) by employing homologous-recombination dependent CRISPR/Cas9 system. Additionally, they have also advocated that the only three streaking rounds result in complete segregation within a week. Again, Li et al. (2016), by employing the same combination, obtained 57% segregation in *Synechococcus elongatus* sp. PCC 7942. They have advocated that CRISPR/Cas9 improve the succinate biological production and can be exploited for the bulk production of numerous industrially relevant chemical. Vividly these findings demonstrated the practicability of the CRISPR-based approaches for cyanobacterial genome editing. Therefore, it can be corroborated that homologous recombination combined with CRISPR/Cas outstandingly improves cyanobacterial genome-editing efficiency. However, the CRISPR/Cas has been employed mainly for cyanobacterial metabolic engineering to produce industrially valuable chemicals.

Aside from the robust genome-editing efficiency, the other cyanobacterial CRISPR-based editing is; (1) Short homologous arms could solve the purpose. A short (400 bp) components would be sufficient for the cyanobacterial homologous recombination (Li et al., 2016). (2) CRISPR/Cas9

TABLE 9.1 CRISPR-based genome engineering in cyanobacteria (Adopted and modified from Behler et al., 2018).

Exploited Cyanobacterial host	Genetic approach and target gene(s)	Demonstrated novelty	Cas protein employed	Vectors used	Selection marker(s)	Expression mode and used promoters	References
Synechococcus elongatus UTEX 2973	Genome editing *nblA*	First markerless Knockout	*Streptomyces* codon-optimized Cas9	pCRISPomyces-2	For the temporary plasmid persistence Apramycin and plasmid curing for markerless Knockout (KO) no antibiotics used.	Trans, Cas9: rpsL (XC)-BbsI gRNA: gapdhp (EL)	Wendt et al. (2016)
Synechococcus elongatus PCC 7942	Genome editing NSI site	Homologous recombination efficiency improvement, template DNA reduction, and segregation acceleration.	*Streptococcus pyogenes* Cas9	dCas9-, tracrRNA-, and crRNA encoding plasmids based on pCas9	Chloramphenicol for dCas9-based plasmids, spectinomycin for NSI targeting	Trans, native promoters for Cas9, tracrRNA, and crRNA	Li et al., (2016), Huang et al. (2016)
Synechococcus elongatus UTEX 2973, *Synechocystis* sp. PCC 6803, *Anabaena* sp. PCC 7120	Genome editing A point mutation in *psbA1*, *isiA*, *nifD* Knockout: *nblA*, *nifH* Knock-in: *eYFP* Gene replacement: *nblA* or *nifH* Knockout and simultaneous *eYFP* knock-in	Markerless point mutation, knockout, knock-in, direct gene replacement using Cas12a	*Francisella novicida* Cas12a (Cpf1)	pSL2680	Kanamycin for *S. elongatus* 2973 and *Synechocystis* sp. PCC 6803, neomycin for *Anabaena* sp. PCC 7120, curing of plasmid to obtain markerless editing without antibiotics	Trans, inducible P_{lac} for cas12a (cpf1), constitutive P_{J23119} for CRISPR array	Ungerer and Pakrasi, (2016)

(*continued*)

TABLE 9.1 (Cont'd)

Exploited Cyanobacterial host	Genetic approach and target gene(s)	Demonstrated novelty	Cas protein employed	Vectors used	Selection marker(s)	Expression mode and used promoters	References
Synechocystis sp. PCC 6803	CRISPRi repression *gfp*, *phaE*, and *glgC*	First gene repression by using CRISPRi, in multiplex mode, documentation of useful promoters for repression	*Streptococcus pyogenes* dCas9	pdCas9-bacteria	dCas9: chromosomal integration into the psbA1 neutral site sgRNA: chromosomal integration into the slr2030-slr2031 site, spectinomycin for dCas9, kanamycin for sgRNA	Inducible pLtetO-1 for dCas9 and sgRNA	Yao et al. (2016)
Synechococcus elongatus PCC 7942	Genome editing Knockout: *glgC* Knock-in: *ppc*, *gltA*	Pathway engineering and simultaneous gene knockout and knock-in	*Streptococcus pyogenes* Cas9	dCas9-, tracrRNA-, and crRNA encoding plasmids based on pCas9	Gentamicin	Trans, pCas9-encoded genes, inducible Ptrc for gltA and ppc	Li et al., (2016), Huang et al. (2016)
Synechocystis sp. PCC 6803	CRISPRi repression *Ado*, *Aar*, *PlsX* (*Slr1510*), *Slr2060*, and *PlsC*	Multiplexing gene repression of six fatty acid metabolism-related genes	*Streptococcus pyogenes* dCas9	pdCas9-bacteria	dCas9: chromosomal integration into the psbA1 neutral site sgRNA: chromosomal integration into the slr2030-slr2031 site, spectinomycin for dCas9, kanamycin for sgRNA	Inducible P_{L22} for dCas9, Ptrc for sgRNA	Kaczmarzyk et al. (2018)

Synechococcus elongatus PCC 7942	CRISPRi repression *eyfp*, *glgC*, *sdhA*, and *sdhB*	Gene repression for redirection of carbon flux for succinate production	*Streptococcuspyogenes* dCas9	pdCas9-bacteria	dCas9: chromosomal integration into the NSI neutral site sgRNA: chromosomal integration into the NSII neutral site, chloramphenicol for dCas9, ampicillin for sgRNA, gentamicin for heterologous eyfp selection	Inducible P_{smt} promoter for dCas9, P_{J23119} for sgRNA	
Synechococcus sp. PCC 7002	CRISPRi repression gene regulation, *yfp*, *cpcB*, *ccmK1*	Increasing central carbon flux by gene repression of a key node in nitrogen assimilation for lactate production	*Streptococcus pyogenes* dCas9	-	dCas9: chromosomal integration at the acsA locus sgRNA: NSI site, kanamycin for sgRNA, gentamicin for eYFP	Inducible EZ3 tet promoter for both dCas9 and sgRNA for best dynamic range, several other promoters tested	
Anabaena sp. PCC 7120	CRISPRi and *glnA* and *devH*	Gene repression in a multicellular cyanobacterium and both in vegetative cells and heterocysts	dCas9	-	dCas9: not mentioned, neomycin sulfate for glnA sgRNA, spectinomycin, and streptomycin for devH sgRNA	Inducible P_{L03} promoter for dCas9, constitutive P_{J23119} promoter for sgRNA	Higo et al. (2018)
Anabaena sp. PCC 7120	Genome editing (CRISPR-Cpf1) and *Pol*, *hetR/hetN*, *pks*	"Two-spacers" strategy for genome editing	*Francisella novicida* dead Cpf1	pSL2680	dCpf1: not mentioned, Neomycin sulfate for genes	Not mentioned	Niu et al. 2019

(*continued*)

TABLE 9.1 (Cont'd)

Exploited Cyanobacterial host	Genetic approach and target gene(s)	Demonstrated novelty	Cas protein employed	Vectors used	Selection marker(s)	Expression mode and used promoters	References
Synechocystis sp. PCC6803	CRISPRi and *psbO*	Single plasmid–based, lesion free and episomal expression	*Streptococcus pyogenes* dCas9	pCRPB1010	dCas9: not mentioned, Chloramphenicol *psbO* sgRNA	dCas9 ORF, a tracrRNA and a crRNA resembling the native *S. pyogenes*	Kirtania et al. (2019)
Synechocystis sp PCC 6803	CRISPRi and multiple gene knockouts	Repression of essential genes	*Streptococcus pyogenes* dCas9	pMD19-T	dCas9: Spectinomycin, Kanamycin resistance gene for sgRNA	Inducible promoter P_{L22} for dCas9, Constitutive promoter psbA1 for sgRNA	Yao et al. (2020)
Synechocystis sp. PCC 6803	CRISPRi and *psbD* (D2 protein)	Promoter to drive the expression of ddCpf1 (DNasedead Cpf1 nuclease)	*Francisella novicida* DNase-dead Cpf1 mutant (ddCpf1)	pCRISPR i-D2	DNasedead Cpf1 nuclease	The chimeric promoter, PrhaBAD-RSW (theophylline responsive riboswitch in rhamnose-inducible Promoter) P_{rhaBAD} to drive transcription of both ddcpf1 and a gRNA targeted to the psbD genes	Liu et al. (2020)

Abbreviations: crRNA, CRISPR RNA; pgRNA, pre-genomic RNA; tracrRNA, trans-activating CRISPR RNA.

reduces the template and plasmid requirement for homologous recombination (Li et al., 2016).

9.4 CRISPR/Cas12a-mediated genome editing in cyanobacteria

A higher expression of Cas9 was toxic for all eukaryotic cells (DiCarlo et al., 2013; Kim et al., 2014) and reduced species' viability under investigation cyanobacteria (Wendt et al., 2016; Li et al., 2016). Ungerer and Pakrasi (2016) explored the feasibility of CRISPR type V-A Cas12a nuclease for cyanobacteria genome editing (Fig. 9.2). Cas12a or Cpf1 is a Class 2, type V-A dual-function nuclease specific for CRISPR RNA (crRNA) maturation from its associated repeat-spacer locus and DNA interference, depends on 50-YTN-30 as PAM on the nontarget DNA strand, and cleaves target DNA leading to a 5-nucleotide staggered cut. Cpf1 is the most suitable and compatible CRISPR system for cyanobacterial genome editing (Ungerer and Pakrasi, 2016). By employing cpf1-based genome editing, Ungerer and Pakrasi (2016) generated knockout and single-point mutations in *S. elongatus UTEX 2973, nblA, psbA* genes and obtained 20% edition efficiency. These genes are well documented to serve as a reporter gene and for knock-in yellow fluorescent protein (eYFP) exploited as a reporter gene to corroborate the transcriptional activities. Likewise, cpf1-based genome editing is also obtained in *Anabaena* sp. PCC 7120 and *Synechocystis* sp. PCC 6803 (Ungerer and Pakrasi, 2016). It warrants mentioning that for editing of the metabolic pathway(s) gene(s), extensive precautions should be taken so that the gene(s) regulatory elements remain undisturbed. In CRISPR-mediated genome engineering, the target sequence depends on the adjacent protospacer motif (PAM) position; hence exploitation of Cas nucleases with different PAM specificities would offer greater editing flexibilities of the cyanobacterial genome. Furthermore, the canonical PAM sequences of 50-YTN-30 and 50-NGG-30 would permit targeting of AT- and GC-rich sequences, respectively, across organisms' genome (Anders et al., 2014; Zetsche et al., 2016; Fonfara et al., 2016). The GC content varies from 30.8% (Prochlorococcus MED4) to 68.4% (*Vulcanococcus limneticus*) on an average across the cyanobacterial genomes, nonetheless with significant exemptions. For example, in *V. limneticus*, the *nif* gene cluster has GC content ~60.5%, encoding the proteins required for the biological dinitrogen fixation (Di Cesare et al. 2018). Therefore, the engineering of biological nitrogen fixation genes into nonnitrogen fixers such as plants could solve the increasing nitrogen demand by plants (Mueller et al., 2016). However, an enormous number of modifications are required in addition to the core (other 22–25 *nif* genes). Also, a tight gene expression regulation is needed, and energy, redox equivalents, and the supply of cofactors have to be protected where and when in need.

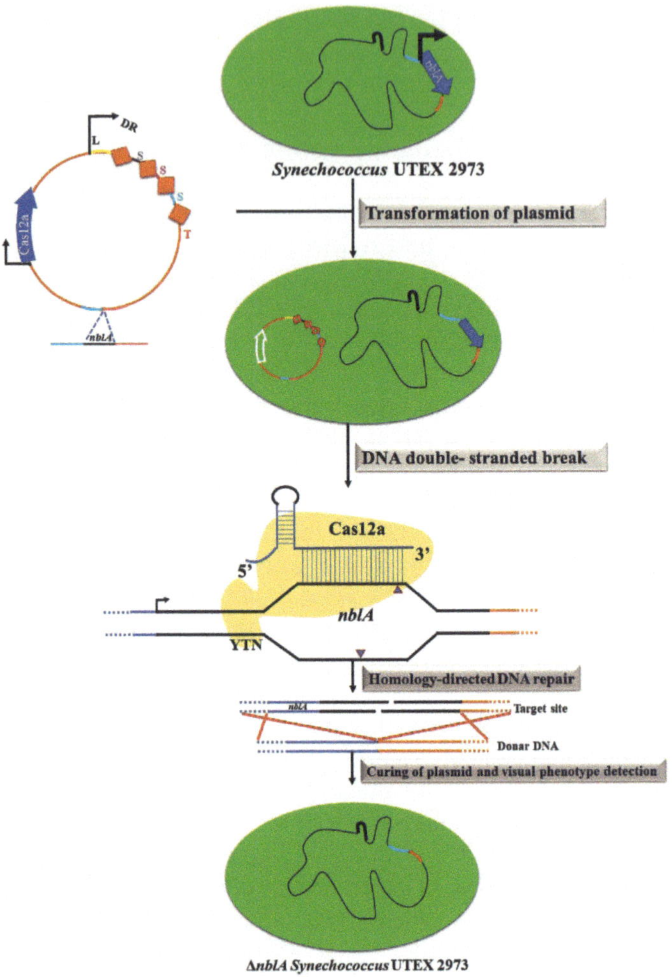

FIG. 9.2 CRISPR/Cas12a-based genome editing in cyanobacteria. A markerless *nblA* deletion mutant obtained by using CRISPR/Cas12a in a unicellular cyanobacterium *Synechococcus elongatus* UTEX 2973 (Ungerer and Pakrasi, 2016). Cas12a ports one split RuvC motif (RuvC I, RuvC II, and RuvC III) and an RNA cleavage domain for cleavage of DNA (Fonfara et al., 2016). Thus, Cas12a sustains a duple function by facilitating both the nblA-targeting crRNA maturation and the entry of a 5-bp staggered double-stranded break (DSB) within the chromosome located downstream PAM site (50-YTN-30). The pSL2680-based vector expresses both Cas12a and the *nblA*-targeting crRNA and also holds 1-kb homology sequences with the target gene. *nblA* crRNA, and Cas12a expressed under the control of constitutive J23119 and inducible lac promoters, respectively. The *nblA* crRNA-encoding array contains the *Francisella novicida* leader sequence (L) followed by direct repeat (DR), spacer (S) sequences, and the *F. novicida* terminator (T). After the DSB introduction, the homologous sequences present within the donor DNA serve as templates for the homology-directed repair mechanisms, and cells can be cured of the plasmid editing. As NblA has been demonstrated to involved in the phycobilisomes degradation during nitrogen starvation, nblA knockout mutants show a nonbleaching phenotype due to the impaired ability of phycobilisomes degradation, compared to wild type. Abbreviations: crRNA, CRISPR RNA; R, repeat; S, spacer.

CRISPR-mediated genome editing offers a promising way to meet the challenges for genome engineering in cyanobacteria. Cas13a is also known as C2c2 single effector protein. It is a Class 2, type VI RNA-guided RNA-targeting CRISPR/Cas effector, a dual nuclease that mediates crRNA maturation and target RNA cleavage.

9.5 Dead Cas9 (dCas9) and cyanobacterial gene expression

The oligo-ploidy and polyploidy in cyanobacterial species makes genome segregation (homozygous knockout mutants) a laborious process on one hand, while on the other Cas9-induced cytotoxicity required preliminary experiments to fine-tune the Cas9 expression (Nogales et al., 2012; Wendt et al., 2016). Therefore, despite exploiting CRISPR-based genome editing, CRISPR interference (CRISPRi) offers an alternative and viable cyanobacterial genome engineering tool that relies on an enzymatically inactive dead Cas9 (dCas9), especially for essential genes. Yao et al. (2016) for the first time established CRISPRi by using green fluorescent protein (GFP) as a reporter and a small set of a single set of guide RNAs (sgRNAs) in *Synechocystis* sp. PCC 6803 for simultaneous editing of two genes (*slr2030* and *slr2031*) by taking their native constitutive promoter. With the use of anhydrotetracycline (aTc)-inducible promoter, P_{L22} a 94% GFP repression was achieved, and this repression was reverted with the removal of inducer aTc (Yao et al., 2016). When the sgRNA targeted the nontemplate strand close to the −35 promoter and transcription start site, the GFP expression's highest repression was observed (Yao et al., 2016). In contrast to the above, Behler et al. (2018) demonstrated when the sgRNA targets +7 to +27 within the nontemplate strand's coding sequence, the RNase E (essential) gene of *Synechocystis* sp. PCC 6803 could be more effectively downregulated. In a separate study, Huang et al. (2016), by exploiting the CRISPRi tool, upheld that the cyanobacterial cells remain repressed for three weeks.

These studies display the pertinence of dCas9 for CRISPRi-based repression in cyanobacteria (Yao et al., 2016; Behler et al., 2016; Huang et al., 2016). It is worth mentioning that irrespective of the promoter (even absence of inducer), the repression in gene expression was observed by Yao et al. (2016). Hence, these responses may be challenging for the regulation of essential genes. Gordon et al. (2016) developed a CRISPRi system with an improved dynamic response that would allow tighter and greater repression control in *Synechococcus* sp. PCC 7002.

Furthermore, they have also assessed fluorescent YFP reporter, various promoters for sgRNA and dCas9 expression. The best observed dynamic range was under the aTc-inducible promoters and when both dCas9 and sgRNA expressed. The transcriptional activity of *YFP* was controlled by CRISPRi. The attempts to improve the dynamic range with alternative promoters or modifications of the sgRNA were unsuccessful. However, by reducing the ribosome's strength-binding site, background levels of repression could be lowered to 30%, resulting in an improved dynamic response (Gordon et al., 2016). Inside the cyanobacterial context of genome editing, tighter induction systems will be crucial for allowing greater dynamic control of metabolic pathways and improving the production of the target chemical's efficiencies.

9.6 dCas9 offers an alternative approach for cyanobacterial metabolic engineering

CRISPRi offers a more realistic cyanobacterial metabolic engineering tool than the Cas9-based genome editing by manipulating the native genes (repressed to reduce) for reducing the carbon flux to flux more carbon through desired metabolic pathways and intensifications target chemical production. For the first time, Yao et al. (2016) introduced a feasible CRISPRi system for cyanobacterial genome editing. They have downregulated the production of glycogen and polyhydroxybutyrate as they serve as natural storage of carbon. Hence, the reduction of flux through their synthetic pathways improves the production of carbon-based target molecules. CRSIPRi-based editing of polyhydroxyalkanoate synthase subunit coding gene (*phaE*) and glycogen biosynthesis coding gene (*glgC*) in *Synechocystis* sp. PCC 6803 eliminated polyhydroxybutyrate and 80% reduction in glycogen synthesis, respectively. Likewise, Huang et al. (2016) also reduced the glycogen content in *S. elongatus* PCC 7942 to 4.8% by repressing glgC and the other two genes: sdhA and *sdhB* genes of glycogen biosynthesis.

It warrants mentioning that to obtain higher photosynthesis by cyanobacterial cells, light penetration in cyanobacterial cultures is mandatory. Kirst et al. (2014) suggested reducing the cyanobacterial light-harvesting antenna's size offers a powerful tool for improved photosynthesis. Gordon et al. (2016) exploited the CRISPRi tool for reducing *cpcB* gene expression (encodes for phycobiliprotein complex subunit) in *Synechococcus* sp. PCC 7002 and showed a severe reduction in the phycobilisomes numbers. Notwithstanding, they also showed *glutamine synthetase I* (*glnA* gene) repression by CRISPRi boosted the carbon metabolism in cyanobacteria. Furthermore, in cyanobacteria, the dinitrogen fixation occurs in heterocyst cells, and mutation in heterocyst cells is challenging to produce nitrogenous compounds. However, Higo et al. (2018) demonstrated that ammonium production could be turned off and on in an inducer (aTc)-dependent manner in both vegetative cells and heterocysts via induction of dCas9 and sgRNA to regulate the *glnA* expression in filamentous cyanobacterium *Anabaena* sp. PCC 7120. Thus, CRISPRi-mediated cyanobacterial genome editing could be exploited for the production of nitrogenous and carbonaceous chemicals.

9.7 Cyanobacterial genome editing offers markerless selection and gene multiplexing

A deep understanding of cyanobacterial metabolic pathways is needed for optimizing the target chemical production. Kanno et al. (2017) evaluated six native cyanobacterial genes independently to enhance the carbon flux toward the 2,3-butanediol a (bioplastics precursor) production. The novelty in the approach was that they modified the genome without using a marker. However, the conventional markerless cyanobacterial mutants' generation techniques are laborious and time-consuming. These marker genes first need to be inserted for cell selection or counterselection and then should be removed from the cell (Berla et al., 2013). However, the CRISPR-mediated editing technologies do not require any selection or counter-selection genes based on cell viability. CRISPR-targeted specificity allows cells with the suitably modified genome to remain viable. Due to the double- DSBs lethality, the cells with nonmodified or partly modified genomes become nonviable and hence markerless cyanobacterial mutants are obtained quickly within a week (Wendt et al., 2016; Cui et al., 2016).

CRISPR multiplexing makes this technology more robust than other genome editing tools (Cobb et al., 2015). For the first time, Yao et al. (2016) knocked down four putative cyanobacterial aldehyde reductases and dehydrogenases coding genes by CRISPR-based multiplexing. Furthermore, recently Kaczmarzyk et al. (2018), by employing the CRISPRi approach, knocked down six native genes of *Synechocystis* sp. PCC 6803 involved in fatty acyl-ACP consumption for enhanced fatty alcohol production. Hence, genome multiplexing by employing CRISPR has opened up the door of metabolic engineering in cyanobacteria. In a nutshell, both the features, markerless and multiplexing of CRISPR, unquestionably can hasten cyanobacterial metabolic engineering.

9.8 Cyanobacterial genome editing: key challenges

CRISPR-mediated genome editing tools have been exploited worldwide for genome editing in various organisms, including cyanobacteria (Wendt et al., 2016; DiCarlo et al., 2013; Shin et al., 2016; Wang et al., 2019; Jiang et al., 2013; Yamano et al., 2017; Wenjing et al., 2020). Even though other genome-cutting tools such as transcription activator-like effector nucleases (TALENs), meganucleases, and zinc finger nucleases have also been exploited but because of their offsite activities and customized engineering requirement, they make them inacceptable for bacterial genomes editing and primarily used for genome editing in eukaryotes (Silva et al., 2011; Gaj et al., 2013; Joang and Sander, 2013) and contrast to the above an increasing number of Cas nucleases CRISPR-mediated genome editing produces DSBs with reasonable specificity and offering sufficient flexibility concerning genome target sites. Also, CRISPR-mediated approaches require very little prior knowledge, genome multiplexing (multiple genetic modifications at a time), cheaper synthesis of sgRNA, and a simple alteration in sgRNA offers numerous rounds of genomic manipulations within a week. These accessibilities have led to the rapid development of markerless and multiplex features of CRISPR-mediated tools and have been exploited for cyanobacterial genome editing. Moreover, CRISPR-mediated off-target frequencies show other unwanted phenotypes and cytotoxic effects (Hendel et al., 2015); however, such undesired effects, if there are any, in cyanobacteria remain elusive. Nevertheless, an improved cleavage specificity could offer the means to reduce the off-target sites, controlled Cas9 induction by employing CRISPR/Cas9 nickases (Shen et al., 2014), or engineered Cas nucleases (Kleinstiver et al., 2015). Given the cytotoxicity, the application of Cas9 needs further investigations and exercised with great attention and tightly controlled induction of Cas9. It is warranted to mentioning that the Cas12a (Cpf1), by contrast, has never shown to have cyanobacterial cell toxicity and hence serves as a better Cas9 alternative. Notwithstanding the above, Cas12a requires only a single CRISPR array, and the Cas enzyme itself processes the RNA into individual sgRNAs. However, Cas9 depends on multiple sgRNA (need to be cloned separately), and each depends on its promoter. Hence, in the CRISPR/Cas12a system, the unpretentious machinery and smaller-sized sgRNA help reduce the cyanobacterial genome engineering costs.

Again, the level of repression significantly influenced CRISPRi approaches due to the complementarity between the target DNA and sgRNA. For obtaining a higher level of repression, sgRNA should be targeted to the transcriptional start site on the nontemplate strand of the DNA (Larson et al., 2013; Qi et al., 2013). Remarkably, Yao et al. (2016) advocated that one of their sgRNAs used in their CRISPRi displayed scrawny repression and conjectured that the sgRNA might target a distal sequence of the transcription start site. Consequently, several different sgRNAs need to be tested to implement the most effective sgRNA for the target gene repression in practice. Another crucial challenge is that cyanobacterial engineering is the choice of the target chemical. Usually, the highest productive target chemicals in cyanobacteria are relatively simple chemical structures from the high-flux pathways (pyruvate or Calvin cycle). Nonetheless, these targets offer excellent fundamental insights into the cyanobacterial pathways metabolic flux, target chemicals diversification, involve complex metabolic routes or poor flux pathways, which will be particularly vital for meeting the industries demands.

9.9 Conclusion and prospects

Atlas of the literature available on CRISPR-mediated cyanobacterial genome editing is summarized in Table 9.1 (Gordon et al., 2016; Huang et al., 2016; Yao et al., 2016, 2020; Shin et al., 2016; Wendt et al., 2016; Li et al., 2016; Ungerer et al., 2016; Higo et al., 2018; Behler et al., 2018; Kaczmarzyk et al., 2018). There is no report providing insights into the application of ZNFs- or TALENs-mediated cyanobacterial genome engineering. Hitherto, the CRISPR technology has been employed successfully in genome engineering of unicellular cyanobacterium *Synechococcus* sp. PCC 7002 (Hunag et al., 2016), *S. elongatus* PCC 7942 (Li et al., 2016; Hunag et al., 2016), *Synechococcus* UTEX 2973 (Wendt et al., 2016; Ungerer and Pakrasi, 2016), *Synechocystis* sp. PCC 6803 (Ungerer and Pakrasi, 2016; Kaczmarzyk et al., 2018; Behler et al., 2018; Yao et al., 2016), and in filamentous cyanobacterium *Anabaena* sp. PCC 7120 (Higo et al., 2018; Kim et al., 2014). Like other microbes, CRISPR-based cyanobacteria systems also assisted with plasmids (Dicarlo et al., 2013; Shin et al., 2016). Conventional (insertion, deletion, or point) mutation in the cyanobacteria genome relies on a gene of interest replacement with a selective marker through double homologous recombination between a suicidal vector and the host genome. At the same time, CRISPR-based approaches feature cyanobacterial genome engineering. CRISPR-based techniques have several aids than conventional strategies.

1. Unnecessary incorporation of selection markers (selective pressure) based on cell viability is integral to the CRISPR approach. Henceforth, CRISPR-based editing allows the production of markerless knockouts and knock-ins, opening up the opportunities of introduction or deletion of an unlimited genes number (multiplexing).
2. Less time-consuming and reduced labor efforts nullify the selection markers application; hence time-consuming downstream protocols omitted out.
3. With CRISPRi, fine-tuning of metabolic pathways and unwanted side products cellular fluxes dipping are made possible without affecting cell viability. Maybe in the coming years, CRISPR-based tools will significantly benefit cyanobacterial metabolic engineering. Knockouts, knock-ins, and target genes transcriptional activity downregulation

are the strategies adopted in cyanobacteria. However, the other possibilities, such as transcriptional activity upregulation by CRISPRi, and transcriptional editing (mRNA regulation/editing), possibly by Cas13a yet to be attempted (Abudayyeh et al., 2016; East-Seletsky et al., 2016). The target genes by using CRISPR-systems limited to the engineering of metabolic pathways for the production of desired chemical biosynthesis or unwanted metabolites generated from competing pathways. Native pathways derepression, by interruption of regulatory pathway genes, has yet not been explored in cyanobacteria. For example, in *Synechocystis* sp. PCC 6803 production of free fatty acid could be improved by deletion of posttranscriptional regulators such as *IsaR1* (Georg et al., 2017) or *PsrR1* (Georg et al., 2014) or *cyAbrB2* transcriptional regulators (Kawahara et al., 2016). Hence, guiding the native metabolic fluxes toward producing desired products will enhance both the production level and target product efficiency. Furthermore, the cyanobacterial genome encodes many hypothetical proteins, and their role in various metabolic pathways is yet to be explored; gene knockout by CRISPR remains unemployed for gene functional analysis (Singh et al., 2017). It warrants mentioning that, to date, very few cyanobacterial investigations have confirmed the CRISPR multiplexing. The highest number of genes that have been edited in parallel is six (Kaczmarzyk et a., 2018). Meanwhile, the cyanobacteria metabolic optimization for producing a target chemical possibly involves several genetic and regulatory factors. Hence, there is an urgent need for extensive investigation of CRISPR multiplexing in future. A deep understanding of several CRISPR systems and their mechanisms will unquestionably encourage further advancement in cyanobacteria genome editing approaches. Bearing in mind the CRISPR-based methods reliabilities for metabolic engineering, the CRISPR components characterization from various organisms would be an excellent direction for researchers in the upcoming years.

The CRISPR-based technologies' potential for cyanobacterial metabolic engineering cannot be overrated as long as researchers with more extensive options of CRISPR-based tools, each with slightly erratic mechanisms and properties, will assist in driving groundbreaking techniques and approaches for not only cyanobacterial engineering but also other biotechnologically germane organisms. We are leaving here with some outstanding questions; What would be the possibility of genome editing in filamentous forms? At which capacity the CRISPRi and multiplexed CRISPR/Cas could be applied for the metabolic pathway's optimization in cyanobacteria? Is it possible to synthesize nonnative chemicals also by employing CRISPR-based engineering in cyanobacteria? Can we also target the various cyanobacterial genes, including regulators, and which genes would be the most operative for CRISPR-based editing in cyanobacteria? What are the possibilities of off targeting by CRISPR in cyanobacteria? Is there is an additional requirement of bioinformatics tools for CRISPRi-dependent engineering in cyanobacteria? Is it possible to employ CRISPRi for the activation of genes rather than suppression of transcriptional activity? Would it be conceivable to exploit the RNA-based CRISPR/Cas13a for cyanobacterial editing?

Acknowledgments

Financial support by the University Grant Commission, New Delhi, India for UGC start-up research grant.

References

Abudayyeh, O.O., Gootenberg, J.S., Konermann, S., Joung, J., Slaymaker, I.M., Cox, D.B., Shmakov, S., Makarova, K.S., Semenova, E., Minakhin, L., Severinov, K., Regev, A., Lander, E.S., Koonin, E.V., Zhang, F., 2016. C2c2 is a single-component programmable RNA-guided RNA-targeting CRISPR effector. Science 5 (353) (6299), aaf5573. doi:10.1126/science.aaf.

Anders, C., Niewoehner, O., Duerst, A., et al., 2014. Structural basis of PAM-dependent target DNA recognition by the Cas9 endonuclease. Nature 513, 569–573. https://doi.org/10.1038/nature13579.

Behler, J., Sharma, K., Reimann, V., Wilde, A., Urlaub, H., Hess, W.R., 2018a. The host-encoded RNase E endonuclease as the crRNA maturation enzyme in a CRISPR-Cas subtype III-Bv system. Nat. Microbiol. 3 (3), 367–377. doi: 10.1038/s41564-017-0103-5. Epub 2018 Feb 5. PMID: 29403013.

Behler, J., Vijay, D., Hess, W.R., Akhtar, M.K., 2018b. CRISPR-based technologies for metabolic engineering in cyanobacteria. Trends Biotechnol 36 (10), 996–1010. doi:10.1016/j.tibtech.2018.05.011.

Berla, B.M., Saha, R., Immethun, C.M., Maranas, C.D., Moon, T.S., Pakrasi, H.B., 2013. Synthetic biology of cyanobacteria: unique challenges and opportunities. Front. Microbiol 4, 246. doi:10.3389/fmicb.2013.00246 2013 Aug 27.

Cobb, R.E., Wang, Y., Zhao, H., 2015. High-efficiency multiplex genome editing of *Streptomyces* species using an engineered CRISPR/Cas system. ACS Synth. Biol 19, 4 (6), 723–728. doi:10.1021/sb500351f.

Cui, L., Bikard, D., 2016. Consequences of Cas9 cleavage in the chromosome of *Escherichia coli*. Nucleic Acids Res. 44 (9), 4243–4251. https://doi.org/10.1093/nar/gkw223.

Di Cesare, A., Cabello-Yeves, P.J., Chrismas, N.A.M., Sánchez-Baracaldo, P., Salcher, M.M., Callieri, C., 2018. Genome analysis of the freshwater planktonic *Vulcanococcus limneticus* sp. nov. reveals horizontal transfer of nitrogenase operon and alternative pathways of nitrogen utilization. BMC Genom 16 (1), 259. doi:10.1186/s12864-018-4648-3 19.

DiCarlo, J.E., Norville, J.E., Mali, P., Rios, X., Aach, J., Church, G.M., 2013. Genome engineering in *Saccharomyces cerevisiae* using CRISPR-Cas systems. Nucleic Acids Res 41 (7), 4336–4343. doi:10.1093/nar/gkt135.

East-Seletsky, A., O'Connell, M.R., Knight, S.C., Burstein, D., Cate, J.H., Tjian, R., Doudna, JA, 2016. Two distinct RNase activities of CRISPR-C2c2 enable guide-RNA processing and RNA detection. Nature 538 (7624), 270–273. doi:10.1038/nature19802.

Fonfara, I., Richter, H., Bratovič, M., et al., 2016. The CRISPR-associated DNA-cleaving enzyme Cpf1 also processes precursor CRISPR RNA. Nature 532, 517–521. https://doi.org/10.1038/nature17945.

Fujisawa, T., Narikawa, R., Maeda, S., Watanabe, S., Kanesaki, Y., Kobayashi, K., et al., 2017. CyanoBase: a large-scale update on its 20th anniversary. Nucleic Acids Res 45, D551–D554. https://doi.org/10.1093/nar/gkw1131.

Gaj, T., Gersbach, C.A., Barbas 3rd., C.F., 2013. ZFN, TALEN, and CRISPR/Cas-based methods for genome engineering. Trends. Biotechnol 31 (7), 397–405. doi:10.1016/j.tibtech.2013.04.004.

Georg, J., Kostova, G., Vuorijoki, L., Schön, V., Kadowaki, T., Huokko, T., Baumgartner, D., Müller, M., Klähn, S., Allahverdiyeva, Y., Hihara, Y., Futschik, M.E., Aro, E.M., Hess, WR., 2017. Acclimation of oxygenic photosynthesis to iron starvation is controlled by the sRNA IsaR1. Curr. Biol 27 (10), 1425–1436. doi:10.1016/j.cub.2017.04.010 2017e7.

Georg, J., Dienst, D., Schürgers, N., Wallner, T., Kopp, D., Stazic, D., Kuchmina, E., Klähn, S., Lokstein, H., Hess, W.R., Wilde, A., 2014. The small regulatory RNA SyR1/PsrR1 controls photosynthetic functions in cyanobacteria. Plant cell 26 (9), 3661–3679. https://doi.org/10.1105/tpc.114.129767.

Golden, S.S., Brusslan, J., Haselkorn, R., 1987. Genetic engineering of the cyanobacterial chromosome. Methods. Enzymol 153, 215–231.

Golden, S.S., Sherman, L.A., 1984. Optimal conditions for genetic transformation of the cyanobacterium *Anacystis nidulans* R2. J. Bacteriol. 158, 36–42.

Gordon, G.C., Korosh, T.C., Cameron, J.C., Markley, A.L., Begemann, M.B., Pfleger, B.F., 2016. CRISPR interference as a titratable, trans-acting regulatory tool for metabolic engineering in the cyanobacterium *Synechococcus* sp. strain PCC 7002. Metab. Eng. 38, 170–179.

Haro, M.A., De Lorenzo, V., 2001. Metabolic engineering of bacteria for environmental applications: construction of *Pseudomonas* strains for biodegradation of 2 chlorotoluene. J. Biotechnol. 85, 103–113.

Hendel, A., Fine, E.J., Bao, G., Porteus, M.H., 2015. Quantifying on- and off-target genome editing. Trends Biotechnol 33 (2), 132–140. doi:10.1016/j.tibtech.2014.12.001.

Higo, A., Isu, A., Fukaya, Y., Ehira, S., Hisabori, T., 2018. Application of CRISPR interference for metabolic engineering of the heterocyst-forming multicellular Cyanobacterium *Anabaena* sp. PCC 7120. Plant Cell Physiol 59 (1), 119–127. doi:10.1093/pcp/pcx166.

Hirokawa, Y., Maki, Y., Tatsuke, T., Hanai, T., 2016. Cyanobacterial production of 1,3-propanediol directly from carbon dioxide using a synthetic metabolic pathway. Metab. Eng 34, 97–103. doi:10.1016/j.ymben.2015.12.008.

Hou, S., Brenes-Álvarez, M., Reimann, V., Alkhnbashi, O.S., Backofen, R., Muro-Pastor, A.M., Hess, W.R., Apr, 2019. CRISPR-Cas systems in multicellular cyanobacteria. RNA Biol. 16 (4), 518–529. doi:10.1080/15476286.2018.1493330. Epub 2018 Aug 15. PMID: 29995583; PMCID: PMC6546389.

Hu, Q., Sommerfeld, M., Jarvis, E., Ghirardi, M., Posewitz, M., Seibert, M., Darzins, A., 2008. Microalgal triacylglycerols as feedstocks for biofuel production: perspectives and advances. Plant. J 54, 621–639.

Huang, C.H., Shen, C.R., Li, H., Sung, L.Y., Wu, M.Y., Hu, Y.C., 2016. CRISPR interference (CRISPRi) for gene regulation and succinate production in cyanobacterium *S. elongatus* PCC 7942. Microb. Cell. Fact. 15, 196.

Jiang, W., Bikard, D., Cox, D., Zhang, F., Marraffini, L.A., 2013. RNA-guided editing of bacterial genomes using CRISPR-Cas systems. Nat. Biotechnol 31 (3), 233–239. doi:10.1038/nbt.2508.

Joung, J.K., Sander, J.D., 2013. TALENs: a widely applicable technology for targeted genome editing. Nat. Rev. Mol. Cell. Biol 14 (1), 49–55. doi:10.1038/nrm3486 Epub 2012 Nov 21.

Kaczmarzyk, D., Cengic, I., Yao, L., Hudson, E.P., 2018. Diversion of the long-chain acyl-ACP pool in Synechocystis to fatty alcohols through CRISPRi repression of the essential phosphate acyltransferase PlsX. Metab. Eng. 45, 59–66.

Kanno, M., Carroll, A., Atsumi, S., 2017. Global metabolic rewiring for improved CO_2 fixation and chemical production in cyanobacteria. Nat. Commun. 8, 14724. https://doi.org/10.1038/ncomms14724.

Kawahara, A., Sato, Y., Saito, Y., Kaneko, Y., Takimura, Y., Hagihara, H., Hihara, Y., 2016. Free fatty acid production in the cyanobacterium *Synechocystis* sp. PCC 6803 is enhanced by deletion of the cyAbrB2 transcriptional regulator. J. Biotechnol 20 (220), 1–11. doi:10.1016/j.jbiotec.2015.12.035.

Keasling, J.D., 2010. Manufacturing molecules through metabolic engineering. Science 330, 1355–1358.

Khanna, N., Lindblad, P., 2015. Cyanobacterial hydrogenases and hydrogen metabolism revisited: recent progress and future prospects. Int. J. Mol. Sci 16 (5), 10537–10561. https://doi.org/10.3390/ijms160510537.

Kim, S., Kim, D., Cho, S.W., Kim, J., Kim, J.S., Jun 2014. Highly efficient RNA-guided genome editing in human cells via delivery of purified Cas9 ribonucleoproteins. Genome Res. 24 (6), 1012–1019. doi: 10.1101/gr.171322.113. Epub 2014 Apr 2. PMID: 24696461; PMCID: PMC4032847.

Kirst, H., Formighieri, C., Melis, A., 2014. Maximizing photosynthetic efficiency and culture productivity in cyanobacteria upon minimizing the phycobilisome light-harvesting antenna size. Biochim. Biophys. Acta 1837 (10), 1653–1664. doi:10.10.16/j.bbabio.2014.07.009.

Kirtania, P., Hódi, B., Mallick, I., et al., 2019. A single plasmid based CRISPR interference in Synechocystis 6803 - A proof of concept. PLoS One 14 (11), e0225375. Published 2019 Nov 26. doi:10.1371/journal.pone.0225375.

Kiyota, H., Okuda, Y., Ito, M., Hirai, M.Y., Ikeuchi, M., 2014. Engineering of cyanobacteria for the photosynthetic production of limonene from CO_2. J. Biotechnol 185, 1–7. doi:10.1016/j.jbiotec.2014.05.025.

Kleinstiver, B.P., Prew, M.S., Tsai, S.Q., Topkar, V.V., Nguyen, N.T., Zheng, Z., Gonzales, A.P., Li, Z., Peterson, R.T., Yeh, J.R., Aryee, M.J., Joung, J.K., 2015. Engineered CRISPR-Cas9 nucleases with altered PAM specificities. Nature 23 (7561), 481–485. doi:10.1038/nature14592 523.

Kumar, K., R.A., Mella-Herrera, Golden, J.W., 2010. Cyanobacterial heterocysts. Cold Spring Harb. Perspect. Biol 2 (4), a000315. doi:10.1101/cshperspect.a000315.

Larson, M.H., Gilbert, L.A., Wang, X., Lim, W.A., Weissman, J.S., Qi, L.S., 2013. CRISPR interference (CRISPRi) for sequence-specific control of gene expression. Nat. Protoc 8 (11), 2180–2196. doi:10.1038/nprot.2013.132.

Li, H., Shen, C.R., Huang, C.H., Sung, L.Y., Wu, M.Y., Hu, Y.C., 2016. CRISPR-Cas9 for the genome engineering of cyanobacteria and succinate production. Metab. Eng. 38, 293–302.

Libis, V., et al., 2016. Expanding biosensing abilities through computer-aided design of metabolic pathways. ACS Synth. Biol. 5, 1076–1085.

Liu, X., Sheng, J., Curtiss, R., 2011. Fatty acid production in genetically modified cyanobacteria. Proc. Natl. Acad. Sci USA 108 (17), 6899–6904. doi:10.1073/pnas.1103014108.

Liu, D., Johnson, V.M., Pakrasi, H.B., Jun 19, 2020. A Reversibly Induced CRISPRi System Targeting Photosystem II in the Cyanobacterium *Synechocystis* sp. PCC 6803. ACS Synth. Biol. 9 (6), 1441–1449. doi:10.1021/acssynbio.0c00106. Epub 2020 May 21. PMID: 32379958.

Mueller, T.J., Welsh, E.A., Pakrasi, H.B., Maranas, C.D., 2016. Identifying regulatory changes to facilitate nitrogen fixation in the nondiazotroph *Synechocystis* sp. PCC 6803. ACS Synth. Biol 18, 5 (3), 250–258. doi:10.1021/acssynbio.0c00106.

Niu, T.C., Lin, G.M., Xie, L.R., Wang, Z.Q., Xing, W.Y., Zhang, J.Y., Zhang, C.C., 2019. Expanding the potential of CRISPR-Cpf1-based genome editing technology in the cyanobacterium Anabaena PCC 7120. ACS Synth. Biol. 8, 170–180. doi:10.1021/acssynbio.8b00437.

Nogales, J., Gudmundsson, S., Knight, E.M., Palsson, B.O., Thiele, I., 2012. Detailing the optimality of photosynthesis in cyanobacteria through systems biology analysis. Proc. Natl. Acad. Sci. USA 109 (7), 2678–2683. doi:10.1073/pnas.1117907109.

Porter, R.D., 1986. Transformation in cyanobacteria. Crit. Rev. Microbiol. 1986 (13), 111–132.

Qi, L.S., Larson, M.H., Gilbert, L.A., Doudna, J.A., Weissman, J.S., Arkin, A.P., Lim, W.A., 2013. Repurposing CRISPR as an RNA-guided platform for sequence-specific control of gene expression. Cell 28 152, (5), 1173–1183. doi:10.1016/j.cell.2013.02.022.

Rath, D., Amlinger, L., Rath, A., Lundgren, M., 2015. The CRISPR-Cas immune system: biology, mechanisms and applications. Biochimie. 117, 119–128. doi:10.1016/j.biochi.2015.03.025.

Rippka, R., Waterbury, J., Cohen-Bazire, G., 1974. A cyanobacterium which lacks thylakoids. Arch. Microbiol. 100, 419–436. https://doi.org/10.1007/BF00446333.

Shen, B., Zhang, W., Zhang, J., Zhou, J., Wang, J., Chen, L., Wang, L., Hodgkins, A., Iyer, V., Huang, X., Skarnes, W.C., 2014 Apr. Efficient genome modification by CRISPR-Cas9 nickase with minimal off-target effects. Nat. Methods 11 (4), 399–402. doi:10.1038/nmeth.2857.

Shih, P.M., 2015. Cyanobacterial evolution: fresh insight into ancient questions. Curr. Biol. 25, R192–R193. doi:10.1016/j.cub.2014.12.046.

Shin, S.E., Lim, J.M., Koh, H.G., Kim, E.K., Kang, N.K., Jeon, S., Kwon, S., Shin, W.S., Lee, B., Hwangbo, K., Kim, J., Ye, S.H., Yun, J.Y., Seo, H., Oh, H.M., Kim, K.J., Kim, J.S., Jeong, W.J., Chang, Y.K., Jeong, B.R., 2016. CRISPR/Cas9-induced knockout and knock-in mutations in *Chlamydomonas reinhardtii*. Sci. Rep 13 (6), 27810. doi:10.1038/srep27810.

Silva, G., Poirot, L., Galetto, R., Smith, J., Montoya, G., Duchateau, P., Pâques, F., 2011. Meganucleases and other tools for targeted genome engineering: perspectives and challenges for gene therapy. Curr. Gene Ther 11 (1), 11–27. doi:10.2174/156652311794520111.

Singh, P.K., Shrivastava, A.K., Singh, S., Rai, R., Chatterjee, A., Rai, L.C., 2017. Alr2954 of Anabaena sp. PCC 7120 with ADP-ribose pyrophosphatase activity bestows abiotic stress tolerance in *Escherichia coli*. Funct. Integr. Genom 17 (1), 39–52. doi:10.1007/s10142-016-0531-y.

Singh, P.K., Wang, W., Shrivastava, A.K., 2018. Cadmium-mediated morphological, biochemical and physiological tuning in three different *Anabaena* species. Aquat. Toxicol 202, 36–45. doi:10.1016/j.aquatox.2018.06.011.

Singh S., Shrivastava A.K., Kumar A. Singh V.K., Yadav D., et al. (2020). Cyanobacterial genome editing toolboxes: recent advancement and future projections for basic and synthetic biology researches. In: P.K. Singh, A. Kumar, V.K. Singh, and A.K. Shrivastava (Eds.). Advances in Cyanobacterial Biology. Elsevier, Amsterdam, pp. 129–149. doi.org/10.1016/B978-0-12-819311-2.00009-7.

Ungerer, J., Pakrasi, H.B., 2016. Cpf1 is a versatile tool for CRISPR genome editing across diverse species of cyanobacteria. Sci. Rep. 6, 39681.

van der Woude, A.D., Perez Gallego, R., Vreugdenhil, A., Puthan Veetil, V., Chroumpi, T., Hellingwerf, K.J., 2016. Genetic engineering of *Synechocystis* PCC6803 for the photoautotrophic production of the sweetener erythritol. Microb. Cell. Fact 8 (15), 60. doi:10.1186/s12934-016-0458-y.

Veetil, V.P., Angermayr, S.A., Hellingwerf, K.J., 2017. Ethylene production with engineered *Synechocystis* sp PCC 6803 strains. Microb. Cell. Fact 16 (1), 34. doi:10.1186/s12934-017-0645-5.

Vermaas, W., 1996. Molecular genetics of the cyanobacterium *Synechocystis* sp. PCC 6803: principles and possible biotechnology applications. J. Appl. Phycol. 8, 263–273.

Wang, Y., Wang, D., Wang, X., Tao, H., Feng, E., Zhu, L., Pan, C., Wang, B., Liu, C., Liu, X., Wang, H., 2019. Highly efficient genome engineering in *Bacillus anthracis* and *Bacillus cereus* using the CRISPR/Cas9 System. Front. Microbiol. 10 (1932). https://doi.org/10.3389/fmicb.2019.01932.

Watanabe, S., Ohbayashi, R., Kanesaki, Y., Saito, N., Chibazakura, T., Soga, T., Yoshikawa, H., 2015. Intensive DNA replication and metabolism during the lag phase in cyanobacteria. PLoS. One 10, e0136800.

Watanabe, S., 2020. Cyanobacterial multi-copy chromosomes and their replication. Biosci. Biotechnol. Biochem 84 (7), 1309–1321. doi:10.1080/09168451.2020.1736983.

Weiss, T.L., Young, E.J., Ducat, D.C., 2017. A synthetic, light-driven consortium of cyanobacteria and heterotrophic bacteria enables stable polyhydroxybutyrate production. Metab. Eng 44, 236–245. doi:10.1016/j.ymben.2017.10.009.

Wendt, K.E., Ungerer, J., Cobb, R.E., Zhao, H., Pakrasi, H.B., 2016. CRISPR/Cas9 mediated targeted mutagenesis of the fast growing cyanobacterium *Synechococcus elongatus* UTEX 2973. Microb. Cell. Fact. 15, 115.

Wenjing, W., Chen, Q., Singh, P.K., Huang, Y., Pei, D., 2020. CRISPR/Cas9 edited HSFA6a and HSFA6b of *Arabidopsis thaliana* offers ABA and osmotic stress insensitivity by modulation of ROS homeostasis. Plant Signal. Behav 15 (12), 1816321. doi:10.1080/15592324.2020.1816321.

Yamano, T., Zetsche, B., Ishitani, R., Zhang, F., Nishimasu, H., Nureki, O., 2017. Structural basis for the canonical and non-canonical PAM recognition by CRISPR-Cpf1. Mol. Cell 2017 Aug 17, 67 (4), 633–645.e3. doi:10.1016/j.molcel.2017.06.035.

Yao, L., Cengic, I., Anfelt, J., Hudson, E.P., 2016. Multiple gene repression in cyanobacteria using CRISPRi. ACS Synth. Biol. 5, 207–212.

Yao, L., Shabestary, K., Björk, S.M., et al., 2020. Pooled CRISPRi screening of the cyanobacterium *Synechocystis* sp PCC 6803 for enhanced industrial phenotypes. Nat. Commun. 11, 1666. https://doi.org/10.1038/s41467-020-15491-7.

Yoshikawa, K., Toya, Y., Shimizu, H., 2017. Metabolic engineering of *Synechocystis* sp. PCC 6803 for enhanced ethanol production based on flux balance analysis. Bioprocess. Biosyst. Eng 40 (5), 791–796. doi:10.1007/s00449-017-1744-8.

Zerulla, K., Ludt, K., Soppa, J., 2016. The ploidy level of *Synechocystis* sp. PCC 6803 is highly variable and is influenced by growth phase and by chemical and physical external parameters. Microbiology 162, 730–739.

CHAPTER

10

Cyanobacteria and salinity stress tolerance

Rahul Prasad Singh, Priya Yadav, Reena Kujur, Kapil Deo Pandey, Rajan Kumar Gupta

Department of Botany, Institute of Science, Banaras Hindu University, Varanasi, Uttar Pradesh, India

10.1 Introduction

Cyanobacteria are primitive oxygenic, photosynthetic prokaryotes distributed generally through all liable environments on the earth (Singh et al., 2015). Cyanobacteria are also called blue-green algae (BGA) due to phycocyanin pigment and are major photosynthetic microorganisms in several habitats, including bare soil rock, freshwater, and ocean (Whittonet al., 2012). Cyanobacteria were recorded as the oldest known fossil, and it is earliest life forms originated very close to 3.5 billion years ago during oxygen evolution on the earth (Kasting and Siefert, 2002). They are found in different forms such as individual cells, colonies, filaments, branched and unbranched, crusts, and blooms (Catherineet al., 2013). Cyanobacteria are generally distributed in all habitats ranging from terrestrial to aquatic environments and springs, hypersaline water, salt pans, and extreme habitats such as deserts hot thermal vents (Whitton and Potts, 2000). They have evolved specific mechanisms for proliferation under different environmental stress conditions. During stress condition, normal biochemical pathways are altered and resulting in changed physiology that provides a reasonable boundary for living in the changing environment (Dembitsky and Rezenka, 2005). The primary reason for cyanobacteria's evolutionary success is the presence of various metabolic pathways in their cell. The photosynthesis and respiration in some cyanobacterial groups occur simultaneously within a single compartment.

Cyanobacterial Lifestyle and its Applications in Biotechnology.
DOI: https://doi.org/10.1016/B978-0-323-90634-0.00003-2

However, several cyanobacterial species can fix atmospheric nitrogen, and they can exist in a wide variety of ecological conditions (Vermaas, 2001).

The high salinity refers to the total constituents of dissolved ions in a system. It is studied that some specific ions exert nonstop toxic effects on cyanobacterial cells, which can decrease the occurrence of some sensitive species from specific environments (Whitton and Potts, 2000). The unicellular cyanobacterial cell keeps a constant cytosolic osmotic and ionic content to bring water uptake by osmosis that generates a turgor pressure that is compulsory for cell growth and enlargement. Hence, cyanobacterial cellular metabolism and their survival challenge change in the internal salt and water content. High salinity favored the presence of heterocystous cyanobacterial cell growth, whereas high salinity content supports the development of nonheterocystous cyanobacterial species. Two problems arise in high cytosolic salt concentrations; one is lower water availability, and the other is high ions content. Salinity is usually inhibiting all processes of cyanobacteria, including photosynthesis, growth, osmotic pressure of cell, and cellular homeostasis (Josetet al., 1996; Singh et al., 2002), resulting in the generation of reactive oxygen species (ROS) (Srivastavaet al., 2005) that leads to change in gene expression and protein synthesis (Srivastava et al., 2008) ultimately causing to programmed cell death (Ning et al., 2002). Immediate effect of high salinity is ionic, and osmotic stress and oxidative stress are the secondary effects (Zhu, 2001). Many cyanobacteria have salt-out-mechanism to acclimate against high salt concentrations. Cyanobacteria have survived successfully in the usar saline soil by the reclamation of saline soil during salt stress because they have salt tolerance mechanisms (Borowitzka, 1986). Salinity also induces some specific salt stress proteins (Hagemann et al., 1990) and increases cell chlorophyll content (Ferreira and Shaw, 1989). Cyanobacteria are highly specialized prokaryotic organisms that can proliferate in high salt concentration due to entire metabolic modification during evolution according to concentrations of salt (Müller and Oren, 2003).

10.2 Distribution of cyanobacteria in the saline ecosystem

Both unicellular and multicellular cyanobacteria evolved during the Precambrian era (about 3.5 billion years ago) and are considered major colonizers during the evolution of aerobic life. Due to their evolutionary history, it ranges from saline soils to the coastal swamps (Amsaveni, 1995; Komarek, 1998; Hoffmann, 1989). Several workers have described cyanobacteria in alkaline soil (Singh, 1950, 1961; Ali and Sandhu, 1972). Cyanobacteria such as *Aphanothece*, *Anabaena*, and *Synechocystis* have been classified into the following three groups based on salt tolerance capacity: salt-sensitive, fairly halotolerant, and extreme halotolerant (Pandhal et al.,

2008). The appearance of cyanobacteria in saline soil depends on environmental conditions and is indicated by their dominant strains in such soil types (Whitton and Potts, 2000; Pandey et al., 2005). Some heterocystous cyanobacteria such as *Aulosira* sp., *Scytonema* sp., *Anabaena* sp., and *Hapalosiphon* sp. as well as non-heterocystous cyanobacteria such as *Phormidium* sp., *Lyngbya* sp., *Oscillatoria* sp., and *Synechocystis* sp. in saline habitat was shown in (Fig.10.1).

The heterocystous nitrogen-fixing cyanobacterium *Nodularia spumigena* lives in the less saline southern basin (Roney et al., 2009). Cyanobacteria such as heterocystous are colonized everywhere with salt concentrations exceeding up to 10%. A filamentous cyanobacterium, *Oscillatoria limnetica*, isolated from Solar Lake nowadays is renamed as *Leptolyngbya hypolimnetica* (Anagnostidis, 2001). Some cyanobacteria are found in brine pools with 10% salt concentration and in the dry salt crusts of Great Salt Plains (Oklahoma). In this location, with its regularly changing salinities, the *Geiterinema/Phormidium* species were primarily encountered. The isolated species *Geitlerinema* can acclimatize in a wide range of NaCl concentration from 1% to 15%. However, *Cyanodictyon* sp. isolate was identified that acclimatize high salt ranges from 5% to 15 %. Some other cyanobacteria were also isolated, such as *Aphanothece* sp., *Chlorogloeopsis* sp., and *Nodularia* sp. Some isolates tolerated up to 15% salt concentration; however, most of them did not tolerate concentrations close to 5% (Kirkwood et al.,2008). The heterocystous cyanobacteria *Scytonema* sp. were inhabited in San Salvador Island with salt concentrations increasing up to 11% in hypersaline lagoonal mats (Paerl et al., 2000). Photosynthetic activity and growth of some cyanobacterium such as *Lyngbya* and *Coleofasciculus* increase by hurricane rain floods after dilution of the mat (Paerl et al., 2003). In the Baja California lagoons named Guerrero Negro, the highest salt production in a cyanobacterial mat has been studied in depth (Lee et al., 2014; Nubel et al., 1999). The carotenoid content of *Coleofasciculus* increases with the increase in salinity from 9% to 12% during oxidative stress. The proliferation of *Aphanothece* sp. increases with increasing concentration of salinity (Bebout et al., 2002). Different types of colorful layers of many cyanobacteria may have occurred on upper crusts of gypsum and Guerrero Negro area. Hence, these cyanobacteria may proliferate and remain active for several months in the mineral deposits (Rothschild et al., 1994) (Table 10.1).

Another attractive lake is Mono Lake, an alkaline lake of California whose pH ranges from 9.7 to 10. Cyanobacterium such as *Prochlorococcus* and *Synechococcus* associated with marine ecosystem forms blooms in the lake during late summer. These cyanobacterial species are growing in up to 10% salt, and their cells are filled with phycoerythrin; however, the absence of phycourobilin therefore does not explain complementary chromatic formation (Budinoff and Hollibaugh, 2007). The coastal salterns are

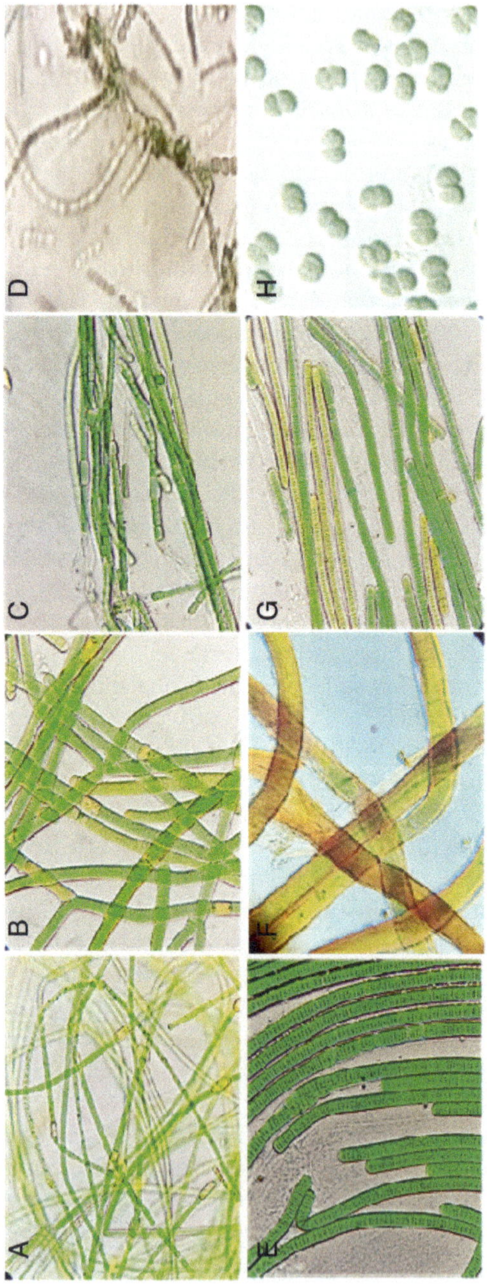

FIG. 10.1 Distribution of some heterocystous (A) *Aulosira* sp., (B) *Scytonema* sp., (C) *Anabaena* sp., and (D) *Hapalosiphon* sp. and nonheterocystous (E) *Phormidium* sp., (F) *Lyngbya* sp., (G) *Oscillatoria* sp., and (H) *Synechocystis* sp. cyanobacteria in saline habitat.

TABLE 10.1 Cyanobacterial species occur in the different hypersaline lakes of the world. Oren, A., 2015. Cyanobacteria in hypersaline environments: biodiversity and physiological properties. Biodiversity and Conservation, 24 (4), 781-798.

Cyanobacterial Species with salt concentration	Bahamas, San Salvador Island	Washington, Hot Lake	California, Solar Salterns worldwide	Utah, Great Salt Lake	Oklahoma, Great Salt Plains	California, Mono Lake
Low (5–15%) salt						
Unicellular						
Chroococcidiopsis	–	–	–	–	–	–
Synechococcus	–	–	+	–	–	+
Pleurocapsa	–	–	–	–	–	–
Gloeothece/Gloeocapsa	–	–	–	–	–	–
Aphanothece/Halothece/Euhalothece	–	–	–	+	+	–
Myxobactron/7Dactylococcopsis	–	–	–	–	–	–
Filamentous Heterocystous	–	–	–	–	–	–
Nodularia	–	–	–	–	–	–
Filamentous nonheterocystous	–	–	–	–	–	–
Coleofasciculus	+	–	–	+	–	–
Halomicronema	–	–	–	–	–	–
Halospirulina/Spirulina	–	–	–	–	–	–
Plectonema	–		–	–	–	–
Leptolyngbya	–	+	–	–	–	–
Lyngbya	+	–	–	–	–	–

(continued)

TABLE 10.1 (Cont'd)

Cyanobacterial Species with salt concentration	Bahamas, San Salvador Island	Washington, Hot Lake	California, Solar Salterns worldwide	Utah, Great Salt Lake	Oklahoma, Great Salt Plains	California, Mono Lake
Oscillatoria	+	+	–	+	–	–
Phormidium/Geitlerinema	–	–	–	+	+	–
Johannes baptistia	+	–	–	–	–	–
Moderate (15–22%) salt						
Unicellular						
Synechococcus	–	–	+	–	–	–
Aphanothece/Halothece/ Euhalothece	–	–	+	+	+	–
Filamentous, nonheterocystous						
Coleofasciculus	–	–	–	–	–	–
Leptolyngbya	–	+	+	–	–	–
Plectonema	–	+	–	–	–	–
Oscillatoria	–	+	–	–	–	–
Halospirulina/Spirulina	–	–	+	–	–	–
Phormidium/Geitlerinema	–	–	+	–	+	–
High (>22%) salt						
Unicellular						
Aphanothece/Halothece/Euhalothece	–	–	–	+	–	–

providing large habitats for cyanobacteria proliferation at high salt concentrations. Benthic cyanobacterial mats form up to 20–22 salinity and, from time to time, higher. Hence, the formation of the cyanobacterial mat is considered beneficial for the salt-making procedure.

10.3 Sensing salinity by cyanobacterial cell

The direct effect in the cytosolic ion content of the cyanobacterial cell and associated changes in existing water is due to change in environmental condition and this may lead to the direct sensing and signaling of salt stress which results in the variation in the macromolecules and influencing protein-protein and lipid-protein interactions (Wood, 2011). Inorganic ions such as KCl/NaCl activate many transport system and enzymes such as the K^+ transport system and glycosyl glycerol (GG) synthesizing enzyme. Most enzymes remain in their inactive state inside the cell at low salt concentration due to high binding affinity to DNA and sometimes RNA. This type of binding occurs in a sequence-independent manner and may be accessible by incorporating inorganic ions such as KCl/NaCl. The quantity of salt concentration can directly induce the gene expression mechanism of cyanobacterium. It has been known that the expression of the gene *glucosyl glycerol-phosphate synthase* (*ggpS)* is transformed in the sigF mutant cyanobacteria *Synechocystis* sp.PCC 6803 causing lower salinity tolerance. In salinity-acclimated cyanobacterial cells, the concentration of ions is somewhat released, and a low quantity of nucleic acids is increased combined with the increased levels of GgpS protein (Proteau, 1993; Rai, 1980). The low amount of GgpS protein is sufficient enough to form glucosyl glycerol for cell growth (Jager et al., 1997).

In the cyanobacterial cells, sensing and signaling salinity is completed by the signal transduction system with a two-component signaling system consisting of a histidine protein kinase (HIK) with a response-regulator protein (RRE). They play a crucial role in sensing and transducing salinity stress from the changing environmental condition—the presence of the entire genome sequence of *Synechocystis* sp. PCC 6803 (Kaneko et al., 1996) has identified many potential transcriptional sensors and transducers during various environmental signals in *Synechocystis* sp. PCC 6803(Suzuki et al., 2000; Paithoonrangsarid et al., 2004; Shoumskaya et al., 2005).

The positive regulation and negative regulation during salinity stress-inducible gene expression involves Hik/Rre systems (Fig. 10.2). Under positive regulation (Fig. 10.2A), the Hik system is inactive with nonstress cells and the resultant inactivation of the corresponding Rre system. The phosphorylation process activates the Hik, and it transfers the various signal to the associated Rre system, enhancing the expression of several genes in salt-stressed cells. This type of regulation occurs in *Synechocystis* sp. during stress-inducible gene expression (Murata et al., 2006). Under

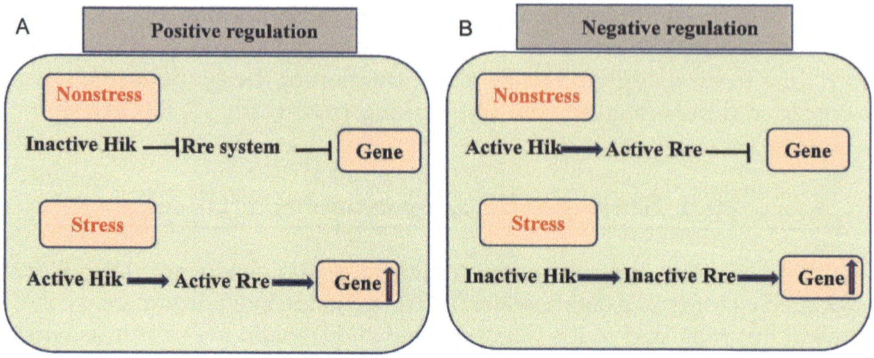

FIG. 10.2 Schematic representation salinity-stress-inducible gene expression under positive (A) and negative (B) regulation of Hik and Rre systems (Modified from Los et al., 2008).

negative control, the Hik and Rre systems are active during nonstress conditions, resulting in suppressing many genes under nonstress conditions. In stressed conditions, the Hik and Rre become inactive, resulting in all formerly repressed genes (Fig. 10.2B).

In *Synechocystis* sp. PCC 6803 there are five two-component signaling systems concerned with the activation of various salinity-induced genes such as HIK2-RRE1, HIK10-RRE3, HIK34 RRE1, HIK16-HIK41-RRE17 and HIK33-RRE31 (Shoumskaya, 2005). The HIK33 is associated with some cyanobacterial cell's thylakoid membrane, which was the recognized as a salt-stress sensor. A mutant of *Nostoc* sp. PCC 7120 faulty in the *rre* gene was osmotic stress sensitive (Schwartz et al., 1998). This Rre also called OrrA controls the salt-stimulated sucrose biosynthesis. Mutated *orrA* accumulates less sucrose, while the *orrA* overexpression induces a higher level of sucrose accumulation (Ehira et al., 2014).

10.3.1 Salinity sensing by SOS pathway

Cyanobacterial cells reduce the effects of high salinity concentration due to the evolved sensing mechanism called salt overly sensitive (SOS) pathway (Fig. 10.3). The plasma membrane receptors sense the extracellular signals from the surroundings and activate a signaling cascade domain. This signaling cascade domain consists of three genes that are SOS1, SOS2, and SOS3. The SOS1 acts as a plasma membrane H^+/Na^+ antiporters, SOS2 acts as protein kinases, and SOS3 acts as a Ca^{2+} sensor. Cyanobacterial plasma membrane regulates the extracellular and intracellular ion homeostasis, and it also plays an essential role in the sensing of high salinity and acclimates against adverse environmental conditions. In *Synechocystis*, changes in the plasma membrane permeability were well known during adverse environmental conditions, leading to

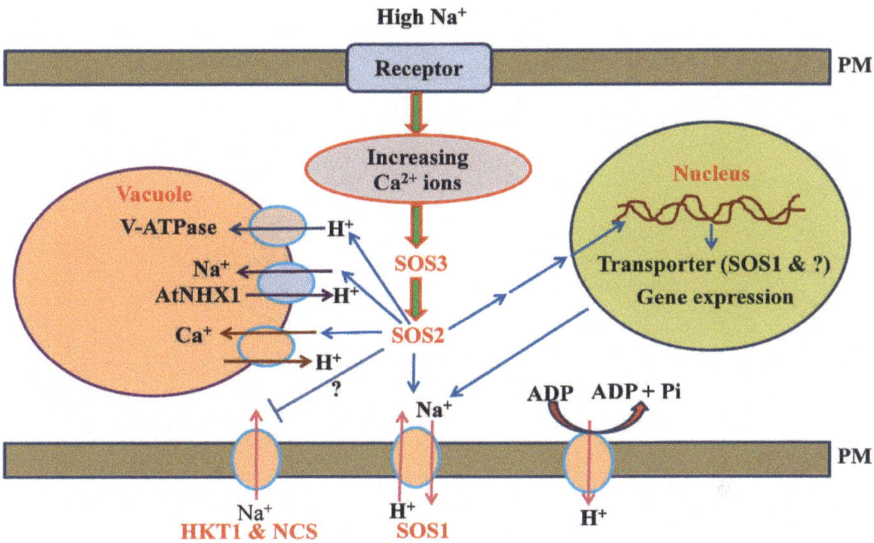

FIG. 10.3 An overview of salt overly sensitive (SOS) pathway.

the proportion of saturated fatty acids and increasing the membrane protein (Huang et al., 2006).

The cyanobacterial plasma membrane sensors sense high salt stress. After that, the cytosolic Ca^{2+} gets disrupted. These are recognized distinctively by the Ca^{2+}-sensing proteins sensed by SOS3 and SOS3 interacting with SOS2. SOS2 activates various antiporters present on the plasma membrane, such as H+/Na+ antiporters, resulting in the efflux of the excess quantity of Na^+ from the cell and helps in their ion homeostasis. SOS2 and SOS3 complex leads to the downregulation of AtHKT1 activity under high salinity stress (Mahajan and Tuteja, 2005). SOS2 also results in ion compartmentalization by influencing the Na^+/H^+ exchange activity.

10.4 Physiological and biochemical responses

Some common physiological and biochemical responses to salinity are addressed next and shown in Fig. 10.4.

10.4.1 Photosynthesis

Generally, in photoautotrophic prokaryotic organisms, salinity stress drastically reduces photosynthetic activity (Marin, 2004 & Allakhverdiev et al., 2000). It has been known that there is a decrease in the quantity of chlorophyll and phycocyanin. It has been studied already that a unicellular freshwater cyanobacterium *Synechocystis* sp. PCC 6803 may tolerate NaCl

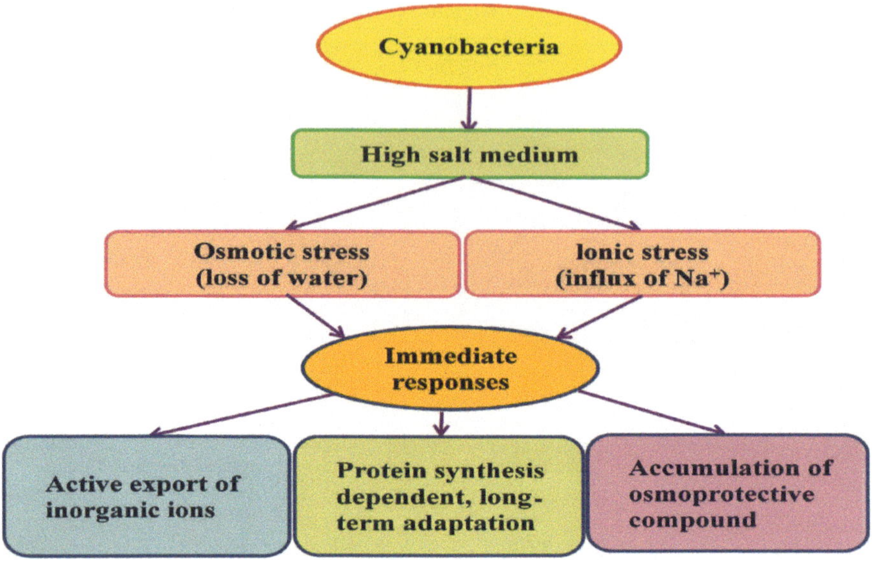

FIG. 10.4 An overview of physiological processes operating in response to salinity in the cell of cyanobacterial species.

concentration up to 1.2 M (Desplats et al., 2005). In cyanobacteria, photosynthesis and respiratory electron transport occur in the thylakoid membrane (Mullineaux, 2014). In cyanobacterial cells, a pigment known as phycobilin is attached with the protein components to form phycobiliproteins (PBPs). There are three major types of PBPs found in cyanobacterial cells: phycoerythrin, phycocyanin, and allophycocyanin. PBPs are served generally as a primary light harvesting antenna for photosystem II (PSII) attached to the thylakoid membranes toward the stromal side. The high salinity decreases the amount of phycocyanin protein in *Synechocystis* sp. resulting in the reduction in the transfer of energy from phycocyanin protein to PSII, which further leads to reduction in the photosynthesis process. High salinity stress also interrupts the transfer of the energy process from PBPs to the (PSII) reaction center. The amount of carotenoid increases, whereas β-carotene remains unchanged (Schubert and Hagemann, 1990; Schubert et al., 1993; Lu and Vonshak, 2002). Increasing the concentration of Na^+ in the cyanobacterial cell generally dissociates two proteins called PsbU and cytochrome C550 from the PSII complex through plastocyanin which decreases the rate of light-induced formation of oxygen (Allakhverdiev et al., 1999). Salt stress also reduces the efficiency of electron transport from Plastoquinone A (QA) to Plastoquinone B (QB), which causes the damage of phycobilisome proteins (Lu and Vonshak, 2002). It was also studied that high salinity in *Synechocystis* cells causes the inhibition of the D1 protein of PSII (Allakhverdiev et al., 2000). The cyanobacterial cell is harmed by high salinity, but it also increases NAD(P)H dehydrogenase function, which is independent of NDH complex during cyclic electron transfer by ferredoxin and can supply extra energy for ion export (Jeanjean et al., 1998).

The cyanobacterial cell shows variations in photosynthetic carbon fixation with different salt-stress tolerance capacity under various salinity conditions. *Aphanothece halophytica,* a halotolerant cyanobacterial species, causes an increase in carbon fixation due to increased NaCl concentration (Takabe et al., 1988); however, no such increment was found in *Aphanothece stagnina,* a freshwater cyanobacterium (Takabe et al., 1988). Hence, different cyanobacterial species respond differently to photosynthetic activity under the various concentration of salinity stress.

10.4.2 Plasma membrane

Cyanobacterial plasma membranes play a crucial role in the adaptation to salinity stress. Plasma membrane permeability allows enough water exchange across the membrane during salinity stress. Cyanobacterial cell membrane fatty acid saturation level is a critical component that plays a crucial role to adapt to various salinity stress condition. The increase in the fatty acids saturation level membrane lipids increases the tolerance of *Synechococcus* photosynthetic and H^+/Na^+ antiport systems to high salt stress condition (Allakhverdiev et al., 1999). The activities of various membrane-associated enzymes depend on the fluidity of cell membranes, and high salt stress reduces the permeability to the Na^+ and H^+ channel of the membrane. Many antiporters system activated by fatty acid saturation level stimulates the synthesis of H^+ ATPase(s) and Na^+/H^+ antiporter(s) to cope up against salinity stress. The cyanobacterial plasma membrane antiporter system reduces the Na^+ levels inside the cytosol and protects the photosystem (PSII and PSI) from salinity induces damages and activates its recovery (Allakhverdiev et al., 2001).

The photosynthetic apparatus is disrupted by high ROS levels inside the cell due to their toxic effect. Salt stress stimulates the high level of unsaturated fatty acids and high quantity of phosphatidylglycerol derivatives. In *A. halophytica,* salt stress produces phosphatidylglycerol derivatives approximately 75% of all cell membrane lipids. An elevated amount of polar membrane lipids enhances the cell membrane surface ions that stimulate the variation in the ion transport properties of the cell membrane (Joset et al., 1996). Salinity stress induces the outflow of water by water channels from the cytosol. It causes cell volume reduction, which increases the cytosolic concentration of salts and inactivation of the oxygen-evolving complex (Allakhverdiev et al., 2000b).

10.4.3 Nitrogen fixation

The assimilation of nitrogen occurs by the two active enzymes: nitrate reductase and nitrite reductase, respectively. Nitrate reductase is found in cyanobacteria as an important ferredoxin-dependent enzyme. Nitrate reductase and glutamine synthetase are two major enzymes for salinity stress tolerance and cyanobacterial cell growth. The maximum inhibition of nitrate reductase and glutamine synthetase was shown in *Nostoc calcicole* after 48 and 24 h under 2000 mM NaCl concentration (Singh et al., 2015). The cyanobacterial cells grown in 170 mM NaCl concentration showed an increased rate of nitrate uptake. In *Anabaena torulosa*, uptake of nitrate enhancement played a significant role in salinity tolerance by the Na^+ influx inhibition. During the salinity stress, a small amount of nitrate increases, acting as an effective osmolyte for the affected cell. In *Anabaena doliolum*, a freshwater nitrogen-fixing cyanobacterium has been reported to have a similar mechanism of salt-stress tolerance (Rai and Abraham, 1995). In *Synechocystis* sp. PCC 6803, the increased concentration of osmolarity generated by NaCl can stimulate nitrate uptake (Baebprasert et al., 2011).

It was studied that Na^+ influx into the cytosol is lowest when nitrate concentration is available in the cells. Cyanobacteria are prokaryotic photoautotrophic organisms; however, diazotrophic filamentous cyanobacterial cells possess specialized structure for nitrogen fixation called heterocysts that defend the nitrogenase from the surrounding oxygen. Some unicellular and nonheterocystous filamentous cyanobacteria such as *Gloeothece, Synechococcus* and *Oscillatoria* sp. perform aerobic nitrogen fixation. It was reported that the nitrogenase activity was decreased in some strains of *Anabaena* sp. due to salt stress (Fernandes et al., 1993).

10.4.4 Formation of ROS and antioxidative defense system

In photosynthesis, the electron transport chain under salt stress condition leads to the formation of ROS. Accumulation of several ROS such as hydrogen peroxide, hydroxyl anion, and superoxide and oxygen-free radical causes oxidative damages to the cyanobacterial cells. Salinity stress can lead to various changes in the photosynthetic electron transport rate, which enhance ROS production and inhibit the scavenging of the antioxidative defense system. This antioxidative resistance system consists of both enzymatic and nonenzymatic machinery. The enzymatic antioxidative part contains catalase (CAT), ascorbate peroxidase (APX), glutathione reductase (GR), and superoxide dismutase (SOD); however, the nonenzymatic component includes ascorbate (ASA), carotenoid (CAR), α-tocopherol (α-TOC), and glutathione reduced (GSH) (Neill et al., 2002). These enzymes involve in both scavenging of ROS and participate in repair mechanism. Superoxide anion (O_2^-) was scavenged by SOD and stimulated the formation of H_2O_2 and O_2; later the H_2O_2 was scavenged by CAT and resulted in H_2O and O_2 (Mckersie and Leshem, 1994). It was shown that under high salinity, antioxidant acts against oxidative damage and increased ROS level resulted in a high level of malondialdehyde.

10.5 Accumulation of compatible solutes

Compatible solutes are highly hydrophilic, low molecular weight organic compounds accumulated in more amounts without interrupting any cellular metabolism and other cellular processes (Brown, 1976). The heteroside 2-O-D-glucopyranosylglycerol (glucosyl glycerol) is the first identified cyanobacterial compatible solute from the marine *Synechococcus* RRIMP N100 (Borowitzka et al., 1980). The other compatible solutes have also determined that can be assigned to different substance classes such as disaccharide: sucrose and trehalose, polyols glycerol, and sorbitol; amino acids: proline and glutamate; quaternary ammonium compounds: glycine betaine and glutamate betaine (Reed et al., 1984, 1986; Hagemann and Pade, 2015). These accumulated compatible solutes provide defense against salinity stress. The preference for the accumulation of different specific compatible solutes at different levels of salinity was associated with the protection and stabilization of macromolecules such as enzymes and membrane proteins (Borges et al., 2002; Hincha and Hagemann, 2004). Marine cyanobacteria such as *Crocosphaera* sp. accumulates trehalose (Pade et al., 2012), and *Prochlorococcus* sp. generally accumulate sucrose (Klähn et al., 2010a). However, some species *Synechococcus* and *Prochlorococcus* also produce glycine betaine and negatively charged glucosyl glycerate (Klähn et al., 2010a). *Trichodesmium erythraeum* is a significant nitrogen-fixing cyanobacterium; this uses the currently identified compatible solutes called homoserine betaine (Pade et al., 2016a).

Salt-tolerated cyanobacterial species that accumulate GG and polyols are considered the dominant compatible solutes in the cell. The cyanobacterial strain accumulating disaccharides such as sucrose and trehalose is deemed inferior and can be least tolerant under salt concentration. In contrast, the cyanobacterial strains accumulating glycine betaine can acquire tolerance maximally under salinity stress (Reed et al., 1986). The betaine and glutamate betaine have been found in a single cyanobacterial isolate, *Calothrix* sp. (Mackay et al., 1984). In the cyanobacterial cell, proline is a widely distributed compatible solute, and its concentration increases continuously during salt stress (Riccardi et al., 1983; Reed and Stewart, 1988; Fulda et al., 1999). Another widely distributed compatible solute is the GG which is the first osmo-protective organic compound reported in a cyanobacterial cell. It has been already reported that in many different cyanobacterial genera, the salt-induced increase of GG is the major compatible solute; however, these strains also accumulated a low quantity of other compatible solutes such as sucrose and trehalose during salt stress. Cyanobacteria containing GG as a major compatible solute can defend against salinity at least two times higher than seawater. Hence, GG accumulation is the characteristic feature of cyanobacterial species that occur in marine habitats.

10.5.1 Biosynthesis of compatible solutes

Most of the cyanobacteria favor the *de novo* synthesis of compatible solutes during salt stress. In cyanobacteria, the biosynthetic pathways with genes encoding enzymes are identified for compatible solutes (Fig. 10.5). Two steps are involved in each compatible solute biosynthetic pathway; the first step is the rate-limiting step, which precedes the polyol phosphate that synthesizes a phosphorylated intermediate by sugar nucleotide condensation process; and in the second step, the polyol phosphate synthesized previously is hydrolyzed by phosphatase enzyme to form the osmolyte.

10.5.1.1. Glucosyl glycerate (GG) synthesis

The formation of GG has been investigated in some cyanobacterial strain isolate such as *Synechocystis* sp. (Reed and Stewart, 1985). This solute is formed via two enzymatic reactions one is by GG-phosphate synthase (GgpS) (Marin et al., 1998), and the second is by GG-phosphate phosphatase (GgpP) (Hagemann et al., 1997a). GgpS is a catalyzing enzyme of glucosyltransferase group that catalyzes the formation of GG3-phosphate from glycerol 3-phosphate and ADP-glucose, then GG3-phosphate intermediate is dephosphorylated through GgpP enzyme (Hagemann and Erdmann, 1994). GG synthesis is not restricted to cyanobacterial strains; however, it also possesses several heterotrophic bacteria, and it synthesizes GG concerning osmotic stress (Pocard et al., 1994; Roder et al., 2005).

10.5.2 Glycine betaine synthesis

Glycine betaine is a well-known compatible solute found in many cyanobacterial species during salinity stress (Chen and Murata,2011). In many cyanobacterial species, glycine betaine is formed by a three-step methylation process with choline or glycine as a precursor found in those strains capable of this osmolyte synthesis (Moore, 1987; Laloknam et al., 2006). These three steps reactions are completed by following enzymes such as glycine/sarcosine-*N*-methyltransferase (GSMT) and dimethylglycine-*N*-methyltransferase (DMT). These enzymes catalyze the

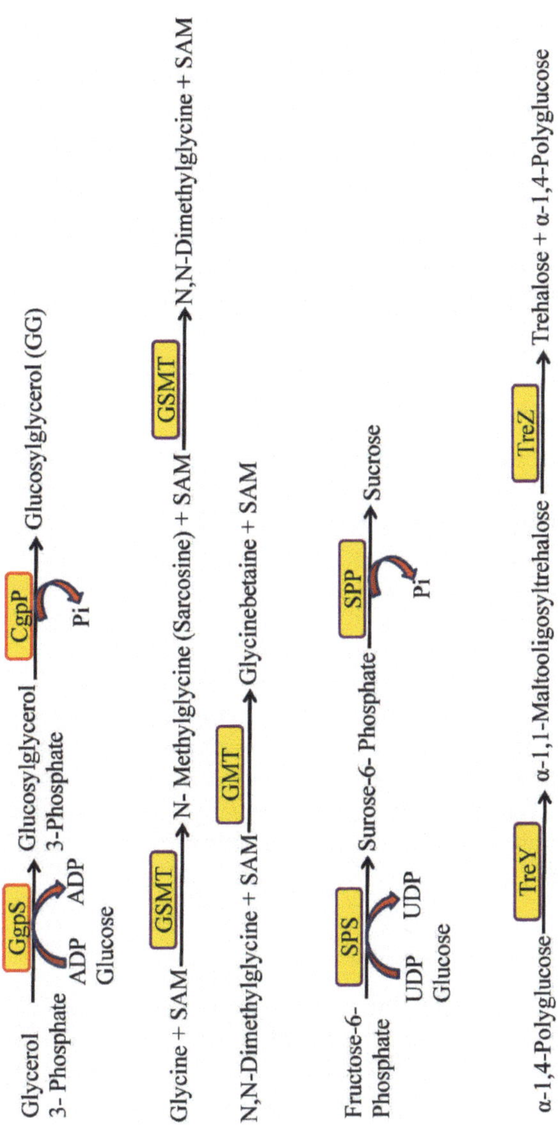

FIG. 10.5 Reactions and enzymes involved in the synthesis of different compatible solutes in cyanobacteria during salt stress.

methylation process of glycine synthesis with methyl-donor S-adenosyl-methionine.

All glycine betaine–accumulating cyanobacteria use direct methylation of glycine during salt stress, and its activity has been confirmed in some hypersaline cyanobacterial model such as *Aphanothece* sp. and *Synechococcus* sp. WH 8102 (Waditee et al., 2003,2005). The cyanobacterial strains, such as *Aphanothece, Lyngbya* sp. PCC 8106 and currently identified species *Moorea* product 3L (Engene et al., 2012), contain the GsmT and Dmt genes in their genome to synthesize glycine betaine. The N_2-fixing cyanobacteria *T. erythraeum* IMS101 are found only the Dmt encoding gene, whereas the GsmT encoding gene lacks the genome. These glycine betaines–encoding genes are also found in *Synechococcus* species. It is known, cyanobacteria live in dynamic ecosystems due to the presence of many enzymes encoding genes involved in the synthesis of many compatible solutes (Mao et al., 2010; Scanlan et al., 2009).

10.5.3 Sucrose synthesis

The synthesis of sucrose is performed by sucrose-6-phosphate synthase, which is recognized as a vital factor against excessive desiccation tolerance. The primary studies suggest that the sucrose synthesis during cyanobacteria evolution was identified as a first osmoprotective compound in primitive cyanobacterial species (Blank, 2013). In the cyanobacterial cell, the widespread sucrose distribution as a compatible solute is proved by cyanobacterial genome sequence (Hagemann, 2013; Kolman et al., 2015). Sucrose is dominant among freshwater cyanobacterial species. It is also found in few moderately salt-tolerant cyanobacterial strains such as *Synechocystis* sp. in the form of secondary compatible solute to GG. Most probably, the genes involved in sucrose metabolism during salt stress were transferred from cyanobacteria to eukaryotic algae and plants during endosymbiotic development of chloroplasts (Lunn, 2002; Salerno and Curatti, 2003).

10.5.4 Trehalose synthesis

Trehalose is a widespread compatible solute protecting against salt stress as well as helps to tolerate several other stresses such as heat, dehydration (Furuki et al., 2009). For the synthesis of trehalose, cyanobacteria utilize two different biosynthetic pathways, such as TreY/TreZ. The presence of the TreYZ pathway among cyanobacterial species was first demonstrated in *Nostoc* 7120 during the study of a gene cluster (Higo et al., 2006). The TreYZ pathway has also been reported in other trehalose-synthesizing cyanobacterial strains such as *Nostoc punctiforme* IAM M-15 and *Nostoc flagelliforme* (Yoshida and Sakamoto, 2009; Wu et al., 2010).

In this pathway, glycogen acts as a pioneer molecule and the α-1,4 sugar bond changed into the α-1,1 bond in the first reaction, and then two last glucose molecules are cleaved off. In the marine cyanobacterium such as *Crocosphaera*, the OtsAB bacterial like pathway was also identified to synthesize trehalose. The trehalose accumulation during salt-induced stress was also detected in *Microcystis aeruginosa* is strong evidence of the trehalose synthesis pathway in cyanobacteria (Tables 10.2 and 10.3).

TABLE 10.2 Different cyanobacterial strains synthesize different compatible solutes based on their salinity tolerance levels. Klähn, S. and Hagemann, M., 2011. Compatible solute biosynthesis in cyanobacteria. Environmental microbiology, 13 (3), 551-562.

Cyanobacterial strains	Tolerance level	Nature of strain	Compatible solutes
Aphanothece, Spirulina	High	Unicellular Non-N_2-fixing	Glycine/glutamate betaine Glycine/glutamate betaine
Synechocystis 6714 *Synechocystis* 6803	Moderate	Unicellular Unicellular	Glucosylglycerol Sucrose Glucosylglycerol Sucrose
Anabaena, *Synechococcus* 6301	Low	N_2-fixing Unicellular	Sucrose Trehalose

TABLE 10.3 High salt concentration enhances modulation of enzyme activities in cyanobacteria.

Cyanobacterial species	Salt increases enzyme activity	Enzymes	Compatible solutes	References
Synechococcus sp. PCC 6301	Three to fourfold increase by NaCl	Sucrose-phosphate synthase	Sucrose	Hagemann and Marin (1999)
Synechocystis sp. PCC 6803	Different salts, optimum at a concentration of 100 mM	Glucosyl glycerol-phosphate synthase	Glucosyl glycerol	Fulda et al. (1999)

10.6 Mechanism of salt tolerance

In cyanobacteria, salt-tolerance mechanisms, including active Na^+ efflux, K^+ transport system, osmotically active molecules, and enzyme activation and deactivation, play a crucial role in stress during salinity stress. The high concentration of Na^+ in the cell can change osmolyte-synthesizing enzymes. The frequently known salt-tolerance mechanism is given next (Table 10.4).

10.6.1 Stress response proteins

High salt concentration can induce polypeptides known as ionic stress proteins. These stress response proteins are usually present for a short period and expressed at low levels (Apte et al., 1998). During various abiotic stresses, cyanobacterial cells showed rapid change in the protein synthesis programme, like in *Anabaena* sp. expression of many stress-induced proteins occurred during transcriptional activation of genes (Apte et al., 1998).

10.6.1.1 Na+ influx

Cyanobacteria limit the entry of sodium ions inside the cell and keep a low cytosolic Na^+ concentration, preventing cell injury during high salinity concentration in the soil (Roychoudhury et al., 1985; Jha et al., 1987). In many cyanobacteria such as *A. halophytica, Nodularia harveyana,* and *Synechocystis* PCC 6803, salt-tolerance mechanism is identified to maintain a high Na^+ / K^+ ratio (Warr et al., 1985). The cyanobacteria surviving in high saline habitat utilize the porins protein in their plasma membrane that allows Na^+ movements inside the cell. The cyanobacterial cells can release

TABLE 10.4 Cyanobacterial species that increase the soil remediation on salt-affected soil.

Cyanobacterial species	Soil pH	Effects	Reference
Anabaena torulosa, Anabaena sp. strain L-31	8.3	Decreased electric conductivity, Increased total nitrogen.	Apte and Thomas (1997)
Anabaena sp., *Calothrix* sp., *Nostoc calcicola*	10.5	Organic carbon increased	Singh and Singh (2015)
Nostoc ellipsosporum HH-205, *Nostoc punctiforme* HH-206	8.0	Increased electrical conductivity	Nisha et al. (2018)
Nostoc commune, Scytonema hyalinum, Tolypothrix distorta	8.57	Improved soil properties	Roman et al. (2018)

exopolysaccharides (EPS), which leads to chelating of a high percentage of Na^+ and immobilizes the excessive Na^+ during the saline environment. A high concentration of intracellular Na^+ may arrest several metabolic processes such as nitrogen metabolism, protein synthesis, and photosynthesis and inhibit cyanobacteria growth.

10.6.1.2 Na^+ efflux

ATP is required for the movement of Na^+ either through pumping of Na^+ or H^+ pumping to generate a proton gradient that drives Na^+/H^+ antiport. During the cell growth, a cytosolic saline atmosphere developed with low salt concentration due to the efflux of Na^+ generally requires an H^+ gradient. In some cyanobacterial species such as *Streptococcus faecalis*, Na^+ efflux occurred by Na^+ efflux channel and was ATPase dependent (Heefner and Harold, 1982).

10.6.1.3 Limitation of K^+ uptake

The Nap family antiporters, such asAp-NapA1-2, help in the uptake of K^+ in *A. halophytica*, and it is functionally is similar to the K^+/Na^+ antiporter (Wutipraditkul et al., 2005).The long-term loss of K^+ from salt-induced cells is complete by accumulating more compatible solutes (Reed and Stewart, 1985).There are generally three types of K^+ uptake systems found in the cell such as Kdp system, which is less critical; Ktr system, which plays a significant role at the time of high concentrations of NaCl; and K^+ channels of the cell membrane (Berry et al., 2003; Nakamura et al., 1998).

10.7 Role of cyanobacteria in the remediation of salt-affected soil

Salt-affected soils that contain high salt/alkaline is less productive for plants because excessive salts content in the crust of soil become rigid and impervious to water. The three types of soils, such as sodic soil, saline soil, and saline-sodic soil, are studied during salt stress (Amini et al., 2016; Nouri et al., 2017). The alkaline soil possesses a high pH, more exchangeable Na^+ ion, electrical conductivity, and measurable amounts of carbonates (Nouri et al., 2017; Meena et al., 2018). This type of soil has less aeration, poor hydraulic conductivity, and high osmotic pressure, reducing the water absorption and uptake of nutrients through plant root. A high amount of salinity in the soil due to high osmotic stress and more accumulation of Na^+ and Cl^- ions negatively impact several plant metabolic processes and their growth (Deinlein et al., 2014). The microbial fauna and flora and carbon cycling are also reduced in the soil due to salt stress (Rath and Rousk, 2015).

The cyanobacteria can be reclaimed by salt-affected soil by removing salts from the soil and enhancing soil fertility in a sustainable way

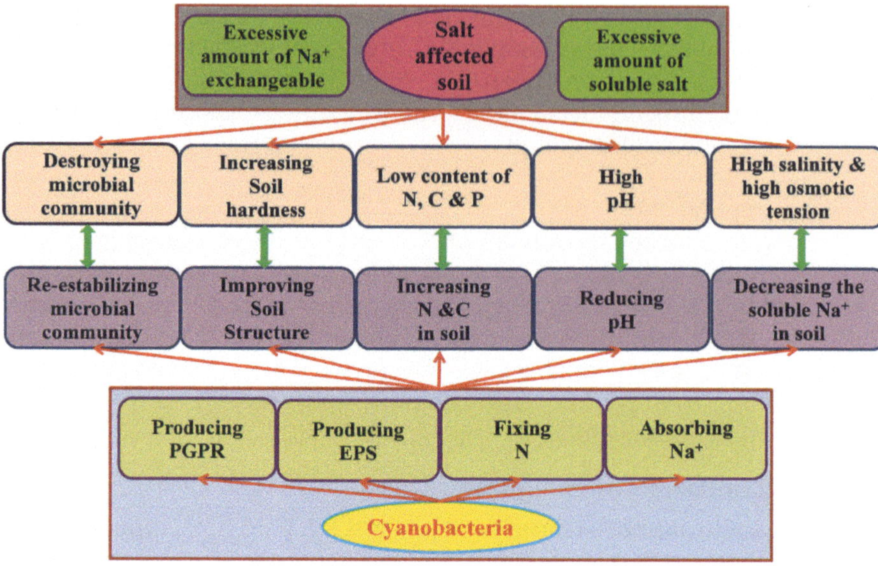

FIG. 10.6 Adverse effects of high salinity and positive impacts of cyanobacteria on salt-affected soils.

(Jesus et al., 2015). Cyanobacteria also improve soil profiling, soil fertility, and re-establish microbial communities similar to physical methods in a sustainable way (Fig. 10.6). Some N_2-fixing cyanobacterial species can enhance phosphate solubilization and increase mineral release to improve nitrogen fixation and soil fertility and play an essential role in soil reclamation to salt stress (Singh, 2014). In the soil microbial community, BGA are not dominant microorganisms; however, they have played an important role in increasing soil fertility and soil microbial community (Rossi et al., 2017). Cyanobacteria such as *Anabaena* sp. release exopolysaccharides (EPS), which bind with sodium ions in soil, it has been proven that cyanobacteria reclaim saline user soil (Singh, 1950). It is reported that the cyanobacterial filament also removes sodium ions from the salt-rich soil (Li et al., 2019). Heterocystous cyanobacteria have a specific network of the trichomes with a mucilage layer in the soil, which binds with the soil particles and also open the way for soil particles at depth, which can improve soil nutrient statuses such as nitrogen and organic carbon in arid soils.

It is known that the phytohormones or growth-promoting substances released in the salt-affected soils by *Phormidium foveolarum* increased the length of the leaves and height of the rice plant. Based on the above information, cyanobacteria can act as bioremediating agents for plant growth and provide a suitable environment for developing other soil microbial communities in the salt-affected soil. The essential features of cyanobacteria were nitrogen fixation, increasing the organic carbon contents, and extracellular polymeric substance production; hence, they played a crucial role in developing soil alteration agents and biofertilizers. Various mechanisms were studied in cyanobacteria, including nitrogen fixation, accumulation of compatible solutes, the release of an extracellular polymeric substance, production of phytohormones, active export of ions, and several defense enzymes are produced during soil reclamation in salt-affected soil. The soil fertility is increased by N_2-fixing cyanobacterium, which released the fixed nitrogen into salt-affected soil. Nitrogen fixation is one of the best potential mechanisms of soil remediation to salt-affected soil.

10.8 Conclusion

This chapter gives us basic information about the salinity-tolerance mechanism of cyanobacterial cells. Cyanobacterium such as *Synechocystis* sp. is generally used as a model organism to study several salt stress tolerance strategies. Their adaptive mechanisms are well studied *in-vitro,* but their salt tolerance adaptation to cyanobacteria in adverse conditions is not well researched. Several adaptive mechanisms against salt tolerance in cyanobacteria such as responsive proteins, extrusion of Na^+, the limited influx of Na^+ and uptake of K^+ maintain the ion homeostasis inside the cytosol. At the time of acclimation against salt stress, cyanobacterial species can synthesize several compatible solutes and many transporters used in the uptake and regulation of many proteins. The plasma membrane also plays an essential role in adaptation against salinity stress. The proteins and lipids of the plasma membrane having several types of antiporters and transporters are crucial agents in maintaining cytosolic ion homeostasis under salinity stress. In future, there is a necessity to focus more on the signaling and regulation system of the cyanobacterial high salinity stress response, which is currently not more known.

Acknowledgments

The authors thank the Head, Department of Botany, Banaras Hindu University for providing basic facilities and the University Grants Commission and Council of Scientific and Industrial Research for fellowship and DST- FIST, CAS- UGC New Delhi and ISLS, ISC, BHU for financial assistance.

References

Ali, S., Sandhu, G.R., 1972. Blue-green algae of the saline soils of the Punjab. Oikos, 268–272.

Allakhverdiev, S.I., Kinoshita, M., Inaba, M., Suzuki, I., Murata, N., 2001. Unsaturated fatty acids in membrane lipids protect the photosynthetic machinery against salt-induced damage in *Synechococcus*. Plant Physiol. 125 (4), 1842–1853.

Allakhverdiev, S.I., Nishiyama, Y., Suzuki, I., Tasaka, Y., Murata, N., 1999. Genetic engineering of the unsaturation of fatty acids in membrane lipids alters the tolerance of *Synechocystis* to salt stress. Proc. Natl. Acad. Sci. USA 96 (10), 5862–5867.

Allakhverdiev, S.I., Sakamoto, A., Nishiyama, Y., Inaba, M., Murata, N., 2000. Ionic and osmotic effects of NaCl-induced inactivation of photosystems I and II in *Synechococcus* sp. Plant Physiol. 123 (3), 1047–1056.

Allakhverdiev, S.I., Sakamoto, A., Nishiyama, Y., Murata, N., 2000b. Inactivation of photosystems I and II in response to osmotic stress in *Synechococcus*. Contribution of water channels. Plant Physiol. 122 (4), 1201–1208.

Amini, S., Ghadiri, H., Chen, C., Marschner, P., 2016. Salt-affected soils, reclamation, carbon dynamics, and biochar: a review. J. Soils Sediments 16 (3), 939–953.

Amsaveni, P., 1995. Effect of Certain Nutrients on the Growth and Ammonia Excretion by the Saline Tolerant Cyanobacteria and Their Role as Biofertilizers for Bioreclamation of Saline and Sodic Soils (Ph. D. Thesis). TNAU, Coimbatore, Tamil Nadu, India.

Anagnostidis, K., 2001. Nomenclatural changes in cyanoprokaryotic order *Oscillatoriales*. Preslia-Praha 73 (4), 359–376.

Apte, S.K., Fernandes, T., Badran, H., Ballal, A., 1998. Expression and possible role of stress-responsive proteins in *Anabaena*. J. Biosci. 23 (4), 399–406.

Apte, S.K., Thomas, J., 1997. Possible amelioration of coastal soil salinity using halotolerant nitrogen-fixing cyanobacteria. Plant Soil 189 (2), 205–211.

Baebprasert, W., Karnchanatat, A., Lindblad, P., Incharoensakdi, A., 2011. Na+-stimulated nitrate uptake with increased activity under osmotic upshift in *Synechocystis* sp. strain PCC 6803. World J. Microbiol. Biotechnol. 27 (10), 2467–2473.

Bebout, B.M., Carpenter, S.P., Des Marais, D.J., Discipulo, M., Embaye, T., Garcia-Pichel, F., Hoehler, T.M., Hogan, M., Jahnke, L.L., Keller, R.M., Miller, S.R., 2002. Long-term manipulations of intact microbial mat communities in a greenhouse collaboratory: simulating earth's present and past field environments. Astrobiology 2 (4), 383–402.

Berry, S., Esper, B., Karandashova, I., Teuber, M., Elanskaya, I., Rögner, M., Hagemann, M., 2003. Potassium uptake in the unicellular cyanobacterium *Synechocystis* sp. strain PCC 6803 mainly depends on a Ktr-like system encoded by slr1509 (ntpJ). FEBS Lett. 548 (1–3), 53–58.

Blank, C.E., 2013. Origin and early evolution of photosynthetic eukaryotes in freshwater environments: reinterpreting proterozoic paleobiology and biogeochemical processes in light of trait evolution. J. Phycol. 49 (6), 1040–1055.

Borges, N., Ramos, A., Raven, N.D., Sharp, R.J., Santos, H., 2002. Comparative study of the thermostabilizing properties of mannosylglycerate and other compatible solutes on model enzymes. Extremophiles 6 (3), 209–216.

Borowitzka, L.J., 1986. Osmoregulation in blue-green algae. Progress Phycol. Res. 4, 243–256.

Borowitzka, L.J., Demmerle, S., Mackay, M.A., Norton, R.S., 1980. Carbon-13 nuclear magnetic resonance study of osmoregulation in a blue-green alga. Science 210 (4470), 650–651.

Brown, A.D., 1976. Microbial water stress. Bacteriol. Rev. 40 (4), 803.

Budinoff, C.R., Hollibaugh, J.T., 2007. Ecophysiology of a Mono Lake picocyanobacterium. Limnol. Oceanogr. 52 (6), 2484–2496.

Catherine, Q., Susanna, W., Isidora, E.S., Mark, H., Aurelie, V., Jean-François, H., 2013. A review of current knowledge on toxic benthic freshwater cyanobacteria–ecology, toxin production and risk management. Water Res. 47 (15), 5464–5479.

Chen, T.H., Murata, N., 2011. Glycinebetaine protects plants against abiotic stress: mechanisms and biotechnological applications. Plant Cell Environ. 34 (1), 1–20.

Deinlein, U., Stephan, A.B., Horie, T., Luo, W., Xu, G., Schroeder, J.I., 2014. Plant salt-tolerance mechanisms. Trends Plant Sci. 19 (6), 371–379.

Dembitsky, V.M., Rezenka, T., 2005. Metabolites produced by nitrogen fixing *Nostoc* species. Folia Microbiol 50, 363–391.

Desplats, P., Folco, E., Salerno, G.L., 2005. Sucrose may play an additional role to that of an osmolyte in *Synechocystis* sp. PCC 6803 salt-shocked cells. Plant Physiol. Biochem. 43 (2), 133–138.

Ehira, S., Kimura, S., Miyazaki, S., Ohmori, M., 2014. Sucrose synthesis in the nitrogen-fixing cyanobacterium *Anabaena* sp. strain PCC 7120 is controlled by the two-component response regulator OrrA. Appl. Environ. Microbiol. 80 (18), 5672–5679.

Engene, N., Rottacker, E.C., Kaštovský, J., Byrum, T., Choi, H., Ellisman, M.H., Komárek, J., Gerwick, W.H., 2012. Moorea producens gen. nov., sp. nov. and *Moorea bouillonii* comb. nov., tropical marine cyanobacteria rich in bioactive secondary metabolites. Int. J. Syst. Evol. Microbiol. 62 (Pt 5), 1171.

Fernandes, T.A., Iyer, V., Apte, S.K., 1993. Differential responses of nitrogen-fixing cyanobacteria Salinity and osmotic stresses. Appl. Environ. Microbiol. 59 (3), 899–904.

Ferreira, R.B., Shaw, N.M., 1989. Effect of osmotic stress on protein turnover in *Lemna minor* fronds. Planta 179 (4), 456–465.

Fulda, S., Huckauf, J., Schoor, A., Hagemann, M., 1999. Analysis of stress responses in the cyanobacterial strains *Synechococcus* sp. PCC 7942, *Synechocystis* sp. PCC 6803, and *Synechococcus* sp. PCC 7418: osmolyte accumulation and stress protein synthesis. J. Plant Physiol. 154 (2), 240–249.

Furuki, T., Oku, K., Sakurai, M., 2009. Thermodynamic, hydration and structural characteristics of alpha, alpha-trehalose. Front. Biosci. 14, 3523–3535.

Hagemann, M., 2013. Genomics of salt acclimation: synthesis of compatible solutes among cyanobacteria. Adv. Bot. Res. 65, 27–55.

Hagemann, M., Erdmann, N., 1994. Activation and pathway of glucosylglycerol synthesis in the cyanobacterium *Synechocystis*sp. PCC 6803. Microbiology 140 (6), 1427–1431.

Hagemann, M., Pade, N., 2015. Heterosides–compatible solutes occurring in prokaryotic and eukaryotic phototrophs. Plant Biol. 17 (5), 927–934.

Hagemann, M., Marin, K., 1999. Salt-induced sucrose accumulation is mediated by sucrose-phosphate-synthase in cyanobacteria. J. Plant Physiol. 155 (3), 424–430.

Hagemann, M., Schoor, A., Jeanjean, R., Zuther, E., Joset, F., 1997a. The *stpA*gene from *Synechocystis*sp. strain PCC 6803 encodes the glucosylglycerol-phosphate phosphatase involved in cyanobacterial osmotic response to salt shock. J. Bacteriol. 179, 1727–1733.

Hagemann, M., Wölfel, L., Krüger, B., 1990. Alterations of protein synthesis in the cyanobacterium *Synechocystis* sp. PCC 6803 after a salt shock. Microbiology 136 (7), 1393–1399.

Heefner, D.L., Harold, F.M., 1982. ATP-driven sodium pump in *Streptococcus faecalis*. Proc. Natl. Acad. Sci. USA 79 (9), 2798–2802.

Higo, A., Katoh, H., Ohmori, K., Ikeuchi, M., Ohmori, M., 2006. The role of a gene cluster for trehalose metabolism in dehydration tolerance of the filamentous cyanobacterium *Anabaena* sp. PCC 7120. Microbiology 152 (4), 979–987.

Hincha, D.K., Hagemann, M., 2004. Stabilization of model membranes during drying by compatible solutes involved in the stress tolerance of plants and microorganisms. Biochem. J. 383 (2), 277–283.

Hoffmann, L., 1989. Algae of terrestrial habitats. Bot. Rev. 55 (2), 77–105.

Huang, F., Fulda, S., Hagemann, M., Norling, B., 2006. Proteomic screening of salt-stress-induced changes in plasma membranes of *Synechocystis* sp. strain PCC 6803. Proteomics 6 (3), 910–920.

Jäger, K.M., Johansson, C., Kunz, U., Lehmann, H., 1997. .Sub-cellular element analysis of a cyanobacterium (*Nostoc* sp.) in symbiosis with *Gunneramanicata* by ESI and EELS. Bot. Acta 110 (2), 151–157.

Jeanjean, R., Bédu, S., Havaux, M., Matthijs, H.C., Joset, F., 1998. Salt-induced photosystem I cyclic electron transfer restores growth on low inorganic carbon in a type 1 NAD (P) H dehydrogenase deficient mutant of *Synechocystis*PCC6803. FEMS Microbiol. Lett. 167 (2), 131–137.

Jesus, J.M., Danko, A.S., Fiúza, A., Borges, M.T., 2015. Phytoremediation of salt-affected soils: a review of processes, applicability, and the impact of climate change. Environ. Sci. Pollut. Res. 22 (9), 6511–6525.

Jha, M.N., Venkataraman, G.S., Kaushik, B.D., 1987. Response of *Westiellopsisprolifica* and *Anabaena* sp. to salt stress. MIRCEN J. Appl. Microbiol. Biotechnol. 3 (3), 307–317.

Joset, F., Jeanjean, R., Hagemann, M., 1996. Dynamics of the response of cyanobacteria to salt stress: deciphering the molecular events. Physiol. Plant. 96 (4), 738–744.

Kaneko, T., Sato, S., Kotani, H., Tanaka, A., Asamizu, E., Nakamura, Y., Miyajima, N., Hirosawa, M., Sugiura, M., Sasamoto, S., Kimura, T., 1996. Sequence analysis of the genome of the unicellular cyanobacterium *Synechocystis* sp. strain PCC6803. II. Sequence determination of the entire genome and assignment of potential protein-coding regions. DNA Res. 3 (3), 109–136.

Kasting, J.F., Siefert, J.L., 2002. Life and the evolution of Earth's atmosphere. Science 296 (5570), 1066–1068.

Kirkwood, A.E., Buchheim, J.A., Buchheim, M.A., Henley, W.J., 2008. Cyanobacterial diversity and halotolerance in a variable hypersaline environment. Microb. Ecol. 55 (3), 453–465.

Klähn, S., Steglich, C., Hess, W.R., Hagemann, M., 2010a. Glucosylglycerate: a secondary compatible solute common to marine cyanobacteria from nitrogen-poor environments. Environ. Microbiol. 12, 83–94.

Kolman, M.A., Nishi, C.N., Perez-Cenci, M., Salerno, G.L., 2015. Sucrose in cyanobacteria: from a salt-response molecule to play a key role in nitrogen fixation. Life 5 (1), 102–126.

Komarek, J., 1998. Studies on the cyanophytes of Cuba. Folia Glo-Bot Phytotaxon 30, 81–90.

Laloknam, S., Tanaka, K., Buaboocha, T., Waditee, R., Incharoensakdi, A., Hibino, T., Tanaka, Y., Takabe, T., 2006. Halotolerant cyanobacterium *Aphanothece halophytica* contains a betaine transporter active at alkaline pH and Salinity. Appl. Environ. Microbiol. 72 (9), 6018–6026.

Lee, J.Z., Burow, L.C., Woebken, D., Everroad, R.C., Kubo, M.D., Spormann, A.M., Weber, P.K., Pett-Ridge, J., Bebout, B.M., Hoehler, T.M., 2014. Fermentation couples *Chloroflexi* and sulfate-reducing bacteria to cyanobacteria in hypersaline microbial mats. Front. Microbiol. 5, p. 61.

Li, H., Zhao, Q., Huang, H., 2019. Current states and challenges of salt-affected soil remediation by cyanobacteria. Sci. Total Environ. 669, 258–272.

Los, D.A., Suzuki, I., Zinchenko, V.V., Murata, N., 2008. Stress responses in Synechocystis: regulated genes and regulatory systems. In: Herrero, A., Flores, E. (Eds.), Stress responses in Synechocystis: regulated genes and regulatory systems. The Cyanobacteria: Molecular Biology, Genomics and Evolution, 117–157.

Lu, C., Vonshak, A., 2002. Effects of salinity stress on photosystem II function in cyanobacterial *Spirulina platensis* cells. Physiol. Planta. 114 (3), 405–413.

Lunn, J.E., 2002. Evolution of sucrose synthesis. Plant Physiol. 128 (4), 1490–1500.

Mackay, M.A., Norton, R.S., Borowitzka, L.J., 1984. Organic osmoregulatory solutes in cyanobacteria. Microbiology 130 (9), 2177–2191.

Mahajan, S., Tuteja, N., 2005. Cold, salinity and drought stresses: an overview. Arch. Biochem. Biophys. 444 (2), 139-158.

Mao, C., Brown, C.R., Falkovskaia, E., Dong, S., Hrabeta-Robinson, E., Wenger, L., Boeger, H., 2010. Quantitative analysis of the transcription control mechanism. Mol. Syst. Biol. 6 (1), 431.

Marin, K., Kanesaki, Y., Los, D.A., Murata, N., Suzuki, I., Hagemann, M., 2004. Gene expression profiling reflects physiological processes in salt acclimation of *Synechocystis* sp. strain PCC 6803. Plant Physiol. 136 (2), 3290–3300.

Marin, K., Zuther, E., Kerstan, T., Kunert, A., Hagemann, M., 1998. The ggpS gene from *Synechocystis*sp. strain PCC 6803 encoding glucosyl-glycerol-phosphate synthase is involved in osmolyte synthesis. J. Bacteriol. 180 (18), 4843–4849.

McKersie, B.D., Leshem, Y.Y., 1994. Stress and stress coping in cultivated plants mechanisms. Trends Plant Sci 19, 371–379.

Meena, M.D., Narjary, B., Sheoran, P., Jat, H.S., Joshi, P.K., Chinchmalatpure, A.R., Yadav, G., Yadav, R.K., Meena, M.K., 2018. Changes of phosphorus fractions in saline soil amended with municipal solid waste compost and mineral fertilizers in a mustard-pearl millet cropping system. Catena 160, 32–40.

Moore, D.J., Reed, R.H., Stewart, W.D.P., 1987. A glycine betaine transport system in *Aphanothece halophytica* and other glycibetaine-synthesizing cyanobacteria. Arch. Microbiol. 147 (4), 399–405.

Müller, V., Oren, A., 2003. Metabolism of chloride in halophilic prokaryotes. Extremophiles 7 (4), 261–266.

Mullineaux, C.W., 2014. Co-existence of photosynthetic and respiratory activities in cyanobacterial thylakoid membranes. Biochim. Biophys. Acta 1837 (4), 503–511.

Murata, N., Suzuki, I., 2006. Exploitation of genomic sequences in a systematic analysis to access how cyanobacteria sense environmental stress. J. Exp. Bot. 57 (2), 235–247.

Nakamura, T., Yuda, R., Unemoto, T., Bakker, E.P., 1998. KtrAB, a new type of bacterial K+-uptake system from *Vibrio alginolyticus*. J. Bacteriol. 180 (13), 3491–3494.

Neill, S., Desikan, R., Hancock, J., 2002. Hydrogen peroxide signalling. Curr. Opin. Plant Biol. 5 (5), 388–395.

Ning, S.B., Guo, H.L., Wang, L., Song, Y.C., 2002. Salt stress induces programmed cell death in prokaryotic organism *Anabaena*. J. Appl. Microbiol. 93 (1), 15–28.

Nisha, R., Kiran, B., Kaushik, A., Kaushik, C.P., 2018. Bioremediation of salt affected soils using cyanobacteria in terms of physical structure, nutrient status and microbial activity. Int. J. Environ. Sci. Technol. 15 (3), 571–580.

Nouri, H., Borujeni, S.C., Nirola, R., Hassanli, A., Beecham, S., Alaghmand, S., Saint, C., Mulcahy, D., 2017. Application of green remediation on soil salinity treatment: a review on halophyte remediation. Process Saf. Environ. Prot. 107, 94–107.

Nübel, U., Garcia-Pichel, F., Kühl, M., Muyzer, G., 1999. Spatial scale and the diversity of benthic cyanobacteria and diatoms in a salina. In: Zehr, J.P., Voytek, M.A. (Eds.), Molecular Ecology of Aquatic Communities. Springer, Dordrecht, pp. 199–206.

Pade, N., Compaoré, J., Klähn, S., Stal, L.J., Hagemann, M., 2012. The marine cyanobacterium *Crocosphaerawatsonii* WH8501 synthesizes the compatible solute trehalose by a laterally acquired OtsAB fusion protein. Environ. Microbiol. 14 (5), 1261–1271.

Pade, N., Michalik, D., Ruth, W., Belkin, N., Hess, W.R., Berman-Frank, I., Hagemann, M., 2016a. Trimethylated homoserine functions as the major compatible solute in the globally significant oceanic cyanobacterium *Trichodesmium*. Proc. Natl. Acad. Sci. USA 113 (46), 13191–13196.

Paerl, H.W., Steppe, T.F., Buchan, K.C., Potts, M., 2003. Hypersaline cyanobacterial mats as indicators of elevated tropical hurricane activity and associated climate change. AMBIO 32 (2), 87–90.

Paerl, H.W., Pinckney, J.L., Steppe, T.F., 2000. Cyanobacterial–bacterial mat consortia: examining the functional unit of microbial survival and growth in extreme environments. Environ. Microbiol. 2 (1), 11–26.

Paithoonrangsarid, K., Shoumskaya, M.A., Kanesaki, Y., Satoh, S., Tabata, S., Los, D.A., Zinchenko, V.V., Hayashi, H., Tanticharoen, M., Suzuki, I., Murata, N., 2004. Five histidine kinases perceive osmotic stress and regulate distinct sets of genes in *Synechocystis*. J. Biol. Chem. 279 (51), 53078–53086.

Pandey, K.D., Shukla, P.N., Giri, D.D., Kashyap, A.K., 2005. Cyanobacteria in alkaline soil and the effect of cyanobacteria inoculation with pyrite amendments on their reclamation. Biol. Fertil. Soil 41 (6), 451–457.

Pandhal, J., Snijders, A.P., Wright, P.C., Biggs, C.A., 2008. A cross-species quantitative proteomic study of salt adaptation in a halotolerant environmental isolate using 15N metabolic labelling. Proteomics 8 (11), 2266–2284.
Pocard, J.A., Smith, L.T., Smith, G.M., Le Rudulier, D., 1994. A prominent role for glucosylglycerol in the adaptation of *Pseudomonas mendocina* SKB70 to osmotic stress. Journal of Bacteriol. 176 (22), 6877–6884.
Proteau, P.J., Gerwick, W.H., Garcia-Pichel, F., Castenholz, R., 1993. The structure of scytonemin, an ultraviolet sunscreen pigment from the sheaths of cyanobacteria. Experientia 49 (9), 825–829.
Rai, A.K., Abraham, G., 1995. Relationship of combined nitrogen sources to salt tolerance in freshwater cyanobacterium *Anabaena doliolum*. J. Appl. Bacteriol. 78 (5), 501–506.
Rai, A.N., Rowell, P., Stewart, W.D.P., 1980. $NH4^{+}$ assimilation and nitrogenase regulation in the lichen *Peltigeraaphthosa* Willd. New Phytol. 85 (4), 545–555.
Rath, K.M., Rousk, J., 2015. Salt effects on the soil microbial decomposer community and their role in organic carbon cycling: a review. Soil Biol. Biochem. 81, 108–123.
Reed, R.H., Stewart, W.P., 1988. The responses of cyanobacteria to salt stress. In: Rogers, L.J., G.R. (Ed.), Biochemistry of the Algae and Cyanobacteria. Clarendon Press, Oxford, 217–231.
Reed, R.H., Stewart, W.D., 1985. Evidence for turgor-sensitive K+ influx in the cyanobacteria *Anabaena variabilis* ATCC29413 and *Synechocystis* PCC6714. Biochim. Biophys. Acta 812 (1), 155–162.
Reed, R.H., Chudek, J.A., Foster, R., Stewart, W.D.P., 1984. Osmotic adjustment in cyanobacteria from hypersaline environments. Arch. Microbiol. 138 (4), 333–337.
Reed, R.H., Warr, S.R., Kerby, N.W., Stewart, W.D., 1986. Osmotic shock-induced release of low molecular weight metabolites from free-living and immobilized cyanobacteria. Enzyme Microb. Technol. 8 (2), 101–104.
Riccardi, G., Cella, R., Camerino, G., Ciferri, O., 1983. Resistance to azetidine-2-carboxylic acid and sodium chloride tolerance in carrot cell cultures and *Spirulina platensis*. Plant Cell Physiol. 24 (6), 1073–1078.
Roder, A., Hoffmann, E., Hagemann, M., Berg, G., 2005. Synthesis of the compatible solutes glucosylglycerol and trehalose by salt-stressed cells of *Stenotrophomonas* strains. FEMS Microbiol. Lett. 243, 219–226. doi:10.1016/j. femsle.2004.12.005.
Román, J.R., Roncero-Ramos, B., Chamizo, S., Rodríguez-Caballero, E., Cantón, Y., 2018. Restoring soil functions by means of cyanobacteria inoculation: importance of soil conditions and species selection. Land Degrad. Dev. 29 (9), 3184–3193.
Roney, H.C., Booth, G.M., Cox, P.A., 2009. Competitive exclusion of cyanobacterial species in the Great Salt Lake. Extremophiles 13 (2), 355–361.
Rossi, F., Li, H., Liu, Y., De Philippis, R., 2017. Cyanobacterial inoculation (cyanobacterisation): perspectives for the development of a standardized multifunctional technology for soil fertilization and desertification reversal. Earth-Science Rev. 171, 28–43.
Rothschild, L.J., Giver, L.J., White, M.R., Mancinelli, R.L., 1994. Metabolic activity of microorganisms in evaporites 1. J. Phycol. 30 (3), 431–438.
Roychoudhury, P., Kaushik, B.D., Venkataraman, G.S., 1985. Response of *Tolypothrixceylonica* to sodium stress. Curr. Sci. 54 (22), 1181–1183.
Salerno, G.L., Curatti, L., 2003. Origin of sucrose metabolism in higher plants: when, how and why? Trends Plant Sci. 8 (2), 63–69.
Scanlan, D.J., Ostrowski, M., Mazard, S., Dufresne, A., Garczarek, L., Hess, W.R., Post, A.F., Hagemann, M., Paulsen, I., Partensky, F., 2009. Ecological genomics of marine picocyanobacteria. Microbiol. Mol. Biol. Rev. 73 (2), 249–299.
Schubert, H., Hagemann, M., 1990. Salt effects on 77K fluorescence and photosynthesis in the cyanobacterium *Synechocystis* sp. PCC 6803. FEMS Microbiol. Lett. 71 (1-2), 169–172.

Schubert, H., Fulda, S., Hagemann, M., 1993. Effects of adaptation to different salt concentrations on photosynthesis and pigmentation of the cyanobacterium *Synechocystis* sp. PCC 6803. J. Plant Physiol. 142 (3), 291–295.

Schwartz, S.H., Black, T.A., Jäger, K., Panoff, J.M., Wolk, C.P., 1998. Regulation of an osmoticum-responsive gene in *Anabaena* sp. strain PCC 7120. J. Bacteriol. 180 (23), 6332–6337.

Shoumskaya, M.A., Paithoonrangsarid, K., Kanesaki, Y., Los, D.A., Zinchenko, V.V., Tanticharoen, M., Suzuki, I., Murata, N., 2005. Identical Hik-Rre systems are involved in perception and transduction of salt signals and hyperosmotic signals but regulate the expression of individual genes to different extents in *Synechocystis*. J. Biol. Chem. 280 (22), 21531–21538.

Singh, R.N., 1950. Reclamation of 'Usar'lands in India through blue-green algae. Nature 165 (4191), 325–326.

Singh, D.P., Kshatriya, K., 2002. NaCl-induced oxidative damage in the cyanobacterium *Anabaena doliolum*. Curr. Microbiol. 44 (6), 411–417.

Singh, R.N., 1961. Role of Blue-green Algae in Nitrogen Economy of Indian Agriculture, Agriculture Indian Council of Agricultural Research, New Delhi.

Singh, R.N., 1950. Reclamation of 'Usar'lands in India through blue-green algae. Nature 165 (4191), 325–326.

Singh, V., Singh, D.V., 2015. Cyanobacteria modulated changes and its impact on bioremediation of saline-alkaline soils. Bangladesh J. Bot. 44 (4), 653–658.

Singh, S., 2014. A review on possible elicitor molecules of cyanobacteria: their role in improving plant growth and providing tolerance against biotic or abiotic stress. J. Appl. Microbiol. 117 (5), 1221–1244.

Singh, V., Pandey, K.D., Mesapogu, S., Singh, D.V., 2015. Influence of NaCl on photosynthesis and nitrogen metabolism of cyanobacterium *Nostoc calcicola*. Appl. Biochem. Microbiol. 51 (6), 720–725.

Srivastava, A.K., Bhargava, P., Thapar, R., Rai, L.C., 2008. Salinity-induced physiological and proteomic changes in *Anabaena doliolum*. Environ. Exp. Bot 64 (1), 49–57.

Srivastava, A.K., Bhargava, P., Rai, L.C., 2005. Salinity and copper-induced oxidative damage and changes in the antioxidative defence systems of *Anabaena doliolum*. World J. Microbiol. Biotechnol. 21 (6), 1291–1298.

Suzuki, I., Los, D.A., Kanesaki, Y., Mikami, K., Murata, N., 2000. The pathway for perception and transduction of low-temperature signals in *Synechocystis*. EMBO J. 19 (6), 1327–1334.

Takabe, T., Incharoensakdi, A., Arakawa, K., Yokota, S., 1988. CO_2 fixation rate and RuBisCO content increase in the halotolerant cyanobacterium, *Aphanothece halophytica*, grown in high salinities. Plant Physiol. 88 (4), 1120–1124.

Vermaas, W.F., 2001. Photosynthesis and respiration in cyanobacteria. *e LS*. https://doi.org/10.1038/npg.els.0001670, Corpus ID: 19016706.

Waditee, R., Bhuiyan, M.N.H., Rai, V., Aoki, K., Tanaka, Y., Hibino, T., Suzuki, S., Takano, J., Jagendorf, A.T., Takabe, T., Takabe, T., 2005. Genes for direct methylation of glycine provide high levels of glycinebetaine and abiotic-stress tolerance in *Synechococcus* and *Arabidopsis*. Proc. Natl. Acad. Sci. USA. 102 (5), 1318–1323.

Waditee, R., Tanaka, Y., Aoki, K., Hibino, T., Jikuya, H., Takano, J., Takabe, T., Takabe, T., 2003. Isolation and functional characterization of N-methyl transferases that catalyze betaine synthesis from glycine in a halotolerant photosynthetic organism *Aphanothece halophytica*. J. Biol. Chem. 278 (7), 4932–4942.

Warr, S.R.C., Reed, R.H., Chudek, J.A., Foster, R., Stewart, W.D.P., 1985. Osmotic adjustment in *Spirulina platensis*. Planta 163 (3), 424–429.

Whitton, B.A., Mateo, P., 2012. Rivulariaceae. In: Whitton, B.A. (Ed.), Ecology of Cyanobacteria II. Springer, Dordrecht, pp. 561–591.

Wood, J.M., 2011. Bacterial osmoregulation: a paradigm for the study of cellular homeostasis. Annu. Rev. Microbiol. 65, 215–238.

Wu, S., Shen, R., Zhang, X., Wang, Q., 2010. Molecular cloning and characterization of maltooligosyl trehalose synthase gene from *Nostoc flagelliforme*. J. Microbiol. Biotechnol. 20 (3), 579–586.

Wutipraditkul, N., Waditee, R., Incharoensakdi, A., Hibino, T., Tanaka, Y., Nakamura, T., Shikata, M., Takabe, T., Takabe, T., 2005. Halotolerant cyanobacterium *Aphanothece halophytica*contains NapA-type Na+/H+ antiporters with novel ion specificity that are involved in salt tolerance at alkaline pH. Appl. Environ. Microbiol. 71 (8), 4176–4184.

Whitton, B.A., Potts, M., 2000. Introduction to the cyanobacteria. In: Whitton, B.A., Potts, M. (Eds.), The Ecology of Cyanobacteria. © Kluwer Academic Publishers 2000, Springer, Dordrecht. https://doi.org/10.1007/0-306-46855-7_1.

Yoshida, T., Sakamoto, T., 2009. Water-stress induced trehalose accumulation and control of trehalase in the cyanobacterium *Nostoc punctiforme* IAM M-15. J. Gen. Appl. Microbiol. 55 (2), 135–145.

Zhu, J.K., 2001. Plant salt tolerance. Trends in Plant Sci. 6 (2), 66–71.

CHAPTER

11

Cyanobacteria as biostimulants in the paddy fields

Prashant Kumar Singh[a], *Ajay Kumar*[b], *Mukesh Kumar Yadav*[a], *Kaushalendra*[d], *Savita Singh*[c]

[a]Department of Biotechnology, Mizoram University (A Central University), Pachhunga University College Campus, Aizawl, Mizoram, India
[b]Agriculture Research Organization (ARO), The Volcani Center, Rishon-Lezion, Israel
[c]Department of Botany, Babu Shivnath Agrawal College, Mathura, Uttar Pradesh, India
[d]Department of Zoology, Mizoram University (A Central University), Pachhunga University College Campus, Aizawl, Mizoram, India

11.1 Introduction

The increasing world population poses severe pressure on land for increased crop production worldwide (Lutz et al., 2008; York et al., 2003). On the other hand, this flooding population has contributed significantly to (1) reduced availability of cultivated land, (2) scarcity and degradation of water resources, (3) soil xenobiotic accumulation, and (4) soil fertility and quality degradation. To overcome the challenges combined with rising global population and anthropogenic pressure and encounter the food surplus, there is an urgent need to improve crop productivity practices in hand with sustainability (Pathak et al., 2018). Higher productions can be accomplished either by boosting crops' yields or reducing agricultural productivity losses due to various stresses (both abiotic and biotic). The scientific school is flooded with approaches to expand and take crop production to the next level. In this regard, various solutions are coming up to boost agricultural production. Though there are many techniques for barring any ill effects, the most scientific community relies on sustainable means of increasing crop production. Sustainable agriculture focuses on using biological means

Cyanobacterial Lifestyle and its Applications in Biotechnology.
DOI: https://doi.org/10.1016/B978-0-323-90634-0.00005-6

and discards using inorganic and chemical fertilizers in agricultural fields that are safe both for humans and the environment. Researchers have engrossed their attention on biological products in the hunt for sustainable and ecologically responsive solutions of improved farm yields, with cyanobacteria and microalgae emerging as a valuable means for their protection and production of stimulating and fertilizing activities (Gonçalves, 2021). It is worthwhile to mention that the exploitation of cyanobacteria and microalgae in agriculture has been in practice since ages; for example, in European coastal areas, farmers used algae for fertilization purposes (Baweja et al., 2019). Later in the twentieth century, algal and worldwide farmers were fascinated by cyanobacterial extracts because of the capacity of extracts to produce biologically active compounds such as carotenoids, phenols, terpenoids, free fatty acids, and polysaccharides, which demonstrated to have promising effects on crops' productivity (Pan et al., 2019; Singh et al., 2017). Algal and cyanobacterial metabolites have been shown to offer (Gorka et al., 2018; Ronga et al., 2019): (1) fertilization and decontamination of soils; (2) biotic and abiotic stress tolerance to crops; and (3) plant growth and development by production of phytohormones, which serves as plant-growth promoters. Both microalgae and cyanobacterial biomass and extracts are commercially available nowadays.

Hence, it is well acknowledged that biostimulants application stimulates plants' growth and protects them by providing strength to cope with different stresses. Biostimulants is a diverse term applicable to any product or organism that has a beneficial effect on crop plants' growth and development. Various biostimulants have been proposed, such as protein hydrolysates and polyamines, seaweed extracts, humic and fulvic acids, arbuscular mycorrhizal fungi, cyanobacteria and plant growth-promoting rhizobacteria. Among the biostimulants where microorganisms serve the purpose, cyanobacteria encompass a very pivotal role (Fig. 11.1). This

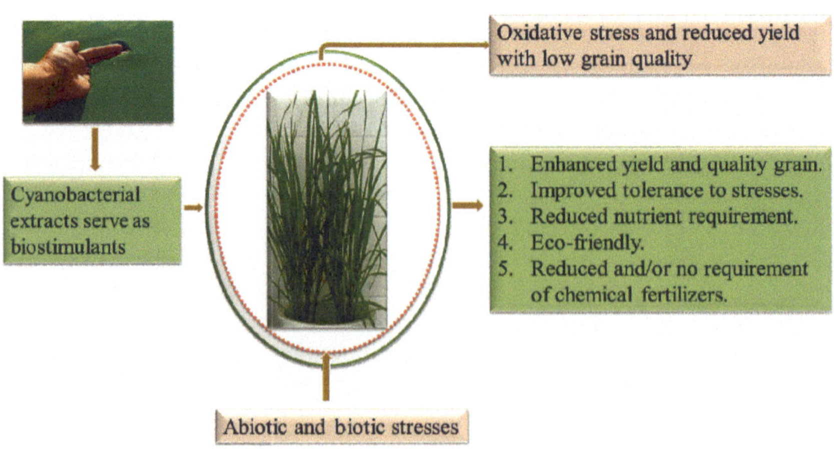

FIG. 11.1 Cyanobacterial extracts serve as biostimulants.

unique ability is because of thier property of being photosynthetic autotrophic prokaryotes, most of which have the unique ability to fix atmospheric dinitrogen nitrogen and are relatively easy to grow. This chapter aims to weave cyanobacteria's concept serving as biostimulants and their comparative pros and cons compared to other types of biostimulants known in the literature.

11.2 Cyanobacterial biostimulants and plant growth–promoting potential

Table 11.1 shows a summary of cyanobacterial biostimulants and their role in plant-growth promotion. For agricultural practices, these compounds serve a crucial role in crop protection against biotic and abiotic stresses because of their antimicrobial and antioxidant properties, respectively. These properties make them valuable resources for biostimulants, biopesticides, and biofertilizers to enhance crop productivity sustainably. Besides, it has also been reported that these compounds also play a very crucial role in plants reproductive success (Gershenzon and Dudareva, 2007).

11.3 Cyanobacteria and their extracts: impacts on crops' productivity

As discussed above, cyanobacteria are an essential source of biologically active compounds that can significantly improve agricultural productivity and exploited as biostimulants. The extensive cyanobacterial active metabolites diversities, collected with their multiple functions in crops' productivity improvement, permit their exploitation as a crude or pure extract. Furthermore, this metabolites diversity impacts crops' production at different levels: (1) some extracts induce higher crops' productivities through soil quality improvement; (2) some through the protection against biotic and abiotic stresses; and (3) some metabolites act directly on plant growth stimulation. The role of these metabolites under the above heads discussed below how different metabolites involved crop productivity improvement with suitable examples.

11.3.1 Role of cyanobacterial metabolites in soil health improvement

Some cyanobacterial extracts are known to affect soil fertility and, hence, positively affect crop productivity. Some well-known activities of cyanobacterial species are to (1) fix atmospheric dinitrogen to make it available

TABLE 11.1 Summary of biostimulants of cyanobacterial origin and their role in plant-growth promotion.

Metabolites	Examples cyanobacterial source	Biological activity	Role in agriculture	Reference
Terpenoids	*Oscillatoria peronata*; *Pseudanabaena articulate*; *Pseudanabaena* sp.; *Synechocystis* sp.; *Thermosynechococcus elongatus*	Antibacterial; antioxidant	Crop protection against abiotic and biotic stress	Betterle and Melis (2019), Pattanaik and Lindberg (2019)
Free fatty acids	*Anabaena* sp. *Spirulina* sp.	Antibiotic; antifungal; antioxidant; antiviral	Crop protection against abiotic and biotic stress	El-Baz et al. (2013)
Polysaccharides	*Aphanothece* sp.; *Arthrospira* sp.; *Nostoc* sp.; *Scytonema* sp.	Antibacterial; anticancer; anticoagulant; antiinflammatory; antioxidant	Soil quality improvement and crops growth stimulation and protection against biotic and abiotic stresses	Campos et al. (2015)
Phytohormones	*Phormidium* sp.	Chemical messengers	Stimulation and regulation of crops growth and crops' response to stresses	Singh et al. (2017)
Carotenoids	*Spirulina* sp.	Antioxidant	Crop protection against abiotic and biotic stress	Singh et al. (2017)
Phenolics	*Spirulina* sp.	Antioxidant	Crop protection against abiotic and biotic stress	Singh et al. (2017)

for the plant's assimilation; (2) offer essential plants nutrients, and (3) improve soil's physical and chemical properties. Table 11.2 shows the exploitation of cyanobacteria and/or their extracts contributions in restoration/improvement of soil health for crop productivities improvement.

TABLE 11.2 Impacts of cyanobacteria and/or their metabolites on the soil.

Mode of action	Cyanobacterial extracts	Target crop species/Soil	Improvements advocated	References
1. Atmospheric nitrogen fixation	*Anabaena sphaerica*, *Aulosira fertilissima*, *Cylindrospermum majus*, *Nostoc hatei*, and *Westiellopsis prolifica* as a cyanobacterial inoculum	*Oryzae*	1. Increase availability of nitrogen in the soil. 2. Increase yields of grain and straw.	Jha and Prasad (2006)
	Vegetative cells of *Nostoc* sp.	*Oryzae* sp.	Increase yields of grain.	Innok et al. (2009)
	Anabaena variabilis wild-type and herbicide-resistant strains	*Oryzae* sp.	1. Increase yields of grain seed and straw. 2. Increase in plant height and leaf length.	Singh and Dutta (2007)
	Biofilms of cyanobacterial species such as *Calothrix* sp., *Anabaena laxa*, *Anabaena torulosa*, *Anabaena doliolum*, *Nostoc carneum*, *Nostoc piscinale* and bacterial biofilms *Trichoderma viride*, *Pseudomonas fluorescens* and *Azotobacter chroococcum*	Soybean and mungbean	Increase in nitrogen availability in the soil Increase in plant fresh weight	Prasanna et al. (2013)
	Biofilm of *Anabaena torulosa*	Wheat	Enhance soil nitrogen.	Swarnalakshmi et al. (2013)
	Nostoc entophytum and *Oscillatoria angustissima*	Pea	1. Increase in nitrogen fixation. 2. Increase in growth parameters, percentage germination and photosynthetic pigments. 3. Increased pea seeds' nutritional values.	Osman et al. (2010)

(continued)

TABLE 11.2 (Cont'd)

Mode of action	Cyanobacterial extracts	Target crop species/Soil	Improvements advocated	References
2. Soil nutrients' availability	Consortia of cyanobacterial species: *Anabaena doliolum*, *Cylindrospermum sphaerica* and *Nostoc calcicola*	Millet and wheat	1. Enhanced nitrogen and phosphorus bioavailability in the soil. 2. Soil density decreases. 3. Water retention capacity improves, 4. Improve grain yields and nutritional values (increased protein content in grain and leaves).	Nisha et al. (2007)
	Calothrix ghosei, *Hapalosiphon intricatus* and *Nostoc* sp.	Wheat	1. Increased soil organic carbon content. 2. Improved grain yield.	Karthikeyan et al., (2007)
	Microalgal and cyanobacterial consortia, including unicellular and filamentous species of *Chlorella*, *Scenedesmus*, *Chlorococcum*, *Chroococcus*, *Phormidium*, *Anabaena*, *Fischerella* and *Spirogyra*	Wheat	1. Enhanced soil availability of nitrogen, phosphorus, and potassium. 2. Increased soil organic carbon content. 3. Product quality improvement.	Renuka et al., (2016)
	Consortia and biofilms of bacterium (*Azotobacter* sp. and *Providencia* sp.) and cyanobacterial species (*Calothrix* sp. and *Anabaena* sp.)	Okra	1. An increased bioavailability of zinc and iron in the soil. 2. Microbiome amendments. 3. Increase in yield of root and weight.	Manjunath et al. (2016)

	Microalgal and cyanobacterial consortia of unicellular and filamentous species of *Chlorella*, *Scenedesmus*, *Chlorococcum*, *Chroococcus*, *Phormidium*, *Anabaena*, *Fischerella*, and *Spirogyra*	Wheat	Increase in zinc, iron, copper and manganese availability in the soil. Increase in organic carbon content in the soil Increase in grain yield Improvement of plant nutritional value (increase in grain micronutrient contents)	Renuka et al. (2017)
3. Soil physical and chemical amendments	*Spirulina platensis* phycocyanin extract and its inactive biomass	Diesel- and biodiesel-contaminated soil	1. Reduction in diesel and biodiesel from soil contaminated with these pollutants.	DeCesaro et al. (2017)
	Nostoc sp.	Poorly aggregated tropical soils	Aggregates' stability increase.	Issa et al. (2007)
	Microcoleus vaginatus, *Phormidium tenue*, *Scytonema javanicum*, *Nostoc* sp., and *Desmococcus olivaceus*	Unconsolidated sand	1. Enhanced crust cohesion 2. Resistance to wind erosion	Hu et al. (2002)

11.3.1.1 Nitrogen fixation

Cyanobacteria are diazotrophic bacteria and can fix atmospheric dinitrogen (N_2) into organic nitrogen, which is easily assimilated by plants (Singh et al., 2018). Consequently, the cyanobacterial utilization in soils increases the accessibility of nitrogen, which is one of the fundamental supplements for plants' development (Singh et al., 2018). The beneficial impacts of cyanobacteria in farming soils (for the creation of various yields) have effectively been shown (Renuka et al., 2016). For instance, utilizing a cyanobacterial inoculum in a rice oar, Jha and Prasad (2006) inferred that this inoculum improved the grain and straw yields and expanded the paddy fields nitrogen accessibility. Additionally, utilizing *Anabaena variabilis* strains in a rice field, Singh and Datta (2007) showed that the preowned strains advanced a general improvement in the rice crops (e.g., expansion in plant tallness and leaf length, and improvement of seeds, grain and straw productivities). While applying *Nostoc* sp. vegetative cells to a rice field, Innok et al. (2009) acquired an expansion in grain yields practically identical to the one achieved with a synthetically based nitrogen compost. Although most of the investigations allude to the utilization of cyanobacteria in rice fields, ongoing studies have indicated the constructive outcome of cyanobacteria in different cultures. Osman et al. (2010) vaccinated the field planned for pea plant creation with the cyanobacterial species *Nostoc entophytum* and *Oscillatoria angustissima* noticing a critical expansion germination rate, photosynthetic shades, and development boundaries of this plant. Likewise, the authors saw that this expansion was related to a higher N_2 obsession movement by cyanobacteria. In the investigation performed by Swarnalakshmi et al. (2013), the utilization of a biofilm formed by the cyanobacterium *Anabaena torulosa* in a wheat crop altogether expanded the accessibility of nitrogen in the field. Prasanna et al. (2013) assessed the capability of a few cyanobacterial and bacterial consortia as manures for soybean and mungbean, introducing positive results in terms of nitrogen accessibility in the soil and plant productivity. Considering the enhancements such as nitrogen accessibility, it is feasible to conclude that cyanobacteria could substitute compound nitrogen fertilizers, diminishing harvests' production costs by 25–40%. Moreover, the utilization of these microorganisms advances nitrogen reusing efficiencies on the one hand while preventing the persistent accumulation of chemical-based nitrogenous fertilizers. At last, cyanobacteria are known for their capacity to frame soil outside layers that stay away from nitrogen filtering/spillover and, hence, the dangers of water tainting with nitrogen (Renuka et al., 2018).

11.3.1.2 Soils nutrients' bioavailability

The utilization of cyanobacteria in soils can likewise expand the accessibility of other fundamental supplements for plants. Photosynthetic cyanobacteria species take up nitrogen and phosphorus from various conditions (even from places where these nutrients are limited) and accumulate them in their biomass. As these supplements are fundamental for advancing higher plants, cyanobacterial biomass can be considered a significant wellspring of these supplements, preventing utilization of conventional compound manures (Sharma et al., 2012). Other essential microelements such as potassium, magnesium, sulfur and iron), which are required for plants' growth and development also present in the cyanobacterial biomass. These components are commonly associated with redox responses, assuming a significant part in plants' digestion.

Consequently, the utilization of cyanobacteria in soils can improve the accessibility of both large scale and micronutrients for plants' development. This improvement in supplements' accessibility has effectively been accounted for in writing (Marks et al., 2019). In an investigation performed by Manjunath et al. (2016), the utilization of cyanobacterial-bacterial biofilms in addition to their consortia to okra crops brought about an expansion in zinc and iron substance in the field. Furthermore, it added to an expanded root yield and weight. Cyanobacteria can be near a wellspring of organic matter to higher plants, which should be provided sufficient focus for a legitimate plants' development (Renuka et al., 2018; Shrama et al., 2012). Initially, these photosynthetic microorganisms fix atmospheric CO_2, changing over it into natural biomass. Besides, in general, these organic entities will discharge extracellular polymeric substances expanding the natural carbon levels in soils and advancing the development of other microorganisms. Finally, cyanobacterial biomass present in farming fields might be corrupted by different living beings and slow eaters, building the natural carbon content in the areas. The increment of natural carbon substance in soils utilizing microalgae or potentially cyanobacteria has effectively been depicted in writing (Mark et al., 2019). For model, Karthikeyan et al. (2007) exhibited that the utilization of cyanobacterial suspensions (*Calothrix ghosei*, *Hapalosiphon intricatus* and *Nostoc* sp.) in wheat creation expanded biological carbon fixation in the fields while advancing higher grain yields. More recently, Renuka et al. (2018) assessed the impact of unicellular and filamentous microalgal cyanobacterial consortia on soil synthesis and grain nature of a wheat crop. Other than an expansion in micronutrient accessibility in the field (basically zinc, iron, copper and manganese), the contemplated consortia additionally upgraded its natural carbon content. This improvement in soil quality brought about an expanded grain yield and a better item, as the acquired grains introduced an improved healthful synthesis.

11.3.1.3 Amendments in soil physical and chemical properties

Additionally, cyanobacteria assume a significant part in the recuperation of soil contaminant, as they can: (1) adjust soils' pH, (2) control soils' salinity or secure plants from high salinity levels, and (3) eliminate heavy metals, metalloids, and different pollutants from soils (Singh et al., 2015, 2018; Pandey et al., 2012; Rai et al., 2014; Pan et al., 2019; Win et al., 2018; Prasanna et al., 2013; Sharma et al., 2012; Priya et al., 2015). Concerning pH control, a few investigations have detailed either an expansion or, on the other hand, a decline in soils' pH when utilizing cyanobacterial biomass. Be that as it may, it is critical to feature the beneficial impact of this biomass in controlling the acidic/essential nature of various soils (Pan et al., 2019; Chatterjee et al., 2017). As far as soils' salinity is concerned, a few investigations have revealed an improvement in plants' development in soils introducing high salinity (Shrama et al., 2017). This improvement might be an aftereffect of (1) a decrease in the soils' salt (by expanding the water maintenance limit of soils), or (2) an expansion in plants' resistance to high salt levels (which may be accomplished by gibberellic corrosive, a phytohormone ordinarily found in cyanobacteria). Finally, cyanobacteria can sequester minor components (e.g., iron, zinc, copper, and manganese), hefty metals (e.g., cadmium, lead, and chromium), hydrocarbons, and different mixtures from the general climate. Hence, cyanobacteria may add to its remediation and expand soils' quality and fruitfulness (Priynka et al., 2020). Decesaro et al. (2017) assessed the capability of dormant *Spirulina platensis* cells and phycocyanin extracts from similar species on hydrocarbons degradation from a soil contaminated with diesel and biodiesel. Contrasted with the control, phycocyanin extricates and inert cells have adequately eliminated hydrocarbons from the defiled soils. The concentrates are more potent in biodiesel expulsion (88.8%) and *S. platensis* dormant cells in diesel expulsion (63.9%). Other than their significance in the advancement of soils with fundamental supplements and the recovery of soil properties, cyanobacteria can likewise improve a few soil qualities such as porosity, aggregation, ventilation, permeability, and humidity (Baweja et al., 2019; Chatzissavvidis and Therios, 2014; Pan et al., 2019; Renuka et al., 2018). Regularly, the improvement of these qualities is related to the polysaccharides discharged by a few cyanobacterial species (Baweja et al., 2019). When released to the soil, these polysaccharides will form a glue and thick sheath that adds to soils' collection and keeps them from disintegration (Baweja et al., 2019). These polysaccharides additionally assume a significant part in soils' air circulation and in their capacity for water maintenance, which is pivotal to keep up soils' temperature, pH, and saltiness (Baweja et al., 2019; Chatzissavvidis and Therios, 2014; Pan et al., 2019; Renuka et al., 2018). The improvement of these soil properties fundamentally upgrades yields' productivities, as they animate root development and lift the soil microbial movement. Accordingly, cyanobacteria have been generally utilized to rebuild soils' properties and improve soil fertility. Hu et al.

(2002) exhibited that the protection from wind disintegration of unconsolidated sand was expanded within sight of desert soil green growth, for example, *Microcoleus vaginatus*, *Phormidium tenue*, *Scytonema javanicum*, *Nostoc* sp., and *Desmococcus olivaceus*, presuming that this expansion was related with higher outsides' union in the examples treated with the alluded green growth. Also, Issa et al. (2007) inoculated inadequately accumulated soil with *Nostoc* strains, accomplishing a more increased level of collection in the inoculated soil when contrasted with the noninoculated one.

11.3.3 Crops direct growth stimulation

Aside from their gainful impacts at the soil level, the utilization of cyanobacteria (and/or their extracts) can straightforwardly animate plants' development and advancement through an improvement in germination rates and plant qualities (e.g., expanded shoot and root length, expanded area leaf and higher nutrient contents). These upgrades are accomplished because of the cyanobacterial metabolite's movement, which can trigger many metabolic processes such as respiration, photosynthesis, nucleic and chlorophyll synthesis, nutrient uptake (Górka et al., 2018). Table 11.3 presents a few models where cyanobacteria (biomass or their extracts or their consortia) have been effectively applied in the direct development incitement of higher plants.

Usually, these metabolites incorporate plant development–promoting components, such as phytohormones, some amino acids, nutrients (e.g., nutrient B12 and biotin), polysaccharides, and polypeptides (Sharma et al., 2012). Saadatnia and Riahi (2009) advocated that paddy field treatment with *Anabaena* enhances the seeds' germination rate and crop growth improvements such as plant height, root length and leaves fresh and dry weight, stems and roots. Elarroussia et al. (2016) assessed the impact of applying all polysaccharides extracts of *S. platensis* on tomato and pepper and recorded improved crop traits for both.

11.3.4 Crops' protection against stresses

Yields security against biotic and abiotic stresses is vital for the improvement of agriculture productivities. Quite a few crop plants have their resilience mechanisms, yet, to accomplish the hostile objective of productivities, outside assurance specialists should utilize food crops. Considering the vast diversity of cyanobacterial bioactive compounds, the utilization of these cyanobacterial species and/or their extracts can advance a sufficient yields' insurance against both biotic and abiotic factors. Table 11.3 presents a few instances of the use of cyanobacteria (biomass and extracts) on yields' security and profitability.

TABLE 11.3 Cyanobacterial and/or their metabolites and their impact on crop biostimulants.

Mode of action	Cyanobacterial extracts	Target crop species/Soil	Improvements advocated	References
1. Direct plants growth stimulation	Strains of *Anabaena such as Anabaena osillarioides, Anabaena variabilis, Anabaena spiroides,* and *Anabaena torulosa*	*Oryzae*	1. Improved rate of germination of seeds. 2. Increase in shoots length and plant height. 3. Increase leaf, stem, and root dry and fresh weights. 4. Soil properties such as soil moisture and porosity improvement.	Saadatnia and Hossein (2009)
	Spirulina platensis total polysaccharides extract	Pepper and tomato	1. Upsurge plants' size and root weight. 2. Improved size and number of nodes.	Elarroussia et al., 2016
	Spirulina platensis and *Chlorella vulgaris*	Maize	1. Improved rate of germination of seeds and plant yield. 2. Increase in shoots length and plant height. 3. Increase in leaf, stem, root and whole plants dry and fresh weights. 4. Increase in the length of the shoot and leaves a number.	DineshKumar et al. (2017)
	Arthrospira platensis	Lettuce	Increase in growth of seedling and spermine content.	Mógor et al. (2018)
2. Protection against abiotic stresses	*Scytonema hofmanni*	*Oryzae*	Salt stress tolerance alleviation	Rodriguez et al. (2006)
	Oscillatoria agardhii	Wheat	1. Drought stress alleviation. 2. Antioxidative defense enzymes such as catalase, superoxide dismutase and peroxidase, and hydrolytic enzymes activities upregulated. 3. Grain yield increases. 4. Plant quality parameters improved.	Haggag et al. (2018)

3. Protection against biotic stresses	*Calothrix elenkinii* and *Anabaena laxa*	Cumin, coriander, and fennel	1. In shoots and roots, and upregulated activity of 1,3 endoglucanase. 2. Antifungal potential. 3. Germination rate alleviation. 4. Shoot and root length increment. 5. Enhanced plant dry weight.	Kumar et al., 2013
	Anabaena sp., *Oscillatoria* sp., *Nostoc* sp., *Calothrix* sp. and *Nodularia* sp.	Pepper	1. Antifungal activity against Fungicidal activity against *Botrytis cinerea* and *Alternaria alternate*	Kim (2006)
	Anabaena oscillarioides and *Anabaena variabilis*	Tomato	1. Fungal disease severity. Improved plant growth (increased height and fresh weight of plants)	Chaudhary et al. (2012)
	Calothrix elenkinii	*Oryzae*	1. Upregulated activities of defense enzymes such as phenylalanine ammonia-lyase, peroxidase, and polyphenol oxidase, and hydrolytic enzymes. 2. Upregulated nitrogenase activity and indole acetic acid (auxin) production 3. Plant growth increment.	Priya et al., (2015)
	Oscillatoria chlorina	Tomato	Antifungal activity against *Meloidogyne arenaria* and improved plant growth.	Khan et al. (2007)
	Anabaena laxa, Anabaena torulosa, Anabaena azollae, Calothrix sp., *and Anabaena oscillarioides*	Wheat	1. Improved nitrogen fixation. 2. Enhance hydrolytic and defense enzymes such as endoglucanase, phenylalanine ammonia-lyase, and peroxidase, polyphenol oxidase activities. 3. Improved plant fresh and dry weight.	Santosh et al., (2015)

11.3.4.1 Protection against abiotic stresses

Plants' openness to outrageous natural conditions (such as freezing and hot temperatures, high salinity, drought, waterlogging, heavy metals) adversely affects plant growth and developments, hence, productivity and quality (Singh et al., 2018). On account of excessive temperatures, low temperatures may reduce the metabolic activity of higher plants and may be detrimental to the cell membrane, as the phospholipidic layer that comprises these layers becomes unstable. Exposure to high temperatures may likewise be hurtful for cell membranes and impact the synthesis and activities of proteins and enzymes (Wang et al., 2020). However, high salinity disturbed the plant osmotic pressure, which influences the plant's nutritional values and, henceforth, plants' growth and development (Górka et al., 2018; Bulagri et al., 2019; Kaya et al., 2002). Dry spell is moreover answerable for the osmotic pressure in plants.

Nonetheless, it additionally influences plants' capacity for gas exchange (stomatal movement), essentially impacting the photosynthetic and respiration rates and, therefore, plant productivity (Górka et al., 2018; Bulagri et al., 2019). Considering the adverse consequences of outrageous ecological conditions on plants' development and advancement, expanding the resistance of higher plants to these unforgiving conditions is pivotal for improving crop productivities. For this reason, the utilization of biostimulants can be an exceptionally encouraging choice, as these extracts are ready to (Bulagri et al., 2019; Kaya et al., 2002): (1) offer accumulation of antioxidants and antioxidative defense system enzymes, which help crops in tolerating abiotic stress; and (2) improve the overall survival of plants under hostile environments to sustain crop productivity. Although a few natural extracts are known for their stimulatory movement, the utilization of the second-degree metabolites produced by cyanobacteria has pulled in researchers worldwide because of their wide range and capacity to offer stress resilience metabolic activity enhancement. Indeed, the utilization of these biostimulants has been effectively applied in food crops under temperature, salt and drought stress (Table 11.3) (Rodriguez et al., 2006; Haggag et al., 2018).

11.3.4.2 Protection against biotic stresses

Biotic stresses that can affect yields' productivities incorporate organisms such as bugs, nematodes, bacteria, fungi, and viruses (Kumar et al., 2020). Insurance against these pathogenic life forms is regularly achieved by polysaccharides, which perceive the flagging particles present in the pathogenic life form cell wall and instigate several defense reactions (Chanda et al., 2019). Usually, the defense pathways such as signaling pathways, induction of specific defense pathways, and expression of genes, which result in the accumulation of secondary metabolites (such as terpenoids, phenolic compounds, and other compounds), serve as antimicrobial, antioxidant, and fungicidal assets (Chanda et al., 2019; Wang et al., 2020). When applied in agriculture fields, cyanobacteria may advance these safeguard reactions, as cyanobacteria are a source of incredible polysaccharides source (Chanda et al., 2019). Table 11.3 shows the application of cyanobacterial extracts applied in crop growth protection from the biotic agents.

11.4 Cyanobacteria as biostimulants in agriculture

Cyanobacteria are ubiquitous oxygenic photosynthetic prokaryotes occurring in almost all realms of life on the earth. They evolved nearly 3.5 billion years ago and were responsible for making the earth's atmosphere oxygenic. The adaptive responses developed by cyanobacteria have helped them to survive under different environmental conditions. They occur in a variety of habitats land, water, or rock surface. They even thrive in extremes of temperature, salinity, and pH. Cyanobacteria range from unicellular forms to highly complex structures. One can quickly draw a line of evolution from the order *Chroococcales*, unicellular forms to highly evolved *stigonematales*. Many cyanobacterial forms could be valued ecologically as they have the unique ability to convert unusable dinitrogen to ammonia. Cyanobacteria contribute to soil fertility by atmospheric nitrogen fixation and act as a safe supplementary biofertilizer in agriculture. Cyanobacterial species such as *Nostoc* and *Anabaena* have outstanding capabilities to treat the soil salinization problem by fixing atmospheric dinitrogen and producing an extracellular matrix and compatible solutes. Cyanobacteria can readily help reduce soil salinity, weed growth and decay, and increase soil biomass.

Cyanobacteria possess a unique attribute of nitrogen fixation in paddy fields and, consequently, increase rice growth and productivity (Tripathi et al., 2008; Mishra and Pabbi 2004; Padhy et al. 2016). Their agricultural importance in rice cultivation generally relies on their nitrogen-fixing ability and other beneficial effects on soil and plants. For example, *Anabaena, Aulosira, Nostoc, Calothrix,* and *Plectonema* are the main nitrogen-fixing species. These species play the role of biofertilizers and help maintain long-term soil fertility and sustainability by fixing atmospheric dinitrogen and allowing for conversion to ammonium, solubilization of fixed nutrients, or the modification of insoluble phosphorus in the soil to plant usable forms (Burns et al. 2013). In a recent study, *Spirulina platensis* filtrates were observed to increase the chlorophyll content and elemental constituents in radish seedlings (Godlewska et al., 2019). These findings proved algal extracts' potential for application in future horticulture and agriculture (Bhowmik et al. 2010). The products derived from *Spirulina* species are already in the market, thus ensuring food production meets the increasing population and environmental protection demands. Cyanobacterial and microalgal biomass applied to soils as fertilizer improve the water-binding capacity and nutrient composition of nutrient-poor soils. Overall, biofertilizers such as cyanobacteria are an environmentally and economically sustainable option for agricultural purposes because of their lesser input of capital and energy than conventional synthetic fertilizers such as urea. Their short life-cycle and high biomass, and the unique property of photosynthesis, just like higher plants, are incomparable to any other life-forms, which could aid in sustainable growth. Furthermore, alkaline phosphatase produced by cyanobacteria help make phosphorous available to plants; along with it, they also produce hydrolytic enzymes, such as protease, carboxymethyl cellulase (CMCase), cellobiose, and filter paperase (Fpase), that have a fungicidal activity and are responsible for nutrient recycling of carbon, nitrogen, and phosphate necessary for plant growth (Grzesik and Romanowska-Duda 2015; Burns et al. 2013).

As biostimulants, their role is magnetic; now and then, new reports are being added to the literature highlighting their importance. Vaishampayan et al. (2001) recommended that the *Azolla-Anabaena* (the free-living cyanobacteria *Anabaena* and the water fern *Azolla*) symbiotic N_2-fixing complex be considered self-renewable natural nitrogen resources to reduce inorganic N requirements to the bare minimum. The cyanobacterium *Tolypothrix* sp. was found to produce bioproducts in tropical regions by using low nitrogen-containing water sources. Besides *Anabaena* and *Tolypothrix*, *Aulosira* sp., *Cylindrospermum* sp. and *Nostoc* sp. have also been used in wetland rice cultivation (Zayadan et al. 2014; Ashok et al. 2017). In another work, the inoculants of *Anabaena laxa* and *Anabaena*-Rhizobium consortium were used to formulate biofilm in Chickpea cultivation. The *A. laxa* inoculation for the biofilm led to a 50% higher grain yield (1724 kg/ha) than the control (847 kg/ha). Foliar application of *S. platensis* has been used for eggplant and other leafy vegetables (Alves Dias et al., 2017; Wuang et al., 2016).

Many of the metabolites from cyanobacteria have been shown to act as biostimulants affecting plants' physiology and production. A combination of bacterial and cyanobacterial inoculants increased the average germination compared to seeds untreated or treated by hormones (Indole-3-acetic acid (IAA), Gibberellic acid (GA3), and cytokinins) in *Lupinus termis* (Tantawy and Atef 2010). Rodrı´guez et al. 2006 reported that the extracellular products of *Scytonema hofmanni* produced gibberellin-like plant growth regulators, which enabled the hormone homeostasis of rice seedlings under salt stress.

Another study by El Arroussi et al. (2018) advocated that blue-green alga *S. platensis* polysaccharide extract on every three-day intervals spraying on tomato and pepper augmented the plant size, root dry weight, the nodes number and size up to 100% and 50% for tomato and pepper, respectively. Barone et al. (2018a) tested the effect of *Scenedesmus quadricauda* extracts on sugar beet seedlings grown hydroponically and demonstrated the changes in root attributes. They advocated a higher total root length, surface, and number of root tips. Further, they supported that the several biological pathways genes involved in the primary and secondary metabolism have been upregulated.

Bioactive molecules of *Arthrospira platensis* are the polyamine content, mainly spermidine, which could stimulate cell multiplication and plant growth. Biostimulation effect has been observed on foliar applications of the aqueous suspension of *A. platensis* biomass on red beet seedlings growth (Mógor et al. 2018).

One of the latest findings makes use of nanoparticles (NPs) and nanomaterials (NMs) as biostimulants to increase crop production (Morales-Díaz et al. 2017; Zuverza-Mena et al. 2017). NPs) and NMs) on application at lesser amounts on foliar spray or in nutrient solutions positively change the metabolism, composition, and nutraceutical quality of food crops, as well as their tolerance to stress (Juárez-Maldonado et al. 2019).

11.5 Other biostimulants and their role in plant growth stimulations

By definition, plant biostimulants include various substances and microorganisms that enhance plant growth and production. The description, however, keeps evolving as new findings and concepts are being added. Thus plant biostimulants positively affect plant growth and development by direct or indirect means affecting the physiology, morphology, or functional aspects of plant metabolism as a whole or partially. The pioneer worker in the field of biostimulants is du Jardin (2012), who for the first time made a bibliographic analysis of plant biostimulants and grouped them into eight categories, namely: seaweed extracts, humic substances and fulvic acids and other complex organic compounds, some elements, inorganic salts (such as phosphite), N-containing substances, chitin and chitosan derivatives, and as well as free amino acids. Humic acids are high-molecular-weight compounds generally formed by microbial decomposition and degradation of dead biota (Asli and Neumann 2010; Schiavon et al. 2010; Simpson et al. 2002; Nardi et al. 2009). Fulvic acids are, on the other hand, low-molecular-weight compounds (Varanini and Pinton 2001). The fundamental structural differences of humic and fulvic acids depend on the organic material source, and the transformation time taken (Muscolo et al. 2007b). The chemical structure also determines the type of function performed. Humic substances have been reported to favor many plants such as wheat, rice, soybean, maize, potato, tomato, cucumber, pepper, lemon, grape, and *Arabidopsis*. Humic substances affect the various ions uptake by plants that usually depend on the concentration and type of the humic substance, pH of the growing media, and the plant species (Muscolo et al. 2007a, Nardi et al. 2009). Humic substances have been known to affect plant physiology by improving the soil structure and fertility, nutrient uptake efficiency, and root architecture (Trevisan et al. 2010).

Fulvic acids (FAs) have more excellent total acidity, more significant numbers of carboxyl groups, and cation exchange capacities and higher adsorption than humic acid (Bocanegra et al., 2006). FAs are responsible for the metal ions (including Fe and Al) chelation and mobilization (Esteves da Silva et al. 1998; Lobartini et al. 1998). At high salt concentrations and broader pH range, small molecular weight FAs remain in soil solution (Zhang and Ervin 2004; Zhang et al. 2006; Zhang and Ervin 2008; Zhang et al. 2010; Zhuang et al. 2007; Zimmerli et al. 2008).

Thus, both humic acid and FA interact with soil nutrients and evoke physiological responses in plants, leading to increased plant growth and ameliorating abiotic stresses.

Protein-based products are broadly divided into two major groups: protein hydrolysates consisting of a mixture of peptides and amino acids that may be of animal or plant origin and individual amino acids such as glutamate, glutamine, proline, and glycine betaine. The uptake of amino acids both root and foliar has been known to ultimately enhance plant growth (Furuya and Umemiya 2002; Miller et al. 2007; Maini 2006; Stiegler et al. 2013). Protein hydrolysates are made by hydrolysis of animal and plant residues. Siapton (amino acid–based biostimulant) contains a high proportion of glycine and proline, while hydrolysate of glutamine and arginine predominates in carob germ (Parrado et al. 2008). The other class of protein-based products includes 20 structural amino acids. Studies with individual amino acids propose that they may play a signaling role in regulating nitrogen acquisition and C and N metabolism regulation. There is a growing number of shreds of evidence suggesting that protein hydrolysates and specific amino acids, including proline, betaine, their derivatives and precursors, can elicit the plant defense responses and enhances plant tolerance to various abiotic stresses such as salinity, drought, temperature, and oxidative conditions (Ashraf and Foolad 2007; Chen and Murata 2008; Kauffman et al. 2007; Apone et al. 2010; Ertani et al. 2013a).

Seaweed extracts are known to act as chelators, thus stimulating root growth particularly (Milton 1964). Seaweed extracts also work as biostimulants, enhancing seed germination and establishment, improving plant growth, yield, flower set and fruit production, increasing resistance to biotic and abiotic stresses, and improving postharvest shelf life (Mancuso et al. 2006; Norrie and Keathley 2006; Hong et al. 2007a, 2007b; Rayorath et al. 2008; Khan et al. 2009; Craigie 2011; Mattner et al. 2013). The biostimulatory effect is due to plant growth hormones and related low-molecular-weight compounds present in the extracts (Stirk and van

Staden 1997; Tarakhovskaya et al. 2007), even the presence of high-molecular-weight polysaccharides and polyphenols aids in enhancing resistance to stress (Klarzynski et al. 2003; Zhang et al. 2006; Rioux et al. 2007; González et al. 2013). There are several reports of beneficial effects of seaweed extract on shoot growth and crop yield (Verkleij 1992; Khan et al. 2009; Craigie 2011).

11.6 Challenges involved in using cyanobacteria as biostimulants

Biostimulants have a positive response on plants by affecting plants' physiology and biology; however, the successful application demands healthy tripartite interactions among biostimulants, the plant, and the environment (du Jardin 2015). Biostimulants are becoming popular in agriculture, but they are not used in routine; more acceptance is needed from the agricultural sector. General setbacks and problems associated with biostimulants are concerned with screening and application.

In plant growth-promoting microbes need to be screened for the successful field applications require plenty of time and thoughtful research experiments. Biostimulants may be used in soil preparations (powders, granules, or solutions added to the soil) or liquid foliar application products. Most commonly, biostimulants are used in an extract or powder form. The timing of the application is also essential. Some biostimulants are applied directly onto harvested fruits (Battacharyya et al., 2015).

Moreover, farmers usually do not readily take up the idea to adopt biostimulant products unless their success is guaranteed, mostly in developing countries.

11.7 Prospects and conclusion

Cyanobacterial biostimulants serve as an effective and sustainable tool to enhance plant growth and productiveness, improving tolerance against abiotic stresses. There is no doubt regarding the implications of the use of cyanobacteria as biostimulants in agriculture. However, more research studies simplifying application and success problems in field conditions should be a priority.

Acknowledgment

Prashant Kumar Singh is thankful to University Grant Commission (UGC), New Delhi, for Start-up Research Grant.

References

Alves Dias, G., Garcı´a-Martı´nez, M., Osornio-Berthet, L., 2017. Growth, yield, and postharvest quality in eggplant produced under different foliar fertilizer (*Spirulina platensis*) treatments. Water Sci. Technol. 37 (6), 1693–1701.

Apone, F., Tito, A., Carola, A., et al., 2010. A mixture of peptides and sugars derived from plant cell walls increases plant defence responses to stress and attenuates ageing-associated molecular changes cultured skin cells. J. Biotechnol. 145, 367–376.

Ashok, A.K., Ravi, V., Saravanan, R., 2017. Influence of cyanobacterial auxin on sprouting of taro (*Colocasia esculenta* var. *antiquorum*) and corm yield. Indian J. Agric. Sci. 87 (11), 1437–1444.

Ashraf, M., Foolad, M.R., 2007. Roles of glycine betaine and proline in improving plant abiotic stress resistance. Environ. Exp. Bot. 59, 206–216.

Asli, S., Neumann, P.M., 2010. Rhizosphere humic acid interacts with root cell walls to reduce hydraulic conductivity and plant development. Plant Soil 336, 313–322.

Barone, V., Baglieri, A., Stevanato, P., Broccanello, C., Bertoldo, G., Bertaggia, M., 2018a. Root morphological and molecular responses induced by microalgae extracts in sugar beet (*Beta vulgaris* L.). J. Appl. Phycol. 30, 1061–1072. doi:10.1007/s10811-017-1283-3.

Battacharyya, D., Babgohari, M.Z., Rathor, P., Prithiviraj, B., 2015. Seaweed extracts as biostimulants in horticulture. Sci. Hortic. 196, 39–48.

Baweja, P., Kumar, S., Kumar, G., 2019. Organic fertilizer from algae: A novel approach towards sustainable agriculture. In: Giri, B., Prasad, R., Wu, Q.S., Varma, A. (Eds.), Biofertilizers for Sustainable Agriculture and Environment. Springer, Cham, Switzerland, pp. 353–370.

Betterle, N., Melis, A., 2019. Photosynthetic generation of heterologous terpenoids in cyanobacteria. Biotechnol. Bioeng. 116, 2041–2051 2019.

Bhowmik, D., Dubey, J., Mehra, S., 2010. Evaluating the potential of *Spirulina* as inoculant for pulses. Acad. J. Plant Sci. 3 (4), 161–164.

Bocanegra, M.P., Lobartini, J.C., Orioli, G.A., 2006. Plant uptake of iron chelated by humic acids of different molecular weights. Commun. Soil Sci. Plant Anal. 37, 1–2.

Bulgari, R., Franzoni, G., Ferrante, A., 2019. Biostimulants application in horticultural crops under abiotic stress conditions. Agronomy 9, 306.

Burns, R.G., DeForest, J.L., Marxsen, J., 2013. Soil enzymes in a changing environment: current knowledge and future directions. Soil Biol. Biochem 58, 216–234.

Campos, E.V.R., De Oliveira, J.L., Fraceto, L.F., Singh, B., 2015. Polysaccharides as safer release systems for agrochemicals. Agron. Sustain. Dev. 35, 47–66.

Chanda, M.J., Merghoub, N., Hicham, E.A., 2019. Microalgae polysaccharides: the new sustainable bioactive products for the development of plant bio-stimulants? World J. Microbiol. Biotechnol. 35, 177 2019.

Chatterjee, A., Singh, S., Agrawal, C., Yadav, S., Rai, R., Rai, L.C., 2017. Role of algae as a biofertilizer. In: Rastogi, R.P., Madamwar, D., Pandey, A. (Eds.). Algal Green Chemistry, 2017. Elsevier BV, Amsterdam, The Netherlands, pp. 189–200.

Chatzissavvidis, C., Therios, I., 2014. Role of algae in agriculture. In: Pomin, V.H. (Ed.), Seaweeds: Agricultural Uses, Biological and Antioxidant Agents. Nova Science Publishers, Inc., Hauppauge, NY, USA, pp. 1–37.

Chaudhary, V., Prasanna, R., Nain, L., Dubey, S.C., Gupta, V., Singh, R., Jaggi, S., Bhatnagar, A.K., 2012. Bioefficacy of novel cyanobacteria amended formulations in suppressing damping-off disease in tomato seedlings. World J. Microbiol. Biotechnol. 28, 3301–3310.

Chen, T.H.H., Murata, N., 2008. Glycinebetaine: an effective protectant against abiotic stress in plants. Trends Plant. Sci. 13, 499–505.

Craigie, J.S., 2011. Seaweed extract stimuli in plant science and agriculture. J. Appl. Phycol. 23, 371–393.

DeCesaro, A., Rampel, A., Machado, T.S., Thomé, A., Reddy, K., Margarites, A.C., Colla, L.M., 2017. Bioremediation of soil contaminated with diesel and biodiesel fuel using biostimulation with microalgae biomass. J. Environ. Eng. 143, 04016091.

Dineshkumar, R., Subramanian, J., Gopalsamy, J., Jayasingam, P., Arumugam, A., Kannadasan, S., Sampathkuma, P., 2017. The impact of using microalgae as biofertilizer in maize (*Zea mays* L.). Waste Biomass Valori. 10, 1101–1110.

Du Jardin, P., 2012. The Science of Plant Biostimulants- A bibliographic analysis. Adhoc Study Report to the European Commission DG ENTR 2012. http://ec.europa.eu/enterprise/sectors/chemicals/files/fertilizers/finalreportbio2012en.pdf.

Du Jardin, P., 2015. Plant biostimulants: Definition, concept, main categories and regulation. Sci. Hortic. 196, 3–14.

El Arroussi, H., Benhima, R., Elbaouchi, A., Sijilmassi, B., El Mernissi, N., Aafsar, A., 2018. Dunaliella salina exopolysaccharides: a promising biostimulant for salt stress tolerance in tomato (*Solanum lycopersicum*). J. Appl. Phycol. 30, 2929–2941. doi:10.1007/s10811-017-1382-1.

Elarroussia, H., Elmernissia, N., Benhimaa, R., El Kadmiria, I.M., Bendaou, N., Smouni, A., Wahbya, I., 2016. Microalgae polysaccharides a promising plant growth biostimulant. J. Algal Biomass Util. 7, 55–63.

El-Baz, F., El-Senousy, W., El-Sayed, A., Kamel, M., 2013. In vitro antiviral and antimicrobial activities of *Spirulina platensis* extract. J. Appl. Pharm. Sci. 3, 52–56.

ErtaniA, PizzeghelioD, Altissimo, A., Nardi, S., 2013a. Use of meat hydrolyzate derived from tanning residues as plant biostimulant for hydroponically grown maize. J Plant Nutr. Soil Sci. 176, 287–296.

Esteves da Silva, J.C.G., Machado, AASC, Oliveira, C.J.S., 1998. Effect of pH on complexation of Fe(III) with fulvic acids. Environ. Toxicol. Chem. 17, 1268–1273.

Furuya, S., Umemiya, Y., 2002. The influence of chemical forms on foliar-applied nitrogen absorption for peach trees. Acta Hortic. 594, 97–103.

Gershenzon, J., Dudareva, N., 2007. The function of terpene natural products in the natural world. Nat. Chem. Biol. 3, 408–414.

Godlewska, K., Michalak, I., Pacyga, P., Baśladyńska, S., Chojnacka, K., 2019. Potential applications of cyanobacteria: *Spirulina platensis* filtrates and homogenates in agriculture. World J. Microbiol. Biotechnol. 35, 80. 0112 33456789. https://doi.org/10.1007/s11274-019-2653-6.

Gonçalves, A.L., 2021. The use of microalgae and cyanobacteria in the improvement of agricultural practices: a review on their biofertilising, biostimulating and biopesticide roles . Appl. Sci. 11 (871). 2021. https://doi.org/10.3390/app11020871.

González, A., Castro, J., Vera, J., Moenne, A., 2013. Seaweed oligosaccharides stimulate plant growth by enhancing carbon and nitrogen assimilation, basal metabolism, and cell division. J. Plant Growth Regul. 32, 443–448.

Górka, B., Korzeniowska, K., Lipok, J., Wieczorek, P.P., 2018. The Biomass of Algae and Algal Extracts in Agricultural Production. In: Chojnacka, K., Wieczorek, P.P., Schroeder, G., Michalak, I. (Eds.), Algae Biomass: Characteristics and Applications. Springer Science and Business Media LLC, Berlin, Germany, pp. 103–114.

Grzesik, M., Romanowska-Duda, Z., 2015. Ability of cyanobacteria and green algae improves willow plants' metabolic activity and development. Pol. J. Environ. Stud 24, 1003–1012 2015.

Haggag, W., Hoballah, M., Ali, R., 2018. Applications of nano biotechnological microalgae product for improve wheat productivity in semiarid areas. Int. J. Agric. Technol. 14, 675–692.

Hong, D.D., Hien, H.M., Son, P.N., 2007. Seaweeds from Vietnam used functional food, medicine and biofertilizer. J. Appl. Phycol. 19, 817–826.

Hu, C., Liu, Y., Song, L., Zhang, D., 2002. Effect of desert soil algae on the stabilization of fine sands. Environ. Biol. Fishes 14, 281–292.

Innok, S., Chunleuchanon, S., Boonkerd, N., Teaumroong, N., 2009. Cyanobacterial akinete induction and its application as biofertilizer for rice cultivation. Environ. Biol. Fishes 21, 737–744.

Issa, O.M., Défarge, C., Le Bissonnais, Y., Marin, B., Duval, O., Bruand, A., D'Acqui, L.P., Nordenberg, S., Annerman, M., 2007. Effects of the inoculation of cyanobacteria on the microstructure and the structural stability of a tropical soil. Plant Soil 290, 209–219.

Jha, M.N., Prasad, A.N., 2006. Efficacy of new inexpensive cyanobacterial biofertilizer including its shelf-life. World J. Microbiol. Biotechnol. 22, 73–79.

Juárez-Maldonado, A., Ortega-Ortíz, H., Morales-Díaz, A.B., González-Morales, S., Morelos-Moreno, A., Cabrera-De la Fuente, M., Sandoval-Rangel, A., Cadenas-Pliego, G., Benavides-Mendoza, A., 2019. Nanoparticles and nanomaterials as plant biostimulants. Int. J. Mol. Sci. 20, 162. doi:10.3390/ijms20010162.

Karthikeyan, N., Prasanna, R., Nain, L., Kaushik, B.D., 2007. Evaluating the potential of plant growth-promoting cyanobacteria as inoculants for wheat. Eur. J. Soil Biol. 43, 23–30.

Kauffman III, G.L., Kneival, D.P., Watschke, T.L., 2007. Effects of biostimulant on the heat tolerance associated with photosynthetic capacity, membrane thermostability, and polyphenol production of perennial ryegrass. Crop Sci. 47, 261–267.

Kaya, C., Higgs, D., Sakar, E., 2002. Response of two leafy vegetables grown at high salinity to supplementary potassium and phosphorus during different growth stages. J. Plant Nutr. 25, 2663–2676.

Khan, W., Rayirath, U.P., Subramanian, S., et al., 2009. Seaweed extracts as biostimulants of plant growth and development. J. Plant Growth Regul. 28, 386–399.

Khan, Z., Kim, Y.H., Kim, S.G., Kim, H.W., 2007. Observations on the suppression of root-knot nematode (*Meloidogyne arenaria*) on tomato by incorporation of cyanobacterial powder (*Oscillatoria chlorina*) into potting field soil. Bioresour Technol 98 (1), 69–73. doi:10.1016/j.biortech.2005.11.029.

Kim, J.D., 2006. Screening of cyanobacteria (blue-green algae) from rice paddy soil for anti-fungal activity against plant pathogenic fungi. Mycobiology 34, 138–142.

Klarzynski, O., Descamps, V., Plesse, B., et al., 2003. Sulfated fucan oligosaccharides elicit defence responses in tobacco and local and systemic resistance against tobacco mosaic virus. Mol. Plant-Microbe Interact. 16, 115–122.

Kumar, M., Prasanna, R., Bidyarani, N., Babu, S., Mishra, B.K., Kumar, A., Adak, A., Jauhari, S., Yadav, K., Singh, R. et al., 2013. Evaluating the plant growth promoting ability of thermotolerant bacteria and cyanobacteria and their interactions with seed spice crops. Sci. Hortic. 164, 94–101.

Kumar, A., Changwal, C., Hada, A., Singh, P.K., 2020. Molecular insight of plant-pathogen interactions. In: Roychowdhury, R., Choudhury, S., Hasanuzzaman, M., Srivastava, S. (Eds.), Sustainable Agriculture in the Era of Climate Change. Springer, Cham. https://doi.org/10.1007/978-3-030-45669-6_21.

Lobartini, J.C., Tan, K.H., Pape, C., 1998. Dissolution of aluminium and iron phosphate by humic acids. Commun. Soil Sci. Plant Anal. 29, 535–544.

Lutz, W., Sanderson, W.C., Scherbov, S., 2008. The coming acceleration of global population ageing. Nature 451 (7179), 716–719.

Maini, P., 2006. The first biostimulant experience, based on amino acids and peptides: a short retrospective review on the laboratory research and the practical results. Fertil. Agrorum 1, 29–43.

Mancuso, S., Azzarello, E., Mugnai, S., Briand, X., 2006. Marine bioactive substances (IPA extract) improve foliar ion uptake and water stress tolerance in potted *Vitis vinifera* plants. Adv. Hortic. Sci. 20, 156–161.

Manjunath, M., Kanchan, A., Ranjan, K., Venkatachalam, S., Prasanna, R., Ramakrishnan, B., Hossain, F., Nain, L., Shivay, Y.S., Rai, A.B., et al., 2016. Beneficial cyanobacteria and eubacteria synergistically enhance the bioavailability of soil nutrients and yield of okra. Heliyon 2, e00066 2016.

Marks, E.A., Montero, O., Rad, J.C., 2019. The biostimulating effects of viable microalgal cells applied to a calcareous soil: increases in bacterial biomass, phosphorus scavenging, and precipitation of carbonates. Sci. Total. Environ. 692, 784–790.

Mattner, S.W., Wite, D., Riches, D.A., Porter, I.J., Arioli, T., 2013. The effect of kelp extract on seedling establishment of broccoli on contrasting soil types in southern Victoria, Australia. Biol. Agric. Hortic. 29, 258–270.

Miller, A.J., Fan, X., Shen, Q., Smith, S.J., 2007. Amino acids and nitrate as signals for nitrogen acquisition regulation. J. Exp. Bot. 59, 11–119.

Milton R.F., 1964. Liquid seaweed as a fertilizer. Proc. Int. Seaweed Symp. 4, 428–431.

Mishra, U., Pabbi, S., 2004. Cyanobacteria: a potential biofertilizer for rice. Resonance 9 (6), 6–10.

Mógor, A.F., Amatussi J de, O., Mógor, G., Lara, G.B., 2018. Bioactivity of cyanobacterial biomass related to amino acids induces growth metabolic changes on seedlings and yield gains of organic red beet. Sci. Res. 9 (5), 1–4.

Mógor, ÁF, Ördög, V., Lima, G.P.P., Molnár, Z., Mógor, G., 2018. Biostimulant properties of cyanobacterial hydrolysate related to polyamines. Environ. Biol. Fishes 30, 453–460.

Morales-Díaz, ABAB, Ortega-Ortíz, H., Juárez-Maldonado, A., Cadenas-Pliego, G., González-Morales, S., Benavides-Mendoza, A., 2017. Application of nanoelements in plant nutrition and its impact in ecosystems. Adv. Nat. Sci. Nanosci. Nanotechnol. 8, 013001.

Muscolo, A., Sidari, M., Attiná, E., Francioso, O., Tugnoli, V., Nardi, S., 2007b. Biological activity of humic substances is related to their chemical structure. Soil Sci. Soc. Am. J. 71, 75–85.

Muscolo, A., Sidari, M., Francioso, O., Tugnoli, V., Nardi, S., 2007a. The auxin-like activity of humic substances is related to membrane interactions in carrot cell cultures. J. Chem. Ecol. 33, 115–129.

Nardi, S., Carletti, P., Pizzeghello, D., Muscolo, A., 2009. Biological activities of humic substances. In: Senesi, N., Xing, B., Huang, P.M. (Eds.), Biophysico-chemical Processes Involving Natural Nonliving Organic Matter in Environmental Systems. Wiley, Hoboken, pp. 305–339.

Nisha, R., Kaushik, A., Kaushik, C., 2007. Effect of indigenous cyanobacterial application on structural stability and productivity of an organically poor semi-arid soil. Geoderma 38, 49–56.

Norrie, J., Keathley, J.P., 2006. Benefits of *Ascophyllum nodosum* marine-plant extract applications to "Thompson seedless' grape production. Acta Hortic. 727, 243–247.

Osman, M.E.H., El-Sheekh, M.M., El-Naggar, A.H., Gheda, S.F., 2010. Effect of two species of cyanobacteria as biofertilizers on some metabolic activities, growth, and yield of pea plant. Biol. Fertil. Soils 46, 861–875.

Padhy, R.N., Nabakishore, N., Dash-Mohini, R.R., et al., 2016. Growth, metabolism and yield of rice cultivated in soils amended with fly ash and cyanobacteria and metal loads in plant parts. Rice Sci. 23 (1), 22–32.

Pan, S., Jeevanandam, J., Danquah, M.K., 2019. Benefits of algal extracts in sustainable agriculture. In: Rampelotto, P.H., Trincone, A. (Eds.), Grand Challenges in Marine Biotechnology. Springer Science and Business Media LLC: Berlin, Germany, pp. 501–534.

Pandey, S., Rai, R., Rai, L.C., 2012. Proteomics combines morphological, physiological and biochemical attributes to unravel the survival strategy of *Anabaena* sp. PCC7120 under arsenic stress. J. Proteom 75 (3), 921–937. doi:10.1016/j.jprot.2011.10.011.

Parrado, J., Bautista, J., Romero, E.F., et al., 2008. Production of a carob enzymatic extract: potential use as a biofertilizer. Bioresour. Technol. 99, 2312–2318.

Pathak, J., Rajneesh, Maurya PK, Singh, S.P., Häder, D.P., Sinha, R.P., 2018. Cyanobacterial farming for environment-friendly sustainable agriculture practices: innovations and perspectives. Front. Environ. Sci. 2018 (6), 7–19.

Prasanna, R., Triveni, S., Bidyarani, N., Babu, S., Yadav, K., Adak, A., Khetarpal, S., Pal, M., Shivay, Y.S., Saxena, A.K., 2013. Evaluating the efficacy of cyanobacterial formulations and biofilmed inoculants for leguminous crops. Arch. Agron. Soil Sci. 60, 349–366.

Priya, H., Prasanna, R., Ramakrishnan, B., Bidyarani, N., Babu, S., Thapa, S., Renuka, N., 2015. Influence of cyanobacterial inoculation on the culturable microbiome and growth of rice. Microbiol. Res 171, 78–89. doi:10.1016/j.micres.2014.12.011.

Priyanka, Kumar C, Chatterjee, A., Wenjing, W., Yadav, D., Singh, P.K., 2020. Cyanobacteria: potential and role for environmental remediation. In: Singh, P., Kumar, A., Borthakur, A. (Eds.), Abatement of Environmental Pollutants. Elsevier, Amsterdam, pp. 193–202.

Rai, S., Agrawal, C., Shrivastava, A.K., Singh, P.K., Rai, L.C., 2014. Comparative proteomics unveils cross-species variations in *Anabaena* under salt stress. J. Proteom 98, 254–270. doi:10.1016/j. jprot.2013.12.020.

Rayorath, P., Jithesh, M.N., Farid, A., Khan, W., Palanisamy, R., Hankins, S.D., Critchley, A.T., Prithiviraj, B., 2008. Rapid bioassays to evaluate the plant growth-promoting activity of *Ascophyllum nodosum* (L.) Le Jol. using a model plant, *Arabidopsis thaliana* (L.) Heynh. J. Appl. Phycol. 20, 423–429.

Renuka, N., Guldhe, A., Prasanna, R., Singh, P., Bux, F., 2018. Microalgae as multi-functional options in modern agriculture: current trends, prospects and challenges. Biotechnol. Adv. 36, 1255–1273.

Renuka, N., Prasanna, R., Sood, A., Ahluwalia, A.S., Bansal, R., Babu, S., Singh, R., Shivay, Y.S., Nain, L., 2016. Exploring the efficacy of wastewater-grown microalgal biomass as a biofertilizer for wheat. Environ. Sci. Pollut. Res. 23, 6608–6620.

Renuka, N., Prasanna, R., Sood, A., Bansal, R., Bidyarani, N., Singh, R., Shivay, Y.S., Nain, L., Ahluwalia, A.S., 2017. Wastewater grown microalgal biomass as inoculants for improving micronutrient availability in wheat. Rhizosphere 3, 150–159.

Rioux, L-E, Turgeon, S.L., Beaulieu, M., 2007. Characterization of polysaccharides extracted from brown seaweeds. Carbohydr. Polym. 69, 530–537.

Rodriguez, A.A., Stella, A.M., Storni, M.M., Zulpa, G., Zaccaro, M.C., 2006. Effects of cyanobacterial extracellular products and gibberellic acid on salinity tolerance in *Oryza sativa* L. Saline Syst 2, 7 2006.

Ronga, D., Biazzi, E., Parati, K., Carminati, D., Carminati, E., Tava, A., 2019. Microalgal biostimulants and biofertilisers in crop productions. Agronomy 9, 192 2019.

Saadatnia, H., Hossein, R., 2009. Cyanobacteria from paddy fields in Iran as a biofertilizer in rice plants. Plant Soil Environ 55, 207–212 10.17221/384-PSE.

Santosh, B., Prasanna, R., Bidyarani, N., Singh, R., 2015. Analyzing the colonization of inoculated cyanobacteria in wheat plants using biochemical and molecular tools. Environ. Boil. Fishes 27, 327–338.

Schiavon, M., Pizzeghello, D., Muscolo, A., Vaccaro, S., Francioso, O., Nardi, S., 2010. High molecular size humic substances enhance phenylpropanoid metabolism in maize (*Zea mays* L.). J. Chem. Ecol. 36, 662–669.

Sharma, R., Khokhar, M., Jat, R., Khandelwal, S., 2012. Role of algae and cyanobacteria in sustainable agriculture system. Wud Pecker J. Agric. Res. 1, 381–388.

Simpson, A.J., Kingery, W.L., Hayes, M.H.B., et al., 2002. Molecular structures and associations of humic substances in the terrestrial environment. Naturwissenschaften 89, 84–88.

Singh, P.K., Shrivastava, A.K., Chatterjee, A., Pandey, S., Rai, S., Singh, S., Rai, L.C., 2015. Cadmium toxicity in diazotrophic *Anabaena* spp. adjudged by hasty up-accumulation of transporter and signalling and severe down-accumulation of nitrogen metabolism proteins. J Proteom 127 (Pt A) (Sep 8), 134–146. doi:10.1016/j. jprot.2015.05.019.

Singh, P.K., Wang, W., Shrivastava, A.K., 2018. Cadmium-mediated morphological, biochemical and physiological tuning in three different *Anabaena* species. Aquat. Toxicol 202, 36–45. doi:10.1016/j.aquatox.2018.06.011.

Singh, R., Parihar, P., Singh, M., Bajguz, A., Kumar, J., Singh, S., Singh, V.P., Prasad, S.M., 2017. Uncovering potential applications of cyanobacteria and algal metabolites in biology, agriculture and medicine: current status and future prospects. Front. Microbiol. 8, 515 2017.

Singh, S., Dutta, P., 2007. Outdoor evaluation of herbicide resistant strains of *Anabaena variabilis* as biofertilizer for rice plants. Plant Soil 296 (1), 95–102. doi: 10.1007/s11104-007-9293-6.

Stiegler, J.C., Richardson, M.D., Karcher, D.E., Roberts, T.L., Norman, R.J., 2013. Foliar absorption of various inorganic and organic nitrogen sources by creeping bentgrass. Crop Sci. 52, 1148–1152.

Stirk, W.A., van Staden, J., 1997. Comparison of cytokinin- and auxin-like activity in some commercially used seaweed extracts. J. Appl. Phycol. 8 (2004), 503–508.

Swarnalakshmi, K., Prasanna, R., Kumar, A., Pattnaik, S., Chakravarty, K., Shivay, Y.S., Singh, R., Saxena, A.K., 2013. Evaluating the influence of novel cyanobacterial biofilmed biofertilizers on soil fertility and plant nutrition in wheat. Eur. J. Soil Biol. 55, 107–116.

Tantawy, S.T., Atef, N.M., 2010. Growth responses of *Lupinus termis* to some plant growth-promoting cyanobacteria and bacteria as biofertilizers. J. Food Agric. Environ., 1178–1183.

Tarakhovskaya, E.R., Maslov Yu, I., Shishova, M.F., 2007. Phytohormones in algae. Rus. J. Plant Physiol. 54 (2), 163–170.

Trevisan, S., Francioso, O., Quaggiotti, S., Nardi, S., 2010. Humic substances biological activity at the plant-soil interface. Plant Signal. Behav. 5, 635–643.

Tripathi, R.D., Dwivedi, S., Shukla, M.K., et al., 2008. Role of blue-green algae biofertilizer in ameliorating the nitrogen demand and fly-ash stress to the growth and yield of rice (*Oryza sativa* L.) plants. Chemosphere 70 (10), 1919–1929.

Vaishampayan, A., Sinha, Rajeshwar P, Hader, Donat, 2001. Cyanobacterial biofertilizers in rice agriculture. Bot. Rev. 67 (4), 453–516.

Varanini, Z., Pinton, R., 2001. Direct versus indirect effects of soil humic substances on plant growth and nutrition. In: Pinton, R., Varanini, Z., Nannipieri, P. (Eds.), The Rhizosphere. Marcel Dekker, Basel, pp. 141–158.

Verkleij, F.N., 1992. Seaweed extracts in agriculture and horticulture: a review. Biol. Agric. Hortic. 8, 309–324.

Wang, W., Chen, Q., Singh, P.K., Huang, Y., Pei, D., 2020. CRISPR/Cas9 edited HSFA6a and HSFA6b of *Arabidopsis thaliana* offers ABA and osmotic stress insensitivity by modulation of ROS homeostasis. Plant Signal. Behav. 15, 12. doi:10.1080/15592324.2020.1816321.

Win, T.T., Barone, G.D., Secundo, F., Fu, P., 2018. Algal Biofertilizers and Plant Growth Stimulants for Sustainable Agriculture. Ind. Biotechnol. 14, 203–211.

Wuang, S.C., Mar, C.K., Chua, Q.D., 2016. Use of Spirulina biomass produced from aquaculture wastewater treatment as agricultural fertilizers. Algal Res. 15, 59–64.

York, R., Rosa, E.A., Dietz, T., 2003. Footprints on the earth: the environmental consequences of modernity. Am. Sociol. Rev. 68, 279–300.

Zayadan, B.K., Matorin, D.N., Baimakhanova, G.B., 2014. Promising microbial consortia for producing biofertilizers for rice fields. Microbiology 83 (4), 391–397.

Zhang, Q., Zhang, J., Shen, J., Silva, A., Dennis, D.A., 2006. A simple 96-well microplate method for estimation of total polyphenol content in seaweeds. J. Appl. Phycol. 18, 445–450.

Zhang, X., Ervin, E.H., 2004. Cytokinin-containing seaweed and humic acid extracts associated with creeping bentgrass leaf cytokinins and drought resistance. Crop Sci. 44, 1737–1745.

Zhang, X., Ervin, E.H., 2008. Impact of seaweed extract-based cytokinins and zeatin riboside on creeping bentgrass heat tolerance. Crop Sci. 48, 364–370.

Zhang, X., Wang, K., Ervin, E.H., 2010. Optimizing dosages of seaweed extract-based cytokinins and zeatin biocide for improving creeping bentgrass heat tolerance. Crop Sci. 50, 316–320.

Zhuang, X., Chen, J., Shim, H., Bai, Z., 2007. New advances in plant growth-promoting rhizobacteria for bioremediation. Environ. Int. 33, 406–413. doi:10.1016/j.envint.2006.12.005.

Zimmerli, L., Hou, B-H, Tsai, C-H, et al., 2008. The xenobiotic betaaminobutyric acid enhances *Arabidopsis* thermotolerance. Plant J. 53, 144–156.

Zuverza-Mena, N., Martínez-Fernández, D., Du, W., Hernandez-Viezcas, J.A., Bonilla-Bird, N., López-Moreno, M.L., Komárek, M., Peralta-Videa, J.R., Gardea-Torresdey, J.L., 2017. Exposure of engineered nanomaterials to plants: insights into the physiological and biochemical responses-A review. Plant Physiol. Biochem. 110, 236–264.

CHAPTER

12

Molecular characterization of local cyanobacterial isolates using *16S rRNA*, *rpo*B, and *nif*H biomarkers

Nidhi Kumari[a,b], *Lal Chand Rai*[b,*]

[a]Department of Biological Sciences, Indian Institute of Science Education and Research Mohali, Mohali, Punjab, India

[b]Centre for Advanced Study in Botany, Institute of Science, Banaras Hindu University, Varanasi, Uttar Pradesh, India

[*]BHU Distinguished Professor, Centre for Advanced Study in Botany, Institute of Science, Banaras Hindu University, Varanasi, Uttar Pradesh, India

12.1 Introduction

Cyanobacteria, also known as blue-green algae and cyanophytes, are an ancient, remarkable, and widespread group of photosynthetic prokaryotes comprising a single phylogenetic group within the domain *Bacteria* (Castenholz, 2001a). The name "cyanobacteria" has been derived from the Greek word "kyanós" (meaning blue) due to the pigment color of the bacteria (cyan; etymonline.com, 2018). They have been defined as "microorganisms that harbor within a typically prokaryotic cell, a photosynthetic apparatus similar in structure and function to that located in the chloroplast of phototrophic eukaryotes" (Stanier, 1977). In evolutionary terms, they represent a link between bacteria and green plants. They are estimated to have occurred approximately 3.6 billion years ago based on cyanobacterial gene phylogenies (Garcia-Pichel et al., 2019), genomic sequence analysis (Hedges et al., 2001), fossil records (Schopf, 2000; 2011), and organic biomarkers (Brocks et al., 1999). These earliest cyanobacteria are believed to have played

Cyanobacterial Lifestyle and its Applications in Biotechnology.
DOI: https://doi.org/10.1016/B978-0-32-390634-0.00004-4

a major role in the conversion of the primitive anaerobic reducing environment to an oxygen-rich atmosphere on the earth about 2.3 million years ago (Blankenship, 1992; Lukow et al., 2000) that is called as the great oxidation event (Garcia-Pichel et al., 2019). Probably endosymbiont event between cyanobacteria and eukaryote gave rise to plastids (photosynthetic organelles), and consequently, algae and plants photosynthesize and possess chlorophyll a (Bhattacharya et al., 2004). Hence, cyanobacteria are considered as ancestors of chloroplasts (Archibald, 2009). They play a major role in primary productivity as well as nitrogen economy of various ecosystems. The contribution of N_2-fixing cyanobacteria as a natural biofertilizer in increasing soil fertility has been appreciated by several workers including Singh, (1961), Venkataraman, (1972), Stewart, (1980), Sinha and Kumar, (1992), Singh et al., (2000), Vaishampayan et al., (2001), Ariosa et al., (2004), Tirkey and Adhikary, (2005), Song et al., (2005), Nayak and Prasanna, (2007), Srivastava et al., (2009) and Kumari et al., (2012). Notwithstanding, cyanobacteria are the primary colonizers of the rock surfaces (Whitton, 1987) and play an important role in preventing soil erosion and preserving water in the soil (Ehling-Schulz et al., 2002). These bacteria are morphologically diverse (Whitton and Potts, 2000). There are both filamentous and unicellular forms, which can aggregate as colonies. In colonies, cells and filaments may be arranged in different ways, for example, radially, in strict planes, or irregularly. Filaments can be branched, coiled, or straight. Some cyanobacteria have evolved specialized cells for nitrogen fixation (heterocyst), survival in stressed conditions (akinetes), and dispersion (hormogonia). They have many fascinating features such as buoyancy, photosynthesis, fixation of atmospheric nitrogen (Castenholz, 2001b), and production of a wide variety of bioactive compounds (Burja et al., 2001). They possess gas vacuoles, allowing buoyancy and facilitating the formation of blooms (mass occurrences) (Walsby, 1994). They are cosmopolitan in distribution and capable of growth and multiplication in a wide range of ecological habitats (Whitton and Potts, 2000). Probably due to their physiological flexibility and long evolutionary history, cyanobacteria inhabit a large variety of terrestrial and aquatic habitats from deserts to lakes as well as hot springs and glaciers (Mur et al., 1999), on shores and surfaces of stones, plants and artificial objects as biofilms (Stal, 2007), temperate lakes (Reynolds, 1987), in the Arctic ice (Torsvik et al., 1996), deserts (Garcia-Pichel et al., 2001), oceans (Thajuddin and Subramanian, 2005), and even in the Arctic hot springs (Rios et al., 2007) and Antarctic granite rocks (Roeselers et al., 2007).

Apart from these diverse habitats, rice fields are also an important niche for cyanobacteria where they contribute a lot for nitrogen and carbon economy of the soil. Flooded rice fields are a privileged niche for biological nitrogen fixation as they possess a variety of N_2-fixing systems (Roger, 1995). Diazotrophic cyanobacteria are the main contributors to photodependent N_2-fixation in rice fields, and among them, heterocystous cyanobacteria are quite ubiquitous (Sinha and Häder, 1996). They play a vital role in soil fertility maintenance and buildup (Roger and Watanabe, 1986). Fristch (1907) was the first algologist to pronounce the abundance of cyanobacteria in paddy fields. The paddy fields are inhabited by blue-green algae that grow significantly in the water-logged field. In India, the Philippines, and China, cyanobacteria are specially grown in colossal mass spread out in the rice paddy fields. Actively growing rice fields harboring a good population of water ferns (*Azolla*) and cyanobacteria have been found to register increased yield up to 38%. The adaptability of cyanobacteria to different environmental conditions is attributed to the plasticity of their genome. Thus evolutionary significance, a wide ranging distribution in diverse habitats, and the immense potential as a biofertilizer make cyanobacteria an important model for study. Keeping in view the fact that cyanobacteria supplement significant amount of carbon and nitrogen to our ecosystems, it becomes very important to have an accurate knowledge of their diversity to unlock their hidden potential for human welfare.

Biological diversity or biodiversity may be defined as a "cluster of concepts" (Contoli, 1991), encompassing many interrelated aspects, from genetics and molecular biology to community structure and habitat heterogeneity (Battaglia, 1996; Boero, 1996). However, the most fundamental meaning of biodiversity probably lies in the concept of species richness (Baltanas, 1992), that is, the number of species occurring in a site, region, or ecosystem. India is one of the mega-biodiversity countries of the world, having almost all possible climatic variations and encompasses a great variety of plants and microbes. It is also known that one gram of soil may contain more than 10 billion microbes of thousands of different species (Torsvik and Ovreas, 2002). Culture-independent estimates of prokaryotic biodiversity fall between millions and billions of species (Dykhuizen 1998; Gans et al. 2005). Estimates of the current cyanobacterial biodiversity range from 2000 (Sant'Anna et al., 2006) to 8000 (Guiry 2012). Nabout et al. (2013) applied a discovery curve to cyanobacteria utilizing the CyanoDB database (http://www.cyanodb.cz/) with three asymptotic models, yielding from 3166 to 6280 species, depending on the model of choice. However, the actual number of species can be barely assessed by statistics. It requires extensive observation of the species diversity and distribution in nature (Foissner, 2006) with subsequent quantification. In this way, genetic and molecular data will be increasingly helpful for cyanobacterial diversity assessment (Dvořák et al., 2015).

Besides enormous cyanobacterial diversity, its molecular assessment remains a neglected area in India. Cyanobacteria have been traditionally identified and classified based on their morphological and physiological characteristics (Gietler, 1932, Desikachary, 1959). Its taxonomy has been formulated under the International Code of Botanical Nomenclature based on phenotypic characters (Bourrelly, 1985; Marsh et al., 1997; Banerjee et al., 2000). Identification of cyanobacteria using the above proposals of classification is primarily based on features such as the morphology of the filament, vegetative cells, heterocyst, and akinete (Komárek and Anagonistidis, 1989). The form of the colony, shape of the terminal cells, the presence or absence of sheath and gas vesicles, and life cycle are additional features used for identifying some genera (Rajaniemi et al., 2005). However, the reliability of these phenotypic characters and their use in the taxonomy has been questioned by Anagnostidis and Komárek (1989) because these characters are prone to environmental changes. For example, nitrogen sources such as NH_4^+, NO_3^- inhibit the formation of heterocyst and the presence of phosphate inhibits the formation of akinetes in cyanobacteria.

Moreover, cyanobacteria in nature may be present in the form of vegetative cells, akinetes, spores, hormogones, and other resting bodies, which may not regenerate in the culture so quickly. Identifying these resting bodies is not possible under the microscope (Komárek and Anagonistidis, 1989). Similarly, the morphology of the strains may change depending on environmental conditions, and the diversity of the strains can be altered by selective culture conditions (Palinska and Marquardt, 2008). For example, the strains of *Nodularia* were found to lose gas vesicles (Lehtimäki et al., 2000), and *Aphanizomenon flos-aquae* formed colony (Gugger et al., 2002a) in their culture conditions. Komárek and Anagnostidis (1989) have estimated that more than 50% of the strains in culture collections are misidentified. During the turbulent history of cyanobacteria classification, several major revisions and changes have been proposed and more or less adopted. Anagnostidis and Komárek (1985), Wilmotte (1994), and Turner (1997) have reviewed the history of botanical classification extensively. Therefore, an overview of the two most commonly adopted classification systems—the bacteriological approach in Bergey's Manual of Systematic Bacteriology (Boone and Castenholz, 2001) and the botanical approach of Anagnostidis and Komárek (1985)—are explained as the most recent proposal for the cyanobacterial classification system. Due to the reasons mentioned above, the assessment of biodiversity using only morphological and physiological variables may provide misleading information; therefore, molecular characterization was thought to be an important alternative. However, it is pertinent here to mention that most of the molecular approaches use 16S rRNA sequences (Woese, 1987), *nif* gene, phycocyanin locus (Neilan et al., 2002), RNA polymerase gene (Neilan et al., 1995) for assessing cyanobacterial diversity. These genes function as information molecules that can be used to reveal certain characteristics of the respective organisms. Hence, these are referred as molecular markers containing information about the genetic and evolutionary history of the source organisms.

12.2 Molecular markers used to assess cyanobacterial biodiversity

Molecular assessment of cyanobacterial diversity has exploited many markers till date. The most frequently used ones are the *16S rRNA* gene, phycocyanin locus gene, RNA polymerase gene, *nif*H gene, PC-IGS (Phycobillin-Intergenic spacer) regions, etc. Some of these may be explained as:

12.2.1 *16S rRNA* gene

The gene coding for 16S ribosomal RNA is referred to as the *16S rRNA* gene that is universally distributed in prokaryotes. Woese and Fox (1977) proposed using a small subunit *16S rRNA* gene for phylogenetic analysis due to the slow rates of evolution of the gene. They are considered pioneer researchers for using *16S rRNA* in phylogenetics (Woese et al., 1990). The *16S rRNA* gene has several properties of a good phylogenetic marker gene, such as universal occurrence in prokaryotes, variable and conserved regions, functional consistency, large size, and the high information content (Woese, 1987, Ludwig and Klenk, 2001). Subsequently, it was found that the *16S rRNA* gene shows a perfect correlation with genome relatedness, that is, denoted with DNA:DNA reassociation kinetics values (Stackebrandt and Goebel, 1994) or as ANI/AAI values (the average nucleotide or amino acid identity of shared genes, Konstantinidis and Tiedje, 2005a, b). These correlations support the robustness of the *16S rRNA* gene-based microbial phylogeny (Konstantinidis and Tiedje, 2005b). *16S rRNA* sequences formed the basis for systematic assignment in the latest edition of Bergey's Manual of Systematic Bacteriology, and it has helped distinguish broad taxonomic groups as well as individual species (Casamatta et al., 2005, Rajaniemi et al., 2005, Litavitis, 2002, Svenning et al., 2005). For broad phylogenetic studies, sequence data from the *16S rRNA* gene are the most commonly used due to their efficacy for distinguishing higher level taxonomic groups and traditional species.

The *16S rRNA* gene, the most commonly used marker gene, has played a crucial role in cyanobacterial diversity assessment, phylogenetic analysis, and molecular identification. At the time of Woese's review, the cyanobacteria were circumscribed by eight rRNA sequences from isolates representing only a minor portion of cyanobacterial diversity. The cyanobacteria and the chloroplasts of higher plants and the prochlorophytes (Turner, 1997) were considered within a common lineage that comprises one of 10 major eubacterial taxa (Woese, 1987). A more detailed study by Giovannoni et al. (1988) analyzed the *16S rRNAs* of 29 strains of cyanobacteria, including key reference strains from each of the five orders. This study permitted its authors to delineate the major phylogenetic patterns of the cyanobacteria currently available in pure culture. They found that the rRNA sequence diversity within the cyanobacteria is considerably less than the diversity that separates other major eubacterial taxa, indicating that relatively close phylogenetic relationships underlie the extensive morphological diversity occurring within the cyanobacteria. Oscillatoriales traditionally includes the well-known genera *Oscillatoria, Phormidium,* and *Lyngbya,* which are primarily differentiated by sheath properties. However, sheath production is only one of the typical diagnostic features subjected to the direct effects of both environmental and culture conditions and, therefore, not phylogenetically informative (Rippka et al., 1979, Sigler et al., 2003). DNA sequence information for the small subunit *rRNA* gene obtained from a cyanobacterial culture has been used to investigate cyanobacteria's presence and their abundance in natural habitat (Rudi et al., 2000). The study of the *16S rRNA* gene has also helped understand the phylogenetic relationship between *Synechocystis trididemni* and *Prochloron didemni*. The cytological similarities suggest that *Prochloron* may have evolved from an ancestral *Synechocystis*. This assumption was verified by Shimada et al. (2003), and it was found that *S. trididemni* and *P. didemni* had a common cyanobacterial ancestor. They inferred it from phylogenetic analysis of these two cyanobacteria using gene fragments of the 16S ribosomal RNA (16S rRNA) and the large subunit of the Rubisco gene (*rbcL*). Nowadays, there are at least four well-known *16S rRNA* gene databases, that is, EzBioCloud (https://www.ezbiocloud.net/, Yoon et al., 2017), Ribosomal Database Project (http://rdp.cme.msu.edu/, Larsen et al., 1993), SILVA (Pruesse et al., 2007), and GreenGenes (http://greengenes.secondgenome.com, DeSantis et al., 2006; McDonald et al., 2012) which are very useful for microbial diversity assessment. Additionally, the *16S rRNA* gene sequences are relatively easy to align and compare, and large databases (EMBL, DDBJ, GeneBank) are available online (Ludwig and Klenk, 2001). However, the *16S rRNA* gene's resolution power is at or above species level (Fox et al., 1992, Stackebrandt and Goebel, 1994).

As the *16S rRNA* molecular analysis was performed for cyanobacteria, it has been clear that morphology alone is not enough for distinguishing species and genera. That is why modern concepts of species have been created using polyphasic characterization, and molecular analysis has become an essential criterion to delimit genera (Johansen and Casamatta 2005). These analyses permitted access to a not yet known biodiversity. For example, in the last 10 years, 15 genera of *Nostocales* alone have been described based on this polyphasic approach (Hentschke and SantÀnna, 2015).

12.2.2 RNA polymerase

First of all, Bergsland and Haselkorn (1991) proposed the DNA-dependent RNA polymerase gene to be an efficient biomarker due to its existence in single copies in prokaryotic genome. Afterward Palenik and Haselkorn (1992) designed specific PCR primers to amplify the fragment of *rpo*C1 gene of cyanobacterial origin that has proved to be very useful for assessing cyanobacterial evolutionary relationship and the results are comparable with that of *16S rRNA* gene (Urback et al., 1992). These RNA polymerase gene sequences were specifically used by Palenik (1994) to assess the cyanobacterial diversity. The cyanobacterial species that are closely related to each other, display a greater degree of divergence in the *rpoC1* nucleotide sequence than *16S rRNA* gene. Two *Synechococcus* isolates WH7805 and WH8103 demonstrated 98.6% identity for *16S rRNA* gene, whereas their respective *rpoC1* gene fragments were only 83% identical with each other. Thus the use of *rpoC1* gene provides an additional advantage for molecular assessment of cyanobacterial diversity in a natural environment (Palenik, 1994; Lyra et al., 2005).

12.2.3 *nif*H gene

Another useful biomarker for N_2-fixing diazotrophic cyanobacterial diversity assessment is *nif*H gene; a highly conserved gene that encodes for dinitrogenase reductase protein of the nitrogenase complex playing an important role in N_2-fixation and it is a common indicator gene for all the N_2-fixers diazotrophic cyanobacteria. Ben-Porath et al. (1993) have used this biomarker for molecular identification of *Cylindrospermopsis raciborskii* isolates from different geographical locations and their subsequent characterization as N_2-fixing cyanobacteria. A group of *Nostoc* sp., *Anabaena* sp., *Anabaenopsis* sp., *Nodularia* sp., and some unidentified cyanobacteria was characterized using *nif*H gene marker by Teaumroong et al. (2002). Thus, *nif*H gene has the potential to distinguish N_2-fixing cyanobacterial isolates at the intraspecies level for phylogenetic analysis.

At all taxonomic levels above species, the sequence analysis of genes encoding small-subunit ribosomal RNA (*16S rRNA*) is currently the most promising approach for biodiversity assessment as well as for phylogenetic classification of cyanobacteria as outlined by Hentschke and Sant'Anna (2015). While discussing cyanobacterial diversity under multifarious environmental conditions, Kumari and Rai (2020) have also mentioned these three biomarkers in details.

12.3 Biodiversity documentation

Biodiversity documentation seems impossible without the help of a statistical analysis. With the advancements in technology, various computational and structural artifacts are also being reported (Sneath and Sokal, 1973; Moreira and Philippe, 2000). There are several methods to illustrate the organismal diversity and decipher their genetic relationship such as UPGMA (Unweighted Pair Group Method with Arithmetic Mean), Neighbor-joining method (Saitou and Nei, 1987), PHYLIP (Phylogeny Inference Package), PAUP (Phylogenetic Analysis Using Parsimony), MEGA4.0 (Molecular Evolutionary Genetic Analysis, http://evolution.genetics.washington.edu/phylip/software.html), etc. Among these, MEGA provides a user-friendly interface with many options to carry out evolutionary analysis independently.

Parsimony is a nonparametric statistical method commonly used in computational phylogenetics for estimating phylogenies. Under parsimony, the preferred phylogenetic tree is the tree that requires the least evolutionary change to explain some observed data. Parsimony is part of a class of character-based tree estimation methods that use a matrix of discrete phylogenetic characters to infer one or more optimal phylogenetic trees for a taxon set, commonly a set of species or reproductively isolated populations of a single species. These methods operate by evaluating candidate phylogenetic trees according to an explicit optimality criterion; the tree with the most favorable score is taken as the best estimate of the phylogenetic relationships of the included taxa. Maximum parsimony (MP) is used with most kinds of phylogenetic data; until recently, it was the only widely used character-based tree estimation method used for morphological data. The input data used in a maximum parsimony analysis are in the form of "characters" for a range of taxa. There is no generally agreed-upon definition of a phylogenetic character. Still, operationally a character can be thought of as an attribute, an axis along which taxa can vary. These attributes can be physical (morphological), molecular, genetic, physiological, or behavioral. The only widespread agreement on characters seems to be that character analysis variation should reflect the heritable variation. Whether it must be directly heritable or indirect inheritance (e.g., learned behaviors) is acceptable, is not entirely resolved (Kolaczkowski and Thornton, 2004).

12.4 Molecular characterization of local cyanobacterial isolates

Keeping the above-mentioned facts in mind, it was decided to characterize local cyanobacterial isolates using three molecular markers namely; *16S rRNA*, *rpo*B, and *nif*H gene. Paddy fields situated in the Indo-Gangetic belt of Eastern Uttar Pradesh and Western Bihar were selected as experimental sites for the study due to the mandate of the AMAAS network project of the Indian Council of Agricultural Research, New Delhi and NBAIM Mau to explore the cyanobacterial diversity of these areas that are preferred fields for rice cultivation from a very long time and crop yield is reportedly high due to abundant cyanobacteria present in these paddy fields (Singh, 1961). A total of 40 rice fields situated in eight districts of Eastern Uttar Pradesh and Western Bihar (Kumari et al., 2012) were selected for sampling (situated from 25°16′ to 28°12′ N and 79°22′ to 85°13′ E). Briefly, a total of 23 strains used in the study were isolated and cultured in BG-11 medium (Rippka et al., 1979) buffered with Tris/HCl (pH 7.5) under continuous cool white fluorescent light 72 μmol photon/m^2/s PAR (Photosynthetically Active Radiation) irradiance with a photoperiod of 14:10 h (light: dark) at 24±2 °C. Cultures were manually shaken two to three times daily. Morphological characterization of 23 cyanobacterial strains was done using a trinocular microscope (Kyowa Getner, Opto-Plan, 2KT, Japan, Model No. 90116) equipped with a high-resolution Nikon digital camera (model no. DXm 1200). Morphological characters were analyzed according to traditional criteria (Desikachary, 1959, Geitler, 1932). Their photomicrographs are given in Fig. 12.1. Total genomic DNA was extracted by Srivastava et al. (2007) using the phenol-free method. The amplification of *16S rRNA* was performed using cyano-specific primers CYA106F (forward primer) and CYA781R (reverse primer) (Nübel et al., 1997). In contrast, *nif* and *rpo* gene amplification was carried out with primers used earlier by Zehr et al. (1997) and Palenik (1994),

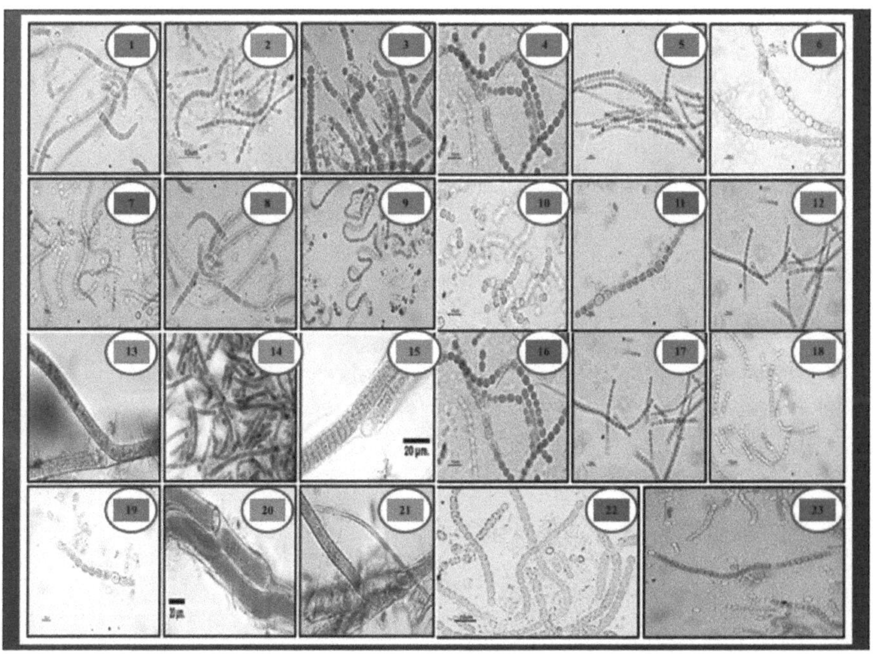

FIG. 12.1 Photomicrograph of 23 local cyanobacterial isolates. (Serial no. 6, 7, 11, 16, 22, 23 in green boxes are *Anabaena* sp.; 2, 4, 5, 9, 10, 12, 17, 18, 19 in gray boxes are *Nostoc* sp.; 1, 3, 8 in blue boxes are *Calothrix* sp.; and 13, 14, 15, 20, 21 are in red boxes *Tolypothrix* sp.)

respectively. These lyophilized PCR products were subjected to purification and sequencing on a commercial basis (Macrogen Inc., Korea). Partial *16S rRNA, nif,* and *rpo* gene sequences of the isolates were deposited in the GenBank database with accession numbers FJ948086-FJ948087 GQ255932-GQ255947, GU055197-GU055198, GU055201, GU055207, GU055211, HQ836192, HQ836196-HQ836199, HQ836204-HQ836221 and HQ836222-HQ836244, respectively. A total of 69 sequences of *16S rRNA, nif*H, and *rpo*B gene (23 each) obtained from 23 local isolates were multiple sequences aligned using CLUSTAL X (2.0) (Larkin et al., 2007) and manually corrected by using Jalview. A phylogenetic analysis was performed using the MEGA 4.0 platform (molecular evolutionary genetics analysis) (Tamura et al., 2007), an analytical package. With the aligned sequences, maximum-parsimony analysis in conjunction with a maximum mini Branch-&-bound option was carried out. To evaluate the robustness of branches in the inferred tree, the bootstrap resampling of Felsen stein was employed using 111 random seeds, and 100 replicates, and a consensus tree was generated.

Of these 23 cyanobacterial isolates, 9 were *Nostoc* species, 6 *Anabaena,* 5 *Tolypothrix,* and 3 *Calothrix* species from different geographical locations of paddy fields of Eastern Uttar Pradesh and Western Bihar. These cyanobacteria have also been previously reported from Eastern Uttar Pradesh paddy fields by Singh (1961) and others (Kaushik, 1994; Nayak et al., 2007, Prasanna et al., 2009, Srivastava et al., 2009; Kumari et al., 2012) which strengthen our observation of cyanobacterial diversity of this area. Apart from this, the BLASTn search also helped in the identification and confirmation of these morphologically identified genera, as shown in Table 12.1 (A)–(C).

TABLE 12.1 (A) Details of cyanobacterial *16S rRNA* gene sequences along with their percentage (%) sequence similarity as revealed by NCBI-BLASTn search.

Serial no.	Accession no.	Length of nucleotides Sequenced (bp)	Most closely related *16S rRNA* gene sequence		
			Organism name	**Accession no.**	**Sequence similarity (%)**
1	FJ948086	787	*Nostoc sp.* Cc2	AM711532	96
2	FJ948087	754	*Anabaena variabilis* NIES23	AF247593	98
3	GQ255932	662	*Nostoc* sp. PCC 7120	BA000019	98

(continued)

Serial no.	Accession no.	Length of nucleotides Sequenced (bp)	Most closely related *16S rRNA* gene sequence		
			Organism name	Accession no.	Sequence similarity (%)
4	GQ255933	610	*Calothrix* sp. PCC 7101	AB325535	99
5	GQ255934	660	*Nostoc carneum*	AB325906	97
6	GQ255935	724	*Calothrix* sp. PCC 7101	AB325535	99
7	GQ255936	780	*Calothrix* sp. PCC 7101	AB325535	99
8	GQ255937	667	*Tolypothrix* sp. PCC 7504	FJ660997	99
9	GQ255938	665	*Tolypothrix* sp. IAM M-259	AB093486	97
10	GQ255939	659	*Nostoc carneum*	AB325906	97
11	GQ255940	663	*Nostoc carneum*	AB325906	98
12	GQ255941	633	*Nostoc* sp. PCC 7120	BA000019	98
13	GQ255942	645	*Tolypothrix* sp. PCC 7504	FJ661002	98
14	GQ255943	639	*Nostoc* sp. PCC 7120	BA000019	100
15	GQ255944	687	*Nostoc carneum*	AB325906	97
16	GQ255945	660	*Anabaena flos-Aquae*	DQ234824	97
17	GQ255946	635	*Tolypothrix* sp. PCC 7504	FJ661002	96
18	GQ255947	669	*Anabaena flos-aquae*	DQ234824	99
19	GU055197	285	*Anabaena* sp. KVJF17	EU022721	97
20	GU055198	579	*Tolypothrix* sp. PCC 7504	FJ660998	95
21	GU055201	612	*Anabaena variabilis*	AF247593	99
22	GU055207	612	*Nostoc* sp. CENA88	GQ259207	96
23	GU055211	490	*Anabaena variabilis*	AF247593	96

TABLE 12.1 (B) Details of cyanobacterial *rpo*B gene sequences along with their percentage (%) sequence similarity as revealed by NCBI-BLASTn search

Serial no.	Accession no.	Length of nucleotides Sequenced (bp)	Most closely related *rpo* gene sequence		
			Organism name	**Accession no.**	**Sequence similarity (%)**
1	HQ836222	608	*Anabaena variabilis*	AB074795	99
2	HQ836223	589	*Calothrix brevissima*	AB074791	75
3	HQ836224	519	*Nostoc* sp. PCC 7120	BA000019	82
4	HQ836225	750	*Anabaena variabilis*	CP000117	94
5	HQ836226	343	*Calothrix brevissima*	AB074791	77
6	HQ836227	507	*Tolypothrix* sp. IAM M-259	AB096730	85
7	HQ836228	588	*Calothrix Brevissima*	AB074791	93
8	HQ836229	930	*Nostoc Punctiforme*	CP001037	80
9	HQ836230	661	*Nostoc* sp. PCC 7120	BA000019	74
10	HQ836231	679	*Nostoc* sp. PCC 7120	BA000019	99
11	HQ836232	913	*Nostoc* sp. PCC 7120	BA000019	97
12	HQ836233	660	*Tolypothrix* sp. IAM M-259	AB096730	98
13	HQ836234	685	*Nostoc* sp. PCC 7120	BA000019	96
14	HQ836235	586	*Nostoc* sp. PCC 7120	BA000019	96
15	HQ836236	535	*Tolypothrix* sp. IAM M-259	AB096730	79
16	HQ836237	819	*Tolypothrix* sp. IAM M-259	AB096730	88
17	HQ836238	588	*Tolypothrix* sp. IAM M-259	AB096730	75
18	HQ836239	940	*Anabaena bergii* ZIE 26	FJ234861	73
19	HQ836240	606	*Nostoc* sp. PCC 7120	BA000019	99
20	HQ836241	455	*Anabaena variabilis*	CP000117	95
21	HQ836242	586	*Nostoc* sp. PCC 7120	BA000019	99
22	HQ836243	618	*Anabaena variabilis*	CP000117	96
23	HQ836244	601	*Anabaena variabilis*	CP000117	96

TABLE 12.1 (C) Details of cyanobacterial *nif*H gene sequences along with their percentage (%) sequence similarity as revealed by NCBI-BLASTn search

Serial no.	Accession no.	Length of nucleotides Sequenced (bp)	Most closely related *nif*H gene sequence		
			Organism name	**Accession no.**	**Sequence similarity (%)**
1	HQ836192	556	*Nostoc* sp. PCC 7120	BA000019	96
2	HQ836196	669	*Tolypothrix* sp. PCC 7601	AY768421	96
3	HQ836197	662	*Nostoc* sp. PCC 7120	BA000019	98
4	HQ836198	585	*Nostoc* sp. PCC 7120	BA000019	86
5	HQ836199	667	*Anabaena* sp. A2	AF124377	80
6	HQ836204	935	*Anabaena Variabilis*	AB074795	88
7	HQ836205	670	*Nostoc* sp. PCC 7120	BA000019	95
8	HQ836206	417	*Nostoc punctiforme*	CP001037	83
9	HQ836207	344	*Calothrix* sp.	U73130	87
10	HQ836208	524	*Calothrix* sp.	U73130	89
11	HQ836209	467	*Calothrix* sp.	U73130	88
12	HQ836210	336	*Nostoc* sp. PCC 7120	BA000019	85
13	HQ836211	404	*Nostoc* sp.	JF923554	96
14	HQ836212	317	*Nostoc* sp. PCC 7120	BA000019	98
15	HQ836213	383	*Anabaena* sp. L-31	LO4499	93
16	HQ836214	322	*Tolypothrix* sp. PCC 7101	AY221817	90
17	HQ836215	275	*Anabaena* sp. L-31	LO4499	94
18	HQ836216	216	*Anabaena Azotica*	DQ294218	78
19	HQ836217	320	*Tolypothrix* sp. PCC 7101	AY221817	90
20	HQ836218	328	*Anabaena* sp. L-31	LO4499	94
21	HQ836219	323	*Tolypothrix* sp. PCC 7601	AY768421	96
22	HQ836220	315	*Tolypothrix* sp. PCC 7101	AY221817	90
23	HQ836221	311	*Nostoc* sp. PCC 7120	BA000019	93

12.5 Phylogenetic analysis of local cyanobacterial isolates using three different biomarkers

12.5.1 *16S rRNA* gene–based phylogenetic tree

The maximum parsimony–based consensus phylogenetic tree of 23 local cyanobacterial isolates, 5 reference sequences from the database, and 1 outgroup *Arthrospira* sp. were constructed using their respective 16SrRNA gene sequences. The consensus tree inferred from the 35 most parsimonious trees has been shown in Fig. 12.2 (Eck and Dayhoff, 1966). Branches corresponding to partitions reproduced in less than 50% of trees are collapsed. The consistency index is 0.508929, the retention index is 0.702703, and the composite index is 0.426641 (0.357625) for all sites and parsimony-informative sites (in parentheses). The percentage of parsimonious trees in

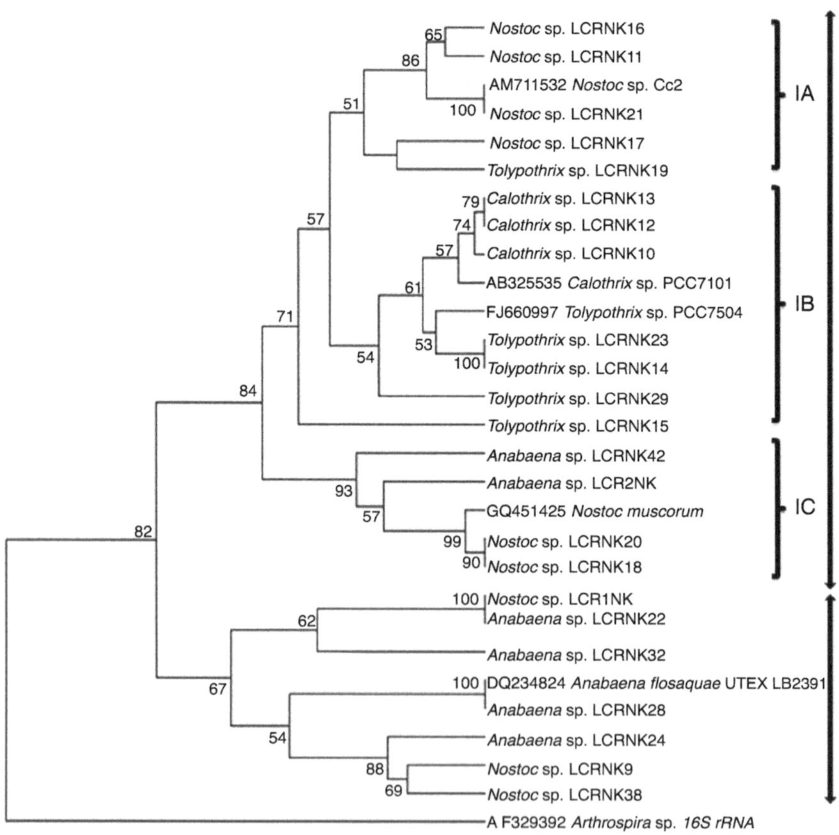

FIG. 12.2 *16S rRNA* gene–based evolutionary relationships of 24 taxa. The evolutionary history was inferred using the maximum parsimony method (Tamura et al., 2007; Nei and Kumar, 2000; Eck and Dayhoff, 1966).

which the associated taxa clustered together is shown next to the branches. The MP (Maximum Parsimony) tree was obtained using the close-neighbor-interchange algorithm with search level 3 (Nei and Kumar, 2000, Tamura et al., 2007). The initial trees were brought with the random addition of sequences (10 replicates). All positions containing gaps and missing data were eliminated from the dataset (Complete deletion option). There were 295 positions in the final dataset, of which 43 were parsimony informative. Phylogenetic analyses were conducted in MEGA4 (Tamura et al., 2007), which divided the phylogenetic tree into two major clusters.

The first cluster (cluster I) consisted of *Nostoc, Anabaena, Tolypothrix,* and *Calothrix* species. In contrast, the second cluster (cluster II) was made up of *Anabaena* and *Nostoc,* and their sequences showed 96–99% homology to their respective database counterparts (Table 12.1A). The cluster I was further segregated into three subclusters, that is, IA, IB, and IC, based on cyanobacteria grouping with each other inside the cluster. Subcluster IA was homogenously (except for one *Tolypothrix* species) comprised four *Nostoc* isolates clubbed together with AM711532 *Nostoc* sp. reference sequence. In contrast, subcluster IB was heterogeneous in composition, with four *Tolypothrix* and three *Calothrix* species grouped with their respective database reference sequences. A common clade formation by one *Tolypothrix* sp. LCRNK19 isolate with *Nostoc* sp. LCRNK17 in subcluster IA can be explained in view of the less than 50% bootstrap support value at the respective node and only 91.2% match in percent similarity matrix of the CLUSTALW multiple sequence alignment (MSA) results. Similarly, IC harbored *Nostoc* and *Anabaena* isolates which showed similarity with heterogeneous cluster II due to *Nostoc* and *Anabaena* isolates forming the same clade. However, these cluster patterns were also supported with their respective database reference sequences and more than 50% bootstrap support.

12.5.2 *rpo*B gene–based phylogenetic tree

The consensus-based parsimony tree of 23 *rpo*B gene of local cyanobacterial isolates and five reference sequences from the database and one outgroup *Arthrospira* sp. suggested that there were two major clusters in Fig. 12.3. The consensus tree inferred from the 23 most parsimonious trees has been shown. Branches corresponding to partitions reproduced in less than 50% of trees are collapsed. The consistency index is 0.696246, the retention index is 0.730303, and the composite index is 0.538004 (0.508470) for all sites and parsimony-informative sites (in parentheses). The percentage of parsimonious trees in which the associated taxa clustered together is shown next to the branches. The MP tree was obtained using the close-neighbor-interchange algorithm with search level 3, in which the initial trees were obtained with the random addition of sequences (10 replicates).

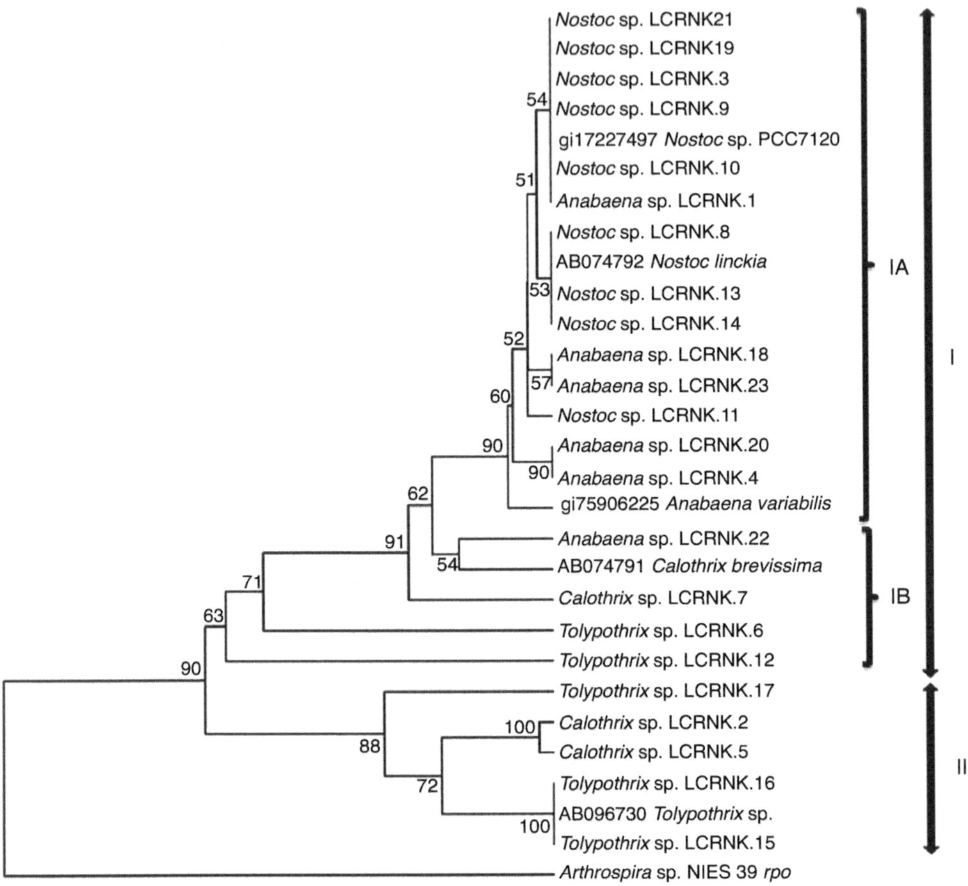

FIG. 12.3 *rpo*B gene–based evolutionary relationships of 24 taxa. The evolutionary history was inferred using the maximum parsimony method (Tamura et al., 2007; Nei and Kumar, 2000; Eck and Dayhoff, 1966).

All positions containing gaps and missing data were eliminated from the dataset (complete deletion option). There were 128 positions in the final dataset, of which 99 were parsimony informative. Phylogenetic analyses were conducted in MEGA4 (Eck and Dayhoff, 1966; Nei and Kumar, 2000; Tamura et al., 2007). The first cluster (cluster I)was heterogeneous, consisting of *Nostoc*, *Anabaena*, *Calothrix*, and *Tolypothrix* species showing 74–97% homology, respectively (Table 12.1B). This cluster can be further separated into two subclusters IA and IB. *Nostoc* and *Anabaena* grouped in subcluster IA and their respective database reference sequences, whereas *Calothrix* and *Tolypothrix* formed the subcluster IB. The second cluster (cluster II) was also of mixed type harboring *Tolypothrix* and *Calothrix* with 85–99% homology (Table 12.1B). The intermingling of *Anabaena* sp. LCRNK.1 and

Anabaena sp. LCRNK.22 with *Nostoc* sp. LCRNK.10 and AB074791 *Calothrix brevissima* can be explained based on 87.8% and 84.3% value in the percentage similarity matrix of CLUSTALW MSA results, respectively.

12.5.3 *nif*H gene–based phylogenetic tree

The most parsimonious consensus phylogenetic analysis resulted in two major clusters (Fig. 12.4) out of 23 local cyanobacterial isolates, 4 reference sequences and 1 *Sinorhizobium meliloti nif*H sequence. The consensus tree inferred from the 11 most parsimonious trees is shown. Branches corresponding to partitions reproduced in less than 50% of trees are collapsed. The consistency index is (0.447761), the retention index is (0.634868), and the composite index is 0.356329 (0.284269) for all sites and parsimony-informative sites (in parentheses). The percentage of parsimonious trees in which the associated taxa clustered together is shown next to the

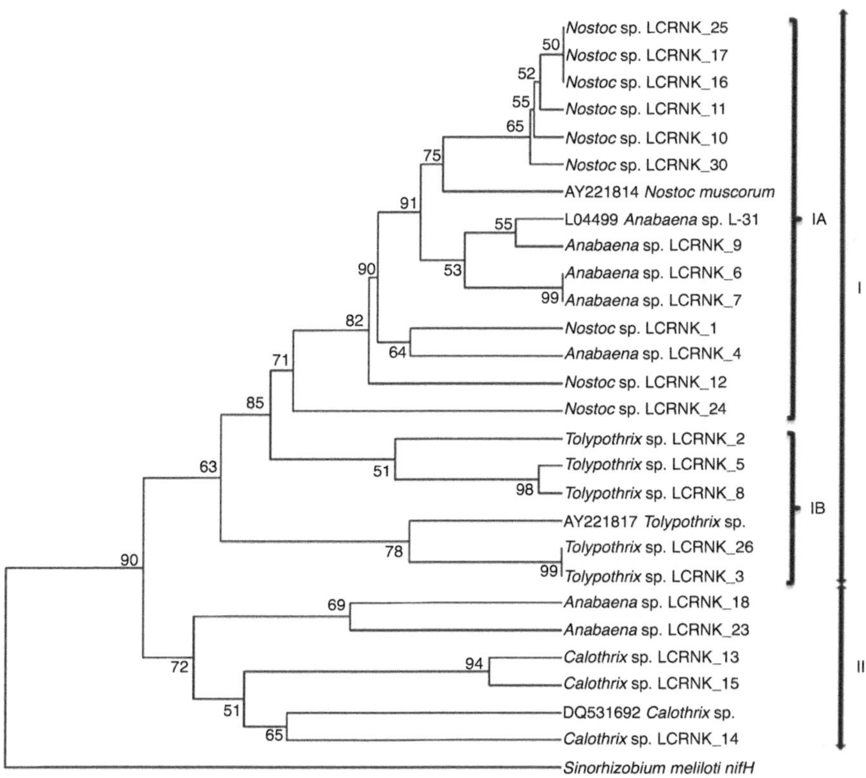

FIG. 12.4 *nif*H gene–based evolutionary relationships of 24 taxa. The evolutionary history was inferred using the maximum parsimony method (Tamura et al., 2007; Nei and Kumar, 2000; Eck and Dayhoff, 1966).

branches. The MP tree was obtained using the close-neighbor-interchange algorithm with search level 3, in which the initial trees were obtained with the random addition of sequences (10 replicates). All positions containing gaps and missing data were eliminated from the dataset (Complete deletion option). There were 119 positions in the final dataset, of which 49 were parsimony informative. Phylogenetic analyses were conducted in MEGA4 (Eck and Dayhoff, 1966; Nei and Kumar, 2000; Tamura et al., 2007). The first Cluster (cluster I) consisted of nine *Nostoc,* four *Anabaena,* and five *Tolypothrix* species isolated from various districts.

In contrast, the second cluster (cluster II) was made up of two *Anabaena* and three *Calothrix* isolates with 90–96% homology (Table 12.1C). The first cluster can be divided into two subclusters IA and IB, comprising *Nostoc, Anabaena,* and *Tolypothrix,* respectively, with significant homology to their database reference sequences. In the second cluster, three *Calothrix* clubbed together with their reference sequence and two *Anabaena* sp. LCRNK_18 and LCRNK_23 separated from the *Calothrix* group with 72% bootstrap support.

All the three phylogenetic trees showed good congruence of mixed clusters, that is, cluster I demonstrated the same cyanobacteria in both *16S rRNA* and *rpo*B gene–based tree. In contrast, cluster II of the *rpo*B gene–based tree was similar to that of the *16S rRNA* tree subcluster IB. Most of the clustering pattern of these three phylogenetic trees has been well supported with earlier database reference sequences (Figs. 12.2–12.4). It has also been strengthened further by BLASTn search results (Table 12.1A–C). This study has resulted in some genetically homogenous clusters of *Tolypothrix* (subcluster IB of *nif*H-based tree) and *Nostoc* and *Anabaena* were clustered in IC, II of *16S rRNA*-based phylogenetic tree, which is in accordance with the findings of *nif*H (subcluster IA) and *rpo*B (subcluster IA) based results.

Despite morphological differences between *Anabaena* and *Nostoc,* they appeared to be the same at the genetic level as suggested by this biomarkers-based phylogenetic analysis. Several reports support this finding that is, many genera were found to be polyphyletic: *Anabaena, Aphanizomenon, Trichormus* (Lyra et al., 2001; Gugger et al., 2002a, b; Iteman et al., 2002; Rajaniemi et al., 2005), *Tolypothrix, Calothrix* (Sihvonen et al., 2007), and the traditional classification needs to be revised (Rajaniemi et al., 2005; Sihvonen et al., 2007; Berrendero et al., 2008). However, the morphology alone seems to be insufficient for distinguishing different *16S rRNA* clusters. *Anabaena, Aphanizomenon, Trichormus,* and *Nostoc* strains examined by Rajaniemi et al. (2005) consistently formed six clusters in the analyses of *16S rRNA*, *rpo*B, and *rbc*LX genes with strain of different genera intermixed. Furthermore, the separation of the genera *Nostoc* and *Anabaena* has also been discussed in recent years (Tamas et al., 2000; Henson et al., 2002). On the other hand, some morphologically defined taxa seem to be in good correspondence with molecular data: *Cylindrospermopsis* (Saker and Neilan, 2001; Dyble et al., 2002; Neilan et al., 2003), *Nodularia* (Lehtimaki et al., 2000; Laamanen et al., 2001; Moffitt et al., 2001), *Anabaena circinalis* (Beltran and Neilan, 2000). In general, it can be concluded that cyanobacterial phylogenetic studies have demonstrated that genetic relationships conflict with morphological classification (Lyra et al., 2001; Iteman et al., 2002; Gugger and Hoffmann, 2004).

In *16S rRNA*- and *rpo*B-based consensus maximum parsimony tree, clustering of *Calothrix* and *Tolypothrix* can be explained in the light of the work of Sihvonen et al. (2007), which contrasts with our clustering results produced by *nif*H gene-based phylogeny of *Tolypothrix*. It can be clarified in view of less than 75% sequence identity of different *Calothrix* and *Tolypothrix* genera (Table 12.1B and C). This viewpoint is strengthened by the report that suggests the existence of 63–19% homology difference among various cyanobacterial species of a particular genus due to the presence of a different form of *nif* gene (i.e., *vnf*H and *anf*H), which may be amplified with the same primers (Ben-Porath and Zehr, 1994). Similarly, two species of *Synechococcus* WH7805 and WH8103 were reported to be only 83% homologous to each other based on *rpoC1* fragment sequence, whereas their *16S rRNA* gene revealed 98.6% sequence identity (Palenik, 1994). The congruence of mixed cluster and cluster containing *Anabaena* and *Nostoc* in the present study based on these three biomarkers attested cyanobacteria's polyphyletic nature. *16S rRNA* gene was considered to be the most suitable for phylogenetic analyses providing information not affected by horizontal gene transfer. However, it was denied by Miller et al. (2005), who found cyanobacterium with the incorporated bacterial loop in its *16S rRNA*. Cyanobacteria can contain multiple rrn operons and intragenomic sequence heterogeneity of the *16S rRNA* genes, which is inconsistent with phylogenetic use of *16S rRNA* for this group (Iteman et al., 2002; Lokmer 2007). Despite these facts, the *16S rRNA* gene remains the most used and a valuable tool in molecular phylogeny at the genus level due to its most extensive database availability. Large phylogenetic studies dealing with the phylogenetic relationships of whole orders or phylum were executed many years ago (Turner, 1997; Honda et al., 1999; Ishida et al., 2001; Litvaitis, 2002; Gugger and Hoffmann, 2004; Henson et al., 2004). Since that time, the number of sequences in GenBank has increased exponentially. The higher number of sequences, the more information, but also the more errors and taxa misidentification, and that is the main reason why scientists focused on smaller analyses dealing with phylogenies involving only single genera or few related taxa (Bolch et al., 1999; Boyer et al., 2002; Casamatta et al., 2005; Rajaniemi et al., 2005; Sihvonen et al., 2007; Marquardt and Palinska, 2007; Palinska and Marquardt, 2008; Papaefthimiou et al., 2008). Progress in phylogeny tends toward sequencing whole genomes and building phylogenetic trees from such large datasets (Zhaxybayeva et al. 2006, Shi and Falkowski 2008). The first-ever complete genome of cyanobacteria was sequenced nearly two decades ago, and CyanoBase (http://genome.microbedb.jp/cyanobase), the first database for cyanobacteria, was simultaneously created (Fujisawa et al., 2017). It has expanded the number of cyanobacterial genomic sequences to 376 species, which consists of 86 complete and 290 draft genomes. This information indicates the advancements of cyanobacteria molecular characterization.

Molecular techniques based on PCR amplification targeting conserved regions inside the *16S rRNA* gene have allowed the determination of phylogenetic affiliations among cyanobacteria and the development of modern cyanobacterial taxonomy (Komárek, 2006). Other genes coding for proteins, such as hetR (Han et al., 2009) and *nif*H (Zehr and McReynolds, 1989), have also been shown to have phylogenetic discriminating potential. However, many discrepancies between traditional morphological classification (Rippka et al., 1979; Castenholz, 1989; Castenholz and Waterbury, 1989) and phylogenetic analysis are still unresolved (Gugger et al., 2002a, b; Komárek, 2010). Therefore, it was felt by the scientific community to reevaluate or revise the cyanobacterial taxonomy. According to Komárek (2010), the modern system of cyanobacteria classification must be based on the molecular definition of genotypes (= basic clusters with a similarity index of ± 95% or less using the *16S rRNA* gene sequencing, considered as the standard method), which corresponds to the traditional taxonomic category of "genus." Obligatory separation of genera must also be done by at least one diacritical phenotypic character (or autapomorphic set of characters) and ecological, physiological, ultrastructural, and biochemical characteristics as an integral part of the generic definition.

Nevertheless, nowadays, the polyphasic approach is considered a unique, modern, unequivocal, and entirely acceptable methodological procedure. The combination of different methods should be based on molecular sequencing as the basic approach. Other criteria (morphological, ecological, etc.) must be added if they are available, distinct, and recognizable in cyanobacterial populations. The use of such characteristics is necessary; hence must be obligatorily included for the final characterization of strains and natural populations (Komárek, 2016; 2018).

12.6 Concluding remarks

In a nutshell, the only method for understanding and exploring knowledge about organismal diversity is taxonomic classification. However, it becomes complicated in phototrophic cyanobacteria, which contain very simple unicellular to multicellular forms with a differentiated and diversified thallus. This phylogenetically old group has various genotypes that are adaptable to various specialized ecosystems (Komárek, 2016). Their diversification occurs through ecological adaptation, facilitated by asexual reproduction, homologous recombination, horizontal gene transfer, and selection pressures on ecotypes leading to speciation, and a clear taxonomy has not yet been universally accepted (Willis and Woodhouse, 2020). Nevertheless, the introduction and combination of modern molecular, cytomorphological, and ecological methods in the taxonomy of cyanobacteria have shown the necessary insights and should be accepted as the only method for the elaboration of their modern systematics. To recognize the extent of biodiversity in the cyanobacteria, the scientific community should look forward to the state-of-the-art polyphasic methodology. It will certainly improve the degree of resolution and provide the broadest picture possible with in-depth information about cyanobacterial diversity.

Acknowledgements

Nidhi Kumari is thankful to UGC, New Delhi, for the Postdoctoral fellowship for women at the Department of Biological Sciences, IISER, Mohali and Centre of Advanced Study in Botany, BHU, Varanasi for facilities. L. C. Rai is thankful to NASI, Allahabad, for Senior Scientist, ICAR-NBAIM, and SERB, New Delhi for the financial support in the form of projects and Centre of Advanced Study in Botany BHU, Varanasi for facilities.

References

Anagnostidis, K., Komárek, J., 1985. Modern approach to the classification system of Cyanophytes 1-Introduction. Arch. Hydrobiol. Suppl. 71, 291–302.

Anagnostidis, K., Komárek, J., 1989. Modern approach to the classification system of cyanophytes 3 Oscillatoriales. Arch. Hydrobiol. Suppl. 80, 327–472.

Archibald, J.M., 2009. The puzzle of plastid evolution. Curr. Biol. 19, R81–R88.

Ariosa, Y., Quesada, A., Aburto, J., Carrasco, D., Carreres, R., Leganes, F., Fernandez Valiente, E., 2004. Epiphytic cyanobacteria on *Chara vulgaris* are the main contributors to N_2 fixation in rice fields. Appl. Environ. Microbiol. 70, 5391–5397.

Baltanas, A., 1992. On the use of some methods for the estimation of species richness. Oikos 65, 484–492.

Banerjee, M., Whitton, B.A., Wynn-Williams, D.D., 2000. Phosphatase activities of endolithic communities in rocks of the Antarctic dry valleys. Microb. Ecol. 39, 80–81.

Battaglia, B., 1996. La biodiversita marina a livello intraspecifico. Biol. Mar. Mediterr. 3, 1–12.

Beltran, E.C., Neilan, B.A., 2000. Geographical segregation of the neurotoxin producing cyanobacterium *Anabaena circinalis*. Appl. Environ. Microbiol. 66, 4468–4474.

Ben-Porath, J., Carpenter, E.J., Zehr, J.P., 1993. Genotypic relationship in *Trichodesmium* (Cyanophyceae) based on *nif*H sequence comparison. J. Phycol. 29, 806–810.

Ben-Porath, J., Zehr, J.P., 1994. Detection and characterization of cyanobacterial *nif*H genes. Appl. Environ. Microbiol. 60, 880–887.

Bergsland, K.J., Haselkorn, R., 1991. Evolutionary relationship among eubacteria, cyanobacteria and chloroplast: evidence from the *rpoC1* gene of *Anabaena* sp. strain PCC-7120. J. Bacteriol. 173, 3446–3455.

Berrendero, E., Perona, E., Mateo, P., 2008. Genetic and morphological characterization of *Rivularia* and *Calothrix* (Nostocales, Cyanobacteria) from running water. Int. J. Syst. Evol. Microbiol. 58, 447–460.

Bhattacharya, D., Yoon, H.S., Hackett, J.D., 2004. Photosynthetic eukaryotes unite: endosymbiosis connects the dots. Bioessays 26, 50–60.

Blankenship, R.E., 1992. Origin and early evolution of photosynthesis. Photosynth. Res. 3, 91–111.

Boero, F., 1996. Specie, biodiversita e comunita. Biol. Mar. Mediterr. 3, 13–20.

Bolch, C.J.S., Orr, P.T., Jones, G.J., Blackburn, S.I., 1999. Genetic, morphological and toxicological variation among globally distributed strains of *Nodularia* (cyanobacteria). J. Phycol. 35, 339–355.

Boone, D.R., Castenholz, R.W., 2001. The Archaea and the Deeply Branching and Phototrophic Bacteria. In: Boone, D.R., Castenholz, R.W., Garrity, G.M. (Eds.), Bergey's Manual of Systematic Bacteriology. Springer-Verlag, New York, USA.

Bourrelly, P., 1985. *Les algues bleues et rouges, les Eugleniens, Peridiniens et Cryptomonadiens,* In: *Les algues da'eau douce III,* 2nd Edn. Societe Nouvelle Des Editions Boubee, Paris, p. 606.

Boyer, S.L., Johansen, J.R., Flechtner, V.R., 2002. Phylogeny and genetic variance in terrestrial *Microcoleus* (Cyanophycae) species based on sequence analysis of the 16S rRNA gene and associated 16S-23S ITS region. J. Phycol. 38, 1222–1235.

Brocks, J.J., Logan, G.A., Buick, R., Summons, R.E., 1999. Archean molecular fossils and the early rise of eukaryotes. Science 285, 1033–1036.

Burja, A.M., Banaigs, B., Abou-Mansour, E., Burgess, J.G., Wright, P.C., 2001. Marine cyanobacteria - a prolific source of natural products. Tetrahedron 57, 9347–9377.

Casamatta, D.A., Johansen, J.R., Vis, M.L., Broadwater, S.T., 2005. Molecular and morphological characterization of ten polar and near-polar strains within the Oscillatoriales (Cyanobacteria). J. Phycol. 41, 421–438.

Castenholz, R.W., 2001a. Oxygenic photosynthetic bacteria. In: Boone, D.R., Castenholz, R.W. (Eds.), Bergey's Manual of Systematic Bacteriology 2nd Edn. Springer-Verlag, New York, USA, pp. 523–561.

Castenholz, R.W., 2001b. General characteristics of the cyanobacteria. In: Boone, D.R., Castenholz, R.W. (Eds.), Bergey's Manual of Systematic Bacteriology2nd Edn. Springer-Verlag, New York, USA, pp. 474–487.

Castenholz, R.W., Waterbury, J.B., 1989. Oxygenic photosynthetic bacteria. Group I. Cyanobacteria. In: Staley, J.T., Bryant, M.P., Pfenning, N., Holt, J.G. (Eds.). Bergey's Manual of Systematic Bacteriology, 3. Williams and Wilkins, Baltimore, pp. 1710–1789.

Contoli, L., 1991. Diversita biologica. SItE Notizie 12, 19–22.

DeSantis, T.Z., Hugenholtz, P., Larsen, N., Rojas, M., Brodie, E.L., Keller, K., Huber, T., Dalevi, D., Hu, P., Andersen, G.L., 2006. Greengenes, a chimaera-checked 16S rRNA gene database, and workbench compatible with ARB. Appl. Environ. Microbiol. 72 (7), 5069–5072.

Desikachary, T.V., 1959. Cyanophyta, ICAR monograph on algae. ICAR, New Delhi.

Dvořák, P., Poulíčková, A., Hašler, P., Belli, M., Casamatta, D.A., Papini, A., 2015. Species concepts and speciation factors in cyanobacteria, with connection to the problems of diversity and classification. Biodivers. Conserv. 24, 739–757.

Dyble, J., Paerl, H.W., Neilan, B.A., 2002. Genetic characterization of *Cylindrospermopsis raciborskii* (Cyanobacteria) isolates from diverse geographical origins based on *nifH* and *cpcBA* IGS nucleotide sequence analysis. Appl. Environ. Microbiol. 68, 2567–2571.

Dykhuizen, D.E., 1998. Santa Rosalia revisited: why are there so many species of bacteria? Antonie Van Leeuwenhoek 73, 25–33.

Eck, R.V., Dayhoff, M.O., 1966. Atlas of Protein Sequence and Structure. National Biomedical Research Foundation, Silver Springs, MD.

Ehling-Schulz, M., Schulz, S., Wait, R., Görg, A., Scherer, S., 2002. The UV-B stimulon of the terrestrial cyanobacterium *Nostoc commune* comprises early shock proteins and late acclimation proteins. Mol. Microbiol. 46, 827–843.

Foissner, W., 2006. Biogeography and dispersal of microorganisms: a review emphasizing protists. Acta Protozool 45, 111–136.

Fox, G.E., Wisotzkey, J.D., Jurtshuk Jr., P, 1992. How close is close: 16S rRNA sequence identity may not be sufficient to guarantee species identity. Int. J. Syst. Bacteriol. 42, 166–170.

Fritsch, F.E., 1907a. The subarial and freshwater algal flora of the tropics. Ann. Bot. 30, 235–275.

Fujisawa, T., Narikawa, R., Maeda, S.I., Watanabe, S., Kanesaki, Y., Kobayashi, K., et al., 2017. CyanoBase: a large-scale update on its 20th anniversary. Nucleic Acids Res 45 (D1), D551–D554.

Gans, J., Wolinsky, M., Dunbar, J., 2005. Computational improvements reveal great bacterial diversity and high metal toxicity in soil. Science 309, 1387–1390.

Garcia-Pichel, F., Lombard, J., Soule, T., Dunaj, S., Wu, S.H., Wojciechowski, M.F., 2019. Timing the evolutionary advent of cyanobacteria and the later great oxidation event using gene phylogenies of a sunscreen. mBio 10 (3), 561–619.

Garcia-Pichel, F., López-Cortés, A., Nübel, U., 2001. Phylogenetic and morphological diversity of cyanobacteria in soil desert crusts from the Colorado Plateau. Appl. Environ. Microbiol. 67, 1902–1910.

Geitler, L., 1932. Cyanophyceae. In: Rabenhorst, L. (Ed.). Kryptogamenflora von Deutschland, Österreich und der Schweiz, 14. Akad Verlag, Leipzig, pp. 916–931.

Giovannoni, S.J., Turner, S., Olsen, G.J., Barns, S., Lane, D.J., Pace, N.K., 1988. Evolutionary relationships among cyanobacteria and green chloroplasts. J. Bacteriol. 170, 3584–3592.

Gugger, M., Lyra, C., Henriksen, P., Coutá, A., Humbert, J.F., Sivonen, K., 2002b. Phylogenetic comparison of the cyanobacterial genera *Anabaena* and *Aphanizomenon*. Int. J. Syst. Evol. Microbiol. 52, 1867–1880.

Gugger, M., Lyra, C., Suominen, I., Tsitko, I., Humbert, J.F., Salkinoja-Salonen, M.S., Sivonen, K., 2002a. Cellular fatty acids as chemotaxonomic markers of the genera *Anabaena, Aphanizomenon, Microcystis, Nostoc* and *Planktothrix* (cyanobacteria). Int. J. Syst. Evol. Microbiol. 52, 1007–1015.

Gugger, M.F., Hoffmann, L., 2004. Polyphyly of true branching cyanobacteria (*Stigonematales*). Int. J. Syst. Evol. Microbiol. 54, 349–357.

Guiry, M.D., 2012. How many species of algae are there? J. Phycol. 48, 1057–1063.

Han, D., Fan, Y., Hu, Z., 2009. An evaluation of four phylogenetic markers in *Nostoc*: implications for cyanobacterial phylogenetic studies at the intrageneric level. Curr. Microbiol. 58, 170–176.

Hedges, S.B., Chen, H., Kumar, S., Wang, D.Y., Thompson, A.S., Watanabe, H., 2001. A genomic timescale for the origin of eukaryotes. BMC Evol. Biol. 1, 4.

Henson, B.J., Hesselbrock, M., Watson, L.E., Barnum, S.R., 2004. Molecular phylogeny of the heterocytous cyanobacteria (subsection IV and V) based on *nif*D. Int. J. Syst. Evol. Microbiol. 54, 493–497.

Henson, B.J., Watson, L.E., Barnum, S.R.O.V., 2002. Molecular differentiation of the heterocystous cyanobacteria, *Nostoc* and *Anabaena*, based on complete *nif*D sequences. Curr. Microbiol. 45, 161–164.

Hentschke, G.S., Santa'Anna, C.L., 2015. Current trends and prospects for cyanobacterial taxonomy: are only cultured populations enough? Algol. Stud. 147, 3–6.

Honda, D., Yokota, A., Sugiyama, J., 1999. Detection of seven major evolutionary lineages in cyanobacteria based on the 16S rRNA gene sequence analysis with new sequences of five marine *Synechococcus* strains. J. Mol. Evol. 48, 723–739.

Ishida, T., Watanabe, M.M., Sugiyama, J., Yokota, A., 2001. Evidence for polyphyletic origin of the members of the orders of *Oscillatoriales* and *Pleurocapsales* as determined by 16S rDNA analysis. FEMS Microbiol. Lett. 201, 79–82.

Iteman, I., Rippka, R., Tandeau de Marsac, N., Herdman, M., 2002. rDNA analyses of planktonic heterocystous cyanobacteria, including members of the genera *Anabaenopsis* and *Cyanospira*. Microbiology 148, 481–496.

Johansen, J.R., Casamatta, D.A., 2005. Recognizing cyanobacterial diversity through adoption of a new species paradigm. Algol. Stud. 117, 71–93.

Kaushik, B.D., 1994. Algalization of rice in salt-affected soils. Ann. Agric. Res. 14, 105–106.

Kolaczkowski, B., Thornton, J.W., 2004. Performance of maximum parsimony and likelihood phylogenetics when evolution is heterogeneous. Nature 431, 980–984.

Komárek, J., 2010. Recent changes (2008) in cyanobacteria taxonomy based on a combination of molecular background with phenotype and ecological consequences (genus and species concept). Hydrobiologia 639, 245–259.

Komárek, J., 2016. A polyphasic approach for the taxonomy of cyanobacteria: principles and applications. Eur. J. Phycol. 51, 346–353.

Komárek, J., 2018. Several problems of the polyphasic approach in the modern cyanobacterial system. Hydrobiologia 811, 7–17.

Komárek, J., Anagnostidis, K., 1989. Modern approach to the classification system of cyanophytes 4 Nostocales. Arch. Hydrobiol. Suppl. 82, 247–345.

Konstantinidis, K.T., Tiedje, J.M., 2005a. Towards a genome-based taxonomy for prokaryotes. J. Bacteriol. 187, 6258–6264.

Konstantinidis, K.T., Tiedje, J.M., 2005b. Genomic insights that advance the species definition for prokaryotes. Proc. Natl. Acad. Sci. USA 102, 2567–2572.

Kumari, N., Narayan, O.P., Rai, L.C., 2012. Cyanobacterial diversity-shifts induced by butachlor in selected Indian rice fields in eastern Uttar Pradesh and western Bihar analyzed with PCR and DGGE. J. Microbiol. Biotechnol. 22 (1), 1–12.

Kumari, N., Rai, L.C., et al., 2020. Cyanobacterial Diversity: Molecular Insights Under Multifarious Environmental Conditions. In: Singh, P.K. et al (Ed.), Advances in Cyanobacterial Biology. Academic Press Elsevier, USA, pp. 17–33.

Laamanen, M.J., Gugger, M.F., Lehtimaki, J.M., Haukka, K., Sivonen, K., 2001. Diversity of toxic and nontoxic *Nodularia* isolates (cyanobacteria) and filaments from Baltic Sea. Appl. Environ. Microbiol. 67, 4638–4647.

Larkin, M.A., Blackshields, G., Brown, N.P., Chenna, R., McGettigan, P.A., McWilliam, H., Valentin, F., Wallace, I.M., Wilm, A., Lopez, R., Thompson, J.D., Gibson, T.J., Higgins, D.G., 2007. Â ClustalW and ClustalX version 2.0. Bioinformatics 23, 2947–2948.

Larsen, N., Olsen, G.J., Maidak, B.L., McCaughey, M.J., Overbeek, R., Macke, T.J., Marsh, T.L., Woese, C.R., 1993. The ribosomal database project. Nucleic Acids Res 21 (13), 3021–3023.

Lehtimäki, J., Lyra, C., Suomalainen, S., Sundman, P., Rouhiainen, L.L., Paulin, L., Salonen, M.S., Sivonen, K., 2000. Characterization of *Nodularia* strains, cyanobacteria from brackish waters by genotypic and phenotypic methods. Int. J. Syst. Evol. Microbiol. 50, 1043–1053.

Litavitis, M.K., 2002. A molecular test of cyanobacterial phylogeny: inference from constraint analysis. Hydrobiologia 468, 135–145.

Lokmer, A., 2007. Polyphasic approach to the taxonomy of the selected oscillatorian strains (Cyanobacteria). (M.Sc. thesis). Faculty of Biological Sciences, University of SouthBohemia, Czech Republic, p. 40.

Lukow, T., Dunfiel, P.F., Liesack, W., 2000. Use of the T-RFLP technique to assess spatial and temporal changes in the bacterial community structure within an agricultural soil planted with transgenic and non-transgenic potato plants. FEMS Microbiol. Ecol. 32, 241–247.

Lyra, C., Laamanen, M., Lehtimaki, J.M., Surakka, A., Sivonen, K., 2005. Benthic cyanobacteria of the genus *Nodularia* are non-toxic, without gas vacuoles, able to glide and genetically more diverse than planktonic *nodularia*. Int. J. Syst. Evol. Microbiol. 55, 555–568.

Lyra, C., Suomalainen, S., Gugger, M., Vezie, C., Sundman, P., Paulin, L., Sivonen, K., 2001. Molecular characterization of planktic cyanobacteria of *Anabaena, Aphanizomenon, Microcystis* and *Planktothrix* genera. Int. J. Syst. Evol. Microbiol. 51, 513–526.

Marquardt, J., Palinska, K.A., 2007. Genotypic diversity of cyanobacteria assigned to the genus *Phormidium* (Oscillatoriales) from different habitats and geographical sites. Arch. Microbiol. 187, 397–413.

Marsh, T.L., Liu, W.T., Cheng, H., Forney, L.J., 1997. Characterization of microbial diversity by determining terminal restriction fragment length polymorphisms of genes encoding 16S rRNA. Appl. Environ. Microbiol. 63, 4516–4522.

McDonald, D., Price, M.N., Goodrich, J., Nawrocki, E.P., DeSantis, T.Z., Probst, A., Andersen, G.L., Knight, R., Hugenholtz, P., 2012. An improved Greengenes taxonomy with explicit ranks for ecological and evolutionary analyses of bacteria and archaea. Int. J. Syst. Evol. Microbiol. 6 (3), 610–618.

Miller, S.R., Augustine, S., Olson, T.L., Blankenship, R.E., Selker, J., Wood, A.M., 2005. Discovery of a free-living chlorophyll d-producing cyanobacterium with a hybrid proteobacterial/cyanobacterial small-subunit rRNA gene. Proc. Natl. Acad. Sci. USA 102, 850–855.

Moffitt, M.C., Blackburn, S.I., Neilan, B.A., 2001. rRNA sequences reflect the ecophysiology and define the toxic cyanobacteria of the genus *Nodularia*. Int. J. Syst. Evol. Microbiol. 51, 505–512.

Moreira, D., Philippe, H., 2000. Molecular phylogeny: pitfalls and progress. Int. Microbiol. 3, 9–16.

Mur, L.R., Skulberg, O.M., Utkilen, H., 1999. Cyanobacteria in the Environment. In: Chorus, I., Bartram, J. (Eds.), Toxic Cyanobacteria in Water. E and FN Spon, London, pp. 15–40.

Nabout, J.C., Rocha, B.D.S., Carneiro, F.M., Santa'Anna, C.L., 2013. How many species of Cyanobacteria are there? Using a discovery curve to predict the species number. Biodivers. Conserv. 22, 2907–2918.

Nayak, S., Prasanna, R., 2007. Soil pH and its role in cyanobacteria abundance and diversity in rice field soils. Appl. Ecol. Environ. Res. 5, 103–113.

Nayak, S., Prasanna, R., Prasanna, B.M., Sahoo, D.B., 2007. Analyzing diversity among Indian isolates of *Anabaena* (Nostocales, Cyanophyta) using morphological, physiological and biochemical characters. World J. Microbiol. Biotechnol. 23, 1575–1584.

Nei, M., Kumar, S., 2000. Molecular Evolution and Phylogenetics. Oxford University Press, New York.

Neilan, B.A., Burns, B.P., Relman, D.A., Lowe, D.R., 2002. Molecular identification of cyanobacteria associated with stromatolites from distinct geographical locations. Astrobiology 2, 271–280.

Neilan, B.A., Jacobs, D., Goodman, A.E., 1995. Genetic diversity and phylogeny of toxic cyanobacteria determined by DNA polymorphisms within the phycocyanin locus. Appl. Environ. Microbiol. 61, 3875–3883.

Neilan, B.A., Saker, M.L., Fastner, J., Törökné, A., Burns, B.P., 2003. Phylogeography of the invasive cyanobacterium *Cylindrospermopsis raciborskii*. Mol. Ecol. 12, 133–140.

Nübel, U., Garcia-Pichel, F., Muyzer, G., 1997. PCR primers for *16S rRNA* genes from cyanobacteria. Appl. Environ. Microbiol. 63, 3327–3332. In this issue.

Palenik, B., 1994. Cyanobacterial community structure as seen from RNA polymerase gene sequence analysis. Appl. Environ. Microbiol. 60, 3212–3219.

Palenik, B., Haselkorn, R., 1992. Multiple evolutionary origins of prochlorophytes, the chlorophyll b-containing prokaryotes. Nature 355, 265–267.

Palinska, K.A., Marquardt, J., 2008. Genotypic and phenotypic analysis of strains assigned to the widespread cyanobacterial morphospecies *Phormidium autumnale* (Oscillatoriales). Arch. Microbiol. 189, 325–335.

Papaefthimiou, D., Van Hove, C., Lejeune, A., Rasmussen, U., Wilmotte, A., 2008. Diversity and host specificity of *Azolla* cyanobionts. J. Phycol. 44, 60–70.

Prasanna, R., Jaiswal, P., Nayak, S., Sood, A., Kaushik, B.D., 2009. Cyanobacterial diversity in the rhizosphere of rice and its ecological significance. Indian J. Microbiol. 49, 89–97.

Pruesse, E., Quast, C., Knittel, K., Fuchs, B.M., Ludwig, W., Peplies, J., Glöckner, F.O., 2007. SILVA: a comprehensive online resource for quality checked and aligned ribosomal RNA sequence data compatible with ARB. Nucleic Acids Res 35 (21), 7188–7196.

Rajaniemi, P, Hrouzek, P, Kaštovská, K, Willame, R, Rantala, A, Hoffmann, L, Komárek, J, Sivonen, K, 2005. Phylogenetic and morphological evaluation of the genera *Anabaena*, *Aphanizomenon*, *Trichormus* and *Nostoc* (Nostocales, Cyanobacteria). Int. J. Syst. Evol. Microbiol. 55, 11–26.

Reynolds, C.S., 1987. Cyanobacterial water-blooms. Adv. Bot. Res. 13, 67–143.

Rios, A.D., Grube, M., Sancho, L.G., Ascaso, C., 2007. Ultrastructural and genetic characteristics of endolithic cyanobacterial biofilms colonizing Antarctic granite rocks. FEMS Microbiol. Ecol. 59, 386–395.

Rippka, R., Deruelles, J., Waterbury, J.B., Herdman, M., Stanier, R.Y., 1979. Generic assignments, strain histories, and properties of pure cultures of cyanobacteria. J. Gen. Microbiol. 111, 1–61.

Roeselers, G., Norris, T.B., Castenholz, R.W., Rysqaard, S., Glud, R.N., Kuhl, M., Muyzer, G., 2007. Diversity of phototrophic bacteria in microbial mats from Arctic hot springs (Greenland). Environ. Microbiol. 9, 126–138.

Roger, P.A., 1995. Biological N_2-fixation and its management in wetland rice cultivation. Fertil. Res. 42, 261–276.

Roger, P.A., Watanabe, I., 1986. Technologies for using biological nitrogen fixation in wetland rice: potentialities, current usage and limiting factors. Fertil. Res. 9, 39–77.

Rudi, K., Skulberg, O.M., Skulberg, R., Jakobsen, K.S., 2000. Application of sequence-specific labeled 16S rRNA gene oligonucleotide probes for genetic profiling of cyanobacterial abundance and diversity by array hybridization. Appl. Environ. Microbiol. 66, 4004–4011.

Saitou, N., Nei, M., 1987. The neighbour-joining method: a new method for reconstructing phylogenetic trees. Mol. Biol. Evol. 4, 406–425.

Saker, M.L., Neilan, B.A., 2001. Varied diazotrophic, morphologies and toxicities of genetically similar isolates of *Cylindrospermopsis raciborskii* (Nostocales, Cyanophyceae) from northern Australia. Appl. Environ. Microbiol. 67, 1839–1845.

Sant'Anna, C.L., Azevedo, M.T.P., Agujaro, L.F., Carvalho, M.C., Carvalho, L.R., Souza, R.C.R., 2006. Manual ilustrado para identificação e contagem de cianobactérias planctônicas de águas continentais brasileiras. Rima, Rio de Janeiro, p. 58.

Schopf, J.W., 2011. The paleobiological record of photosynthesis. Photosynth. Res. 107, 87–101.

Schopf, W.J., 2000. The fossil record: tracing the roots of the cyanobacterial lineage. In: Whitton, B.A., Potts, M. (Eds.), The Ecology of Cyanobacteria: Their Diversity in Time and Space. Kluwer Academic Publishers, Dordrecht, The Netherlands, pp. 13–35.

Shi, T., Falkowski, P.G., 2008. Genome evolution in cyanobacteria: the stable core and the variable shell. Proc. Natl. Acad. Sci. USA 105, 2510–2515.

Shimada, A., Yano, N., Kanai, S., Lewin, R.A., Maruyama, T., 2003. Molecular phylogenetic relationship between two symbiotic photo-oxygenic prokaryotes, *Prochloron* sp. and *Synechocystis trididemni*. Phycologia 42 (2), 193–197.

Sigler, W.B., Bachofen, R., Zeyer, J., 2003. Molecular characterization of endolithic cyanobacteria inhabiting exposed dolomite in central Switzerland. Environ. Microbiol. 5, 618–627.

Sihvonen, L.M., Lyra, C., Fewer, D.P., Rajaniemi-Wacklin, P., Lehtimäki, J.M., Wahlsten, M., Sivonen, K., 2007. Strains of the cyanobacterial genera *Calothrix* and *Rivularia* isolated from the Baltic Sea display cryptic diversity and are distantly related to *Gloeotrichia* and *Tolypothrix*. FEMS Microbiol. Ecol. 61, 74–84.

Singh, P.K., Dhar, D.W., Pabbi, S., Prasanna, R., Arora, A., 2000. Biofertilizers: Blue-Green Algae and Azolla. IARI, New Delhi Venus Printers and Publishers, New Delhi.

Singh, R.N., 1961. The Role of Blue-Green Algae in the Nitrogen Economy of Indian agriculture. Indian Council Agricultural Research, New Delhi, p. 175.

Sinha, R.P. Häder, D.P., 1996. Photobiology and ecophysiology of rice field cyanobacteria. Photochem. Photobiol. 64, 887–896. In this issue.

Sinha, R.P., Kumar, A., 1992. Screening of blue-green algae for biofertilizer. In: Patil, P.L. (Ed.), Proc. National seminar on Organic Farming. Pune, India, 95–97.

Sneath, P.H.A., Sokal, R.R., 1973. Numerical Taxonomy. Freeman, San Fransisco.

Song, T., Mårtensson, L., Eriksson, T., Zheng, W., Rasmussen, U., 2005. Biodiversity and seasonal variation of the cyanobacterial assemblage in a rice paddy field in Fujian, China. FEMS Microbiol. Ecol. 54, 131–140.

Srivastava, A.K., Ara, A., Bhargava, P., Mishra, Y., Rai, S.P., Rai, L.C., 2007. A rapid and cost-effective method of genomic DNA isolation from cyanobacterial mat and soil suitable for genomic fingerprinting and community analysis. J. Appl. Phycol. 19, 373–382.

Srivastava, A.K., Bhargava, P., Kumar, A., Rai, L.C., Neilan, B.A., 2009. Molecular characterization and the effect of salinity on cyanobacterial diversity in the rice fields of Eastern Uttar Pradesh. India. Saline Syst. 5, 4.

Stackebrandt, E., Goebel, B.M., 1994. Taxonomic note: a place for DNA: DNA reassociation and 16S rRNA sequence analysis in the present species definition in bacteriology. Int. J. Syst. Bacteriol. 44, 846–849.

Stal, L.J., 2007. Cyanobacteria: Diversity and versatility, clues to life in extreme environment. In: Seckbach, J. (Ed.), Algae and Cyanobacteria in Extreme Environment. Springer, The Netherlands, pp. 659–680.

Stanier, R.Y., 1977. The position of cyanobacteria in the world of phototrophs. Carlsberg Res. Commun. 42, 77.

Stewart, W.D.P., 1980. Some aspects of structure and function in N_2-fixing cyanobacteria. Annu. Rev. Microbiol 34, 497–536.

Svenning, M.M., Eriksson, T., Rasmussen, U., 2005. Phylogeny of symbiotic cyanobacteria within the genus *Nostoc* based on 16S rDNA sequence analyses. Arch. Microbiol. 183, 19–26.

Tamas, I., Svircev, Z., Andersson, S.G.E., 2000. Determinative value of a portion of the *nif*H sequence for the genera *Nostoc* and *Anabaena* (Cyanobacteria). Curr. Microbiol. 41, 197–200.

Tamura, K., Dudley, J., Nei, M., Kumar, S., 2007. MEGA4: Molecular Evolutionary Genetics Analysis (MEGA) software version 4.0. Mol. Biol. Evol. 24, 1596–1599.

Teaumroong, N., Innok, S., Chunleuchanon, S., Boonkerd, N., 2002. Diversity of nitrogen-fixing cyanobacteria under various ecosystems of Thailand: I. morphology, physiology and genetic diversity. World J. Microbiol. Biotechnol. 18, 673–682.

Thajuddin, N., Subramanian, G., 2005. Cyanobacterial biodiversity and potential applications in biotechnology. Curr. Sci. 89, 47–57.

Tirkey, J., Adhikary, S.P., 2005. Cyanobacteria in biological soil crusts of India. Curr. Sci. 89, 515–521.

Torsvik, V., Ovreas, L., 2002. Microbial diversity and function in soil: from genes to ecosystem. Curr. Opin. Microbiol. 5, 240–245.

Torsvik, V., Sorheim, R., Goksoyr, J., 1996. Total bacterial diversity in soil and sediment communities-a review. J. Ind. Microbiol. 17, 170–178.

Turner, S., 1997. Molecular systematics of oxygenic photosynthetic bacteria. Plant Syst. Evol. 11, 13–52.

Urback, E., Robertson, D.L., Chisholm, S.W., 1992. Multiple evolutionary origins of prochlorophytes within the cyanobacterial radiation. Nature 355, 267–270.

Vaishampayan, A., Sinha, R.P., Häder, D.P., Dey, T., Gupta, A.K., Bhan, U., Rao, A.L., 2001. Cyanobacterial biofertilizers in rice agriculture. Bot. Rev. 67, 453–516.

Venkataraman, G.S., 1972. Algal Biofertilizers and Rice Cultivation. Today and Tomorrow Printers and Publishers, Faridabad, India 1972.

Walsby, A.E., 1994. Gas vesicles. Microbiol. Rev. 58, 94–144.

Whitton, B.A., 1987. Survival and dormancy of blue green algae. In: Henis, Y. (Ed.), Survival and dormancy of Microorganisms. Wiley, New York, pp. 109–167.

Whitton, B.A., Potts, M., 2000. Introduction to the cyanobacteria. In: Whitton, B.A., Potts, M. (Eds.), The ecology of cyanobacteria: Their diversity in time and space. Kluwer Academic Publishers, Dordrecht, The Netherlands, pp. 1–11.

Willis, A., Woodhouse, J.N., 2020. Defining cyanobacterial species: diversity and description through genomics. Crit. Rev. Plant Sci. 39 (2), 101–124.

Wilmotte, A., 1994. Molecular evolution and taxonomy of the cyanobacteria. In: Bryant, D.A. (Ed.), The Molecular Biology of Cyanobacteria. Kluwer Academic Publishers, Dordrecht, The Netherlands, p. 881.

Woese, C.R., 1987. Bacterial evolution. Microbiol. Rev. 51, 221–271.

Woese, C.R., Fox, G.E., 1977. Phylogenetic structure of the prokaryotic domain: the primary kingdoms. Proc. Natl. Acad. Sci. USA 74 (11), 5088–5090.

Woese, C.R., Kandler, O., Wheelis, M.L., 1990. Towards a natural system of organisms: proposal for the domains Archaea, Bacteria, and Eucarya. Proc. Natl. Acad. Sci. USA 87 (12), 4576–4579.

Yoon, S.H., Ha, S.M., Kwon, S., Lim, J., Kim, Y., Seo, H., Chun, J., 2017. Introducing EzBioCloud: a taxonomically united database of 16S rRNA and whole-genome assemblies. Int. J. Syst. Evol. Microbiol. 67, 1613–1617.

Zehr, J.P., McReynolds, L.A., 1989. Use of degenerate oligonucleotides for amplification of the nifH gene from the marine cyanobacterium *Trichodesmium thiebautii*. Appl. Environ. Microbiol. 55, 2522–2526.

Zehr, J.P., Mellon, M.T., Hiorns, WD, 1997. Phylogeny of cyanobacterial *nifH* genes: evolutionary implications and potential applications to natural assemblages. Microbiology 143, 1443–1450.

Zhaxybayeva, O., Gogarten, J.P., Charlebois, R.L., Doolittle, W.F., Papke, R.T., 2006. Phylogenetic analyses of cyanobacterial genomes: quantification of horizontal gene transfer events. Genome Res 16, 1099–1108.

CHAPTER

13

Cyanobacteria-mediated heavy metal and xenobiotics bioremediation

Shweta Pandey[a], *Sumit Kumar Dubey*[b], *Arun Kumar Kashyap*[c], *Buddhi Prakash Jain*[d]

[a]Govt VYT PG Autonomous College, Durg, Chhattisgarh, India

[b]Department of Microbiology, D. L. S. Postgraduate College, Bilaspur, Chhattisgarh, India

[c]Department of Biotechnology, Govt. E Raghavendra Rao Postgraduate Science College, Bilaspur, Chhattisgarh, India

[d]Department of Zoology, Mahatma Gandhi Central University Motihari, Bihar, India

13.1 Introduction

The population of the world is continuously increasing. Today, there are 7.7 billion people in the world globally, and it is estimated to increase to 9.7 billion people by 2030. This population growth results from advancement in medical facilities, the industrial revolution, and increased agricultural productivity. This resulted in various technological advancements to benefit humanity and meet the ever-increasing demand for their necessities. These developments, in turn, have resulted in increased pollution, contamination as well as the exploitation of natural resources. It has further led to the increase in the concentration of heavy metals and xenobiotics in nature.

The word "xenobiotics" depicts a foreign chemical substance to which an organism is subjected. It also includes the substances that are present in higher than the average concentration in the environment. The metals with atomic number more than 20 and specific gravity greater than 5 are termed heavy metals (Ali and Khan, 2018; Csuros and Csuros, 2002). Both heavy metals and xenobiotics of them are a considerable challenge to the

Cyanobacterial Lifestyle and its Applications in Biotechnology.
DOI: https://doi.org/10.1016/B978-0-323-90634-0.00001-9

ecosystem. The source of heavy metals can be natural (e.g., natural weathering of rocks and volcanic eruption) as well as anthropogenic (e.g., industrialization, urbanization, mining, and element extraction process), while that of xenobiotics is mainly anthropogenic (exception: natural chemical formed in defense against predator by a prey can be xenobiotic for the predator). Both heavy metals and xenobiotics affect the ecosystem adversely, threatening all the living organisms (Pimentel et al., 2005). There is an urgent need to maintain and restore its balance, which is the main driving force of life.

There are several methods to eliminate the chemicals from the pollution site and have been practiced for a long time. These remediation methods include traditional approaches such as their transportation to a landfill or cap and contain method, incineration, and various chemical technologies and bioremediation. There are various drawbacks associated with these approaches, for example, handling and transport of hazardous material to landfill is unsafe; the efficiency of cap and contain method, which involves isolation and monitoring of the site, is less and the process of chemical treatment is costly, complex as well as hazardous. However, applying various microorganisms such as bacteria, algae, fungi, etc., to remove contaminants and pollutants from the environment is much more economical and promising. This process of introduction of microbes for remediation is known as bioremediation. Bioremediation is a pretty cost-effective method and can be done on-site so that no transportation is required, which reduces the risk of exposure of hazardous materials to living beings. There are various advantages and disadvantages of bioremediation, which are listed as follows (Dobos and Puia, 2009):

Advantages:

1. Safe for the environment.
2. Highly cost-effective.
3. Does not generate waste byproducts.
4. It eliminates the cost of transportation and can be done directly on sites. The exposure of harmful material to the living system is minimal, and disruption of the ecosystem is minimal.
5. It may be combined with other treatment technologies.

Disadvantages:

1. A few materials are not fully degradable, although they are converted to simpler forms.
2. The results observed in the laboratory may vary while it is done in field conditions.
3. It is not very effective in compact and clay soil, where nutrients and oxygen are complex.
4. Hydrological and geochemical conditions of the site may change over time.

In 2001, Vidali considered this as a noninvasive technique that keeps the ecosystem intact. Several factors affect the rate of bioremediation, such as the supply of nutrients, type of pollutants, structure of soil, pH and temperature, microbial diversity and moisture content (Vidali, 2001). Algae, bacteria, and fungi mainly mediate the process of bioremediation. Recently cyanobacteria have been identified as potential agents for bioremediation. Cyanobacteria are Gram-negative prokaryotes that evolve oxygen and are omnipresent in the environment. They may exist as free-living organisms as well as in symbiotic relationship with fungi, plants, and animals. They play an essential role in maintaining soil fertility by sustaining the availability of nutrients, balancing the pH and porosity, increasing the water holding capacity, and reducing the salinity of the soil. They also can degrade, bioaccumulate, and biotransform various heavy metals and xenobiotics (Chaillan et al., 2004; El-Sheekh and Hamouda, 2014). Tripathi et al. and Yin et al., in 2008 and 2012 respectively, have found them suitable to treat the contaminated aquatic ecosystem. In this chapter, we will look at the role of cyanobacteria as the potential candidate to carry out bioremediation (Tripathi et al., 2008; Yin et al., 2012).

13.2 Cyanobacteria and heavy metals

Heavy metals enter the ecosystem through various processes such as mining (Zhuang et al., 2013), industrial activity and weathering of rocks (Gautam et al., 2016), agricultural practices (Naveedullah et al., 2013), sewage water (Wei et al., 2019). When heavy metals enter the living system and exceed the threshold level (differ for each heavy metal and ranges from 1 to 2000 ppm), it gets bioaccumulated, inhibits enzymes, and disrupts various biological functions. Heavy metals such as nickel, cobalt, chromium, arsenic, and vanadium cause damage to biomolecules, as they produce free radical by acting as redox catalysts. To remove these metals from wastewater, different chemical methods such as ion exchange filtration, nanofiltration, catalytic coagulation and lime precipitation, and adsorption methods are used (Arora, 2019; Budak, 2013). These processes are expensive, energy-intensive, and often produce sludge, which needs further treatment. Meanwhile, microbes have a high surface area to volume ratio due to the small size, enabling them to provide extensive contact areas to bind metals.

Cyanobacteria are the largest photosynthetic phylum that sustains diverse habitat, ranging from water bodies to land. They are even found in extreme environment such as Antarctic rocks. They are photoautotrophs and can fix atmospheric nitrogen (Singh et al., 2016). Their cell wall contains various functional groups such as carboxyl and hydroxyl group and possesses multiple binding sites for metals inside the cells. It has high retention capacity and high metal removal efficiency, which enable them to detoxify water contaminated with heavy metal by sequestering the metal ions by the process of adsorption and accumulation (Bilal et al., 2018).

13.2.1 Heavy metal sequestration by cyanobacteria

Microbes remove or biotransform the pollutants using two different uptakes of metals: namely absorption and adsorption. Absorption is a metabolically driven active process and is also known as bioaccumulation. Adsorption is not driven metabolically and is a passive process known as biosorption (De Philippis et al., 2011; Jacob et al., 2018; Kaplan, 2013; R and C, 2017).

13.2.1.1 Biosorption of heavy metals

At physiological pH, bacteria's cell wall (both live and dead) is negatively charged, which forms ionic bonds with the positively charged heavy metals. Thus, biosorption of heavy metals occurs (Fomina and Gadd, 2014; Michalak et al., 2013). These heavy metals are translocated inside the cell and get transform into an innocuous form. These innocuous forms are formed by the process of extracellular precipitation, volatilization, or valance conversion (Igiri et al., 2018). The peptidoglycan receptors can also participate in the biosorption process by any of the following mechanisms (Arief et al., 2008; Chojnacka et al., 2005):

1. The carboxyl group of peptidoglycans is a good surface for the adsorption of heavy metals. They can bind to the membrane proteins, lipopolysaccharides, lipids, and most importantly, exopolysaccharides (EPS) bind to the metal ions by the action of uronic acid.
2. When there exists a gradient of concentration of metal ions between intra- and extracellular environment transporter proteins aid in the active transport of metal ions (Van Ho et al., 2002; Zalups and Ahmad, 2003).
3. Cyanobacteria produce transportation by binding metal ions to a chelating protein such as metallothionein polyphosphates and phytochelatin in their cytoplasm. Following the binding of metal ions with the chelating agents, they are stored in various cellular compartments (Diep et al., 2018; Gardea-Torresdey et al., 1998; Kaplan, 2013).

Among the above mechanisms, the metal sequestration by EPS is found to be very high. EPS is a high-molecular-weight polymer formed naturally in cyanobacteria. It comprises various proteins, polysaccharides, nucleotides, lipids, and secondary metabolites carrying negatively charged functional groups such as carboxylate, amide, amine, sulfate, sulfhydryl, and uronic acid. These groups provide binding sites for various metal ions. Four enzymes, namely, hexokinase, uridine-5′-diphosphate-glucose pyrophosphorylase, glycosyltransferases, and polymerase, are responsible for production. The uptake of metal ions mainly depends on the density and distribution of charges on the cell wall and how accessible are these charges to the metal ions. Carboxylic, hydroxyl, and phosphoric group predominantly take part in the sequestration of metal ions, while uronic acid and sulfate ions impart the highest negative charge so that positively charged metal ions can bind to them (De Philippis et al., 2011; Kanamarlapudi and Muddada, 2017; Ozturk et al., 2009). A few studies have shown that EPS works better with a single metal solution; the reason may be the interaction between metal ions. While some cyanobacterial species show selectivity toward some metals, others do not display any (Pereira et al., 2009). It is always suggested to select an optimum level of metal uptake, using EC50, which is half-maximal effective concentration, for an organism. This value makes organism sustainable at that particular amount of heavy metal (Ozturk et al., 2009). The wild-type cyanobacteria have certain limitations, and they need to be genetically modified for removing all types of metal ions. Most commonly, the approaches are employed to increase anionic moiety in the cell surface and EPS production (Igiri et al., 2018).

13.2.1.2 *Bioaccumulation of heavy metals*

Bioaccumulation occurs when the extracellular metal ion concentration is relatively higher than that of the intracellular ions. This leads to the transport of metal ions inside the cells leading to bioaccumulation (Arunakumara and Zhang, 2008). The metal ions are translocated from the outer membrane to periplasm with the help of porins. Three different proteins, namely, channels, primary active transporters, and secondary carriers, are involved in transporting ions from periplasm to cytoplasm. The transport via channels depends on the gradient of metal ions and is passive transport, while secondary carriers depend on energy and protein motive force. The primary active transporter is a complex of proteins and depends on ATP hydrolysis. In the cytoplasm, metal-binding entities such as metallothionein, polyphosphates, and phytochelatins, bind the metal ions. Inside the cells, the metabolic activities of the cells turn the metal ions into a less toxic form (Pereria et al., 2011). The cell may experience harmful effects of the metal ions if this mechanism fails.

Most wild-type cyanobacteria are less efficient in bioaccumulation due to less efficient mechanism of metal sequestration, low level of expression of genes and proteins, or less efficient mechanism of stress recovery. Multiple transporting systems comprising channels, primary active transporters, and secondary career can be recombined by genetic engineering (Saier, 2016). This improvement of the heavy metal ion storage system by expressing metal-binding proteins (e.g., metallothionein) in the cytoplasm can be achieved (Diep et al., 2018).

13.2.2 Cyanobacterial defense mechanism against toxicity induced by metals

A lot of physiological stress occurs due to metal toxicity, and metal ions can block functional groups of many molecules, leading to the enhanced production of reactive oxygen species, thereby inducing cell damage (Kerfeld and Kirilovsky, 2013). To manage this, cyanobacteria develop various defense mechanisms: the formation of EPS (De Philippis et al., 2011); synthesis of metal-binding proteins such as metallothionein and phytochelatin (Balzano et al., 2020); expression of antiporter proteins in the cells, which transport metal ions out of the cells (Alhadeff and Warshel, 2015; Ito et al., 2017).

13.2.3 Factors affecting metal sequestration by cyanobacteria

Various factors affect metal sequestration by cyanobacteria. These are biomass (Gardea-Torresdey et al., 1998), carrying capacity (Rai and Tripathi, 2007), growth rate (El-Enany and Issa, 2000), contact time (De Philippis et al., 2003), temperature (Goswami et al., 2015), and pH (Harris and Ramelow, 1990). The composition of the cellular wall and the capacity of biosorption are also affected by light intensity and glucose concentration (Chojnacka et al., 2005). The design of the cell wall is affected by light and the concentration of glucose. Therefore, when the same cyanobacterial species is grown under different light and varied glucose concentrations, it showed different metal biosorption characteristics (Chojnacka et al., 2005). It has been found that at different pH, diverse functional group participate in metal binding. For instance, at pH 2-5 carboxyl group mainly participate in the metal-binding process, while at pH 5–9, carboxyl and phosphate group, while at pH 9–12 carboxyl, phosphate hydroxyl, and amine group are involved in the process of metal sequestration (Chojnacka et al., 2005). Similarly, the time of contact to reach equilibrium for removing metal ion from water is variable for different strains. For example, the contact time for *Nostoc* is 5–6 h, while with *Spirulina platensis* and *Synechococcus* sp. PCC 7942, it is 1–2 h.

13.2.4 Various species of cyanobacteria in metal ion sequestration

Cyanobacteria have been widely explored for their efficiency in removing and minimizing harmful industrial wastes (such as metal ions, biocides and xenobiotics) from the ecosystem (El-Bestawy, 2008). Cyanobacteria are known to tolerate and remove heavy metals such as lead, cadmium, zinc, and copper (Nagase et al., 2005). Shing et al. (2008) revealed that the cadmium concentrations (up to 8.0 mg/L) stimulate the rate of oxygen production in *Anabaena torulosa* (Shing et al., 2008). Additionally, biosensors were developed to measure lead toxicity in water samples using immobilized *A. torulosa* (Chia Chay et al., 2008). When *Synechocysis* sp. PCC6803, a unicellular blue-green alga, was treated with 0.5 mM of arsenate for 14 days, it accumulated 1 g/kg dry weight (dw) of arsenic.

Similarly, when it was treated with the same amount of arsenite for the same period, it accumulated 0.9 g/kg DW of arsenic. It can rapidly oxidize arsenite to arsenate and accumulate arsenic through cellular oxidation (Yin et al., 2012). This shows its ability to remove arsenic from the aquatic environment.

Cyanobacterial strains have also been explored to remove heavy metals from water reservoirs and decrease the excess nitrate and phosphate compounds present in the agricultural crop fields (Kesaano and Sims, 2014; Kumar et al., 2016). Consortia of cyanobacteria and bacteria have also been investigated for application in wastewater treatment. The active cultures of *Aphanocapsa* sp. BDU 16, *Oscillatoria* sp. BDU 30501, and *Halobacterium* US 101 reduce the calcium and chloride content from the wastewater and generate the fish-friendly water ecosystem in water reservoirs (Uma and Subramanian, 1990).

Additionally, *Phormidium valderianum* BDU 30501 and *Oscillatoria boryana* BDU 92181 have been revealed for phenol and melanoidin degradation from the distillery effluents (Francisca Kalavathi et al., 2001; Shashirekha et al., 1997). The indigenous microbial consortia are most often utilized for bioremediation of contaminated regions of the soil and water bodies. Therefore, the cyanobacterial strains could be employed to remove diverse groups of contaminants present in the ecosystem. *Chlamydomonas* sp., *Agrobacter* sp., *Oscillatoria* sp., *Chlorella* sp., sulfate-reducing bacteria and microalgae have been isolated from chromate-contaminated wastewater. These microbial strains have been evaluated as sorbing agents (such as adsorption and absorption) for heavy metal ions using continuous flow bioreactors. The use of microbial strains for metal removal has sought attention to better performance in a cost-effective manner than traditional wastewater management technologies. Recent biosorbent technology can remove toxic metals from wastewater, for example *Spirulina fusiformis* has been efficient for extracting chromium from the tannery effluent (Dönmez and Aksu, 2002).

It was demonstrated that Oedogonium, *Hydrocotyle* and *Rhizoclonium* can accommodate heavy metals as much as 8% of their dry mass (especially arsenic and vanadium) (Saunders et al., 2012). A list of cyanobacterial species with the ability to sequester the heavy metal ions has been shown in Table 13.1.

TABLE 13.1 List of various species of cyanobacteria used in the removal of heavy metals from water (adopted from (Al-Amin et al., 2021).

S no.	Name of species	Ion of heavy metal	Type of process
1	*Cyanospira capsulate* and *Nostoc PCC7936*	Cu^{2+}	Biosorption
2	*Tolypothrix ceylonica*	Cu^{2+} Zn^{2+}	Biosorption
3	*Anabaena doliolum*	Cd^{2+}	Biosorption
4	*Nostoc muscorum*	Cu^{2+}, Zn^{2+}, Cd^{2+}, Pb^{2+}	Biosorption and bioaccumulation
5	*Spirulina fusiformis*	Cr^{4+}	Bioaccumulation
6	*Spirulina platensis, Nostoc muscorum* and *Anabaena oryzae*	Hg^{2+}, Cu^{2+}, Pb^{2+}	
7	*Scenedesmus quadricauda*	Ag^{2+}, Cd^{2+}, Cu^{2+}, Zn^{2+}	
8	*Cyanospira capsulata* ATCC43193, *Cyanothece* (ET5, TI4, PE14, VI22, CE4), and *Nostoc PCC7936*	Cr^{6+}	Biosorption
9	*Gloeothece magna*	Cd^{2+}, Mn^{2+}	Biosorption
10	*Limnococcus* sp.	Cr^{6+}	Biosorption and bioaccumulation
11	Microcystis sp.	Cd^{2+}, Cu^{2+}, Cr^{6+}, Sb^{3+}	Biosorption
12	*Nitzschia closterium, Navicula incerta*	Pb^{2+}	Biosorption
13	*Spirulina platensis*	Zn^{2+}	Biosorption
14	*Spirulina fusiformis*	Cr^{6+}	Bioaccumulation
15	*Synechococcus* sp. *PCC 7942*	Cu^{2+}, Ni^{2+}, Pb^{2+}, Cd^{2+}, Cr^{3+}, Cr^{6+}, Zn^{2+}, Cd^{2+}, Cu^{2+}, Co^{2+}, Hg^{2+}, Ag^{2+},	Biosorption
16	*Synechococcus* sp. *PCC 7942*	Hg^{2+}, Pb^{2+}, Cd^{2+}	Bioaccumulation

13.3 Cyanobacteria and xenobiotics

Bioremediation is most often interceded by bacteria, algae, and fungi. Cyanobacteria is used as bioremediators for several contaminants such as heavy metals, crude oil, pesticides, phenol, and xenobiotics (Hamouda et al., 2016; Kesaano and Sims, 2014; Kumar et al., 2016; Singh et al., 2016). *Synechococcus elongatus, Microcystis aeruginosa,* and *Anacystics nidulans* have been identified to eliminate pesticide such as organochlorine and organophosphorus from the contaminated soil and water (Vijayakumar, 2012). Cyanobacterial genera such as *Synechococcus, Oscillatoria, Nodularia, Anabaena, Microcystis,* and *Nostoc* have been reported to degrade organochlorine insecticide lindane (gammaxene) (El-Bestawy, 2008). *Anabaena, Lyngya, Microcystis, Nostoc,* and *Spirulina* have been revealed to metabolize weedicide such as glyphosate from soil (Forlani et al., 2008; Lipok et al., 2009). Therefore, cyanobacterial strains can potentially be explored for their application in sustainable agricultural practice to remediation a wide range of biocides from soil and water bodies.

The microbial consortia, that is, cyanobacteria and chemotrophic bacterial strains together, could potentially be helpful for the wastewater treatments and removal of oil-contaminated water bodies and soil systems (Abed and Köster, 2005). Cyanobacterial strains have been documented to oxidize oil components, herbicides, and surfactants (El-Rahman Mansy and El-Bestawy, 2002; Pathak et al., 2018; Subashchandrabose et al., 2013). Similarly, *Oscillatoria*-Gammaproteobacteria can apply to breakdown phenanthrene, n-octadecane, and dibenzothiophene (Abed and Köster, 2005). Moreover, the group of *Microcoleus chthonoplastes*—organotrophic bacteria—has been reported to fix nitrogen and metabolize organo-sulfur derivatives and cyclic hydrocarbons (Sánchez et al., 2005). Hydrocarbon-degrading bacteria and cyanobacteria develop an artificial biofilm over the gravel particles and glass plates for sanitation of crude oil contaminate seawater (Al-Awadhi et al., 2003). *Anabaena oryzae* and *Chlorella kessleri* have been shown to degrade crude oil in mixotrophic condition (Hamouda et al., 2016). The modern approach of improvement in bioremediation efficiency of microbial cells can be accomplished by genetic engineering. Recently, the cyanobacterial strains have been engineered for commercial and maintenance-free remediation technology toward the high throughput biomass to get valuable products (Cuellar-Bermudez et al., 2017; Kuritz and Wolk, 1995).

13.3.1 Biotransformation of aromatic hydrocarbons by cyanobacteria

Various cyanobacterial species can grow in the presence of xenobiotics, for example, *P. valderianum,* a marine cyanobacterium, can grow in the presence of phenol and eliminate it from the medium by the activity of polyphenol oxidase and laccase enzyme (Shashirekha et al., 1997). Similarly, phenol can be used as a carbon source by various cyanobacterial species such as *Anabaena, Oscillatoria, Nostoc, and Spirulina* (Samanthakamani and Thangaraju, 2015). However, *Synechococcus* PCC 7002 can degrade phenol extracellularly in darkness under nonphotosynthetic conditions and is inhibited by heat, proteases, and metal ions. It leads to the formation of cis-muconic acid (Wurster et al., 2003). The exudates from cyanobacteria can affect the metabolism of bacteria used in paper industries by acting as a source of growth substrate and aid in the degradation of aromatic substances. *Agmenellum quadruplicatum* and *Oscillatoria* sp. have been found to grow photoautotrophically in the presence of aniline. The cyanobacteria metabolized this aniline into less toxic form such as formanilide, acetanilide and 4-aminophenol (Cerniglia et al., 1981). *Microcystis aeruginosa* can adsorb nitrobenzene and reduce it to aniline using enzyme nitrobenzene reductase (Liu et al., 2014). *Anabaena cylindrica* has a high capacity to minimize 2,4-dinitrophenol to 2-amino-4-nitrophenol in the presence of light. A mixed culture of *A. variabilis and A. cylindrica* can reduce 2-amino-4-nitrophenol too. 2,4-dinitrophenol is a precursor of many pesticides, herbicides, and synthetic dyes (Hirooka et al., 2006). Some genera of cyanobacteria, such as, *Nodularia, Arthrospira, Leptolyngbya,* and *Synechococcus,* can transform acetophenone by the activity of 3-ketoacyl-(acyl-carrier-protein) reductase (Hölsch et al., 2008). Acetophenone and its derivative are used as building blocks of pharmaceutical and cosmetic industries (Cao et al., 2012). *Anabaena flos-aquae* and *M. aeruginosa* can degrade phthalate esters, commonly used as plasticizers and significant pollutants. The degradation of phthalate esters occurs through transesterification (Babu and Wu, 2010).

Some cyanobacteria species can also transform organophosphorus pesticide, fenamiphos. They oxidize it to fenamiphos sulfoxide, followed by hydrolysis to fenamiphos sulfoxide phenol (Cáceres et al., 2008). Diaromatic hydrocarbons as naphthalene are oxidized by *Oscillatoria* sp. to 1-naphthol, cis-1,2-dihydroxy-1,2- dihydro naphthalene and 4-hydroxy-1-tetralone (Cerniglia et al., 1980a). *Anabaena, Microcoleus, Coccochloris, Agmenellum, Aphanocapsa,* and *Nostoc* can oxidize naphthalene to 1-naphthol. In 2012, El-Skeekh et al. found that the cyanobacterial strains *Lyngbya lagerlerimi, Nostoc linckia,* and *Oscillatoria rubescens* can degrade α and β naphthol (El-Sheekh et al., 2012). Similarly, *Oscillatoria* sp. can grow in the presence of light and biphenyl, where it gets oxidize to form 4-hydroxybiphenyl (Cerniglia et al., 1980b). *Microcystis aeruginosa* promotes photodegradation of antimicrobial agent triclosan in the aquatic environment (Huang et al., 2016).

Anabaena laxa, Aphanizomenon klebahnii, Nodularia moravica, and *Synechocystis aquatilis* can catalyze the bioreduction of chalcone in the presence of light. Chalcones are aromatic ketones that form the central core of various biological compounds (Żyszka et al., 2017). *Phormidium tenue* strain of cyanobacteria is known to degrade anthracene, and a polycyclic aromatic hydrocarbon produces various dyes. It can degrade anthracene to anthracene-1,2-dione, 8- hydroxy-anthracene-1,2-dione. *Anabaena fertilissima* can degrade anthracene to form 2,4-dimethyl-1-heptene, tetradecanoic acid, 9-octadecane, and benzene ring compounds (Patel et al., 2015). Phenanthrene, a compound used to make dyes, pesticides, plastics, explosives, and drugs, is highly distributed in the environment due to pyrolytic processes. *Agmenellum quadruplicatum* PR-6, a marine cyanobacterium, can metabolize phenanthrene to phenanthrene trans-9,10-dihydrodiol 1-methoxyphenanthrene (Narro et al., 1992). *Anabaena fertilissima* can degrade pyrene to 2,3,4-trimethylhexane (Patel et al., 2015).

13.4 Conclusion

Cyanobacteria have been known as a tool for bioremediation to balance the geochemical cycle in nature. We have seen in detail how they sequester the heavy metals by the process of bioaccumulation and biosorption. The metal binding capacity of cyanobacteria can further be increased by genetic engineering of the EPS pathway. The species *Synechococcus* PCC 7942 is found to be very efficient in removing a variety of metals. However, the detailed process of biotransformation and biodegradation of xenobiotics is the less explored area than that of heavy metals. However, cyanobacteria have been shown to biotransform these exceptionally, the lower molecular weight metals are more efficiently biotransformed than the large aromatic compounds. Cyanobacteria have intensive metabolism, high ability of proliferation, and are easy to manipulate genetically; therefore are quite great area of interest for the processing of xenobiotics and heavy metals. Further understanding of the biotransformation of these compounds is needed to utilize them more extensively.

References

Abed, R.M.M., Köster, J., 2005. The direct role of aerobic heterotrophic bacteria associated with cyanobacteria in the degradation of oil compounds. Int. Biodeterior. Biodegrad. 55, 29–37. https://doi.org/10.1016/j.ibiod.2004.07.001.

Al-Amin, A., Parvin, F., Chakraborty, J., Kim, Y.-I., 2021. Cyanobacteria mediated heavy metal removal: a review on mechanism, biosynthesis, and removal capability. Environ. Technol. Rev. 10, 44–57. https://doi.org/10.1080/21622515.2020.1869323.

Al-Awadhi, H., Al-Hasan, R.H., Sorkhoh, N.A., Salamah, S., Radwan, S.S., 2003. Establishing oil-degrading biofilms on gravel particles and glass plates. Int. Biodeterior. Biodegrad. 51, 181–185. https://doi.org/10.1016/S0964-8305(02)00140-3.

Alhadeff, R., Warshel, A., 2015. Simulating the function of sodium/proton antiporters. Proc. Natl. Acad. Sci. USA 112, 12378–12383. https://doi.org/10.1073/pnas.1516881112.

Ali, H., Khan, E., 2018. What are heavy metals? Long-standing controversy over the scientific use of the term 'heavy metals' – proposal of a comprehensive definition. Toxicol. Environ. Chem. 100, 6–19. https://doi.org/10.1080/02772248.2017.1413652.

Arief, V.O., Trilestari, K., Sunarso, J., Indraswati, N., Ismadji, S., 2008. Recent progress on biosorption of heavy metals from liquids using low cost biosorbents: characterization, biosorption parameters and mechanism studies. CLEAN - Soil Air Water 36, 937–962. https://doi.org/10.1002/clen.200800167.

Arora, R., 2019. Adsorption of heavy metals–a review, Mater. Today Proc., 9th International Conference of Materials Processing and Characterization, ICMPC-2019, 18, 4745–4750. https://doi.org/10.1016/j.matpr.2019.07.462.

Arunakumara, K.K.I.U., Zhang, X., 2008. Heavy metal bioaccumulation and toxicity with special reference to microalgae. J. Ocean Univ. China 7, 60–64. https://doi.org/10.1007/s11802-008-0060-y.

Babu, B., Wu, J.-T., 2010. Production of phthalate esters by nuisance freshwater algae and cyanobacteria. Sci. Total Environ. 408, 4969–4975. https://doi.org/10.1016/j.scitotenv.2010.07.032.

Balzano, S., Sardo, A., Blasio, M., Chahine, T.B., Dell'Anno, F., Sansone, C., Brunet, C., 2020. Microalgal metallothioneins and phytochelatins and their potential use in bioremediation. Front. Microbiol. 11, 517. https://doi.org/10.3389/fmicb.2020.00517.

Bilal, M., Rasheed, T., Sosa-Hernández, J.E., Raza, A., Nabeel, F., Iqbal, H.M.N., 2018. Biosorption: an interplay between marine algae and potentially toxic elements-a review. Mar. Drugs 16. https://doi.org/10.3390/md16020065.

Budak, T.B., 2013. Removal of heavy metals from wastewater using synthetic ion exchange resin. Asian J. Chem. 25, 4207–4210. https://doi.org/10.14233/ajchem.2013.13902.

Cáceres, T.P., Megharaj, M., Naidu, R., 2008. Biodegradation of the pesticide fenamiphos by ten different species of green algae and cyanobacteria. Curr. Microbiol. 57, 643–646. https://doi.org/10.1007/s00284-008-9293-7.

Cao, Y., Zhuang, Y., Yao, C., Wu, B., He, B., 2012. Purification and characterization of an organic solvent-stable lipase from *Pseudomonas stutzeri* LC2-8 and its application for efficient resolution of (R, S)-1-phenylethanol. Biochem. Eng. J. 64, 55–60. https://doi.org/10.1016/j.bej.2012.03.004.

Cerniglia, C.E., Freeman, J.P., Van Baalen, C., 1981. Biotransformation and toxicity of aniline and aniline derivatives of cyanobacteria. Arch. Microbiol. 130, 272–275. https://doi.org/10.1007/BF00425939.

Cerniglia, C.E., Van Baalen, C., Gibson, D.T., 1980a. Metabolism of naphthalene by the cyanobacterium *Oscillatoria* sp., strain JCM. Microbiology 116, 485–494. https://doi.org/10.1099/00221287-116-2-485.

Cerniglia, C.E., Van Baalen, C., Gibson, D.T., 1980b. Oxidation of biphenyl by the cyanobacterium, *Oscillatoria* sp., strain JCM. Arch. Microbiol. 125, 203–207. https://doi.org/10.1007/BF00446877.

Chaillan, F., Le Flèche, A., Bury, E., Phantavong, Y.-H., Grimont, P., Saliot, A., Oudot, J., 2004. Identification and biodegradation potential of tropical aerobic hydrocarbon-degrading microorganisms. Res. Microbiol. 155, 587–595. https://doi.org/10.1016/j.resmic.2004.04.006.

Chia Chay, T., Surif, S., Yook Heng, L., 2008. The behavior of immobilized cyanobacteria *Anabaena torulosa* as an electrochemical toxicity biosensor. Asian J. Biol. Sci. 2, 14–20. https://doi.org/10.3923/ajbs.2009.14.20.

Chojnacka, K., Chojnacki, A., Górecka, H., 2005. Biosorption of Cr3+, Cd2+ and Cu2+ ions by blue-green algae *Spirulina* sp.: kinetics, equilibrium and the mechanism of the process. Chemosphere 59, 75–84. https://doi.org/10.1016/j.chemosphere.2004.10.005.

Csuros, M., Csuros, C., 2002. Environmental Sampling and Analysis for Metals. Lewis, Boca Raton, Fla.

Cuellar-Bermudez, S.P., Aleman-Nava, G.S., Chandra, R., Garcia-Perez, J.S., Contreras-Angulo, J.R., Markou, G., Muylaert, K., Rittmann, B.E., Parra-Saldivar, R., 2017. Nutrients utilization and contaminants removal. A review of two approaches of algae and cyanobacteria in wastewater. Algal Res 24, 438–449. https://doi.org/10.1016/j.algal.2016.08.018.

De Philippis, R., Colica, G., Micheletti, E., 2011. Exopolysaccharide-producing cyanobacteria in heavy metal removal from water: molecular basis and practical applicability of the biosorption process. Appl. Microbiol. Biotechnol. 92, 697–708. https://doi.org/10.1007/s00253-011-3601-z.

De Philippis, R., Paperi, R., Sili, C., Vincenzini, M., 2003. Assessment of the metal removal capability of two capsulated cyanobacteria, *Cyanospira capsulata* and *Nostoc* PCC7936. J. Appl. Phycol. 15, 155–161. https://doi.org/10.1023/A:1023889410912.

Diep, P., Mahadevan, R., Yakunin, A.F., 2018. Heavy metal removal by bioaccumulation using genetically engineered microorganisms. Front. Bioeng. Biotechnol. 6, 157. https://doi.org/10.3389/fbioe.2018.00157.

Dobos, L., Puia, C., 2009. The most important methods for depollution of hydrocarbons polluted soils. Bull. Univ. Agric. Sci. Vet. Med. Cluj-Napoca Agric. 66. https://doi.org/10.15835/buasvmcn-agr:3734.

Dönmez, G., Aksu, Z., 2002. Removal of chromium(VI) from saline wastewaters by *Dunaliella* species. Process Biochem 38, 751–762. https://doi.org/10.1016/S0032-9592(02)00204-2.

El-Bestawy, E., 2008. Treatment of mixed domestic–industrial wastewater using cyanobacteria. J. Ind. Microbiol. Biotechnol. 35, 1503–1516. https://doi.org/10.1007/s10295-008-0452-4.

El-Enany, A.E., Issa, A.A., 2000. Cyanobacteria as a biosorbent of heavy metals in sewage water. Environ. Toxicol. Pharmacol. 8, 95–101. https://doi.org/10.1016/s1382-6689(99)00037-x.

El-Rahman Mansy, A., El-Bestawy, E., 2002. No title found. World J. Microbiol. Biotechnol. 18, 125–131. https://doi.org/10.1023/A:1014490811121.

El-Sheekh, M.M., Ghareib, M., EL-Souod, G.A., 2012. Biodegradation of phenolic and polycyclic aromatic compounds by some algae and cyanobacteria. J. Bioremediation Biodegrad. 03, 133. https://doi.org/10.4172/2155-6199.1000133.

El-Sheekh, M.M., Hamouda, R.A., 2014. Biodegradation of crude oil by some cyanobacteria under heterotrophic conditions. Desalin. Water Treat 52, 1448–1454. https://doi.org/10.1080/19443994.2013.794008.

Fomina, M., Gadd, G.M., 2014. Biosorption: current perspectives on concept, definition and application. Bioresour. Technol. 160, 3–14. https://doi.org/10.1016/j.biortech.2013.12.102.

Forlani, G., Pavan, M., Gramek, M., Kafarski, P., Lipok, J., 2008. Biochemical bases for a widespread tolerance of cyanobacteria to the phosphonate herbicide glyphosate. Plant Cell Physiol 49, 443–456. https://doi.org/10.1093/pcp/pcn021.

Francisca Kalavathi, D., Uma, L., Subramanian, G., 2001. Degradation and metabolization of the pigment—melanoidin in distillery effluent by the marine cyanobacterium *Oscillatoria boryana* BDU 92181. Enzyme Microb. Technol. 29, 246–251. https://doi.org/10.1016/S0141-0229(01)00383-0.

Gardea-Torresdey, J.L., Arenas, J.L., Francisco, N.M.C., Tiemann, K.J., Webb, R., 1998. Ability of immobilized cyanobacteria to remove metal ions from solution and demonstration of the presence of metallothionein genes in various strains. J. Hazard. Subst. Res. 1. https://doi.org/10.4148/1090-7025.1001.

Goswami, S., Syiem, M.B., Pakshirajan, K., 2015. Cadmium removal by *Anabaena doliolum* Ind1 isolated from a coal mining area in Meghalaya, India: associated structural and physiological alterations. Environ. Eng. Res. 20, 41–50. https://doi.org/10.4491/eer.2014.059.

Hamouda, R.A.E.F., Sorour, N.M., Yeheia, D.S., 2016. Biodegradation of crude oil by *Anabaena oryzae*, *Chlorella kessleri* and its consortium under mixotrophic conditions. Int. Biodeterior. Biodegrad. 112, 128–134. https://doi.org/10.1016/j.ibiod.2016.05.001.

Harris, P.O., Ramelow, G.J., 1990. Binding of metal ions by particulate biomass derived from *Chlorella vulgaris* and *Scenedesmus quadricauda*. Environ. Sci. Technol. 24, 220–228. https://doi.org/10.1021/es00072a011.

Hirooka, T., Nagase, H., Hirata, K., Miyamoto, K., 2006. Degradation of 2,4-dinitrophenol by a mixed culture of photoautotrophic microorganisms. Biochem. Eng. J., 29, 157–162. https://doi.org/10.1016/j.bej.2005.03.018.

Hölsch, K., Havel, J., Haslbeck, M., Weuster-Botz, D., 2008. Identification, cloning, and characterization of a novel ketoreductase from the cyanobacterium *Synechococcus* sp. strain PCC 7942. Appl. Environ. Microbiol. 74, 6697–6702. https://doi.org/10.1128/AEM.00925-08.

Huang, X., Tu, Y., Song, C., Li, T., Lin, J., Wu, Y., Liu, J., Wu, C., 2016. Interactions between the antimicrobial agent triclosan and the bloom-forming cyanobacteria Microcystis aeruginosa. Aquat. Toxicol. 172, 103–110. https://doi.org/10.1016/j.aquatox.2016.01.002.

Igiri, B.E., Okoduwa, S.I.R., Idoko, G.O., Akabuogu, E.P., Adeyi, A.O., Ejiogu, I.K., 2018. Toxicity and bioremediation of heavy metals contaminated ecosystem from tannery wastewater: a review. J. Toxicol. 2018, 2568038. https://doi.org/10.1155/2018/2568038.

Ito, M., Morino, M., Krulwich, T.A., 2017. Mrp antiporters have important roles in diverse bacteria and Archaea. Front. Microbiol. 8, 2325. https://doi.org/10.3389/fmicb.2017.02325.

Jacob, J.M., Karthik, C., Saratale, R.G., Kumar, S.S., Prabakar, D., Kadirvelu, K., Pugazhendhi, A., 2018. Biological approaches to tackle heavy metal pollution: a survey of literature. J. Environ. Manage. 217, 56–70. https://doi.org/10.1016/j.jenvman.2018.03.077.

Kanamarlapudi, S.L.R.K., Muddada, S., 2017. Characterization of exopolysaccharide produced by *Streptococcus thermophilus* CC30. BioMed. Res. Int., 1–11. https://doi.org/10.1155/2017/4201809.

Kaplan, D., 2013. Absorption and adsorption of heavy metals by microalgae. In: Richmond, A., Hu, Q. (Eds.), Handbook of Microalgal Culture. John Wiley & Sons, Ltd, Oxford, UK, pp. 602–611. https://doi.org/10.1002/9781118567166.ch32.

Kerfeld, C.A., Kirilovsky, D., 2013. Structural, Mechanistic and Genomic Insights into OCP-Mediated Photoprotection. In: Chauvat, F., Cassier-Chauvat, C. (Eds.), Advances in Botanical Research. Elsevier, pp. 1–26. https://doi.org/10.1016/B978-0-12-394313-2.00001-9.

Kesaano, M., Sims, R.C., 2014. Algal biofilm based technology for wastewater treatment. Algal Res 5, 231–240. https://doi.org/10.1016/j.algal.2014.02.003.

Kumar, D., Pandey, L.K., Gaur, J.P., 2016. Metal sorption by algal biomass: from batch to continuous system. Algal Res 18, 95–109. https://doi.org/10.1016/j.algal.2016.05.026.

Kumar, G.P., Kumar, G.R., Chattopadhyaya, B.S., Pandey, M.C., 2016. Heavy metals in the environment: Fate, transport, toxicity and remediation technologies. In: Pathania (Ed.), Heavy Metals: Sources, Toxicity and Remediation Techniques. Nova.

Kuritz, T., Wolk, C.P., 1995. Use of filamentous cyanobacteria for biodegradation of organic pollutants. Appl. Environ. Microbiol. 61, 234–238. https://doi.org/10.1128/AEM.61.1.234-238.1995.

Lipok, J., Wieczorek, D., Jewgiński, M., Kafarski, P., 2009. Prospects of in vivo 31P NMR method in glyphosate degradation studies in whole cell system. Enzyme Microb. Technol. 44, 11–16. https://doi.org/10.1016/j.enzmictec.2008.09.011.

Liu, Z., Cui, F., Ma, H., Fan, Z., Zhao, Z., Hou, Z., Liu, D., 2014. The transformation mechanism of nitrobenzene in the present of a species of cyanobacteria *Microcystis aeruginosa*. Chemosphere 95, 234–240. https://doi.org/10.1016/j.chemosphere.2013.08.083.

Michalak, I., Chojnacka, K., Witek-Krowiak, A., 2013. State of the art for the biosorption process–a review. Appl. Biochem. Biotechnol. 170, 1389–1416. https://doi.org/10.1007/s12010-013-0269-0.

Nagase, H., Inthorn, D., Oda, A., Nishimura, J., Kajiwara, Y., Park, M.-O., Hirata, K., Miyamoto, K., 2005. Improvement of selective removal of heavy metals in cyanobacteria by NaOH treatment. J. Biosci. Bioeng. 99, 372–377. https://doi.org/10.1263/jbb.99.372.

Narro, M.L., Cerniglia, C.E., Van Baalen, C., Gibson, D.T., 1992. Metabolism of phenanthrene by the marine cyanobacterium *Agmenellum quadruplicatum* PR-6. Appl. Environ. Microbiol. 58, 1351–1359. https://doi.org/10.1128/AEM.58.4.1351-1359.1992.

Naveedullah, null, Hashmi, M.Z., Yu, C., Shen, H., Duan, D., Shen, C., Lou, L., Chen, Y., 2013. Risk assessment of heavy metals pollution in agricultural soils of siling reservoir watershed in Zhejiang Province, China. BioMed. Res. Int., 590306. https://doi.org/10.1155/2013/590306.

Ozturk, S., Aslim, B., Suludere, Z., 2009. Evaluation of chromium(VI) removal behaviour by two isolates of *Synechocystis* sp. in terms of exopolysaccharide (EPS) production and monomer composition. Bioresour. Technol. 100, 5588–5593. https://doi.org/10.1016/j.biortech.2009.06.001.

Patel, J.G., Kumar, J.I.N., Khan, S.R., 2015. Consequences of environmentally hazardous polycyclic aromatic hydrocarbon- anthracene treatment on cyanobacteria. Int. J. Appl. Sci. Biotechnol. 3, 381–386. https://doi.org/10.3126/ijasbt.v3i3.11654.

Pathak, J., Rajneesh, Maurya, P.K., Singh, S.P., Häder, D.-P., Sinha, R.P., 2018. Cyanobacterial farming for environment friendly sustainable agriculture practices: innovations and perspectives. Front. Environ. Sci. 6, 7. https://doi.org/10.3389/fenvs.2018.00007.

Pereira, S., Zille, A., Micheletti, E., Moradas-Ferreira, P., De Philippis, R., Tamagnini, P., 2009. Complexity of cyanobacterial exopolysaccharides: composition, structures, inducing factors and putative genes involved in their biosynthesis and assembly. FEMS Microbiol. Rev. 33, 917–941. https://doi.org/10.1111/j.1574-6976.2009.00183.x.

Pereria, S., Micheletti, E., Zille, A., 2011. Using extracellular polymeric substances (EPS)-producing cyanobacteria for the bioremediation of heavy metals: Do cations compete for the EPS functional groups and also accumulate inside the cell? Microbiology (Reading) 157. https://doi.org/10.1099/mic.0.041038-0.

Pimentel, D., Hepperly, P., Hanson, J., Douds, D., Seidel, R., 2005. Environmental, energetic, and economic comparisons of organic and conventional farming systems. BioScience 55, 573–582. https://doi.org/10.1641/0006-3568(2005)055[0573:EEAECO]2.0.CO;2.

R, P., C, V., 2017. Biosorption of heavy metals by *Spirulina platensis* from electroplating industrial effluent. Environ. Sci. Indian J. 13 (4), 139.

Rai, P.K., Tripathi, B.D., 2007. Removal of heavy metals by the nuisance cyanobacteria *Microcystis* in continuous cultures: an eco-sustainable technology. Environ. Sci. 4, 53–59. https://doi.org/10.1080/15693430601164956.

Saier, M.H., 2016. Transport protein evolution deduced from analysis of sequence, topology and structure. Curr. Opin. Struct. Biol. 38, 9–17. https://doi.org/10.1016/j.sbi.2016.05.001.

Samanthakamani, D., Thangaraju, N., 2015. Potential of freshwater microalgae for degradation of phenol. Ind. J. Sci. Res. and Tech. 3, 9–12. http://www.indjsrt.com.

Sánchez, O., Diestra, E., Esteve, I., Mas, J., 2005. Molecular characterization of an oil-degrading cyanobacterial consortium. Microb. Ecol. 50, 580–588. https://doi.org/10.1007/s00248-005-5061-4.

Saunders, R.J., Paul, N.A., Hu, Y., de Nys, R., 2012. Sustainable sources of biomass for bioremediation of heavy metals in waste water derived from coal-fired power generation. PLoS ONE 7, e36470. https://doi.org/10.1371/journal.pone.0036470.

Shashirekha, S., Uma, L., Subramanian, G., 1997. Phenol degradation by the marine cyanobacterium *Phormidium valderianum* BDU 30501. J. Ind. Microbiol. Biotechnol. 19, 130–133. https://doi.org/10.1038/sj.jim.2900438.

Shing, W.L., Surif, S., Heng, L.Y., 2008. Toxicity Biosensor for the evaluation of cadmium toxicity based on photosynthetic behavior of cyanobacteria *Anabaena torulosa*. Asian J. Biochem. 3, 162–168. https://doi.org/10.3923/ajb.2008.162.168.

Singh, J.S., Kumar, A., Rai, A.N., Singh, D.P., 2016. Cyanobacteria: a precious bio-resource in agriculture, ecosystem, and environmental sustainability. Front. Microbiol. 7, 529. https://doi.org/10.3389/fmicb.2016.00529.

Subashchandrabose, S.R., Ramakrishnan, B., Megharaj, M., Venkateswarlu, K., Naidu, R., 2013. Mixotrophic cyanobacteria and microalgae as distinctive biological agents for organic pollutant degradation. Environ. Int. 51, 59–72. https://doi.org/10.1016/j.envint.2012.10.007.

Tripathi, R.D., Dwivedi, S., Shukla, M.K., Mishra, S., Srivastava, S., Singh, R., Rai, U.N., Gupta, D.K., 2008. Role of blue green algae biofertilizer in ameliorating the nitrogen demand and fly-ash stress to the growth and yield of rice (*Oryza sativa* L.) plants. Chemosphere 70, 1919–1929. https://doi.org/10.1016/j.chemosphere.2007.07.038.

Uma, L., Subramanian, G., 1990. Effective use of cyanobacteria in effluent treatment. Proceedings of the National Symposium on Cyanobacterial N2 fixation IARI, New Delhi.

Van Ho, A., Ward, D.M., Kaplan, J., 2002. Transition metal transport in yeast. Annu. Rev. Microbiol. 56, 237–261. https://doi.org/10.1146/annurev.micro.56.012302.160847.
Vidali, M., 2001. Bioremediation. An overview. Pure Appl. Chem. 73, 1163–1172. https://doi.org/10.1351/pac200173071163.
Vijayakumar, S., 2012. Potential Applications of Cyanobacteria in Industrial Effluents-A Review. J. Bioremediation Biodegrad. 03, 154. https://doi.org/10.4172/2155-6199.1000154.
Wei, J., Duan, M., Li, Y., Nwankwegu, A.S., Ji, Y., Zhang, J., 2019. Concentration and pollution assessment of heavy metals within surface sediments of the Raohe Basin, China. Sci. Rep. 9, 13100. https://doi.org/10.1038/s41598-019-49724-7.
Wurster, M., Mundt, S., Hammer, E., Schauer, F., Lindequist, U., 2003. Extracellular degradation of phenol by the cyanobacterium *Synechococcus* PCC 7002. J. Appl. Phycol. 15, 171–176. https://doi.org/10.1023/A:1023840503605.
Yin, X.-X., Wang, L.H., Bai, R., Huang, H., Sun, G.-X., 2012. Accumulation and transformation of arsenic in the blue-green alga *Synechocysis* sp. PCC6803. Water. Air. Soil Pollut. 223, 1183–1190. https://doi.org/10.1007/s11270-011-0936-0.
Zalups, R.K., Ahmad, S., 2003. Molecular handling of cadmium in transporting epithelia. Toxicol. Appl. Pharmacol. 186, 163–188. https://doi.org/10.1016/s0041-008x(02)00021-2.
Zhuang, P., Li, Z., McBride, M.B., Zou, B., Wang, G., 2013. Health risk assessment for consumption of fish originating from ponds near Dabaoshan mine. South China. Environ. Sci. Pollut. Res. Int. 20, 5844–5854. https://doi.org/10.1007/s11356-013-1606-0.
Żyszka, B., Anioł, M., Lipok, J., 2017. Highly effective, regiospecific reduction of chalcone by cyanobacteria leads to the formation of dihydrochalcone: two steps towards natural sweetness. Microb. Cell Factories 16, 136. https://doi.org/10.1186/s12934-017-0752-3.

CHAPTER

14

Cyanobacteria: As a promising candidate for nanoparticles synthesis

Dnyaneshwar K. Kulal[a], *Dinesh N. Navale*[b], *Santosh W. Zote*[c], *Prasanna B. Ranade*[b], *Pramod K. Kalambate*[d]

[a]Ramnarain Ruia College, Mumbai, Maharashtra, India
[b]Vivekanand Education Society's College of Arts, Science & Commerce, Mumbai, Maharashtra, India
[c]Department of Chemistry, Sathaye College, Mumbai, Maharashtra, India
[d]State Key Laboratory of Materials Processing and Die & Mould Technology, School of Materials Science and Engineering, Huazhong University of Science and Technology, Wuhan, Hubei, PR China

14.1 Introduction

The importance of nanotechnology is been rising day by day and with it, various means of synthesizing nanomaterials. The utilization of nanoparticles in the various fields is being increased that results in a large demand for nanoparticles. A demand for nanoparticles has been increased due to their applications in numerous fields such as catalysis, target drug delivery, biosensors, medicine, bioimaging, cosmetic and personal care, food coloring and packaging, nanofabric, water remediation, optical engineering, paint and surface coating, sports equipment, agriculture, etc. (Stark et al., 2015; Chen et al., 2018). The synthesis of nanoparticles by the physical and chemical method is not economically and environmentally favorable; therefore, researchers are trying to find alternative routes for nanoparticle synthesis. One of them is the use of natural sources for nanoparticle synthesis, which is economically favorable, cleaner, and nontoxic

Cyanobacterial Lifestyle and its Applications in Biotechnology.
DOI: https://doi.org/10.1016/B978-0-323-90634-0.00006-8

(Zikalala et al., 2018; Dahoumane et al., 2017). Biosynthesized metal nanoparticles are found to be efficient and more suitable for medicinal applications in comparison to nanoparticles synthesized by other means.

The natural sources include a lot of materials such as bacteria (Dahoumane et al., 2012), fungi (Guilger-Casagrande et al., 2019), and plants (Mittal et al., 2013), which eliminates the generation of hazardous products; therefore, such synthesis is included in green chemistry. The use of certain microalgae has been reported for the synthesis of nanoparticles of metal such as cadmium, silver, gold, and platinum (Brayner et al., 2007; Parial et al., 2012). One of the most important principles in green chemistry is waste utilization. Ease of handling and genetic manipulation makes the bacteria a promising candidate for the biosynthesis of nanoparticles than the fungi. There are several reports in the literature on the biosynthesis of nanoparticles by using different bacteria. By using bacteria, nanoparticles of many metals are synthesized such as gold, silver, etc. In the agriculture industry, the increased concentration of nitrogen and phosphorous in water enhances the growth of algae. These algae commonly grow on the surface of the water and resist the sun rays to reach into the water which affects the growth of the aquatic animals lying below the surface. However, if such cyanobacteria are utilized for nanoparticle synthesis then it is considered as waste utilization. While using cyanobacteria for nanoparticle synthesis one point should be noted that some cyanobacteria are toxic. Therefore, the selection of cyanobacteria for the preparation of nanoparticles is based on the application of nanoparticles. For example applications such as biosensor, medicine, bioimaging, cosmetic and personal care, food coloring, and packaging, and water remediation requires nontoxic cyanobacteria.

As it is well known that nanoparticles are synthesized by two routes, one is bottom-up and the second one is top-down. Usually, the biosynthesis of metal nanoparticles involves the bottom-up approach. Likewise, cyanobacteria are living organisms; therefore a synthesis of nanoparticles using living things is again divided into two ways that is *in-vitro* and *in-vivo*. *In-vitro* is nothing but the use of cell extracts and *in-vivo* comprises the use of living cells as bioreactors. Furthermore, *in-vivo* method is more preferable over *in-vitro* as it involves intercellular and intracellular self-reproduction of bacteria. Self-reproduction increases the number of bacteria that are directly proportional to the rate of the reaction. Cyanobacterial species are capable to produce nanoparticles continuously (Deplanche et al., 2012). Various cyanobacterial species and environmental conditions produce different sizes and shapes of nanoparticles (Karnani et al., 2013). Furthermore, several experimental parameters responsible for various shapes and sizes are pH, temperature, oxidation state, and concentration of metal ions (Gericke et al., 2006; Mohanpuria et al., 2008; Ramanathan et al., 2011). Before the nanoparticles synthesis, one needs to isolate, purify, and characterize cyanobacteria using appropriate solvents, then it can be utilized for nanoparticle synthesis. The cyanobacteria play an excellent role as a reducing agent in the synthesis of the nanoparticles. Such synthesized nanoparticles can be characterized by numerous important methods including X-ray powder diffraction, transmission electron microscopy (Rosken et. al., 2014), UV-Visible spectroscopy, and X-ray photoelectron spectroscopy (Lengke et. al, 2007).

14.2 Synthesis of nanoparticles using cyanobacteria

Of the known metal nanoparticles, silver nanoparticles have received more attention from researchers, compared to other metals nanoparticles as silver ions are easily reduced by photoreduction. Nowadays, the great attention and demand in the biosynthesis of nanoparticles are due to the increased awareness toward the development of a sustainable environment by reducing or replacing the use of toxic and unsafe materials with clean, easily available, and nontoxic materials (Rai et al., 2011). Along with the above advantages, the biosynthesis method effectively and efficiently controls the growth of the crystal and makes the process economically favorable. We have restricted our study to the synthesis of metallic nanoparticles using cyanobacteria. As it is being one of the most diversified and ecologically more evolved prokaryotic groups. Further, the studies found that the released organic matter from cyanobacteria that is, secondary metabolites, is responsible for reducing the metal ions to metal nanoparticles (Lengke et al., 2007). Moreover, cyanobacteria possess a high growth rate and productivity of biomass that makes it further useful in the synthesis of metallic nanoparticles. Importantly the use of cyanobacteria in the aforementioned synthesis is considered a greener route of synthesis. A lot of studies involving the use of cyanobacterial extracellular and intracellular extracts for the biosynthesis of nanoparticles do exist (Lengke et al., 2007). The possible pathway of the biosynthesis of the nanoparticle has also been discussed by Kannan and Subbalaxmi, 2011. Thus, looking at the advantages and huge applications of nanomaterials, this field of biosynthesizing nanomaterials continues to gain enormous attention. Table 14.1 summarizes the synthesis of different metal nanoparticles such as Ag, Au, Pd, Se, Zn, and Cd employing various species of cyanobacteria. Biosynthesis of metal nanoparticles using cyanobacteria most commonly occurs by following general procedure.

The cyanobacterial extract is prepared in an aqueous/organic solvent either by heating or by boiling, depending upon the nature of cyanobacteria utilized. Production of metal nanoparticles by cyanobacteria was performed by preparing a metal solution in aqueous medium to which an aliquot of cyanobacteria filtrate is added under suitable reaction conditions. Negative control was prepared by taking the same amount of aqueous metal solution;

TABLE 14.1 Biosynthesis of nanoparticles employing various cyanobacteria

S. No	Nanoparticles	Cyanobacteria	Size (nm)	Reference
1	Ag	*Desertifilum* sp.	4.5–26	Hamida et al. (2020)
2	Ag	*Arthrospira platensis* (Spirulina)	4–42	Cicci et al. (2017)
3	Au	*Lyngbya majuscula*	<20	Chakraborty et al. (2009)
4	Pd	*Plectonema boryanum*	≤30	Lengke et al. (2007)
5	Ag	*Arthrospira platensis*	11.6	Mahdieha et al. (2012)
6	Ag	*Dolichospermum* sp.	10	Gabriela et al. (2016)
7	ZnO	*Anabaena* sp. L31	80	Singh et al. (2014)
8	Ag	*Microcoleus* sp.	44–79	Sudha et al. (2013)
9	Ag	*Cylindrospermum stagnale*	38–4	Husain et al. (2015)
10	Au	*Anabaena* sp. (SAG 12.82)	9	Rosken et al. (2014)
11	Se	*Nostoclinckia*	10–80	Zinicovscaia et al. (2016)
12	Ag	*Arthrospira indica* PCC7940	48	Husain et al. (2015)
13	Ag	*Arthrospira indica* SAE-85	67	
14	Ag	*Arthrospira maxima* SAE-49-88	61	
15	Ag	*Arthrospira platensis* NEERI	46	
16	Ag	*Cylindrospermum stagnale* NCCU	38–40	
17	Ag	*Calothrix brevissema* NCCU-65	42	
18	Ag	*Spirulina platensis* NCCU-S5	46	
19	Ag	*Anabaena ambigua* NCCU-1	48	
20	Ag	*Anabaena variabilis* NCCU-441	50	
21	Ag	*Aulosira fertilissma* NCCU-443	58	
22	Ag	*Nostoc muscorum* NCCU-442	42	
23	Ag	*Tolypothrix tenuis* NCCU-122	44	
24	Ag	*Westiellopsis prolifica* NCCU-331	52	
25	Ag	*Oscillatoria* sp. NCCU-369	80	
26	Au	*Spirulina platensis*	6–10	Govindaraju et al. (2008)
27	Cd	*Phormidium tenue*	5	Ali et al. (2012)
28	Au	*Synechocystis* sp. PCC 6803	13 ± 2	Focsan et al. (2011)
29	Au	*Lyngbya majuscula*	41.7 ± 0.2	Bakir et al. (2018)
30	Ag	*Chroococcus* NCCU-207	48	Husain et al. (2015)

(continued)

S. No	Nanoparticles	Cyanobacteria	Size (nm)	Reference
31	Au	*Arthrospira platensis*	17–25	Govindaraju et al. (2008)
32	Ag	*Gloeocapsa gelatinosa* NCCU-430	88	Husain et al. (2015)
33	ZnO	*Nostoc* sp. EA03	50–80	Ebadi et al. (2019)
34	ZnO	*Limnothrix* sp. KO05	35	Davaeifar et al. (2019)
35	Au	*Calothrix pulvinata*	5.4 ± 0.6	Brayner et al. (2007)
36	Au	*Anabaena flos-aquae*	12.0 ± 0.7	
37	Au	*Leptolyngbya foveolarum*	6.0 ± 0.6	
38	Pd	*Calothrix pulvinata*	3.5 ± 0.3	
39	Pt	*Calothrix pulvinata*	3.2 ± 0.3	
40	CuO	*Spirulina platensis*	30-40	Saran et al. (2017)
41	Ag	*Lyngbya majuscula*	16	Duraisamy et al. (2020)

however, instead of cyanobacteria filtrate, the distilled water was employed. Both the solutions were centrifuged at 10,000 rpm for sufficient time by maintaining a particular temperature, and the resultant pellets were thoroughly washed with water at least three times to discard any free biomass remnants or uncapped ligands. The pellets were spread on a sterilized 1/1 petri dish and dried in an oven at 30 °C for 24 h. The dried particles were scraped with the help of a scalpel, and the fine powder was stored for further use. Fig. 14.1 depicts the general process of nanoparticle synthesis using cyanobacteria.

14.3 Advantages and disadvantages of cyanobacteria used in nanoparticle synthesis

Cyanobacteria are considered as one of the best biological nanoparticle synthesizers. Cyanobacteria are microalgae that have been used for

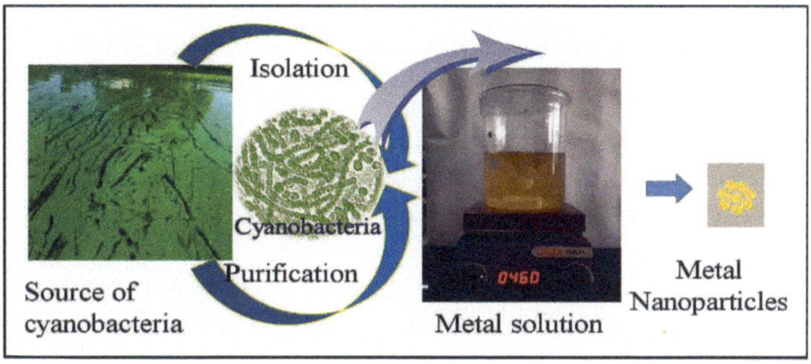

FIG. 14.1 General Scheme for nanoparticle synthesis using cyanobacteria.

bioremediation of toxic metals such as gold and platinum. This property was exploited to synthesize nanoparticles of these metals. The advantages of using cyanobacteria are as follows:

1. They have high biomass yield and are tolerant of different stress types such as nitrates, phosphates, temperature, and heavy metals. This property for resistance to high levels of stress will help in the production process of nanoparticles by making the process of use of such bacteria cheap and sustainable (Patel et al., 2018).
2. They have low-energy requirements and short generation time as compared to traditional chemical methods.
3. Cyanobacterium species can survive in contaminated and polluted environments. As the production of nanoparticles would need that the bacteria used to be robust and survive in harsh synthesis conditions, this property can be used in this context.
4. Cyanobacterium is an oxygenic photosynthesizer that is they release oxygen like plants. This property can be exploited to oxidize metallic nanoparticles.
5. The main advantage of using cyanobacterium is that it converts carbon dioxide to other forms of carbon in the process of photosynthesis. This would help to reduce greenhouse gases and lower production costs (Rai et al., 2019).
6. In the production of silver nanoparticles, it is observed that cyanobacterial cells were encrusted with silver nanoparticles, and the separation of some filaments was observed. The nanoparticles were stabilized by capping agents. As mentioned above, cyanobacteria use nitrate as a major source of nitrogen. The presence of silver nanoparticles may probably kill the bacteria and during death, they may release silver particles stored in the cells. The dead cyanobacterium also releases organic moieties such as proteins that cause further precipitation of silver nanoparticles. Thus, giving high yields of synthesized nanoparticles (Ali et al., 2011).

Thus, the use of cyanobacteria in the production of nanoparticles is a new alternative to traditional chemical methods. This is a new energy-efficient and green alternative; however, it is accompanied by a few limitations as follows:

1. The key disadvantage is the limitation of the size and shape of nanoparticles produced. As the nanoparticles are produced *in-vitro*, it is difficult to control the process and optimize it to get the desired shape nanoparticles.
2. The production would have to take place at ambient temperatures and pH thus it would eliminate major industrial hazards and would be safe.

3. The major challenge is the scaling-up process as the production of tones of nanoparticles will require hundreds of cyanobacteria in the process, thus making the handling of such strains of bacteria difficult and will require sophisticated instruments for handling, thus increasing cost.
4. Another challenge would be to select a proper strain as different nanoparticles depending on their shape and properties will require different strains of bacteria for its production as it would not be a one strain fits all situations. Thus, it would need considerable time to explore the full potential of cyanobacterium in the production of nanoparticles. The use of biological agents for production is still in the laboratory phase and needs optimization for robust scale-up processes.

There is a need for more research in understanding the biological mechanisms by which these bacteria produce nanoparticles to exploit it and use it to scale up to the industry level. With the help of biotechnology and genetic engineering, we can explore the mechanisms by which cyanobacteria produce nanoparticles to get good quality and diverse use of nanoparticles. Thus, we can say that it is still in the developmental stage and needs time for exploiting the full potential of organisms. If this is optimized we would indeed have a method for synthesis of nanoparticles which is green, energy-efficient, and free from harsh chemical processes (Ovais et al., 2018). If the huge library of biosystems that can produce nanoparticles is exploited it would indeed be a potential game-changer in the foreseen future.

14.4 Conclusion

In recent years, cyanobacteria-mediated nanoparticles synthesis has shown a great promise for the development of novel routes as it enables eco-friendly, energy-saving, and inexpensive production of nanoparticles with varied shape and size. Cyanobacteria-mediated nanoparticles possess unique physical, chemical, and biological features that provide applications in various fields such as anticancer, antimicrobial, photocatalytic agents, etc. Similar to various nonbiological chemical routes of synthesis, the shapes and sizes of the cyanobacteria-mediated nanomaterials can be also be controlled by altering reaction conditions such as pH, temperature, time, and reactant ratio. Of the nanoparticles known to be formed by algae, noble metals including silver, gold, and platinum are the most studied. Considering recent progress and ongoing efforts in improving the synthesis of a wider range of nanomaterials and their applications; we expect in the future, implementation of this technology at the industrial scale and their applications at the commercial scale.

Acknowledgment

Author DKK thanks University of Mumbai for Minor Research Grand Project 2019-2020/691. DNN and PR are thankful for FIST-DST and DBT Star College Scheme funding to VES College. The authors thank Ms. Criss Dcosta for her kind support.

References

Ali, D.M., Sasikala, M., Gunasekaran, M., Thajuddin, N., 2011. Biosynthesis and characterization of silver nanoparticles using marine cyanobacterium, *Oscillatoria willei* NTDM01. Dig. J. Nanomater. Biostruct. 6 (2), 385–390.

Ali, D.M., Gopinath, V., Gopinath, N., Rameshbabu, N., Thajuddin, N., 2012. Synthesis and characterization of CdS nanoparticles using C-phycoerythrin from the marine cyanobacteria. Mater. Lett. 74, 8–11.

Bakir, E.M., Younis, N.S., Mohamed, M.E., El Semary, N.A., 2018. Cyanobacteria as nanogold factories: chemical and anti-myocardial infarction properties of gold nanoparticles synthesized by *Lyngbya majuscule*. Mar. Drugs 16 (6), 217.

Brayner, R., Barberousse, H., Hemadi, M., Djedjat, C., Yepremian, C., Coradin, T., Livage, J., Fievet, F., Coute, A., 2007. Cyanobacteria as bioreactors for the synthesis of Au, Ag, Pd, and Pt nanoparticles via an enzyme-mediated route. J. Nanosci. Nanotechnol. 7, 2696–2708.

Chakraborty, N., Banerjee, A., Lahiri, S., Panda, A., Ghosh, A.N., Pal, R., 2009. Biorecovery of gold using cyanobacteria and an eukaryotic alga with special reference to nanogold formation – a novel phenomenon. J. Appl. Phycol. 21, 145–152.

Chen, Y., Fan, Z., Zhang, Z., NiuW., Li C., Yang, N., Chen, B., Zhang, H., 2018. Two-dimensional metal nanomaterials: synthesis, properties, and applications. Chem. Rev. 118 (13), 6409–6455.

Cicci, A., Sed, G., Tirillo, J., Stoller, M., Bravi, M., 2017. Production and characterization of silver nanoparticles in cultures of the cyanobacterium *A. Platensis* (Spirulina). Chem. Eng. Trans. 57, 1405–1410.

Dahoumane, S.A., Djediat, C., Yepremian, C., Coute, A., Fievet, F., Coradin, T., Brayner, R., 2012. Recycling and adaptation of *Klebsormidium flaccidum* microalgae for the sustained production of gold nanoparticles. Biotechnol. Bioeng 109, 284–288.

Dahoumane, S.A., Mechouet, M., Wijesekera, K., Filipe, C.D.M., Sicard, C., Bazylinski, D.A., Jeffryes, C., 2017. Algae-mediated biosynthesis of inorganic nanomaterials as a promising route in nanobiotechnology –a review. Green Chem 19, 552–587.

Davaeifar, S., Modarresi, M.-H., Mohammadi, M., Hashemi, E., Shafiei, M., Maleki, H., Vali, H., Zahiri, H.S., Noghabi, K.A., 2019. Synthesizing, characterizing, and toxicity evaluating of Phycocyanin-ZnO nanorod composites: a back to nature approaches. Colloid. Surf. B Biointerf. 175, 221–230.

Deplanche, K., Merroun, M.L., Casadesus, M., Tran, D.T., Mikheenko, I.P., Bennett, J.A., Zhu, J., Jones, I.P., Attard, G.A., Wood, J., Selenska- Pobell, S., Macaskie, L.E., 2012. Microbial synthesis of core/shell gold/palladium nanoparticles for applications in green chemistry. J. R. Soc. Interf. 9, 1705–1712.

Duraisamy, N., Muthu, S., Krishnan, K., 2020. Synthesis of silver nanoparticles using Lyngbya majuscula extract and their antibacterial effects. Int. J. Pharm. Bio. Sci 11 (2), 83–91.

Ebadi, M., Zolfaghari, M.R., Aghaei, S.S., Zargar, M., Shafiei, M., Zahiri, H.S., Noghabi, K.A., 2019. A bio-inspired strategy for the synthesis of zinc oxide nanoparticles (ZnO NPs) using the cell extract of cyanobacterium *Nostoc* sp. EA03: from biological function to toxicity evaluation. RSC Adv. 9, 23508–23525.

Focsan, M., Ardelean, I.I., Craciun, C., Astileanet, S., 2011. Interplay between gold nanoparticle biosynthesis and metabolic activity of cyanobacterium *Synechocystis* Sp.PCC.6803. Nanotechnology 22 (48), 485101.

Gabriela, K., Zuzana, K., Ivo, V., Eliska, Z., Rai, M., Adam, S., 2016. Noble metal nanoparticles synthesis mediated by the genus Dolichospermum: perspective of green approach in the nanoparticles preparation. Adv. Sci. Lett. 22 (5), 637–641.

Gericke, M., Pinches, A., 2006. Biological synthesis of metal nanoparticles. Hydrometallurgy 83, 132–140.

Govindaraju, K., Basha, S.K., Kumar, V.G., Singaravelu, G., 2008. Silver, gold and bimetallic nanoparticles production using single-cell protein (*Spirulina platensis)* Geitler. J. Mater. Sci. 43, 5115–5122.

Guilger-Casagrande, M., de Lima, R., 2019. Synthesis of silver nanoparticles mediated by fungi: a review. Front. Bioeng. Biotechnol. 7, 287.

Hamida, R.S., Abdelmeguid, N.E., Ali, M.A., Mohammed, M., Meferij, B., Khalil, M.I., 2020. Synthesis of silver nanoparticles using a novel cyanobacteria *Desertifilum* sp. extract: their antibacterial and cytotoxicity effects. Int. J. Nanomed. 15, 49–63.

Husain, S., Sardar, M., Fatma, T., 2015. Screening of cyanobacterial extracts for synthesis of silver nanoparticles. World J. Microbiol. Biotechnol. 31 (8), 1279–1283.

Kannan, N., Subbalaxmi, S., 2011. Biogenesis of nanoparticles – a current prospective. Rev. Adv. Mater. Sci. 27, 99–114.

Karnani, R.L., Chowdhary, A., 2013. Biosynthesis of silver nanoparticle by eco-friendly method. Indian J. Nano Sci. 1, 25–31.

Lengke, M.F, Fleet, M.E., Southarn, G., 2007. Synthesis of palladium nanoparticles by reaction of filamentous cyanobacterial biomass with a palladium(II) chloride complex. Langmuir 23, 8982–8987.

Mahdieha, M., Zolanvari, A., Azimeea, A.S., Mahdiehc, M., 2012. Green biosynthesis of silver nanoparticles by *Spirulina platensis*. Sci. Iran. F 19 (3), 926–929.

Mittal, A.K., Chisti, Y., Banerjee, U.C., 2013. Synthesis of metallic nanoparticles using plant extracts. Biotechnol. Adv. 31 (2), 346–356.

Mohanpuria, P., Rana, N.K., Yadav, S.K., 2008. Biosynthesis of nanoparticles: technological concepts and future applications. J. Nanopart. Res. 10, 507–517.

Ovais, M., Khalil, A.T., Ayaz, M., Ahmad, I., Nethi, S.K., Mukherjee, S., 2018. Biosynthesis of metal nanoparticles via microbial enzymes: a mechanistic approach. Int. J. Mol. Sci. 19, 4100.

Parial, D., Patra, H.K., Dasgupta, A.K.R., Pal, R., 2012. Screening of different algae for green synthesis of gold nanoparticles. Eur. J. Phycol. 47, 22–29.

Patel, V.K., Sundaram, S., Patel, A.K., Kalra, A., 2018. Characterization of seven species of cyanobacteria for high-quality biomass production. Arab. J. Sci. Eng. 43, 109–121.

Rai, M., Gade, A., Yadav, A., 2011. Biogenic nanoparticles: an introduction to what they are, how they are synthesized and their applications. In: Rai, M., Duran, N. (Eds.), Metal Nanoparticles in Microbiology. Springer, Berlin, Heidelberg, pp. 1–14.

Rai, S., Wenjing, W., Shrivastava, A.K., Singh, P.K., 2019. Cyanobacteria as a source of nanoparticles and their applications. In: Kumar, A., Singh, A.K., Choudhary, K.K. (Eds.), Role of Plant Growth Promoting Microorganisms in Sustainable Agriculture and Nanotechnology. Elsevier, Amsterdam, pp. 183–198.

Ramanathan, R., O'Mullane, A.P., Parikh, R.Y., Smooker, P.M., Bhargava, S.K., Bansal, V., 2011. Bacterial kinetics-controlled shape-directed biosynthesis of silver nanoplates using *Morganella psychrotolerans*. Langmuir 27, 714–719.

Rosken, L.M., Korsten, S., Fischer, C.B., Schonleber, A., Smaalen, S.v., Geimer, S., Wehner, S., 2014. Time-dependent growth of crystalline Au0-nanoparticles in cyanobacteria as self-reproducing bioreactors:1. *Anabaena* sp. J. Nanopart. Res. 16, 2370.

Saran S., Sharma G., Kumar M. and Ali M. I., 2017. Biosynthesis of copper oxide nanoparticles using cyanobacteria *Spirulina platensis* and its antibacterial activity. 34 3887–3892.

Singh, G., Babele, P.K., Kumar, A., Srivastava, A., Sinha, R.P., Tyagi, M.B., 2014. Synthesis of ZnO nanoparticles using the cell extract of the cyanobacterium, *Anabaena* strain L31 and its conjugation with UV-B absorbing compound shinorine. J. Photochem. Photobiol. B Biol. 138, 55–62.

Stark, W.J., Stoessel, P.R., Wohlleben, W., Hafner, A., 2015. Industrial applications of nanoparticles. Chem. Soc. Rev. 44, 5793–5805.

Sudha, S.S., Rajamanickam, K., Rengaramanujam, J., 2013. Microalgae mediated synthesis of silver nanoparticles and their antibacterial activity against pathogenic bacteria. Indian J. Exp. Biol. 52, 393–399.

Zikalala, N., Matshetshe, K., Parani, S., Oluwafemi, O.S., 2018. *Biosynthesis* protocols for colloidal metal oxide *nanoparticles*. Nano-Structures Nano-Objects 6, 288–299.

Zinicovscaia, I., Rudi, L., Valuta, A., Cepoi, L., Vergel, K., Frontasyev, M.V., et al., 2016. Biochemical changes in Nostoclinckia associated with selenium nanoparticles biosynthesis. Ecol. Chem. Eng. 23 (4), 559–569.

Index

Page numbers followed by "*f*" and "*t*" indicate, figures and tables respectively.

D

MIX
Papier | Fördert gute Waldnutzung
FSC® C083411

Kolibri 360 GmbH
Ferdinand-Jühlke-Straße 7
99095 Erfurt, Deutschland
produktsicherheit@kolibri360.de